# The Cyclostomata
## An Annotated Bibliography

by

G. Tandler, M. A. Jones and F. W. H. Beamish
Department of Zoology, College of Biological Science,
University of Guelph, Guelph, Ontario, Canada

## Supplement 1973-1978

DR. W. JUNK. B.V. – PUBLISHERS – THE HAGUE – BOSTON – LONDON 1979

ISBN-13: 978-94-009-9633-5        e-ISBN-13: 978-94-009-9631-1
DOI: 10.1007/978-94-009-9631-1
© Dr. W. Junk b.v. – Publishers – The Hague 1979
Cover design Max Velthuijs
Copyright © 1979
Softcover reprint of the hardcover 1st edition 1979

# Table of Contents

# Introduction (How to Use the Bibliography)

# How to use the Bibliography

The bibliography consists of 5616 references given in a MASTER file according to call number (0001-5616). Reference to these articles can be made through AUTHOR, SUBJECT, and SOURCE indices.

## MASTER FILE

All data for each reference is given in the MASTER file. For example,

```
REMBISZEWSKI, J.M.            1967                          0002F
MATERIALY DO POZNANIA MINOGOW (PETROMYZONIDAE) RODZAJO
LAMPETRA GRAY W. POLSCE. I. LAMPETRA (EUDONTOMYZON)
MARIAE BERG. [CONTRIBUTION TO THE KNOWLEDGE OF THE
LAMPREYS (PETROMYZONIDAE) OF THE GENUS LAMPETRA GRAY
IN POLAND. I. LAMPETRA (EUDONTOMYZON) MARIAE BERG.]
FRAGM. FAUN., 13 (14): 249-259.
(POL.)
*EUDONTOMYZON VLADYKOVI POLAND MORPHOMETRY ECOLOGY
SPAWNING ADULTS AMMOCOETES
```

The above reference illustrates the type of information that might be included in this file. Each reference is identified by a call number to the right of the citation.

The "F" signifies the paper has been reviewed and verified. The date of publication is to the immediate right of the author(s). The title and source follow on the next lines. Titles or sources in square brackets i.e. [    ] have been translated, or were available only in translation. Titles in round brackets i.e. (    ) appear in the literature, but could not be verified. If a paper is written in a language other than English, the language is indicated in brackets i.e. (    ) at the end of each citation. The next line(s) contain(s) a list of keywords, an explanation of which is given below under the heading SUBJECT INDEX.

Citations of journals are standardized according to the National Clearinghouse for Periodical Title Word Abbreviations compiled by the Standards Committee Z39 of the American National Standards Institute (1971 edition). The format of each citation is listed according to Library of Congress cataloguing rules. The existence of each journal was verified in one of the following: List of Scientific Serials in Canadian Libraries, (1967), 2nd Edition, 2 Vol., National Science Library, National Research Council of Canada, Ottawa; Library of Congress Catalogue of Printed Cards, (1958), Pageant Books, Inc., New York; Library Congress and National Union Catalogue Author Lists, (1942-1973); A Master Cumulation, compiled by the Gale Research Co., Detroit, Michigan; The National Union Catalogue Pre-1956 Imprints, (1972) compiled and edited by the Library of Congress and the National Union Catalogue Subcommittee of the Resources and Technical Services Division, American Library Association, Mansell Information Publishing Ltd., London, England. When a journal's existence could not be verified it was nevertheless listed according to Library of Congress rules. If a volume number was not available, the date of publication was used in its place.

With few exceptions the existence of each reference was verified, but where it was not possible the source where the reference was cited in the literature is indicated in round brackets.

The contents of some references are restricted. These, indicated by the word RESTRICTED following the citation, were not indexed by keywords. Permission to review these references must be obtained from the government agency concerned.

Books consisting of a number of relevant contributed chapters are listed as a main entry according to the editor (see example 1 below) and each chapter is listed as a citation in the book under the author(s) of that chapter (see example 2 below).

*Example 1*

```
HARDISTY, M.W. (ED.)          1972                    4139F
POTTER, I.C. (ED.)
THE BIOLOGY OF THE LAMPREY. VOL. 2
LONDON: ACADEMIC PRESS.
(MAIN ENTRY, SEE ALSO NO. 1190F, 1330F, 1331F,
1352F, 1353F, 1354F, 1644F)
```

*Example 2*

```
RIGGS, A.                     1972                    1190F
THE HAEMOGLOBINS. PP. 261-286
IN HARDISTY, M.W. AND I.C. POTTER, EDS., THE
BIOLOGY OF THE LAMPREYS. VOL. 2. LONDON:
ACADEMIC PRESS.
```

All other books are listed only by author or editor as for example,

```
SIMEK, Z.                     1959                    2342
RYBY NASICH VOD.
PRAHA: ORBIS. 142 pp.
(CZECH.)
```

For some rare or old books the Library of Congress (L.C.) or Union Catalogue (U.C.) number is included in the MASTER FILE.

Symposia are listed according to the editor, contributors or sponsor.

## AUTHOR INDEX

Authors of articles in the bibliography are listed alphabetically and catalogued according to the Anglo-American Cataloguing Rules (American Library Association, 1967).

The following example illustrates the indexing according to author,

| Author | Date | Call Number |
|---|---|---|
| FAHRENHOLZ, C. | 1936 | 3954* |
| FAIDHERBE, J. | 1950 | 3827 |
| FAIR, E. | 1933 | 0490F |

To the immediate right of the author's name is the date of publication. To the far right is the call number. An asterisk beside the call number indicates the first author is FAHRENHOLZ, C. Lack of an asterisk, as in FAIDHERBE, J. above, indicates a co-author. The call number refers to the MASTER file. An "F" in the call number, as in FAIR, J. above, indicates the article has been reviewed and the citation verified.

## SUBJECT INDEX

The subject index is intended to facilitate the search for reference material dealing with a particular biological area. Each significant word in the title of a reference is included in the subject index along with the following 60 keywords that may also be applicable. Some words are self explanatory while others require some comment.

**Ammocoete**  articles concerning the larval stage of lampreys.

Adult  papers dealing with the post macropthalmia stage of lampreys (for transforming stages see METAMORPHOSIS).

Anatomy  articles about gross morphology and the skeletal system.

Animal  papers referring to animals in addition to members of the Cyclostomata.

Biology  publications dealing in a general way with the Cyclostomata.

Behaviour  articles on the reaction of the lampreys and hagfishes to the external and internal stimuli.

Biochemistry

Blood  papers referring to the composition, chemical and physical properties, and physiology of blood.

Chemistry  used primarily as cross reference with management to identify chemical control (i.e. the larvicide TFM).

Circadian  publications on daily fluctuations in movement and physiology.

Circulatory System  articles on the morphology and function of the circulatory system.

Culture  papers dealing with the rearing of lampreys and hagfish.

Cytology

Digestion

Distribution  articles about the geographical distribution of lampreys and their spatial distribution in a body of water.

Ecology

Egg  papers referring to the unfertilized egg of the Cyclostomata.

Embryology

Endocrinology

Excretion  publications dealing with the morphology and function of kidneys and other excretory tissue as well as defaecation.

Evolution  articles involving phylogenetic and ontogenetic considerations in the evolution of the Cyclostomata.

Fecundity  papers concerning oocyte numbers.

Feeding  papers in which the morphology and function of the branchial filtration system of larval lampreys is discussed, including also feeding preferences, and diet of the Cyclostomata.

Genetics

Growth  papers dealing with changes in body length, weight, and chemical composition.

Gonadogenesis

Hearing  articles referring to the morphology and function of the auditory system.

Histochemistry

Histology

History  research in the Great Lakes on problems and management associated with the invasion of *Petromyzon marinus*.

Immunology  papers dealing with the immunologic processes in the Cyclostomata.

Integument articles dealing with morphology and physiology, including sense receptors, pigmentation, and respiration of the integument.

Ionic Regulation

Life Cycle general articles on the life cycle of the Cyclostomata.

Locomotion papers on the mechanics of movement and swimming performance.

Management publications dealing with methods of lamprey control (stream barriers, chemicals and biological control).

Metabolism

Metamorphosis articles dealing with the process of transformation from larva to adult.

Migration articles on the downstream movement of young adults and the upstream migration of mature adults.

Mortality of papers on death of lampreys either from disease or from management programmes.

Mortality by articles on the effects of lamprey parasitism on fishes.

Morphometry

Muscle papers dealing with the morphology and function.

Mouth publications concerning the form and function (teeth, tongue, buccal glands, etc.).

Nervous System

Olfaction

Osmoregulation

Parasitism by articles dealing with the feeding of lampreys on fish.

Parasitism of papers on the parasites of lampreys.

Pathology publications dealing with the diseases of the Cyclostomata.

Physiology

Pigmentation

Pisces comparative papers in which members of the class PISCES were also studied.

Reproduction articles on the morphology and physiology of the reproductive system.

Respiration papers on the morphology and physiology of the respiratory system.

Sense Receptors articles dealing with chemoreceptors, taste receptors, and the pineal body.

Spawning papers on the act of spawning and on the selection and construction of nests.

Systematics publications dealing with identification and classification of the Cyclostomata.

Techniques papers describing specific chemical, biochemical, or physiological techniques in the study of the Cyclostomata.

Vision papers dealing with the morphology and physiology of the pineal body and eyes, including visual acuity and threshold.

In addition to the keywords, the species (identified by an asterisk), the morphological stage (lampreys), and the geographic location are treated as keywords, providing this information was mentioned in the paper. Moreover, significant words in the title of each citation are included in the subject index. For example in the following article,

```
LARSEN, L.O.                    1965              0075F
EFFECTS OF HYPOPHYSECTOMY IN THE CYCLOSTOME,
LAMPETRA FLUVIATILUS (L.) GRAY.
GEN. COMP. ENDOCRINOL., 5:16-30
```

the significant words in the title are underlined. The paper could also be located in the subject index by each of the assigned keywords.

```
ENDOCRINOLOGY, ADULT, GONADOGENESIS, EGG, LIFE CYCLE,
DENMARK (geographic location)
```

If the title includes a keyword or some form of one, the keyword is omitted. For example in the paper

```
COONFIELD, B.R.                 1940              0301F
THE PIGMENT IN THE SKIN OF MYXINE GLUTINOSA LINN.
AM. MICROSC. SOC., TRANS., 59(3):398-403.
```

PIGMENT is listed in the subject index, at the expense of the keyword PIGMENTATION.

When references were not read, title words only are listed in the subject index. Special effort was made to separate *Petromyzon marinus* into anadramous (ANAD) or landlocked (L-L) forms. The source of landlocked *Petromyzon marinus* is indicated by U.S. (Great Lakes), U.S. (Finger Lakes), or Canada (Great Lakes).

Titles of papers written in a language other than English were subject indexed in the vernacular and English. Many non-English papers were reviewed and assigned keywords.

In searching a topic one or more keywords should be selected from the list. In addition a list should be made of related words. For example the topic — *microscopical examination of the retina* — might be examined under the keyword *VISION* or any of the following words which might appear in the subject index as title words: RETINA, EYE, AUGE (German), L'OEIL (French), MICROSCOPIC, ULTRA-STRUCTURE, etc.

## SOURCE INDEX

This index is of particular value to individuals commencing work on the Cyclostomata as it includes those journals which provide fairly regular publications on the Cyclostomata, as well as to those preparing references for publication. Journals and books are arranged alphabetically in the index.

# Acknowledgements

Financial aid for compilation and publication of the Bibliography was made available by the Great Lakes Fishery Commission, and Dr. W. Junk B.V., Publishers.

The preparation of the Bibliography was possible only with the cooperation of many individuals. Thanks are extended to E. Thomas, D. Rogers, A. Tandler, S. Farringer, T. Watson, N. Weinstein, J. Selley, A. K. Kumaraguru, and F. Zapalac. We would like to also thank A. Galina, B. Campbell, and S. McNeill for administrative assistance.

The Institute of Computer Science, University of Guelph, provided invaluable assistance, and for this we thank J. Demain, S. Yu, and W. Davidson.

The efforts of S. Shield, F. Payer and B. Evans of Alphatext Limited, are greatly appreciated.

Special thanks are extended to all of the contributing authors, for without their cooperation in sending relevant reprints, the Cyclostomata would not have its high measure of validity.

# Master Index

4536

PIGON, A.                    1974
MORITA, M.
BEST, J.B.
CEPHALIC MECHANISM FOR SOCIAL CONTROL OF FISSIONING IN
PLANARIANS PART 2: LOCALIZATION AND IDENTIFICATION OF THE
RECEPTORS BY ELECTRON MICROGRAFIC AND ABLATION STUDIES.
J. NEUROBIOL., 5:443-462.


4537

HOPSON, J.A.                 1974
THE FUNCTIONAL SIGNIFICANCE OF THE HYPOCERCAL TAIL AND
LATERALFIN FOLD OF ANASPID OSTRACODERMS.
CHICAGO NAT. HIST. MUS., FIELIANA: GEOL., 33:83-93.


4538F

TSUNEKI, K.                  1975
ADACHI, T.
ISHII, S.
OHTA, Y.
[MORPHOMETRIC CLASSIFICATION OF NEUROSECRETORY GRANULES IN
THE NEUROHYPOPHYSIS OF THE HAGFISH.]
ZOOL. MAG. (TOKYO), 84:390
(JAP.)


4539

OOI, E.C.                    1976
YOUSON, J.H.
GROWTH OF THE OPISTHONEPHRIC KIDNEY DURING LARVAL LIFE IN
THE ANADROMOUS SEA LAMPREY, PETROMYZON MARINUS L.
CAN. J. ZOOL., 54:1449-1458.


4540F

KARAMYAN, A.I.               1975
ZAGORULKO, T.M.
BELEKHOVA, M.G.
VESELKIN, N.P.
KOSAREVA, A.A.
O KORTIKALIZATSII DVUKH OTDELOV ZRITEL'NOI SISTEMY V
EVOLYUTSII POZVONOCHNYKH. [ON THE CORTICALIZATION OF 2
DIVISIONS OFTHE VISUAL SYSTEM IN THE VERTEBRATES EVOLUTION.]
NEIROFIZIOLOGIYA, 7:12-20.
(RUSS.)
ENGL. SUMM.
*LAMPETRA FLUVIATILIS ADULT NERVOUS SYSTEM EVOLUTION


4541F

HOLCIK, J.                   1974
NALBANT, T.
NOTE ON THE OCCURRENCE OF THE BROOK LAMPREY, LAMPETRA
PLANERI (BLOCH, 1784) (CYCLOSTOMATA) IN ROMANIA.
VESTN. CESK. SPOL. ZOOL., 38:95-97.
*EUDONTOMYZON DANFORD *EUDONTOMYZON MARIAE *EUDONTOMYZON
VLADYKOV ANATOMY DISTRIBUTION


4542F

ROVAINEN, C.M.               1974
SYNAPTIC INTERACTIONS OF IDENTIFIED NERVE CELLS IN THE
SPINAL CORD OF THE SEA LAMPREY.
J. COMP. NEUROL., 154:189-206
*PETROMYZON MARINUS U.S.A. NERVOUS SYSTEM CYTOLOGY
PHYSIOLOGY


4543F

HOLMBERG, K.                 1973
LUNDIN, V.
THE OLFACTORY SYSTEM IN THE HAGFISH MYXINE GLUTINOSA PART
2: FINE STRUCTURE OF OLFACTORY NERVES.
ACTA ZOOL., (STOCKH.) 54:285-295.
SWEDEN HISTOLOGY NERVOUS SYSTEM OLFACTION


4544

CROWLEY, T.E.                1973
GREEN, J.R.
MCMILIAN, N.F.
SOME MOLLUSCA FROM BORNU PROVINCE NORTHERN NIGERIA WITH
APPENDIX STATISTICAL ANALYSES OF 2 SPECIES PILA WERNEI AND
ASPATHRIA COMPLANATA.
J. CONCHOL., 28:81-94.


4545

TELFORD, M.                  1974
BLOOD GLUCOSE IN CRAYFISH PART 2: VARIATIONS INDUCED BY
ART1FICIAL STRESS.
COMP. BIOCHEM. PHYSIOL., 48A:555-560.
ANIMAL BLOOD BIOCHEMISTRY PHYSIOLOGY


4546

PETTER, A.J.                 1974
ATTEMPT AT CLASSIFICATION OF THE FAMILY CUCULLANIDAE.

PARIS. MUS. NATL. HIST. NAT., BULL., ZOOL., 177:1469-1490.

4547F
GOOLD, R.J.                    1976
SEA LAMPREY SURVEYS OF STREAMS TRIBUTARY TO THE CANADIAN
SIDE OF LAKE ONTARIO, 1975. PP. 12-14.
IN  TIBBLES, J.J. ED., ANNUAL REPORT 1975 OF THE SEA LAMPREY
CONTROL CENTRE SAULT STE. MARIE, ONTARIO, TO THE GREAT
LAKES FISHERY COMMISSION. GREAT LAKES FISH. COMM., ANNU.
REP.
RESTRICTED
CANADA AMMOCOETE DISTRIBUTION MANAGEMENT

4548F
MAZIN, A.L.                    1973
VANYUSHIN, B.F.
BELOZERSKII, A.N.
O KHARAKTERE NUKLEOTIDNOI POSLOVATEL'NOSTI DNK NEKOTORYKH
RYB. [CHARACTER OF THE NUCLEOTIDE SEQUENCE OF DNA OF CERTAIN
FISH.]
AKAD. NAUK SSSR, LENINGR., DOKL., SER. BIOL., 210:232-235.
(RUSS.)
ACAD. SCI. USSR, PROC., 210:184-187
(ENG.)
(RUSS.)
ENG. TRANS.
BIOCHEMISTRY GENETICS

4549
SCHWARTZ, M.L.                 1973
PIZZO, S.V.
SULLIVAN, J.B.
HILL, R.L.
MCKEE, P.A.
A COMPARATIVE STUDY OF CROSSLINKED AND NONCROSSLINKED FIBRIN
FROM THE MAJOR CLASSES OF VERTEBRATES.
THROMB. DIATH. HAEMORRH., 29:313-338.
ANIMAL BLOOD BIOCHEMISTRY

4550F
LINNA, T.J.                    1975
FINSTAD, J.
GOOD, R.A.
CELL PROLIFERATION IN EPITHELIAL AND LYMPHO-HEMATOPOIETIC
TISSUES OF CYCLOSTOMES.
AM. ZOOL., 15:29-38.
*EPTATRETUS STOUTII *PETROMYZON MARINUS U.S.A. ADULT BLOOD
CYTOLOGY EVOLUTION IMMUNOLOGY

4552
HERZFELD, J.                   1974
STANLEY, H.E.
A GENERAL APPROACH TO COOPERATIVITY AND ITS APPLICATION
TO THE OXYGEN EQUILIBRIUM OF HEMOGLOBIN AND ITS EFFECTORS.
J. MOL. BIOL., 82:231-265.
BIOCHEMISTRY BLOOD BIOLOGY

4553F
WEISE, J.G.                    1976
SEA LAMPREY SURVEYS OF STREAMS TRIBUTARY TO LAKE ONTARIO,
NEW YORK STATE, 1975. PP.14-16.
IN  TIBBLES, J.J. ED., ANNUAL REPORT 1975 OF THE SEA LAMPREY
CONTROL CENTRE SAULT STE. MARIE, ONTARIO, TO THE GREAT
LAKES FISHERY COMMISSION. GREAT LAKES FISH. COMM., ANNU.
REP.
RESTRICTED
U.S.A. AMMOCOETE DISTRIBUTION MANAGEMENT

4554F
SCHLEEN, L.P.                  1976
INDIVIDUAL STREAM TREATMENT REPORTS, LAKE SUPERIOR, 1975.
PP.16-30.
IN  TIBBLES, J.J. ED., ANNUAL REPORT 1975 OF THE SEA LAMPREY
CONTROL CENTRE SAULT STE. MARIE, ONTARIO TO THE GREAT
LAKESFISHERY COMMISSION. GREAT LAKES FISH. COMM., ANNU. REP.
RESTRICTED
CANADA AMMOCOETE CHEMISTRY MANAGEMENT

4555F
SIVAK, J.G.                    1974
HISTORICAL NOTE: THE VERTEBRATE MEDIAN EYE.
VISION RES., 14:137-140.
ANIMAL EVOLUTION CYTOLOGY SENSE RECEPTORS

4556F
POLENOV, A.L.                  1973
BELEN'KII, M.A.
A NEKOTERYKH ZAKONOMERNOSTYAKH STANOVLENIYA NEIROGEMAL'NYKH
OTDELOV GIPOTALAMO-GIPOFIZAROI NEIROSEKRETORNOI SISTEMY ONTO
I FILOGENEZE POZVONOCHNYKH. [SOME REGULARITIES OF THE
DEVELOPMENT OF NEUROHEMAL PARTS OF THE

HYPOTHALAMO-HYPOPHYSIAL NEUROSECRETORY SYSTEM IN ONTO- AND
PHYLOGENESIS OF VERTEBRATES.]
ZH. EVOL. BIOKHIM. FIZIOL., 9:355-363.
(RUSS.)
ENG. SUMM.
EVOLUTION ANIMAL EMBRYOLOGY ENDOCRINOLOGY NERVOUS SYSTEM


4557F

CHANDLER, J.H.                    1975
MARKING, L.L.
TOXICITY OF THE LAMPRICIDE
3-TRIFLUOROMETHYL-4-NITROPHENOL(TFM) TO SELECTED AQUATIC
INVERTEBRATES AND FROG LARVE.
U.S. DEP. INT., FISH WILDL. SERV., INVEST. FISH CONTROL,
62:3-7.
MANAGEMENT


4558F

RURAK, D.W.                    1974
PERKS, A.M.
THE PHARMACOLOGICAL CHARACTERIZATION OF ARGININE VASOTOCIN
IN THE PITUITARY OF THE PACIFIC HAGFISH POLISTOTREMA
STOUTII.
GEN. COMP. ENDOCRINOL., 22;480-488.
*EPTATRETUS STOUTI *MYXINE GLUTINOSA CANADA ADULT
ENDOCRINOLOGY TECHNIQUES NERVOUS SYSTEM BIOCHEMISTRY


4559F

SCHWAB, M.E.                    1973
SOME NEW ASPECTS ABOUT THE PROSENCEPHALON OF LAMPETRA
FLUVIATILIS (L.): A CYTOARCHITECTURAL AND COMPARATIVE STUDY.
ACTA ANAT., 86:353-375
AMMOCOETE ADULT ANATOMY NERVOUS SYSTEM EVOLUTION


4560

HOMMA, S.                    1973
GIANT INTERNEURONS OF LAMPREY SPINAL CORD.
PHYSIOL. SOC. JAP., J., 35:505.


4561F

PLISETSKAYA, E.M.                    1973
ZHELUDKOVA, Z.P.
VLIYNIE ADRENALINA NA AMNLAZNUYU AKTIVNOST' PECHENI I MYSHTS
MINOGI LAMPETRA FLUVIATILIS. [THE EFFECT OF EPINEPHRINE ON
AMYLASE ACTIVITY IN THE LIVER AND MUSCLES OF THE LAMPREY,
LAMPETRA FLUVIATILIS.]
ZH. EVOL. BIOKHIM. FIZIOL., 9:611-613.
(RUSS.)
ENGL. SUMM.
U.S.S.R. ADULT BIOCHEMISTRY ENDOCRINOLOGY


4562F

LUNDIN, L-G.                    1973
TEGELSTROM, H.
WAHREN, H.
INDICATIONS FOR A DIMERIC STRUCTURE OF AN ACID PHOSPHATASE
IN THE LAMPREY PETROMYZON FLUVIATILIS.
BIOCHEM. GENET., 10:57-67.
SWEDEN ADULT BIOCHEMISTRY TECHNIQUES GENETICS


4563F

KORNELUISSEN, H.                    1973
ULTRASTRUCTURE OF MYOTENDINOUS JUNCTIONS IN MYXINE AND RAT.
SPECIALIZATIONS BETWEEN THE PLASMA MEMBRANE AND THE LAMINA
DENSA.
Z. ANAT. ENTWICKLUNGSGESCH., 142:91-101.
*MYXINE GLUTINOSA L. ATLANTIC ADULT ANATOMY CYTOLOGY MUSCLE


4564F

SUZUKI, S.                    1974
GORBMAN, A.
PROPERTIES OF AN IODOPROTEIN FROM THYROID TISSUE OF THE
PACIFIC HAGFISH, EPTATRETUS STOUTI.
GEN. COMP. ENDOCRINOL., 22:312-314.
CANADA ENDOCRINOLOGY HISTOCHEMISTRY


4565F

POTTER, I.C.                    1975
BEAMISH, F.W.H.
LETHAL TEMPERATURES IN AMMOCOETES OF 4 SPECIES OF LAMPREYS.
ACTA ZOOL., (STOCKH.), 56:85-91.
*LAMPETRA PLANERI *PETROMYZON MARINUS *LAMPETRA
(LETHENTERON) LAMOTTENII *ICHTHYOMYZON FOSSOR AMMOCOETE PHYS


4566F

VON ESCHEN, K.B.                    1974
RUDBACH, J.A.
INACTIVATION OF ENDOTOXIN BY SERUM:A PHYLOGENETIC STUDY.
J. INFECT. DIS., 129:21-27.
*POLISTOTREMO STOUTII *PETROMYZON MARINUS U.S.A. IMMUNOLOGY
ANIMAL PISCES

4567F

MEDNIKOV, B.M.                1974
ANTONOV, A.S.
O STATUSE DVOYAKODYSHASHCHIKH RYB (DIPNOI) I IKH POLOZHENIE
V SISTEME. [STATUS OF LUNGFISHES (DIPNOI) AND THEIR
SYSTEMATIC POSITION.]
AKAD. NAUK SSSR, LENINGR., DOKL., SER. BIOL., 218:474-476.
(RUSS.)


4568

HEATH-EVES, M.J.             1974
MCMILLAN, D.B.
THE MORPHOLOGY OF THE KIDNEY OF THE ATLANTIC HAGFISH, MYXINE
GLUTINOSA.
AM. J. ANAT., 139:309-333.
ATLANTIC ADULT EXCRETION


4569

GONCHAREVSKAYA, O.A.         1977
MICROPUNCTURE STUDY OF THE RESORPTION OF SODIUM CHLORIDE AND
BICARBONATE IN THE PROXIMAL TUBULES OF THE KIDNEY IN THE
LAMPREY.
ZH. EVOL. BIOKHIM. FIZIOL., 13:642-644.
(RUSS.)
J. EVOL. BIOCHEM. PHYSIOL., 13:455-456
(ENG.)


4570F

DAHL, D.                     1973
BIGNAMI, A.
IMMUNOCHEMICAL AND IMMUNOFLUORESCENCE STUDIES OF THE GLIAL
FIBRILLARY ACIDIC PROTEIN IN VERTEBRATES.
BRAIN RES., 61:279-293.
*LAMPETRA TRIDENTATA BIOCHEMISTRY


4571

DAVISON, P.F.                1973
BERMAN, M.
CORNEAL COLLAGENASE SPECIFIC CLEAVAGE OF TYPES ALPHA-1 2,
ALPHA-2 AND ALPHA-1 3 COLLAGENS.
CONNECT. TISSUE. RES., 2:57-64.
ANIMAL BIOCHEMISTRY TECHNIQUES VISION


4572

HSU, D-S.                    1974
HOFFMAN, P.
MASHBURN, T.A. JR.
FRACTIONATION OF CHONDROITIN SULPHATE AND DETERMINATION OF
MOLECULAR WEIGHT BY POLY ACRYLAMIDE GEL ELECTROPHORESIS.
BIOCHIM. BIOPHYS. ACTA, 338:254-264.


4573F

ARLOCK, P.                   1975
ELECTRICAL ACTIVITY AND MECHANICAL RESPONSE IN THE SYSTEMIC
HEART AND THE PORTAL VEIN HEART OF MYXINE GLUTINOSA.
COMP. BIOCHEM. PHYSIOL., 51A:521-522.
SWEDEN TECHNIQUES CIRCULATORY SYSTEM PHYSIOLOGY


4574F

BUUS, O.                     1975
LARSEN, L.O.
ABSENCE OF KNOWN CORTICOSTEROIDS IN BLOOD OF RIVER LAMPREYS
LAMPETRA FLUVIATILIS AFTER TREATMENT WITH MAMMALIAN
CORTICOTROPIN.
GEN. COMP. ENDOCRINOL., 26:96-99.
*PETROMYZON MARINUS DENMARK BIOCHEMISTRY TECHNIQUES BLOOD


4575F

SMITH, K.L. JR.             1974
HESSLER, R.R.
RESPIRATION OF BENTHOPELAGIC FISHES: IN SITU MEASUREMENTS AT
1230 METERS.
SCIENCE, 184:72-73.
*EPTATRETUS DEANI METABOLISM


4576

LEWIS, J.H.                  1973
WILSON, J.H.
VARIATIONS IN ABILITIES OF ANIMAL FIBRINOGENS TO CLUMP
STAPHYLOCOCCI.
THROMB. RES., 3:419-424.


4577F

TSUNEKI, K.                  1975
GORBMAN, A.
ULTRASTRUCTURE OF THE ANTERIOR NEUROHYPOPHYSIS AND THE PARS
DISTALIS OF THE LAMPREY, LAMPETRA TRIDENTATA.
GEN. COMP. ENDOCRINOL., 25:487-508
U.S.A. ADULT HISTOLOGY NERVOUS SYSTEM ENDOCRINOLOGY


4578

JANVIER, P.                    1973
ANATOMIE ET SYSTEMATIQUE DU GENRE BOREASPIS (CYCLOSTOMI,
OSTEOSTRACI), CEPHALASPIDE DU DEVONIEN INFERIEUR DU
SPITSBERG.
P. & M. CURIE UNIV., THESIS: 157 PP.


                                                    4579
BUSTOS-VALDES, S.E.            1974
TREBILCOCK, M.A.
WARD, P.H.
CHANDA, S.K.
A COMPARATIVE STUDY OF HAGFISH AND RAT LIVER HISTONES.
COMP. BIOCHEM. PHYSIOL., 48B:329-342.
ANIMAL BIOCHEMISTRY TECHNIQUES


                                                    4580F
MARTIN, A.R.                   1975
RINGHAM, G.L.
SYNAPTIC TRANSFER AT A VERTEBRATE CENTRAL NERVOUS SYSTEM
SYNAPSE.
J. PHYSIOL., 251:409-426.
*LAMPETRA BIOCHEMISTRY


                                                    4581
SUZUKI, S.                     1975
KAWABATA, I.
STRUCTURE AND FUNCTION OF THE THYROID IN EPTATRETUS BURGERI.
PART 2: LIGHT MICROSCOPY AND ELECTRON MICROSCOPY.
ZOOL. MAG. (TOKYO), 84:405.


                                                    4582F
FANGE, R.                      1973
LARSSON, A.
LIDMAN, U.
LIVER FUNCTION AND BILE COMPOSITION IN MYXINE. PP. 89-92.
IN  FANGE,R. ED. MYXINE GLUTINOSA. BIOCHEMISTRY, PHYSIOLOGY
AND STRUCTURE. REPORT FROM A SYMPOSIUM IN GOTEBORG 28-29
APRIL, 1972. GOTEBORG:KUNGL. VETENSKAPS-OCH VITTERHETS-
SAMHLLT.
PHYSIOLOGY HISTOLOGY METABOLISM ENDOCRINOLOGY


                                                    4583
MASHANSKII, V.F.               1973
RIZHAMADZE, N.A.
A COMPARAIVE STUDY OF THE SARCOPLASMIC RETICULUM AND THE
SYSTEM IN MUSCLES OF SOME MARINE AND FRESH WATER FISHES.
TSITOLOGIYA, 15:1338-1344.


                                                    4584F
POTTER, I.C.                   1975
BROWN, I.D.
CHANGES IN HAEMOGLOBIN ELECTROPHEROGRAMS DURING THE LIFE
CYCLE OF 2 CLOSELY RELATED LAMPREYS.
COMP. BIOCHEM. PHYSIOL., 51B:517-520.
*LAMPETRA PLANERI *LAMPETRA FLUVIATILIS ENGLAND AMMOCOETE
ADULT BLOOD EVOLUTION LIFE CYCLE


                                                    4585F
GORBMAN, A.                    1975
TSUNEKI, K.
A TECHNIQUE FOR HYPOPHYSECTOMY OF THE PACIFIC HAGFISH: 1ST
OBSERVATIONS.
GEN. COMP. ENDOCRINOL., 26:420-422.
*EPTATRETUS STOUTI CANADA ENDOCRINOLOGY REPRODUCTION
PHYSIOLOGY HISTOLOGY


                                                    4586
POLENOV, A.L.                  1974
BELENKY, M.A.
KONSTANTINOVA, M.S.
THE HYPOTHALAMO-HYPOPHYSIAL SYSTEM OF THE LAMPREY LAMPETRA
FLUVIATILIS.
CELL TISSUE RES., 150:505-520.


                                                    4587F
FUJITA, H.                     1975
X-RAY MICROANALYSIS ON THE THYROID FOLLICLE OF THE HAGFISH,
EPTATRETUS BURGERI AND LAMPREY LAMPETRA JAPONICA.
HISTOCHEMISTRY, 43:283-290.
JAPAN ADULT ENDOCRINOLOGY HISTOLOGY


                                                    4588
KORNELIUSSEN, H.                1973
NICOLAYSEN, K.
ULTRASTRUCTURE OF 4 TYPES OF STRIATED MUSCLE  FIBERS IN THE
ATLANTIC HAGFISH MYXINE GLUTINOSA.
Z. ZELLFORSCH. MIKROSK ANAT., 143:273-290.


                                                    4589F
PATZNER, R.A.                  1974
DIE FRUHEN STADIEN DER OOGENESE BEI MYXINE GLUTINOSA L.

(CYCLOSTOMATA). LICHT-UND ELEKTRONENMIKROSKOPISCHE
UNTERSUCHUNGEN. [THE EARLY STAGES OF THE OOGENESIS IN MYXINE
GLUTINOSA (CYCLOSTOMATA). LIGHT- AND ELECTRONMICROSCOPIC
INVESTIGATIONS.].
NORW. J. ZOOL., 22:81-93.
GONADOGENESIS HISTOLOGY


4590F

HITCH, R.K.                     1974
ETNIER, D.A.
FISHES OF THE HIWASSEE RIVER SYSTEM TENNESSEE USA
ECOLOGICAL AND TAXONOMIC CONSIDERATIONS.
TENN. ACAD. SCI., J. 49:81-87.
*ICHTHYOMYZON HUBBSI *LAMPETRA LAMOTTEI *ICHTHYOMYZON
BDELLIUM *ICHTHYOMYZON CASTANEUS U.S.A. ADULT AMMOCOETE
DISTRIBUTION ECOLOGY


4591F

TSUNEKI, K.                     1974
URANO, A.
KOBAYASHI, H.
MONOAMINE OXIDASE AND ACETYLCHOLINESTERASE IN THE
NEUROHYPOPHYSIS OF THE HAGFISH EPTATRETUS BURGERI.
GEN. COMP. ENDOCRINOL., 24:249-256.
JAPAN ADULT HISTOLOGY BIOCHEMISTRY ENDOCRINOLOGY NERVOUS
SYSTEM PHYSIOLOGY


4592F

LARSEN, L.O.                    1974
EFFECTS OF TESTOSTERONE AND OESTRADIOL ON GONADECTOMIZED
AND INTACT MALE AND FEMALE RIVER LAMPREYS LAMPETRA
FLUVIATILIS (L.) GRAY.
GEN. COMP. ENDOCRINOL., 24:305-313.
DENMARK ADULT ENDOCRINOLOGY HISTOLOGY GONADOGENESIS
REPRODUCTION


4593F

BEAMISH, F.W.H.                 1974
SWIMMING PERFORMANCE OF ADULT SEA LAMPREY PETROMYZON MARINUS
IN RELATION TO WEIGHT AND TEMPERATURE.
AM. FISH. SOC., TRANS., 103:355-358.
CANADA ADULT LOCOMOTION MIGRATION


4594F

MURTAUGH, P.A.                  1974
HALVER, J.E.
LEWIS, M.S.
GLADNER, J.A.
CROSS LINKING REACTIONS OF LAMPREY FIBRINOGEN AND FIBRIN.
BIOCHIM. BIOPHYS. ACTA, 359:415-420.
*LAMPETRA TRIDENTATUS U.S.A. BLOOD ANIMAL PHYSIOLOGY


4595F

PFENNINGER, K.H.                1974
ROVAINEN, C.M.
STIMULATION DEPENDENCE AND CALCIUM DEPENDENCE OF VESICLE
ATTACHMENT SITES IN THE SYNAPTIC MEMBRANE A FREEZE CLEAVE
STUDYON THE LAMPREY SPINAL CORD.
BRAIN RES., 72:1-23
*PETROMYZON MARINUS U.S.A. AMMOCOETE ADULT MORPHOLOGY
NERVOUS SYSTEM


4596F

THOMAS, N.W.                    1973
OSTBERG, Y.
FALKMER, S.
A 2ND GRANULAR CELL IN THE ENDOCRINE PANCREAS OF THE HAGFISH
MYXINE GLUTINOSA.
ACTA ZOOL., (STOCKH.), 54:201-207.
SWEDEN ADULT ENDOCRINOLOGY HISTOLOGY CYTOLOGY


4597F

HELLE, K.B.                     1973
LONNING, S.
SARCOPLASMIC RETICULUM  IN THE PORTAL VEIN, HEART AND
VENTRICLE OF THE CYCLOSTOME MYXINE GLUTINOSA.
J. MOL. CELL CARDIOL., 5:433-439.
NORWAY CIRCULATORY SYSTEM CYTOLOGY HISTOLOGY PHYSIOLOGY


4598F

PLISETSKAYA, E.M.               1973
LEIBSON, L.G.
VLIYNIE GORMONOV NA GLIKOGEN-SINTETAZNUYU AKTIVIOST' PECHENI
I MYSHTS U MINOG I SKORPEM. [EFFECT OF HORMONES ON GLYCOGEN
SYNTHETASE ACTIVITY OF THE LIVER AND MUSCLES OF THE LAMPREY
AND SCORPIONFISH.]
AKAD. NAUK SSSR, LENINGR., DOKL., SER. BIOL.,
210:1230-1232..
(RUSS.)
*LAMPETRA FLUVIATILIS U.S.S.R. BEHAVIOUR BIOCHEMISTRY
ENDOCRINOLOGY MUSCLE

4599F

MATHERS, J.S.                    1974
BEAMISH, F.W.H.
CHANGES IN SERUM OSMOTIC AND IONIC CONCENTRATION IN
LANDLOCKED PETROMYZON MARINUS.
COMP. BIOCHEM. PHYSIOL., 49A:677-688.
*LAMPETRA PLANERI *LAMPETRA FLUVIATILIS CANADA
OSMOREGULATION AMMOCOETE ADULT SPAWNING LIFE CYCLE
PHYSIOLOGY METAMORPHOSIS IONIC REGULATION


4600F

VLADYKOV, V.D.                   1973
NORTH AMERICAN NONPARASITIC LAMPREYS OF THE FAMILY
PETROMYZONIDAE MUST BE PROTECTED.
CAN. FIELD NAT., 87:235-239.
U.S.A. AMMOCOETE ADULT ANIMAL BIOLOGY CHEMISTRY ECOLOGY
FEEDING MANAGEMENT MORTALITY OF


4601F

BATUEVA, I.V.                    1974
SHAPOVALOV, A.I.
SINAPTICHESKIE VOZDEYSTVIYA VYZYVAEMYE VMOTONEIRONAKH MINOGI
PRI SUPRASPINAL'NOM 1 INTRASPINAL'NOM RAZDRAZHENII.
[SYNAPTIC ACTIONS EVOKED IN LAMPREY MOTONEURONS BY
SUPRASPINAL AND INTRA SPINAL STIMULATION.]
NEIROFIZIOLOGIYA, 6:629-635.
(RUSS.)
ENGL. SUMM.
*LAMPETRA FLUVIATILIS U.S.S.R. NERVOUS SYSTEM PHYSIOLOGY


4602F

CARTON, Y.                       1974
PARENTES ENTRE LES HEMAGGLUTININES NATURELLES D'ECHINODERMES
ET LES CHAINES DES IMMUNOGLOBULINES DE VERTEBRES. [THE
RELATIONSHIP BETWEEN NATURALLY OCCURING ECHINODERM
HAEMOGGLUTININS AND VERTEBRATE IMMUNOGLOBULINS CHAINS.]
ANN. IMMUNOL., 125C:731-745.
(FR.)
ENG. SUMM.
*PETROMYZON MARINUS FRANCE BIOLOGY EVOLUTION BIOCHEMISTRY


4603F

NAKAO, T.                        1974
UCHINOMIYA, K.
INTRACISTERNAL TUBULES OF LAMPREY CHLORIDE CELLS.
J. ELECTRONMICROSC., 23:51-55.
*LAMPETRA JAPONICA JAPAN HISTOLOGY CYTOLOGY


4604F

POHL, R.J.                       1974
BEND, J.R.
GUARINO, A.M.
FOUTS, J.R.
HEPATIC MICROSOMAL MIXED FUNCTION OXIDASE ACTIVITY OF
SEVERAL MARINE SPECIES FROM COASTAL MAINE USA.
DRUG METAB. DISPOS., 2:545-555.
ANIMAL ADULT BIOCHEMISTRY HISTOLOGY METABOLISM EXCRETION


4605F

GOOLD, R.J.                      1976
SEA LAMPREY SURVEYS OF STREAMS TRIBUTARY TO LAKE HURON,
1975. PP. 7-12.
IN  TIBBLES, J.J. ED., ANNUAL REPORT 1975 OF THE SEA LAMPREY
CONTROL CENTRE, SAULT STE. MARIE, ONTARIO, TO THE GREAT
LAKES FISHERY COMMISSION. GREAT LAKES FISH. COMM., ANNU.
REP.
RESTRICTED
CANADA AMMOCOETE DISTRIBUTION MANAGEMENT


4606F

KONSTANTINOVA, M.S.              1975
MONOAMINY VLIKVOR-KOTAKTNYKH NERVNYKH KLETKAKH GIPOTALAMUSA
U POZVONOCHNYKH. [MONOAMINES IN THE LIQUOR CONTACTING NERVE
CELLS OF THE HYPOTHALAMUS IN VERTEBRATES.]
ZH. EVOL. BIOKHIM. FIZIOL., 11:187-190.
(RUSS.)


4607F

WEISBART, M.                     1975
INVITRO INCUBATIONS OF PRESUMPTIVE ADRENOCORTICAL CELLS FROM
THE OPISTHONEPHROS OF THE ADULT SEA LAMPREY PETROMYZON
MARINUS.
GEN. COMP. ENDOCRINOL., 26:368-373.
U.S.A. ADULT ENDOCRINOLOGY METABOLISM REPRODUCTION


4608

OOTA, Y.                         1974
ELECTRON MICROSCOPIC STUDIES ON THE ADENOHYPOPHYSIS OF THE
HAGFISH EPTATRETUS BURGERI.
SHIZUOKA. UNIV. FAC. SCI., REP., 9:67-78.

4609F

HIROSE, K.                    1975
TAMAOKI, B-I.
FERNHOLM, B.
KOBAYASHI, H.
IN VITRO BIOCONVERSIONS OF STEROIDS IN THE MATURE OVARY OF
THE HAGFISH, EPTATRETUS BURGERI.
COMP. BIOCHEM. PHYSIOL., 51B:403-408.
BIOCHEMISTRY ENDOCRINOLOGY


4610F

KORNELIUSSEN, H.               1973
ULTRASTRUCTURE OF MOTOR NERVE TERMINALS ON DIFFERENT TYPES
OF MUSCLE FIBERS IN THE ATLANTIC HAGFISH MYXINE GLUTINOSA
Z. ZELLFORSCH. MIKROSK. ANAT., 147:87-106.
ADULT HISTOLOGY MUSCLE


4611

HALSTEAD, L.B.                1973
AFFINITIES OF THE HETEROSTRACI AGNATHA.
LINN. SOC. LOND., J. (BIOL.),


4612

KARAMYAN, A.I.                1973
AGAYAN, A.L.
VESELKIN, N.P.
EVOKED POTENTIALS IN VARIOUS REGIONS OF THE LAMPREY BRAIN
FOLLOWING STIMULATION OF DORSAL PARTS OF THE SPINAL CORD.
BIOL. ZH. ARM., 26:56-63.


4613F

SAVINA, M.V.                  1975
WRONISZEWSKA, A.
WOJTCZAK, L.
MITOCHONDRIA FROM THE LAMPREY LAMPETRA FLUVIATILIS
OXIDATIVEPHOSPHORYLATION AND RELATED PROCESSES.
ACTA BIOCHIM. POL., 22:229-238.
BALTIC SEA ADULT METABOLISM HISTOLOGY


4614F

MATHIS, B.J.                  1975
KEVERN, N.R.
DISTRIBUTION OF MERCURY CADMIUM LEAD AND THALLIUM IN A
EUTROPHIC LAKE.
HYDROBIOLOGIA, 46:207-222.


4615

LEHTOLA, K.A.                 1973
ORDOVICIAN VERTEBRATES FROM ONTARIO CANADA.
UNIV. MICH., MUS. PALEONTOL., CONTRIB., 24:23-30.


4616F

KRAYUSHKINA, L.S.             1974
KHLORIDSEKRETIRUYUSHCHIE KLETKI RYB. [CHLORIDE SECRETING
CELLS OF FISHES.]
ARKH. ANAT. GISTOL. EMBRIOL., 67:92-99.
(RUSS.)
METABOLISM


4617F

OSTBERG, Y.                   1975
VAN NOORDEN, S.
PEARSE, A.G.E.
CYTOCHEMICAL, IMMUNOFLUORESCENCE AND ULTRASTRUCTURAL
INVESTIGATIONS ON POLYPEPTIDE HORMONE LOCALIZATION IN THE
ISLET PARENCHYMA AND BILE DUCT MUCOSA OF A CYCLOSTOME,
MYXINE GLUTINOSA.
GEN. COMP. ENDOCRINOL., 25:274-291.
SWEDEN ENDOCRINOLOGY HISTOLOGY DIGESTION HISTOCHEMISTRY
CYTOLOGY


4618F

TOMONAGA, S.                  1975
SAKAI, K.
TASHIRO, J.
AWAYA, K.
HIGH WALLED ENDOTHELIUM IN THE GILLS OF THE HAGFISH.
ZOOL. MAG. (TOKYO), 84:151-155.
*EPTATRETUS BURGERI JAPAN RESPIRATION HISTOLOGY
HISTOCHEMISTRY CYTOLOGY


4619F

CIACCIO, G.                   1972
DELL'AGATA, M.
RICERCHE SULLA ISTOLOGIAE SULLA ISTOCHIMICA DI FIBRE
MUSCOLARI STRIATE A MIOTUBO IN LAMPETRA PLANERI (BLOCH).
[RESEARCH ON THE HISTOLOGY AND HISTOCHEMISTRY OF MYOTUBE
STRIATED MUSCLE IN LAMPETRA PLANERI (BLOCH).]
RIV. BIOL., 65:331-345.
(ITAL.)
ENGL. TRANSL.

ITALY AMMOCOETE ADULT CIRCULATORY SYSTEM HISTOLOGY CYTOLOGY
HISTOCHEMISTRY

4620

READ, L.J.                        1975
ABSENCE OF UREOGENIC PATHWAYS IN LIVER OF THE HAGFISH
BDELLOSTOMA CIRRHATUM.
COMP. BIOCHEM. PHYSIOL., 51B:139-142.
*BDELLESTOMA CIRRHATUM BLOOD EXCRETION


4621

MCINERNEY, J.E.                   1974
RENAL SODIUM REABSORPTION IN THE HAGFISH EPTATRETUS STOUTI.
COMP. BIOCHEM. PHYSIOL., 49:273-280.


4622F

KAMYSHNAYA, M.S.                  1973
TSEPKIN, YE. A.
MATERIALY K EKOLOGII SHCHUKI ESOX LUCIUS L. NIZOV'EV R.
UMBY. [MATERIAL CONCERNING THE ECOLOGY OF THE PIKE ESOX
LUCIUS IN THE LOWER PART OF THE RIVER UMBRA.
J. ICHTHYOL., 13:929-933.
(RUSS.)
*LAMPETRA JAPONICA (MARTENS) CYTOLOGY METABOLISM


4623F

POLMORYKHINA, A.N.                1974
MORFOLOGICHESKIE OSOBENNOSTI I IZMENCHIVOST' SIBIRSKOI
MINOGI LAMPETRA JAPONICA KESSLERI (ANIKIN) VODOEMOV
VERKHNEGO IRTYSHA. [MORPHOLOGICAL PECULIARITIES AND
CHANGEABILITY OF SIBERIAN LAMPREY LAMPETRA JAPONICA KESSLERI
(ANIKIN) OF UPPER IRTYSH BASIN.]
VOPR. IKHTIOL., 14:218-230.
(RUSS.)
J. ICHTHIOL., 14:192-202.
(ENG.)
RUSSIA ADULT ANATOMY DISTRIBUTION ECOLOGY MORPHOMETRY
SYSTEMATICS


4624

HENDERSON, I.W.                   1972
JONES, I.C.
HORMONES AND OSMOREGULATION IN FISHES.
ANN. INST. MICHEL PACHA, 5:69-235.


4625F

ROVAINEN, C.M.                    1974
SYNAPTIC INTERACTIONS OF RETICULOSPINAL NEURONS AND NERVE
CELLS IN THE SPINAL CORD OF THE SEA LAMPREY.
J. COMP. NEUROL., 154:207-224.
*PETROMYZON MARINUS U.S.A. ADULT AMMOCOETE NERVOUS SYSTEM
CYTOLOGY


4626F

WU, T.T.                          1973
KABAT, E.A.
ATTEMPT TO EVALUATE THE INFLUENCE OF NEIGHBOURING AMINO
ACIDS (N-1) AND (N+1) ON THE BACKBONE CONFORMATION OF AMINO
ACIDS IN PROTEINS. USE IN PREDICTING THE 3-DIMENSIONAL
STRUCTUREOF THE POLY PEPT...
J. MOL. BIOL., 75:13-31.
BIOCHEMISTRY ANIMAL


4627F

CSERR, H.F.                       1974
OSTRACH, L.H.
ON THE PRESENCE OF SUBARACHNOID FLUID IN THE MUDPUPPY
NECTURUS MACULOSUS.
COMP. BIOCHEM. PHYSIOL., 48A:145-151.
*MYXINE GLUTINOSA BIOCHEMISTRY


4628F

HIRABAYASHI, T.                   1974
PERRY, S.V.
IMMUNOCHEMICAL STUDY OF THE CALCIUM ION-BINDING PROTEIN
(TROPONIN-C) AND THE INHIBITORY PROTEIN (TROPONIN-I) OF THE
TROPONIN COMPLEX AND THEIR INTERACTION.
BIOCHIM. BIOPHYS. ACTA, 351:273-289.
BIOCHEMISTRY MUSCLE PHYSIOLOGY


4629

CUTFIELD, J.F.                    1974
CUTFIELD, S.M.
DODSON, E.J.
DODSON, G.G.
SABESAN, M.N.
LOW RESOLUTION CRYSTAL STRUCTURE OF HAGFISH INSULIN.
J. MOL. BIOL., 87:23-30.
*MYXINE GLUTINOSA UNITED KINGDOM BIOCHEMISTRY ENDOCRINOLOGY


4630F

SMITH, B.R.                    1974
TIBBLES, J.J.
JOHNSON, B.G.H.
CONTROL OF THE SEA LAMPREY PETROMYZON MARINUS IN LAKE
SUPERIOR 1953-1970.
GREAT LAKES FISH. COMM., TECH. REP., 26:1-60.
RESTRICTED
NORTH AMERICA AMMOCOETE ADULT MANAGEMENT HISTORY MORTALITY
OF MORTALITY BY

                                                          4631F
FERNHOLM, B.                   1975
HOLMBERG, K.
THE EYES IN THREE GENERA OF HAGFISH (EPTATRETUS, PARAMYXINE
AND MYXINE) - A CASE OF DEGENERATIVE EVOLUTION.
VISION RES., 15:253-2599
VISION HISTOLOGY PIGMENTATION

                                                          4632F
KOZITSYN, S.A.                 1974
PTITSYN, O.B.
STRUKTURA GIDROFOBNYKH YADERGLOBINOV. [STUCTURE OF
HYDROPHOBIC CORES OF GLOBINS.]
MOL. BIOL., 8:536-542.
(RUSS.)
MOL. BIOL., 8:427-433.
(ENG.)
HISTOCHEMISTRY HISTOLOGY

                                                          4633F
BEAMISH, R.J.                  1976
WILLIAMS, N.E.
PRELIMINARY REPORT ON THE EFFECTS OF RIVER LAMPREY (LAMPETRA
AYRESII) PREDATION ON SALMON AND HERRING STOCKS.
ENVIRON. CAN., FISH. MAR. SERV., TECH. REP., 611:25PP.
ADULT DISTRIBUTION FEEDING GROWTH

                                                          4634F
BRONSHTEIN, A.A.               1974
ELEKTRONNOMIKROSKOPICHESKIE ISSELEDOVANIYA ORGANA OBONYANIYA
POZVONOCHNYKH. [ELECTRON MICROSCOPIC STUDY OF THE OLFACTORY
ORGAN IN VERTEBRATES.]
ARKH. ANAT. GISTOL. EMBRIOL., 67:22-40.
(RUSS.)
ENG. SUMM.
*CYCLOSTOMATA ANIMAL CYTOLOGY BIOCHEMISTRY

                                                          4635
KOSMACH, P.I.                  1974
GORDIENKO, V.M.
INSULIN IMMUNOGENICITY.
PROBL. ENDOKRINOL., 20:104-111.

                                                          4636F
LUKOMSKAYA, N.YA.              1974
MAGAZANIK, L.G.
'LOKIRUYUSHCHEE DEISTVIE POLIPEPTIDOV ZMEINOGO YADA NA
KHOLINORETSEPTIVNYE MEM'RANY RECHNOI MINOGI LAMPETRA
FLUVIATILIS.[THE INHIBITORY EFFECT OF SNAKE VENOM
POLYPEPTIDES ON CHOLINORECEPTIVE MEMBRANES OF THE LAMPREY
LAMPETRA FLUVIATILIS.]
ZH. EVOL. BIOKHIM. FIZIOL., 10:524-526.
(RUSS.)
ENGL. SUMM.
USSR NERVOUS SYSTEM PHYSIOLOGY ANIMAL

                                                          4637F
GARTMAN, D.K.                  1974
SOME OBSERVATIONS MADE DURING A 1 DAY SCUBA INVESTIGATION
OF THE MILLER BLUE HOLE SANDUSKY COUNTY OHIO USA.
OHIO J. SCI., 74:330-331.
U.S.A. CHEMISTRY

                                                          4638
JONES, A.N.                    1975
A PRELIMINARY STUDY OF FISH SEGREGATION IN SALMON SPAWNING
STREAMS.
J. FISH BIOL., 7:95-104.

                                                          4639F
YOUSON, J.H.                   1973
A COMPARISON OF PRESUMPTIVE INTERRENAL TISSUE IN THE
OPISTHONEPHRIC KIDNEY AND DORSAL VESSEL REGION OF THE LARVAL
SEA LAMPREY, PETROMYZON MARINUS.
CAN. J. ZOOL., 51:769-799.
CANADA LAKE ONTARIO AMMOCOETE ENDOCRINOLOGY EXCRETION

                                                          4640F
CASLEY-SMITH, J.R.             1975
THE FINE STRUCTURE OF THE BLOOD CAPILLARIES OF SOME
ENDOCRINE GLANDS OF THE HAGFISH EPTATRETUS STOUTI:

IMPLICATIONS FOR THE EVOLUTION OF BLOOD AND LYMPH VESSELS.
REV. SUISSE ZOOL., 82:35-40.
HISTOLOGY ENDOCRINOLOGY EVOLUTION CIRCULATORY SYSTEM

4641

WAECHTLER, K.                    1974
THE DISTRIBUTION OF ACETYLCHOLINESTERASE IN THE CYCLOSTOME
BRAIN. PART I: LAMPETRA PLANERI.
CELL TISSUE RES., 152:259-270.


4642F

KARAMYAN, A.I.                   1975
VZGLYADY A.A. UKHTOMSKOGO NA IERARKHICHESKUYU ORGANIZATSIYU
TSENTRL'NOI NERVNOI SISTEMY I SOVREMENNYE DOSTIZHENIYA
EVOLYUTSIONNOI NEIROFIZIOLOGII. [THE IDEAS OF A.A.
UKHTOMSKII ON THE HIERARCHICAL ORGANIZATION OF THE CENTRAL
NERVOUS SYSTEM AND MODERN ACHIEVEMENTS OF EVOLUTIONARY
NEUROPHYSIOLOGY.]
ZH. EVOL. BIOKHIM. FIZIOL., 11:218-224.
(RUSS.)


4643F

RINGHAM, G.L.                    1975
LOCALIZATION AND ELECTRICAL CHARACTERISTICS OF A GIANT
SYNAPSE IN THE SPINAL CORD OF THE LAMPREY.
J. PHYSIOL., 251:395-407.
*PETROMYZON MARINUS ADULT NERVOUS SYSTEM SENSE RECEPTORS
PHYSIOLOGY


4644F

NAKAO, T.                        1975
FINE STRUCTURE OF THE MYOTENDINOUS JUNCTION AND 'TERMINAL
COUPLING' IN THE SKELETAL MUSCLE OF THE LAMPREY, LAMPETRA
JAPONICA.
ANAT. REC., 182:321-338.
JAPAN MUSCLE HISTOLOGY


4645F

SHERA, W.P.                      1975
SEA LAMPREY SURVEYS OF STREAMS TRIBUTARY TO LAKE
HURON,1974.PP. 10-14.
IN  TIBBLES,J.J. ED. ANNUAL REPORT 1974 OF THE SEA LAMPREY
CONTROL CENTRE SAULT STE. MARIE, ONTARIO, TO THE GREAT LAKES
FISHERY COMMISSION. GREAT LAKES FISH. COMM., ANNU. REP.
RESTRICTED
*PETROMYZON MARINUS *ICHTHYOMYZON SP. *LAMPETRA LAMOTTEI
CANADA AMMOCOETE DISTRIBUTION GROWTH


4646F

BAGE, G.                         1975
FERNHOLM, B.
ULTRASTRUCTURE OF THE PRO-ADENOHYPOPHYSIS OF THE RIVER
LAMPREY, LAMPETRA FLUVIATILIS, DURING GONAD MATURATION.
ACTA ZOOL., 56:95-118
SWEDEN ADULT GONADOGENESIS HISTOLOGY PHYSIOLOGY CYTOLOGY


4647F

KAWATSKI, J.A.                   1975
BITTNER, M.A.
UPTAKE ELIMINATION AND BIOTRANSFORMATION OF THE LAMPRICIDE
3-TRIFLUOROMETHYL-4-NITROPHENOL (TFM) BY LARVAE OF THE
AQUATIC MIDGE CHIRONOMUS TENTANS.
TOXICOLOGY, 4:183-194.
CHEMISTRY MANAGEMENT


4648F

JANVIER, P.                      1975
LES YEUX DES CYCLOSTOMES FOSSILES ET LE PROBLEME DE
L'ORIGINE DES MYXINOIDES. [THE EYES OF THE FOSSIL
CYCLOSTOMES AND THE PROBLEM OF THE ORIGIN OF MYXINOIDS.]
ACTA ZOOL., 56:1-9.
(FR.)
ENGL. SUMM.
*EPTATRETUS BURGERI *MYXINE GLUTINOSA VISION EVOLUTION


4649F

POTTER, I.C.                     1974
ROBINSON, E.S.
BROWN, I.D.
STUDIES ON THE ERYTHROCYTES OF LARVAL AND ADULT LAMPREYS
LAMPETRA FLUVIATILIS.
ACTA ZOOL., (STOCKH.) 55:173-177.
GREAT BRITIAN AMMOCOETE ADULT BLOOD CYTOLOGY


4650F

KUZNETSOV, N.V.                  1974
GOROKHOV, YU.A.
POSMNOV, I.E.
SPISOK RYB GOR'KOVSKOI OBLASTI. [LIST OF FISHES IN THE
GORKII REGION.]
VOPR. IKHTIOL., 14:34-40.

(RUSS.)
*PETROMYZONIDAE U.S.S.R. DISTRIBUTION

4651F

CLARIDGE, P.N.                    1974
POTTER, I.C.
HEART RATIOS AT DIFFERENT STAGES IN THE LIFE CYCLE OF
LAMPREYS.
ACTA ZOOL., 55:61-69.
*PETROMYZON MARINUS *LAMPETRA FLUVIATILIS *LAMPETRA PLANERI
GREAT BRITIAN CANADA ADULT AMMOCOETE CIRCULATORY SYSTEM


4652F

LECH, J.J.                        1974
GLUCURONIDE FORMATION IN RAINBOW TROUT - EFFECT OF
SALICYLAMIDE ON THE ACUTE TOXICITY CONJUGATION AND
EXCRETIONOF 3-TRIFLUOROMETHYL-4-NITROPHENOL.
BIOCHEM. PHARMACOL., 23:2403-2410.
BIOCHEMISTRY CHEMISTRY MANAGEMENT


4653F

BRIGHAM, W.U.                     1973
NEST CONSTRUCTION OF THE LAMPREY, LAMPETRA AEPYPTERA.
COPEIA, 1973:135-136.
U.S.A. SPAWNING ADULT


4654F

PETERSON, J.D.                    1975
STEINER, D.F.
EMDIN, S.O.
FALKMER, S.
THE AMINO ACID SEQUENCE OF THE INSULIN FROM A PRIMITIVE
VERTEBRATE THE ATLANTIC HAGFISH MYXINE GLUTINOSA.
J. BIOL. CHEM.,250:5183-5191.
SWEDEN METABOLISM BIOCHEMISTRY


4655F

OHMAN, P.                         1974
FINE STRUCTURE OF THE RETINAL PIGMENT EPITHELIUM OF THE
RIVER LAMPREY LAMPETRA FLUVIATILIS CYCLOSTOMI.
ACTA ZOOL., 55:245-254
SWEDEN VISION HISTOLOGY ADULT


4656F

HENDERSON, N.E.                   1975
LORSCHEIDER, F.L.
THYROXINE AND PROTEIN BOUND IODINE CONCENTRATIONS IN PLASMA
OF THE PACIFIC HAGFISH EPTATRETUS STOUTI CYCLOSTOMATA.
COMP. BIOCHEM. PHYSIOL., 51A:723-726.
CANADA BLOOD BIOCHEMISTRY


4657

DODD, J.M.                        1971
FOLLETT, B.K.
SHARP, P.J.
HYPOTHALMIC CONTROL OF PITUITARY FUNCTION IN SUBMAMMALIAN
VERTEBRATES. PP. 113-223.
IN  LOWENSTEIN,O. ED. ADVANCES IN COMPARATIVE PHYSIOLOGY
AND BIOCHEMISTRY, VOL.4. NEW YORK: ACADEMIC PRESS.


4658

NAKAZAWA, T.                      1974
THE SCHIFF REACTION OF THE LAMPREY RETINA ENTOSPHNUS
JAPONICUS.
ACTA HISTOCHEM. CYTOCHEM., 7:56.


4659F

WEISBART, M.                      1974
YOUSON, J.H.
INVITRO INCUBATIONS OF PRESUMPTIVE ADRENOCORTICAL CELLS AND
TESTICULAR TISSUE OBTAINED FROM THE SEA LAMPREY PETROMYZON
MARINUS DURING THEIR PARASTIC LIFE STAGE.
AM. ZOOL., 14:1296.
(ABSTR.)
HISTOCHEMISTRY GONADOGENESIS ADULT


4660F

HUNN, J.B.                        1975
ALLEN, J.L.
RESIDUE DYNAMICS OF QUINALDINE AND TFM IN RAINBOW TROUT.
GEN. PHARMACOL., 6:15-18.
CHEMISTRY MANAGEMENT


4661

BIRNBERGER, K.L.                  1974
NEURO PHYSIOLOGICAL MECHANISMS OF HABITUATION.
ELECTROENCEPHALOGR. CLIN. NEUROPHYSIOL., 36:438.


4662

HOFFMAN, G.L.                     1973
THE EFFECT OF CERTAIN PARASITES ON NORTH AMERICAN FRESH

WATER FISHES. PP. 1622-1627.
IN   SLADECEK,V. ED., PROCEEDINGS OF THE INTERNATIONAL
ASSOCIATION OF THEORETICAL AND APPLIED LIMNOLOGY. VOL. 18.
LENINGRAD: NAUKA PUBLISHING HOUSE.


4663F

HOSHINO, T.                    1975
AN ELECTRON MICROSCOPE STUDY OF THE OTOLITHIC MACULAE OF
THE LAMPREY ENTOSPHENUS JAPONICUS.
J. ELECTRONMICROSC., 23:299.
ADULT HEARING ANATOMY HISTOLOGY CYTOLOGY


4664

GETSEVICHYUTE, S.              1974
PARASITE FAUNA OF THE RIVER LAMPREY LAMPETRA FLUVIATILIS
FROM THE KURSKI.
ACTA PARASITOL. LITH., 12:59-62.


4665

CHEZE, G.                      1973
GAS, N.
RELATIONS BETWEEN THE PIGMENTARY RETINAL EPITHELIUM AND
VISUALCELLS IN A TELEOST CYPRINUS CARPIO.
SOC. ZOOL. FR., PARIS, BULL., 98:532-539.


4667F

PATZNER, R.A.                  1975
DIE FORTSCHREITENDE ENTWICKLUNG UND REIFUNG DER EIER VON
MYXINE GLUTINOSA L. (CYCLOSTOMATA). LICHT- UND
ELEKTRONENMIKROSKOPISCHE UNTERSUCHUNGEN. [THE PROGRESSIVE
DEVELOPMENT AND MATURATION OF THE EGGS IN MYXINE GLUTINOSA
L. (CYCLOSTOMATA.) LIGHT AND ELECTRON MICROSCOPICAL
INVESTIGATIONS.]
NORW. J. ZOOL., 23:111-120.
(GER.)
ENGL. SUMM.
REPRODUCTION EGG CYTOLOGY


4668

MIKHEL'SON, M. YA.            1975
ONE ASPECT OF EVOLUTIONARY PHARMACOLOGY. OLIGOMERIZATION OF
CHOLINORECEPTORS DURING DEVELOPMENT OF THE ANIMAL KINGDOM.
ZH. EVOL. BIOKHIM. FIZIOL., 11:20-27.
(RUSS.)
J. EVOL. BIOCHEM. PHYSIOL., 11:14-19.
(ENG.)
*LAMPETRA FLUVIATILIS BIOCHEMISTRY MUSCLE EVOLUTION


4669F

TSUNEKI, K.                    1975
FERNHOLM, B.
EFFECT OF THYROTROPIN RELEASING HORMONE ON THE THYROID OF A
TELEOST, CHASMICHTHYS DOLICHOGNATHUS, AND A HAGFISH
EPTATRETUS BURGERI.
ACTA ZOOL., (STOCKH.), 56:61-66.
JAPAN ADULT ANIMAL ENDOCRINOLOGY METABOLISM


4670F

NATOCHIN, YU.V.                1975
LEONT'EV, V.G.
MASLOVA, M.N.
POMAZANSKAYA, L.F.
REABSORBTSIYA NATRIYA I MEMBRANNYE SISTEMY POCHEK
POZVONOCHNYKH. [SODIUM REABSORPTION AND MEMBRANE SYSTEMS IN
THE VERTEBRATE KIDNEY.]
ZH. EVOL. BIOKHIM. FIZIOL., 11:45-52.
(RUSS.)
ENGL. SUMM.
*LAMPETRA FLUVIATILIS ANIMAL IONIC REGULATION


4671F

HIRABAYASHI, T.                1974
PERRY, S.V.
IMMUNOCHEMICAL STUDIES OF TROPONIN-C. PP. 65-67.
IN   DRABIKOWSKI,H., H. STRZELECKA-GOLASZEWSKA AND E.
CARAFOLI, EDS., CALCIUM BINDING PROTEINS. WARSAW: PWN-POLISH
SCIENTIFIC PUBLISHERS.
ANIMAL BIOCHEMISTRY


4672F

SJOBERG, K.                    1973
FISKATANDE FAGLAR I ETT STROMEKOSYSTEM. [FISH-EATING BIRDS
IN A STREAM ECOSYSTEM.]
ZOOL. REVY., 35:131-134.
(SWED.)
ENGL. SUMM.
*LAMPETRA FLUVIATILIS


4673F

FARMER, G.J.                   1975
BEAMISH, F.W.H.

ROBINSON, G.A.
FOOD CONSUMPTION OF THE ADULT LANDLOCKED SEA LAMPREY
PETROMYZON MARINUS.
COMP. BIOCHEM. PHYSIOL., 50A:753-757.
CANADA DIGESTION EXCRETION FEEDING GROWTH PARASITISM BY
PHYSIOLOGY

4674F

ROVAINEN, C.M.              1974
RESPIRATORY MOTONEURONS IN LAMPREYS.
J. COMP. PHYSIOL., 94A:57-68.
*PETROMYZON MARINUS *ICHTHYOMYZON CASTANEUS U.S.A AMMOCOETE
MUSCLE NERVOUS SYSTEM PHYSIOLOGY ADULT

4675F

JANVIER, P.                1974
THE STRUCTURE OF THE NASO-HYPOPHYSIAL COMPLEX AND THE
MOUTH IN FOSSIL AND EXTANT CYCLOSTOMES WITH REMARKS ON
AMPHIASPIFORMS.
ZOOL. SCR., 3:193-200.
*MYXINE ANATOMY EVOLUTION EMBRYOLOGY

4676F

ROHDE, F.C.                1974
ARNDT, R.G.
WANG, J.C.S.
ADDITIONAL RECORDS OF THE LEAST BROOK LAMPREY OKKELBERGIA
AEPYPTERA (ABBOTT), FROM THE DELMARVA PENINSULA.
CHESAPEAKE SCI., 15:154-155.
*PETROMYZON MARINUS U.S.A. AMMOCOETE
ADULT METAMORPHOSIS DISTRIBUTION ECOLOGY SYSTEMATICS
MORPHOMETRY

4678F

BIRD, D.J.                 1976
LUTZ, P.L.
POTTER, I.C.
OXYGEN DISSOCIATION CURVES OF THE BLOOD OF LARVAL AND ADULT
LAMPREYS LAMPETRA FLUVIATILIS.
J. EXP. BIOL., 65:449-458.
*ICHTHYOMYZON UNICUSPIS *ICHTHYOMYZON HUBBSI *LAMPETRA
(ENTOSPENUS) TRIDENTATA *LAMPETRA PLANERI UNITED KINGDOM
ADULT AMMOCOETE BIOCHEMISTRY METABOLISM METAMORPHOSIS
PHYSIOLOGY RESPIRATION

4679

TURNER, P.                 1974
MARINE CALCARENITES FROM THE RINGERIKE GROUP STAGE 10 OF
SOUTHERN NORWAY.
NOR. GEOL. TIDSSKR., 54:1-12.

4680

FRANKENBERG, R.            1974
NATIVE FRESH WATER FISH.
IN  WILLIAMS,W.D. ED. MONOGRAPHIAE BIOLOGICAE, VOL. 25.
BIOGEOGRAPHY AND ECOLOGY IN TASMANIA. NETHERLANDS,THE HAGUE.

4681F

PLISETSKAYA, E.M.          1971
KUZ'MINA, V.V.
UROVEN' GLIKEMII U KRUGLOROTYKH CYCLOSTOMATA I RYB PISCES.
[LEVEL OF GLYCEMIA IN CYCLOSTOMES (CYCLOSTOMATA) AND IN FISH
(PISCES)].
VOPR. IKHTIOL., 11:1077-1087.
(RUSS.)
J. ICHTHYOL., 11:1061-1070.
(ENG.)
BLOOD

4682

MURTAUGH, P.A.             1975
HALVER, J.E.
GLADNER, J.A.
COMPARISON OF THE CROSS LINKING PATTERN OF LAMPREY FIBRIN
WITH LAMPREY FACTOR XIII AND HUMAN FACTOR XIII.
FED. PROC., 34:345.
(ABSTR.)

4683

DOOLITTLE, R.F.           1975
COTTRELL, B.A.
AMINO ACID SEQUENCE STUDIES ON LAMPREY FIBRINOGEN FIBRIN
AND FIBRINO PEPTIDES.
FED. PROC., 34: 281.

4684

NATOCHIN, YU.V.           1974
COMPARATIVE STUDY OF METABOLISM FUNCTION AND ULTRASTRUCTURE
OF THE VERTEBRATE KIDNEY
INST. NATL. SANTE RECH. MED., PARIS, COLLOQ., 30:182.

4685

SMITH, C.G.                    1975
JOHNSON, J.A.
LAKOSKI, J.M.
PHYLOGENETIC DISTRIBUTION OF 6 NEUROTRANSMITTERS AMONG THE
VERTEBRATES.
FED. PROC., 34:305.
(ABSTR.)
ANIMAL BIOLOGY BIOCHEMISTRY EVOLUTION NERVOUS SYSTEM


4686F

WENT, A.E.J.                    1974
SOME INTERESTING FISHES TAKEN FROM IRISH WATERS IN 1973.
IR. NAT. J., 18:57-65.
*PETROMYZON MARINUS IRELAND ADULT


4687F

MCKEOWN, B.A.                    1975
HAZLETT, C.A.
THE EFFECT OF SALINITY ON PITUITARY THYROID AND INTERRENAL
CELLS IN IMMATURE ADULTS OF THE LANDLOCKED SEA LAMPREY
PETROMYZON MARINUS.
COMP. BIOCHEM. PHYSIOL., 50A:379-382.
CANADA ENDOCRINOLOGY HISTOLOGY OSMOREGULATION


4688F

GRABDA, J.                    1971
KATALOG FAUNY PASOZYTNICZEJ POLSKI. CZESC II. PASOZYTY
KRAGLOUSTYCH I RYB.
WARSAW: PANSTW. WYDAWN. NAUK.
(POL.)
POLAND BIBLIOGRAPHY


4689

LOWENSTEIN, O.ED.                    1971
ADVANCES IN COMPARATIVE PHYSIOLOGY AND BIOCHEMISTRY. VOL.4.
NEW YORK: ACADEMIC PRESS.
NEW YORK: ACADEMIC PRESS.


4690F

FUJITA, H.                    1975
SHINKAWA, Y.
ELECTRON MICROSCOPIC STUDIES ON THE THYROID GLAND OF THE
HAGFISH EPTATRETUS BUGERI A PART OF PHYLOGENETIC STUDIES OF
THE THYROID GLAND.
ARCH. HISTOL. JAP., 37:277-290.
JAPAN ADULT  HISTOLOGY CYTOLOGY ENDOCRINOLOGY


4691F

MARKING, L.L.                    1975
BILLS, T.D.
CHANDLER, J.H.
TOXICITY OF THE LAMPRICIDE 3-TRIFLUOROMETHYL-4-NITROPHENOL
(TFM) TO NONTARGET FISH IN FLOW-THROUGH TESTS.
U.S. DEP. INT., FISH WILDL. SERV., INVEST. FISH CONTROL,
61:3-9.
MANAGEMENT


4692F

JANVIER, P.                    1971
PALEONTOLOGIE - LA POSITION ET LA FORME DU SAC NASAL CHEZ
LES OSTEOSTRACI.
ACAD. SCI. PARIS, C.R. HEBD. SEANCES, 272:2434-2436.
(FR.)
*PETROMYZON SPP. ANATOMY OLFACTION


4693F

FAHRENBACH, W.H.                    1975
KNUTSON, D.D.
SURFACE ADAPTATIONS OF THE VERTEBRATE EPIDERMIS TO FRICTION.
J. INVEST. DERMATOL., 65:39-44.
*ENTOSPENUS TRIDENTATUS ANIMAL INTEGUMENT


4694F

GONCHAROVSKAYA, O.A.                    1975
MIKRODISSEKTSIONNE ISSLEDOVANIE NEFRONA MINOGI LAMPETRA
FLUVIATILIS. [MICRO DISSECTION STUDY ON THE NEPHRON IN THE
LAMPREY LAMPETRA FLUVIATILIS.]
ZH. EVOL. BIOKHIM. FIZIOL., 11:88-90.
(RUSS.)
ENGL. SUMM.
U.S.S.R. ANATOMY EXCRETION


4695

YOUSON, J.H.                    1977
WRIGHT, G.M.
OOI, E.C.
THE TIMING OF CHANGES IN SEVERAL INTERRENAL ORGANS DURING
METAMORPHOSIS OF THE ANADROMOUS LARVAL LAMPREY, PETROMYZON
MARINUS L.
CAN. J. ZOOL., 55:469-473.

4696

ROBINSON, E.S.                    1975
POTTER, I.C.
ATKIN, N.B.
THE NUCLEAR DNA CONTENT OF LAMPREYS.
EXPERIENTIA, 31:912-913.


4697F

NIEUWENHUYS, R.            1974
TOPOLOGICAL ANALYSIS OF THE BRAIN STEM, A GENERAL
INTRODUCTION.
J. COMP. NEUROL., 156:255-276.
*LAMPETRA FLUVIATILIS ANATOMY ANIMAL NERVOUS SYSTEM


4698

OOI, E.C.                    1977
MORPHOGENESIS AND GROWTH OF THE OPISTHONEPHRIC KIDNEY IN
THEANADROMOUS SEA LAMPREY, PETROMYZON MARINUS L., DURING
LARVALDEVELOPMENT AND METAMORPHOSIS.
UNIV. TORONTO, PH.D. THESIS: 184 PP.


4699F

FERNHOLM, B.                    1974
DIURNAL VARIATIONS IN THE BEHAVIOUR OF THE HAGFISH
EPTATRETUS BURGERI.
MAR. BIOL., 27:351-356.
JAPAN BEHAVIOUR FEEDING MIGRATION CIRCADIAN


4700F

PFISTER, C.                    1974
DANNER, H.
QUALITATIVE UNTERSUCHUNGEN DER DNS IN DEN KERNEN VON
MACRO-NEUROVEN VON SALMO IRIDEUS (GIBBONS 1855) (TELEOSTEI)
UND LAMPETRA PLANERI (BLOCH 1784) (CYCLOSTOMATA).
[QUALITATIVE INVESTIGATIONS OF THE NUCLEAR DNA IN
MACRONEURONS OF SALMO IRIDEUS TELEOST AND LAMPETRA PLANERI
CYCLOSTOMATA.]
Z. ZELLFORSCH., 88:67-79.
(GER.)
ENGL. SUMM.
HISTOCHEMISTRY NERVOUS SYSTEM PHYSIOLOGY CYTOLOGY PISCES


4701F

GRIGOR'EV, N.I.                    1975
STABLE CYTOLOGICAL AND HISTOLOGICAL INDICES AND THEIR
SIGNIFICANCE IN ANALYZING PROBLEMS IN VERTEBRATE
PHYLOGENESIS AND SYSTEMATICS.
USP. SOVREM. BIOL., 79:302-310.
(RUSS.)


4702F

TIJSKENS, J.                    1974
THE GEOMERICAL PATTERN OF TOOTH IMPLANATION ON THE ORAL DISC
OF THE SEA LAMPREY PETROMYZON MARINUS.
ACTA ZOOL. PATHOL. ANTVERPIENSIA., 58:51-56.
ADULT AMMOCOETE


4703F

FERNHOLM, B.                    1975
OVULATION AND EGGS OF THE HAGFISH EPTATRETUS BURGERI.
ACTA ZOOL., 56:199-204.
JAPAN ADULT EMBRYOLOGY ENDOCRINOLOGY FECUNDITY


4704F

JESPERSEN, A.                    1975
FINE STRUCTURE OF SPERMIOGENESIS IN EASTERN PACIFIC SPECIES
OF HAGFISH MYXINIDAE.
ACTA ZOOL., (STOCKH.), 56:189-198.
*EPTATRETUS STOUTII *EPTATRETUS DEANI *EPTATRETUS SP.
*MYXINE GLUTINOSA *EPTATRETUS BURGERI GONADOGENESIS CYTOLOGY


4705F

ROBERTSON, J.D.                    1974
OSMOTIC AND IONIC REGULATION IN CYCLOSTOMES.149-193PP.
IN  FLORKIN,M. AND B.T.SCHEER EDS. CHEMICAL ZOOLOGY.
VOL.VIII: PRIMITIVE DEUTEROSTOMIANS, CYCLOSTOMATA, FISHES.
NEW YORK:ACADEMIC PRESS.
*PETROMYZON MARINUS *LAMPETRA PLANERI *LAMPETRA FLUVIATILIS
*LAMPETRA TRIDENTATA *MYXINE GLUTINOSA *EPTATRETUS STOUTI
*PARAMYXINE ATAMI ADULT AMMOCOETE EXCRETION INTEGUMENT
ENDOCRINOLOGY PHYSIOLOGY


4706F

FALKMER, S.                    1974
THOMAS, N.W.
BOQUIST, L.
ENDOCRINOLOGY OF THE CYCLOSTOMATA. PP. 195-257.
IN  FLORKIN,M. AND SCHEER,B.T. EDS. CHEMICAL ZOOLOGY. VOL.
VIII: PRIMATIVE DEUTEROSTOMIANS, CYCLOSTOMATA, FISHES. NEW
YORK, ACADEMIC PRESS.
*MYXINE GLUTINOSA *LAMPETRA FLUVIATILIS *EPTATRETUS BURGERI

*PETROMYZON MARINUS *MORDACIA MORDAX *GEOTRIA *LAMPETRA
PLANERI ENDOCRINOLOGY HISTOCHEMISTRY EVOLUTION

4707F

FLORKIN, M. EDS.          1974
SCHEER, B.T.
CHEMICAL ZOOLOGY. VOL. VIII.
NEW YORK: ACADEMIC PRESS.

4708F

CORBEL, M.J.          1975
THE IMMUNE RESPONSE IN FISH. A REVIEW.
J. FISH BIOL., 7:539-564.
*EPTATRETUS STOUTII ANIMAL BIOCHEMISTRY BLOOD ENDOCRINOLOGY
IMMUNOLOGY

4709F

JANVIER, P.          1975
ANATOMIE ET POSITION SYSTEMATIQUE DES GALEASPIDES
(VERTEBRATA, CYCLOSTOMATA) CEPHALASPIDOMORPHES DU DEVONIEN
INTERIEUR DU YUNNAN (CHINE). [ANATOMY AND TAXONOMIC POSITION
OF THE GALEASPIDES VERTEBRATA CYCLOSTOMATA
CEPHALASPIDOMORPHES OF THE LOWER DEVONIAN OF YUNNAN CHINA.]
PARIS. MUS. NATL. HIST. NAT., BULL., SCI. TERRE, 41:1-16.
(FR.)
ANATOMY EVOLUTION NERVOUS SYSTEM EVOLUTION

4710F

BEAMISH, F.W.H.          1975
POTTER, I.C.
THE BIOLOGY OF THE ANADROMOUS SEA LAMPREY PETROMYZON
MARINUSIN NEW BRUNSWICK.
J. ZOOL., 177:57-72.
CANADA ADULT AMMOCOETE BIOLOGY FECUNDITY FEEDING GROWTH
LIFECYCLE METAMORPHOSIS MIGRATION PARASITISM BY MORPHOMETRY

4711

PYCHA, R.L.          1975
KING, G.R.
CHANGES IN THE LAKE TROUT POPULATION OF SOUTHERN LAKE
SUPERIOR IN RELATION TO THE FISHERY THE SEA LAMPREY AND
STOCKING  1950-1970.
GREAT LAKES FISH. COMM., TECH. REP., 28:34 PP.

4712F

MARKING, L.L.          1975
OLSON, L.E.
TOXICITY OF THE LAMPRICIDE 3-TRIFLUOROMETHL-4-NITROPHENOL
(TFM) TO NONTARGET FISH IN STATIC TESTS.
U.S. DEP. INT., FISH WILDL. SERV., INVEST. FISH CONTROL,
60:1-27.
MANAGEMENT

4713

POTTER, I.C.          1975
OSBORNE, T.S.
THE SYSTEMATICS OF BRITISH LARVAL LAMPREYS.
J. ZOOL., 176:311-329.
*PETROMYZON MARINUS *LAMPETRA FLUVIATILIS *LAMPETRA PLANERI
*ICHTHYOMYZON *LAMPETRA LAMOTTENI *LAMPETRA LETHENTERON
BRITAIN AMMOCOETE SYSTEMATICS MORPHOMETRY PIGMENTATION MUSCL

4714

MARTIN, A.R.          1974
RINGHAM, G.
SYNAPTIC TRANSFER CURVES AT A VERTEBRATE CENTRAL SYNAPSE.
J. PHYSIOL., 242:84P-86P.
NERVOUS SYSTEM PHYSIOLOGY

4715F

DOOLITTLE, R.F.          1976
COTTRELL, B.A.
RILEY, M.
AMINO ACID COMPOSITIONS OF THE SUBUNIT CHAINS OF LAMPREY
FIBRINOGEN. EVOLUTIONARY SIGNIFICANCE OF SOME STRUCTURAL
ANOMALIES.
BIOCHIM. BIOPHYS. ACTA, 453:439-452.
ADULT BIOCHEMISTRY BLOOD EVOLUTION TECHNIQUES

4716F

PERCY, R.          1976
POTTER, I.C.
BLOOD CELL FORMATION IN RIVER LAMPREY, LAMPETRA FLUVIATILIS.
J. ZOOL., 178:319-340
ADULT AMMOCOETE BIOLOGY BLOOD HISTOLOGY LIFE CYCLE
PHYSIOLOGY

4717

YONEZAWA, S.          1977
HORI, S.H.
STUDIES ON PHOSPHORYLASE ISOZYMES IN LOWER VERTEBRATES PURIF

ICATION AND PROPERTIES OF LAMPREY PHOSPHORYLASE.

ARCH. BIOCHEM. BIOPHYS., 181:447-453.

4718F

GLOVER, C.J.M.          1976
VERTEBRATE TYPE SPECIMENS IN THE SOUTH AUSTRALIAN MUSEUM.
PART 1: FISHES.
SOUTH AUST. MUS., ADELAIDE, REC., 17:169-175.
ANTARCTICA ANIMAL SYSTEMATICS


4719F

CLARIDGE, P.N.          1973
POTTER, I.C.
HUGHES, G.M.
CIRCADIAN RHYTHMS OF ACTIVITY VENTILATORY FREQUENCY AND
HEART RATE IN THE ADULT RIVER LAMPREY LAMPETRA FLUVIATILIS.
J. ZOOL., 171:239-250.
ADULT BEHAVIOR CIRCULATORY SYSTEM RESPIRATION CIRCADIAN


4720F

ROHDE, F.C.          1975
ARNDT, R.G.
WANG, J.C.S.
RECORDS OF THE FRESH WATER LAMPREYS LAMPETRA LAMOTTENII AND
OKKELBERGIA AEPYPTERA, FROM THE DELMARVA PENINSULA (EAST
COAST, UNITED STATES).
CHESAPEAKE SCI., 16:70-72.
ADULT DISTRIBUTION


4721F

PLISETSKAYA, E.M.          1974
LEIBUSH, B.N.
RADIOIMMUNOLOGICHESKOE OPREDELENIE INSULINA U NIZSHIKH
POZVONOCHNYKH. [RADIOIMMUNOLOGICAL DETERMINATION OF INSULIN
IN LOWER VERTEBRATES.]
ZH. EVOL. BIOKHIM. PIZIOL., 10:623-625.
(RUSS.)
J. EVOL. BIOCHEM. PHYSIOL., 10:567-569.
(ENG.)
*LAMPETRA FLUVIATILIS BALTIC SEA ADULT ENDOCRINOLOGY
TECHNIQUES


4722

NAKAO, T.          1973
FINE STRUCTURE OF AGRANULAR CYTOPLASMIC TUBULES OF LAMPREY
CELLS.
HLORIDE CELLS.
KAIBOGAKA ZASSHI. ACTA ANAT. NIPPON, 48:406.


4723F

HENDERSON, N.E.          1976
THYROXINE CONCENTRATIONS IN PLASMA OF NORMAL AND
HYPOPHYSECTOMIZED HAGFISH, MYXINE GLUTINOSA (CYCLOSTOMATA).
CAN. J. ZOOL., 54:180-184.
CANADA BLOOD ENDOCRINOLOGY HISTOLOGY


4724

CRAWFORD, R.R. JR.          1974
LAWRENCE, A.L.
ABSORPTION OF AMINO ACIDS BY THE GUT OF THE HAGFISH
POLISTOTREME STOUTII.
TEX. J. SCI., 25:122-123.
(ABSTR.)
EVOLUTION PHYSIOLOGY METABOLISM DIGESTION


4725F

FALKMER, S.          1974
EDMIN, S.O.
HAVU, N.
BIUW, L.W.
SUNDBY, F.
CUTFIELD, J.F.
SOME EVOLUTIONARY ASPECTS OF THE MOLECULAR STRUCTURE AND
BIOSYNTHESIS OF INSULIN.
DIABETOLOGIA, 10:364.
(ABSTR.)
*MYXINE GLUTINOSA *LAMPETRA FLUVIATILIS ADULT EVOLUTION
ENDOCRINOLOGY BIOCHEMISTRY


4726

YOUSON, J.H.          1974
HANSEN, S.J.
CAMPBELL, I.M.
A QUANTITATIVE COMPARISON OF THE KIDNEYS OF THE LANDLOCKED
SEA LAMPREY OF THE GREAT LAKES AND THE ANADROMOUS SEA
LAMPREY OF THE ATLANTIC, PETROMYZON MARINUS L.
CAN. J. ZOOL., 52:1447-1455.

4727F
NIIMI, A.J.                    1974
RELATIONSHIP BETWEEN ASH CONTENT AND BODY WEIGHT IN LAMPREY
PETROMYZON MARINUS, TROUT SALMO GAIRDNERI AND BASS
MICROPTERUS SASMOIDES.
COPEIA, 1974:794-795.
CANADA PISCES MIGRATION ADULT ANIMAL GROWTH


4728
SUZUKI, N.                     1975
INTRA CELLULAR POTENTIALS FROM LAMPREY OLFACTORY RECEPTORS.
ZOOL. MAG. (TOKYO), 84:340.


4729F
SUZUKI, S.                     1975
GORBMAN, A.
ROLLAND, M.
MONTFORT, M-F.
LISSITZKY, S.
THYROGLOBULINS OF CYCLOSTOMES AND AN ELASMOBRANCH.
GEN. COMP. ENDOCRINOL., 26:59-69.
*ENTOSPHENUS TRIDENTATUS U.S.A. ANIMAL BIOCHEMISTRY ADULT


4730
KUHLENBECK, H.                 1975
THE CENTRAL NERVOUS SYSTEM OF VERTEBRATES: A GENERAL
SURVEY OF ITS COMPARATIVE ANATOMY WITH AN INTRODUCTION TO
THE PERTIN....
NEW YORK; ACADEMIC PRESS, 3V.
NEW YORK: ACADEMIC PRESS.


4731F
EAKIN, R.M.                    1973
THE PACIFIC LAMPREY (ENTOSPHENUS TRIDENTATUS). PP. 5-7.
IN  EAKIN, R.M. THE THIRD EYE. BERKELEY: UNIV. OF CALIFORNIA
PRESS.
*PETROMYZON MARINUS AMMOCOETE ANATOMY SENSE RECEPTORS


4732
HOLMES, R.L.                   1974
BALL, J.N.
THE PITUITARY GLAND; A COMPARATIVE ACCOUNT.
CAMBRIDGE: UNIV. PRESS. 397P.
(BIOLOGICAL STRUCTURE AND FUNCTION 4.).


4733F
LELEK, A.                      1973
OCCURRENCE OF THE SEA LAMPREY IN MIDWATER OFF EUROPE.
COPEIA, 1975:136-137.
*PETROMYZON MARINUS GERMANY DISTRIBUTION ADULT


4734F
DE VOS, R.                     1973
DE WOLF-PEETERS, C.
DESMET, V.
A MORPHOLOGIC AND HISTOCHEMICAL STUDY OF BILIARY ATRESIA IN
LAMPREY LIVER.
Z. ZELLFORSCH., 136:85-96.
*LAMPETRA PLANERI BELGIUM HISTOCHEMISTRY AMMOCOETE ADULT


4735F
BAUMGARTEN, H.G.               1973
BJORKLUND, A.
LACHENMAYER, L.
NOBIN, A.
ROSENGREN, E.
EVIDENCE FOR THE EXISTENCE OF SEROTONIN-,DOPAMIN-, AND
NORADRENALINE-CONTAINING NEURONS IN THE GUT OF
LAMPETRA FLUVIATILIS.
Z. ZELLFORSCH., 141:33-54.
SWEDEN ADULT HISTOCHEMISTRY BIOCHEMISTRY NERVOUS SYSTEM


4736F
KEMPE, L.L.                    1973
MICROBIAL DEGRADATION OF THE LAMPREY LARVICIDE 3-
TRIFLUOROMETHYL-4-NITROPHENOL IN SEDIMENT-WATER SYSTEMS.
GREAT LAKES FISH. COMM., TECH. REP., 18:1-16.
RESTRICTED
*PETROMYZON MARINUS CANADA AMMOCOETE HISTORY.


4737
ROMERO-HERRERA, A.             1973
LEHMANN, H.
JOYSEY, K.A.
FRIDAY, A.E.
MOLECULAR EVOLUTION OF MYOGLOBIN AND THE FOSSIL RECORD A
PHYLOGENETIC SYNTHESIS.
NATURE (LOND.), 246:389-395.


4738
MURTAUGH, P.A.                 1974

HALVER, J.E.
GLADNER, J.A.
STRUCTURAL CHARACTERIZATION OF LAMPREY FIBRINOGEN AND
FIBRIN: CROSS-LINKING OF THE FIBRIN BY PLASMA
TRANSGLUTAMINASE, FACTOR XIII.
FED. PROC., 33:1474.
(ABSTR.)

4739

BRUKMOZER, P.                 1972
DOBRYLKO, A.K.
EVOKED POTENTIALS IN THE TELENCEPHALON OF THE LAMPREY
LAMPETRA FLUVIATILIS FROM STIMULATION OF OLFACTORY NERVE.
ZH. EVOL. BIOKHIM. FIZIOL., 8:558-560.
(RUSS.)
J. EVOL. BIOCHEM. PHYSIOL., 8:494-560.
(ENG.)
NERVOUS SYSTEM PHYSIOLOGY

4740

RYBAK, B.                     1974
SIMON, M.
DES MECHANISMES BIOCHIMIQUES DE L'AUTOMATISME CARDIAQUE
EXAMEN DE LA MYXINE. [ON THE BIOCHEMICAL MECHANISMS OF THE
HEART AUTOMATICITY STUDY ON MYXINE GLUTINOSA.]
ACTA PHYSIOL. SCAND., 90:501-504.
(FR.)
ENG. SUMM.
CIRCULATORY SYSTEM PHYSIOLOGY ADULT

4741F

PRAVDINA, N.I.                1974
CHEBOTAREVA, M.A.
KRUGLOVA, E.E.
KHIRNYE KISLOTY SFINGOMELINA MIELINA 1 MITOKHONDRII V RYADU
POZVONOCHNYKH. [SPHINGOMYELIN FATTY ACIDS IN BRAIN MYELIN
AND MITOCHONDRIA IN VERTEBRATES.]
ZH. EVOL. BIOKHIM. FIZIOL., 10:325-329.
(RUSS.)
ENGL. SUMM.
*LAMPETRA FLUVIATILIS BIOCHEMISTRY EVOLUTION NERVOUS SYSTEM

4742F

MANUSOVA, N.B.                1973
TSITOFOTOMETRICHESKOE ISSLEDOVANIE AKTIVNOSTI DEGIDROGENAZ V
KLETKAKH NEFRONA POZVONOCHNYKH. [CYTOPHOTOMETRIC STUDIES OF
DEHYDROGENASE ACTIVITIES IN CELLS OF THE NEPHRON IN
VERTEBRATES.]
ZH. EVOL. BIOKHIM. FIZIOL., 9:209-211.
(RUSS.)
ENGL. SUMM.
*LAMPETRA FLUVIATILIS BIOCHEMISTRY EXCRETION ANIMAL

4743F

GEORGE, J.C.                  1974
BEAMISH, F.W.H.
HAEMOCYTOLOGY OF THE SUPRANEURAL MYELOID BODY IN THE SEA
LAMPREY DURING SEVERAL PHASES OF LIFE CYCLE.
CAN. J. ZOOL., 52:1585-1589.
*PETROMYZON MARINUS AMMOCOETE ADULT HISTOCHEMISTRY EVOLUTION
BLOOD HISTOLOGY

4744F

FALKMER, S.                   1975
CUTFIELD, J.F.
CUTFIELD, S.M. ET.AL.
COMPARATIVE ENDOCRINOLOGY OF INSULIN AND GLUCAGON
PRODUCTION.
AM. ZOOL. SUPPL., 15:255-270.
*MYXINE GLUTINOSA ENDOCRINOLOGY DIGESTION EVOLUTION
BIOCHEMISTRY

4745F

YAMADA, Y.                    1973
FINE STRUCTURE OF THE ORDINARY LATERAL LINE ORGAN .I. THE
NEUROMAST OF LAMPREY,ENTOSPHENUS JAPONICUS.
J. ULTRASTRUCT. RES., 43:1-17.
ADULT ANIMAL NERVOUS SYSTEM SENSE RECEPTORS HISTOLOGY

4746F

MOORE, H.H                    1974
DAHL, F.H.
LAMSA, A.K.
MOVEMENT AND RECAPTURE OF PARASITIC PHASE SEA LAMPREYS
PETROMYZON MARINUS TAGGED IN THE ST.MARYS RIVER AND LAKES
HURON AND MICHIGAN, 1963-67.
GREAT LAKES FISH. COMM., TECH. REP., 27:1-19.
RESTRICTED
NORTH AMERICA ADULT DISTRIBUTION PARASITISM BY FEEDING
MIGRATION MANAGEMENT

4747F

STERBA, G.                    1973
HOHEISEL, G.
RUHLE, H.J.
ENGLEMANN, W.E.
EXTRAHYPOTHALAMISCHE PEPTIDERGE NEUROSEKRETION.
[EXTRAHYPOTHALAMIC PEPTIDERGIC NEUROSECRETION.]
Z. ZELLFORSCH., 142:329-345.
(GERM.)
ENGL. SUMM.
*LAMPETRA PLANERII ADULT CYTOLOGY ENDOCRINOLOGY HISTOLOGY
METAMORPHOSIS NERVOUS SYSTEM


4748F

VAN NOORDEN, S.              1974
PEARSE, G.E.
IMMUNOREACTIVE POLYPEPTIDE HORMONES IN THE PANCREAS AND GUT
OF THE LAMPREY.
GEN. COMP. ENDOCRINOL., 23:311-324.
*LAMPETRA FLUVIATILIS GREAT BRITIAN ENDOCRINOLOGY HISTOLOGY
HISTOCHEMISTRY


4749F

YOUSON, J.H.                 1976
FINE STRUCTURE OF GRANULATED CELLS IN THE POSTERIOR
CARDINALAND VEINS OF AMIA CALVA L..
CAN. J. ZOOL., 54:843-851.
ANIMAL CIRCULATORY SYSTEM HISTOLOGY


4750F

VLADYKOV, V.D.               1976
KOTT, E.
A SECOND NONPARASITIC SPECIES OF ENTOSPHENUS GILL,1862
PETROMYZONIDAE FROM KLAMATH RIVER SYSTEM, CALIFORNIA.
CAN. J. ZOOL., 54:974-989.
*ENTOSPHENUS FOLLETTI *ENTOSPHENUS LETHOPHAGUS *ENTOSPHENUS
TRIDENTATUS *ENTOSPHENUS MINIMUS U.S.A. ADULT AMMOCOETE
ANATOMY MORPHOMETRY


4751F

LECH, J.J.                   1975
STATHAM, C.N.
ROLE OF GLUCURONIDE FORMATION IN THE SELECTIVE TOXICITY OF
3-TRIFLUOROMETHYL-4-NITROPHENOL (TFM) FOR THE SEA LAMPREY
;COMPARATIVE ASPECTS OF TFM UPTAKE AND CONJUGATION IN SEA
LAMPREY AND RAINBOW TROUT.
TOXICOL. & APPL. PHARMACOL., 31:150-158.
*PETROMYZON MARINUS U.S.A. AMMOCOETE ADULT ANIMAL CHEMISTRY
BIOCHEMISTRY MORTALITY OF


4752F

VLADYKOV, V.D.               1972
PHARAND, S.
TENTACULES DU VELUM ("VELAR TENTACLES") CHEZ LES LAMPROIES
DE L'AMERIQE DU NORD. [THE VELAR TENTACLES IN LAMPREY OF
NORTH AMERICA.]
ASSOC. CAN.-FR. AV. SCI., MONT., 39:148.
(ABSTR.)
ABSTRACT
*PETROMYZON MARINUS *ICHTHYOMYZON UNICUSPIS *ENTOSPHENUS
TRIDENTATUS *TETRAPLEURODON SPADICEUS *LAMPETRA *LETHENTERON
ANATOMY SYSTEMATICS


4753F

VLADYKOV, V.D.               1972
SOUS-DIVISION EN TROIS SOUS FAMILLES DES LAMPROIES DE
L'HEMISPHERENORD DE LA FAMILLE PETROMYZONIDE. [SUBDIVISION
OF NORTHERN HEMISPHERE LAMPREYS OF THE FAMILY PETROMYZONIDEA
INTO  3 SUBFAMILIES.]
ASSOC. CAN.-FR. AV. SCI., MONT., 39:148.
(FR.)
(ABSTR.)
*PETROMYZONINAE *ENTOSPHENINAE *LAMPETRINAE SYSTEMATICS
DISTRIBUTION


4754F

FARMER, G.J.                 1973
BEAMISH, F.W.H.
SEA LAMPREY PETROMYZON MARINUS PREDATION ON FRESHWATER
TELEOST.
FISH. RES. BOARD CAN., J., 30:601-605.
CANADA ADULT PARASITISM BY BEHAVIOUR


4755F

BEAMISH, F.W.H.              1973
OXYGEN CONSUMPTION OF THE ADULT PETROMYZON MARINUS
IN RELATION TO BODY WEIGHT AND TEMPERATURE.
FISH. RES. BOARD CAN., J., 30:1367-1370
CANADA LOCOMOTION RESPIRATION


4756F

FETTEROLF, C.M. ED.          1975
TECHNICAL REVIEW OF GREAT LAKES FISHERY COMMISSION FUNDED
SEA LAMPREY RESEACH AT HAMMOND BAY BIOLOGICAL STATION
(HBBS), OCTOBER 29-30, 1975.
GREAT LAKES FISH. COMM., TECH. REV., 18PP.
RESTRICTED
*PETROMYZON MARINUS U.S.A. MANAGEMENT CHEMISTRY ANIMAL
DISTRIBUTION CULTURE

                                                        4757F

NAKAO, T.                    1975
"TERMINAL COUPLINGS" IN SOME VERTEBRATE SKELETAL MUSCLES.
INT. CONG. ANAT., 10TH., TOKYO, 327P.
ADULT AMMOCOETE ANIMAL HISTOLOGY

                                                        4758F

PERCY, R.                    1975
LEATHERLAND, J.F.
BEAMISH, F.W.H.
STRUCTURE AND ULTRASTRUCTURE OF THE PITUITARY GLAND IN THE
SEA LAMPREY PETROMYZON MARINUS AT DIFFERENT STAGES IN ITS
LIFE CYCLE.
CELL TISSUE RES., 157:141-164.
CANADA U.S.A. ADULT AMMOCOETE ENDOCRINOLOGY HISTOLOGY
METAMORPHOSIS

                                                        4759F

POTTER, I.C.                 1974
BEAMISH, F.W.H.
JOHNSON, B.G.H.
SEX RATIOS AND LENGTHS OF ADULT SEA LAMPREYS PETROMYZON
MARINUS FROM A LAKE ONTARIO TRIBUTARY.
FISH. RES. BOARD CAN., J., 31:122-123.
CANADA BIOLOGY

                                                        4760F

LAMPE, J.                    1973
PFEIL, W.
BLANCK, J.
BEHLKE, J.
MOHNIKE, W.
SCHELER, W.
UNTERSUCHUNGEN ZUR REAKTION VON ALKYLISOZYANIDEN MIT
NEUNAUGEN HAMOGLOBIN.
[STUDIES ON THE REACTION OF ALKYLISOCYANIDES WITH LAMPREY
HEMOGLOBIN.]
ACTA BIOL. MED. GER., 30:607-615.
(GER.)
ENG. SUMM.
*PETROMYZON MARINUS *LAMPETRA FLUVIATILIS L. ADULT BLOOD
CIRCULATORY SYSTEM BIOCHEMISTRY

                                                        4761F

NURSALL, J.R.                1972
BUCHWALD, D.
LIFE HISTORY AND DISTRIBUTION OF THE ARCTIC LAMPREY
LENTHENTERON JAPONICUM MARTENS OF GREAT SLAVE LAKE, N.W.T.
FISH. RES. BOARD CAN., TECH. REP., 304:28P.
CANADA AMMOCOETE ADULT LIFECYCLE DISTRIBUTION PARASITISM BY
PARASITISM OF BIOLOGY

                                                        4762F

WINBLADH, L.                 1975
HORSTEDT, P.
FOLLICLES IN THE ENDOCRINE PANCREAS OF MYXINE GLUTINOSA
STUDIED BY SCANNING ELECTRON MICROSCOPY.
ACTA ZOOL., 56:213-216.
SWEDEN ADULT ENDOCRINOLOGY HISTOLOGY BIOCHEMISTRY

                                                        4763F

WINBLADH, L.                 1976
FOLLICLES IN THE ENDOCRINE PANCREAS OF SOME MYXINOID SPECIES
ACTA ZOOL., 57:7-11.
*MYXINE GLUTINOSA *EPTATRETUS STOUTI *EPTATRETUS BURGERI
SWEDEN CANADA JAPAN ADULT HISTOLOGY ENDOCRINOLOGY

                                                        4764F

LOHNISKY, K.                 1975
A CONTRIBUTION TO THE KNOWLEDGE OF BIOLOGY OF BROOK LAMPREY,
LAMPETRA PLANERI (BLOCH,1784).
PRIMER CENTENARIO DE LA R.SOC. ESPANOLA DE HIST.NAT.,313-
323.
AMMOCOETE ADULT BIOLOGY GROWTH SPAWNING

                                                        4765F

FALKMER, S.                  1973
EMDIN, S.O.
HAVU, N.
LUNDGREN, G. ET.AL.
INSULIN IN INVERTEBRATES AND CYCLOSTOMES.
AM. ZOOL., 13:625-638.

4766F

VLADYKOV, V.D.                    1974
REQUEST FOR A RULING ON THE STEM OF FAMILY GROUP NAMES BASED
ON THE TYPE GENUS PETROMYZON LINNAEUS,1758. Z.N.(S.)
INT. COMM. ZOOL NOMENCL., BULL., 30:198-199.
SYSTEMATICS

4767F

CHOLETTE, C.                      1973
GAGNON, A.
ISOSMOTIC ADAPTATION IN MYXINE GLUTINOSA L.-II.VARIATIONS
OF THE FREE AMINO ACIDS, TRIMETHYLAMINE OXIDE AND POTASSIUM
OF THE BLOOD AND MUSCLE CELLS.
COMP. BIOCHEM. PHYSIOL., 45A:1009-1021.
PHYSIOLOGY OSMOREGULATION IONIC REGULATION

4768F

VLADYKOV, V.D.                    1976
KOTT, E.
IS OKKELBERGIA CREASER AND HUBBS, 1922 (PETROMYZONIDAE) A
DISTINCT TAXON?
CAN. J. ZOOL., 54:421-425
*LAMPETRA AEPYPTERA U.S.A. SYSTEMATICS

4769F

LETT, P.F.                        1975
BEAMISH, F.W.H.
FARMER, G.J.
SYSTEM SIMULATION OF THE PREDATORY ACTIVITIES OF SEA
LAMPREYS, PETROMYZON MARINUS ON LAKE TROUT SALVELINUS
NAMAYCUSH.
FISH. RES. BOARD CAN., J., 5:623-631.
CANADA MORTALITY BY PARASITISM BY

4770F

VLADYKOV, V.D.                    1973
PETROMYZONIDAE.
IN  HUREAU,J.C. AND MONOD,TH. EDS. CHECK LIST OF THE FISHES
OF THE NORTH EASTERN ATLANTIC AND OF THE MEDITERRANEAN.
PARIS:UNESCO. AGNATHA 1-6.
*PETROMYZON MARINUS *LAMPETRA FLUVIATILIS * LETHENTERON
JAPONICUM (LAMPETRA JAPONICA) *MYXINE GLUTINOSA DISTRIBUTION
SYSTEMATICS BIOLOGY

4771F

NOVITSKAYA, L.I.                  1975
SUR LA STRUCTURE INTERNE ET LES LIENS PHYLOGENETIQUES DES
HETEROSTRACI.
IN  COLLOQUE INTERNATIONAL C.N.R.S., PARIS 4-9 JUIN,
PROBLEMES ACTUELS DE PALEONTOLOGIE - EVOLUTION DES
VERTEBRES. FR., CENT. NAT. RECH. SCI.
(FR.)
ENGL. SUMM.
*MYXINE GLUTINOSA *PETROMYZON ANATOMY ANIMAL BIOLOGY
EVOLUTION OLFACTION NERVOUS SYSTEM

4772F

PLISETSKAYA, E.M.                 1975
GORMONAL'NAYA REGULYATSIYA UGLEVODNOGO OBMENA U NIZSHIKH
POZVONOCHNYKH. [HORMONAL REGULATION OF CARBOHYDRATE
METABOLISM IN LOWER VERTEBRATES.]
LENINGRAD: NAUKA. 215PP.
(RUSS.)

4773F

EISENBACH, G.M.                   1971
WEISE, M.
WEISE, R.
HANKE, K.
STOLTE, H.
BOYLAN, J.W.
RENAL HANDLING OF PROTEIN IN THE HAGFISH, MYXINE GLUTINOSA.
MT. DESERT ISL. BIOL. LAB., BULL., 11:11-15.
ADULT EXCRETION HISTOLOGY IONIC REGULATION

4774F

OSTBERG, Y.                       1976
THE ENTERO-INSULIN ENDOCRINE ORGAN IN A CYCLOSTOME, MYXINE
GLUTINOSA.
UMEA UNIV. MED. DISS., NEW SER., 15:41 PP.
SWEDEN ENDOCRINOLOGY ADULT HISTOLOGY CYTOLOGY

4775F

BLANCK, J.                        1972
LAMPE, J.
SCHELER, W.
KINETISCHE UNTERSUCHUNGEN ZUR DISSOZIATION VON
ALKYLISOZYANID-KOMPLEXEN DES NEUNAUGEN-HAMOGLOBINS
(LAMPETRA FLUVIATILIS L.)

ACTA BIOL. MED. GER., 28:K27-K30.
(GER.)
BLOOD BIOCHEMISTRY CIRCULATORY SYSTEM.

4776F

MOORE, J.W.                    1976
POTTER, I.C.
A LABORATORY STUDY OF THE FEEDING OF LARVAE OF THE BROOK
LAMPREY LAMPETRA PLANERI.
J. ANIM. ECOL., 45:81-90
ENGLAND DIGESTION FEEDING GROWTH

4777F

CZECZUGA, B.                   1973
ASTAXANTHIN-THE DOMINANT XANTHOPHYLL IN LAMPETRA PLANERI
(BLOCH) LARVAE (CYCLOSTOMATA, PETROMYZONTIDAE).
ZOOL. POL., 23:263-267.
(POL.)
POLAND PIGMENTATION

4778F

GRODZINSKI, Z.                 1974
FURTHER STUDIES ON THE YOLK MORPHOLOGY OF THE RIVER LAMPREY,
LAMPETRA FLUVIATILIS L.
ACTA BIOL. CRACOV., SER. ZOOL., 17:263-274.
POLAND ADULT HISTOCHEMISTRY

4779F

LEATHERLAND, J.F.             1976
PERCY, R.
STRUCTURE OF THE NONGRANULATED CELLS IN THE HYPOPHYSEAL
ROSTRAL PARS DISTALIS OF CYCLOSTOMES AND ACTINOPTERYGIANS.
CELL TISSUE RES., 166:185-200.
*PETROMYZON MARINUS CANADA ADULT AMMOCOETE ANIMAL
ENDOCRINOLOGY HISTOLOGY CYTOLOGY

4780F

PIAVIS, G.W.                   1975
DUBIN, N.H.
NARDELL, B.
THE "ESTRADIOL FRACTION" IN PERIPHERAL BLOOD OF THE
LAMPREY,PETROMYZON MARINUS.
ANAT. REC., 181:449-450.
(ABSTR.)
ADULT USA ENDOCRINOLOGY

4781F

WEISBART, M.                   1975
YOUSON, J.H.
STEROID FORMATION IN THE LARVAL AND PARASITIC ADULT SEA
LAMPREY, PETROMYZON MARINUS L.
GEN. COMP. ENDOCRINOL., 27:517-526.
CANADA ENDOCRINOLOGY HISTOLOGY

4782F

NOZAKI, M.                     1975
FERNHOLM, B.
KOBAYASHI, H.
EPENDYMAL ABSORPTION OF PEROXIDASE INTO THE THIRD VENTRICLE
OF THE HAGFISH EPTATRETUS BURGERI (GIRARD).
ACTA ZOOL., 56:265-269.
JAPAN ENDOCRINOLOGY HISTOCHEMISTRY NERVOUS SYSTEM

4783F

TIBBLES, J.J. ED.              1974
ANNUAL REPORT 1973 OF THE SEA LAMPREY CONROL CENTRE SAULT
STE. MARIE, ONTARIO TO THE GREAT LAKES FISHERY COMMISSION.
GREAT LAKES FISH. COMM., ANNU. REP., 91 PP.
RESTRICTED

4784F

TIBBLES, J.J. ED.              1975
ANNUAL REPORT 1974 OF THE SEA LAMPREY CONTROL CENTRE SAULT
STE. MARIE, ONTARIO, TO THE GREAT LAKES FISHERY COMMISSION.
GREAT LAKES FISH. COMM., ANNU. REP., PP.1-93.
RESTRICTED

4785F

STONE, D.J.                    1975
KENNEDY, M.C.
RUBINSON, K.
RETINAL PROJECTIONS IN THE LARVAL LAMPREY,PETROMYZON
MARINUS.
AM. ASSOC. ANAT., LOS ANGELES.
(ABSTR.)
U.S.A. NERVOUS SYSTEM VISION HISTOLOGY

4786F

OSTBERG, Y.                    1976
BOQUIST, L.
ULTRASTRUCTURAL AND FLUORESCENCE MICROSCOPICAL

CHARACTERIZATION OF THE INTESTINAL ENDOCRINE CELLS IN A
CYCLOSTOME, MYXINE GLUTINOSA.
ACTA ZOOL., (STOCKH.), 57:41-51.
SWEDEN ENDOCRINOLOGY HISTOCHEMISTRY CYTOLOGY DIGESTION

4787F

ROVAINEN, C.M.                    1975
SCHIEBER, M.H.
VENTILATION OF LARVAL LAMPREYS.
J. COMP. PHYSIOL., 104:185-203.
*PETROMYZON MARINUS U.S.A. RESPIRATION

4788F

HOMMA, S.                         1975
VELAR MOTONEURONS OF LAMPREY LARVAE.
J. COMP. PHYSIOL., 104:175-183.
*PETROMYZON MARINUS *LAMPETRA AEPYPTERA U.S.A. RESPIRATION
NERVOUS SYSTEM

4789F

HELLE, K.B.                       1972
LONNING, S.
BLASCHKO, H.
OBSERVATIONS ON THE CHROMAFFIN GRANULES OF THE VENTRICLE
AND THE PORTAL VEIN HEART OF MYXINE GLUTINOSA L.
SARSIA, 51:97-106.
NORWAY CIRCULATORY SYSTEM ENDOCRINOLOGY

4790F

BOQUIST, L.                       1975
OSTBERG, Y.
ANNULATE LAMELLAE AND CRYSTALLINE INCLUSIONS IN GRANULAR
ENDOPLASMIC RETICULUM OF THE ISLET ORGAN AND ASSOCIATED
TISSUES OF A CYCLOSTOME, MYXINE GLUTINOSA.
CELL TISSUE RES., 158:75-87.
SWEDEN ADULT HISTOLOGY ENDOCRINOLOGY EVOLUTION

4791F

FERNHOLM, B.                      1972
PITUTARY AND OVARY OF THE ATLANTIC HAGFISH; AN
ENDOCRINOLOGICAL INVESTIGATION.
STOCKHOLM: UNIV. OF STOCKHOLM. 11P.
*MYXINE GLUTINOSA *CYCLOSTOMATA ADULT CYTOLOGY
ENDOCRINOLOGYEVOLUTION PHYSIOLOGY

4792F

HELLE, K.B.                       1975
STORESUND, A.
ULTRASTRUCTURAL EVIDENCE FOR A DIRECT CONNECTION BETWEEN
THE MYOCARDIAL GRANULES AND THE SARCOPLASMIC RETICULUM IN
THE CARDIAC VENTRICLE OF MYXINE GLUTINOSA L.
CELL TISSUE RES., 163:353-363.
NORWAY CIRCULATORY SYSTEM HISTOLOGY HISTOCHEMISTRY MUSCLE

4793F

FANGE, R. ED.                     1973
MYXINE GLUTINOSA. BIOCHEMISTRY, PHYSIOLOGY AND STRUCTURE.
REPORT FROM A SYMPOSIUM IN GOTEBORG 28-29 APRIL, 1972.
GOTEBORG:KUNGL. VETENSKAPS-OCH VITTERHETS-SAMHALLET. 104 PP.

4794F

GUARINO, A.M.                     1972
PRITCHARD, J.B.
BRILEY, P.M.
ANDERSON, J.B.
KINTER, M.A.
RALL, D.P.
COMPARATIVE ASPECTS OF RENAL AND HEPATIC HANDLING OF PHENOL
RED INDOCYANINE GREEN IN THE DOGFISH, FLOUNDER AND HAGFISH.
MT. DESERT ISL. BIOL. LAB., BULL., 12:43-45.
EXCRETION METABOLISM

4795F

TOMONAGA, S.                      1973
HIROKANE, T.
SHINOHARA, H.
AWAYA, K.
THE PRIMITIVE SPLEEN OF THE HAGFISH.
ZOOL. MAG. (TOKYO), 82:215-217.
*EPTATRETUS BURGERI JAPAN BLOOD HISTOLOGY CYTOLOGY

4796F

TOMONAGA, S.                      1973
SHINOHARA, H.
AWAYA, K.
FINE STRUCTURE OF THE PERIPHERAL BLOOD CELLS OF THE HAGFISH

ZOOL. MAG. (TOKYO), 82:211-214.
*EPTATRETUS BURGERI JAPAN BLOOD HISTOLOGY CYTOLOGY

4797F

PLISETSKAYA, E.M.                1972
LEIBUSH, B.N.
INSULIN-LIKE ACTIVITY AND IMMUNOREACTIVE INSULIN IN THE
BLOOD OF THE LAMPREY LAMPETRA FLUVIATILIS.
ZH. EVOL. BIOKHIM. FIZIOL., 8:499-505.
(RUSS.)
J. EVOL. BIOCHEM. PHYSIOL., 8:440-444.
(ENG.)
EUROPE ADULT ENDOCRINOLOGY


                                                        4798F

ROVAINEN, C.M.                1976
VESTIBULO-OCULAR REFLEXES IN THE ADULT SEA LAMPREY.
J. COMP. PHYSIOL., 112:159-164.
*PETROMYZON MARINUS ADULT VISION NERVOUS SYSTEM


                                                        4799F

STOLTE, H.                1973
EISNBACH, G.M.
SINGLE NEPHRON FILTRATION RATE IN THE HAGFISH MYXINE
GLUTINOSA.
MT. DESERT ISL. BIOL. LAB., BULL., 13:120-121
ADULT PHYSIOLOGY TECHNIQUES EXCRETION


                                                        4800F

OSTBERG, Y.                1976
FANGE, R.
MATTISSON, A.
THOMAS, N.W.
LIGHT AND ELECTRON MICROSCOPICAL CHARACTERIZATION OF
HETEROPHILIC GRANULOCYTES IN THE INTESTINAL WALL AND ISLET
PARENCHYMA OF THE HAGFISH, MYXINE GLUTINOSA, CYCLOSTOMATA.
ACTA ZOOL., (STOCKH.), 57:89-102.
SWEDEN HISTOLOGY


                                                        4801F

THOENES, G.H.                1972
THE HAGFISH AT THE PHYLOGENETICAL JUNCTURE TOWARDS
IMMUNOLOGICAL RESPONSE. PP. 69-74.
IN  L'ETUDE PHYLOGENIQUE ET ONTOGENIQUE DE LA REPONSE
IMMUNITAIRE ET SON APPORT A LA THEORY., PARIS,12-14 OCTOBRE.
(PHYLOGENIC AND ONTOGENIC STUDY OF THE IMMUNE RESPONSE AND
ITS CONTRIBUTION TO THE IMMUNOLOGICAL THEORY.) PARIS: SOC.
FRAN. IMMUNOL.
FRANCE ANIMAL IMMUNOLOGY


                                                        4802F

MEDNIKOV, B.M.                1973
ANTONOV, A.S.
POPOV, L.S.
GENOSISTEMATIKA I EVOLYUTSIYA GENOMOV RYB. [GENOSYTEMATICS
AND EVOLUTION OF FISH GENOMES.]
IN  TROSHIN, A.S., ED., BIOKHIMICHESKAYA GENETIKA RYB.
MATERIALY 1-GO VSESOYUZ. SOVESHCH. LENINGRAD, 6-9 FEVR. 1973
G. LENINGRAD: VSESOYUZ. SOVESHCH. PO BIOKHIM. GENET. RYB.
(RUSS.)
ENGL. SUMM.
U.S.S.R. ANIMAL GENETICS


                                                        4803F

SEREBRENIKOVA, T.P.                1969
FILOSOFOVA, E.M.
FRAKTSIONIROVANIE BELKOV I ISSLEDOVANIE FERMENTATISNYKH
AKTIVNOSTEI V SARKOPLAZME MYSHTS NIZSHIKH POZVONOCHNYKH.
PP.50-58. [FRACTIONATION OF PROTEINS AND AN ASSAY OF ENZYMIC
ACTIVITIES IN MUSCLE SARCOPLASM IN LOWER VERTEBRATES.]
IN  KREPSA, E.M. ED. FERMENTY V EVOLYUTSII ZIVOTNYKH,
LENINGRAD: NAUKA.
(RUSS.)
ENGL. SUMM.
*LAMPETRA FLUVIATILIS U.S.S.R. ANIMAL BIOCHEMISTRY
HISTOCHEMISTRY PISCES


                                                        4804F

FANGE, R.                1973
EDSTROM, A.
INCROPORATION OF THYMIDINE, URIDINE AND LEUCINE INTO BLOOD
CELLS OF MYXINE GLUTINOSA. PP. 49-50.
IN  FANGE,R. ED. MYXINE GLUTINOSA. BIOCHEMISTRY, PHYSIOLOGY
AND STRUCTURE. REPORT FROM A SYMPOSIUM IN GOTEBORG 28-29
APRIL,1972. GOTEBORG: KUNGL. VETENSKAPS- OCH
VITTERHETS-SAMHLLT.
CIRCULATORY SYSTEM PHYSIOLOGY


                                                        4805F

MURRAY, M.                1975
JONES, H.
CSERR, H.F.
RALL, D.P.
THE BLOOD BRAIN BARRIER AND VENTRICULAR SYSTEM OF MYXINE
GLUTINOSA.

BRAIN RES., 99:17-33
U.S.A. NERVOUS SYSTEM HISTOLOGY TECHNIQUES

4806F

**OSTBERG, Y.**                    1976
BOQUIST, L.
VAN NOORDEN, S.
PEARSE, A.G.E.
ON THE ORIGIN OF ISLET PARENCHYMAL CELLS IN A CYCLOSTOME,
MYXINE GLUTINOSA.
GEN. COMP. ENDOCRINOL., 28:228-246.
SWEDEN ENDOCRINOLOGY HISTOLOGY CYTOLOGY ADULT TECHNIQUES

4807F

**DELL'AGATA, M.**                  1973
CIACCIO, G.
RICERCHE SULLA BIOLOGIA DI LAMPETRA PLANERI (BLOCH) IN
ABRUZZO. OSSERVAZIONI SULLA PROLUNGATA SOPRAVVIVENZA DI
ESEMPLARI A DIVERSO STADIO DI SVILUPPO CON PARTICOLARE
RIGUARDO ALLA TRASFORMAZIONE DELLA LARVA IN
ADULTO.[RESEARCHES ON THE BIOLOGY OF LAMPETRA PLANERI
(BLOCH) IN ABRUZZO. OBSERVATION ON THE PROLONGED SURVIVAL OF
SPECIMENS AT DIFFERENT STAGES OF DEVELOPMENT WITH
PARTICULAR REFERENCE TO THE TRANSFORMATION OF THE LARVA INTO
THE ADULT.]
RIV. BIOL., 66:270-291.
(ITAL.)
ENGL. TRANSL.
ITALY METAMORPHOSIS LIFE CYCLE SPAWNING ADULT

4808F

**THEISEN, B.**                     1973
THE OLFACTORY SYSTEM IN THE HAGFISH MYXINE GLUTINOSA. I.FINE
STRUCTURE OF THE APICAL PART OF THE OLFACTORY EPITHELIUM.
ACTA ZOOL., 54:271-284
SWEDEN NORWAY OLFACTION SENSE RECEPTORS HISTOLOGY

4809F

**SCHULTZ, H.J.**                   1975
ADAM, H.
ELEKTRONENOPTISCHE HINWEISE AUF EINE NERVOSE VERBINDUNG
ZWISCHEN NEURO- UND ADENOHYPOPHYSE BEI MYXINE GLUTINOSA L.
(CYCLOSTOMATA.) [ELECTRON OPTICAL INDICATIONS OF CONNECTIONS
BETWEEN NEURO- AND ADENOHYPOPHYSIS IN MYXINE GLUTINOSA L.
CYCLOSTOMATA.]
NORW. J. ZOOL., 23:297-306.
(GER.)
NORWAY NERVOUS SYSTEM HISTOLOGY

4810F

**JANVIER, P.**                     1971
PALEONTOLOGIE - NOUVEAU MATERIEL D'ANDREOLEPIS HEDEI GROSS,
ACTINOPTERYGIEN ENIGMATIQUE DU SILURIEN DE GOTLAND (SUEDE).
ACAD. SCI. PARIS, C.R. HEBD. SEANCES, 273:2223-2224.
(FR.)
PISCES SYSTEMATICS

4811F

**ROBERTSON, J.D.**                 1976
CHEMICAL COMPOSITION OF THE BODY FLUIDS AND MUSCLE OF THE
HAGFISH MYXINE GLUTINOSA AND THE RABBIT-FISH CHIMAERA
MONSTROSA.
J. ZOOL., 178:261-277.
NORWAY PHYSIOLOGY BLOOD IONIC REGULATION OSMOREGULATION

4812F

**TIMMONS, T.J.**                   1976
NOTES ON THE FISHES OF NICKAJACK RESERVOIR, TENNESSEE, NEAR
THE RACOON MOUNTAIN PUMPED-STORAGE PLANT
TENN. ACAD. SCI., J., 51:66-67.
*ICHTHYOMYZON CASTANEUS U.S.A. ADULT ANIMAL DISTRIBUTION
MORTALITY BY

4813

**FEDOROV, B.A.**                   1976
LONG RANGE ORDER IN GLOBULAR PROTEINS.
FEBS LETT., 62:139-141.
BIOCHEMISTRY TECHNIQUES

4814

**RYAPOLOVA, N.I.**                 1974
SPAWNING STOCK AND FISHERY OF RIVER LAMPREY IN THE EASTERN
BALTIC.
ANN. BIOL., 29:177

4815

**QUERO, J.C.**                     1974
OBSERVATIONS ON RARE FISH IN FRANCE IN 1972.
ANN. BIOL., 29:182-183.

4816

BLACKER, R.W.                    1974
ENGLISH OBSERVATIONS ON RARE FISH IN 1972.
ANN. BIOL., 29:180-181.

4817F

GOVARDOVSKII, V.I.               1975
METOD OPREDELENIYA ZERITEL'NYKH PIGMENTOV IN SITU I IKH
IZUCHENIE U NEKOTORYKH POZVONOCHNYKH. [A METHOD FOR
DETERMINATION OF VISUAL PIGMENTS IN SITU AND STUDY OF SUCH
PIGMENTS IN CERTAIN VERTEBRATES.]
ZH. EVOL. BIOKHIM. FIZIOL., 11:346-352.
(RUSS.)
ENG. SUMM.
J. EVOL. BIOCHEM. PHYSIOL., 11:304-309.
(ENG.)
*LAMPETRA FLUVIATILIS ANIMAL BIOCHEMISTRY VISION

4818

SMITH, D.S.                      1975
JARLFORS, U.
CAMERON, B.F.
MORPHOLOGICAL EVIDENCE FOR THE PARTICIPATION OF MICRO
TUBULES IN AXONAL TRANSPORT.
N.Y. ACAD. SCI., ANN., 253:472-506.

4819F

KIER, P.M.                       1975
THE ECHINOIDS OF CARRIE-BOW CAY BELIZE.
SMITHSONIAN CONTRIB. ZOOL., 206:1-45
ANIMAL

4821

VLADYKOV, V.D.                   1976
KOTT, E.
A NEW NONPARASITIC SPECIES OF LAMPREY OF THE GENUS
ENTOSPENUS PETROMYZONIDAE FROM SOUTH CENTRAL CALIFORNIA
U.S.A.
SOUTH. CALIF. ACAD. SCI., LOS ANGELES, BULL., 75:60-67.

4822

DATHE, H.                        1975
POCKET BOOK OF ZOOLOGY, V.4, VERTEBRATES: 1: PISCES,
AMPHIBIA, REPTILIA. 244 PP.
JENA: GUSTAV FISHER VERLAG.
(GER.)

4823

OBRUCHEV, D.V.                   1968
ESSAYS ON THE PHYLOGENY AND SYSTEMATICS OF FOSSIL FISH AND
AGNATHA.
MOSCOW: NAUKA PRESS, 214 PP.

4825

NILSSEN, H.                      1975
DE GROOT, S.J.
THE FRESH WATER FISHES OF THE NETHERLANDS.
WET. MEDED. K.N.N.V.(K.NED. NATUURHIST. VER.), 108:1-44.

4826F

JOHNSON, B.G.H.                  1974
SURVEY OF THE INTERNATIONAL RAPIDS AREA, ST. MARY'S RIVER.
PP. 88-91.
IN  TIBBLES,J.J. ANNUAL REPORT 1973 OF THE SEA LAMPREY
CONTROL CENTRE SAULT STE. MARIE, ONTARIO, TO THE GREAT LAKES
FISHERY COMMISSION. GREAT LAKES FISH. COMM., ANNU. REP.
RESTRICTED
*PETROMYZON MARINUS CANADA AMMOCOETE BIOLOGY

4827

BLOCH, W.W.                      1976
AUSTIN, J.C.
RESPONSE OF THE EXPOSED DENTAL PULP TO A CORTICO STEROID
PASTE.
J. DENT. RES., 55:538

4828

KRIPPNER, M.                     1973
RADIOACTIVE CONTAMINATION OF THE MARINE ENVIRONMENT.
NEW YORK: UNIPUB, 786 PP.

4829

LEAKE, L.D.                      1975
COMPARATIVE HISTOLOGY. AN INTRODUCTION TO THE MICROSCOPIC
STRUCTURE OF ANIMALS.
LONDON: ACADEMIC PRESS, 738 PP.

4830

DONNER, P.J.                     1972
ROTIFERA FROM THE WATER SEDIMENT BORDER LAYER FROM THE
NEUSIEDLER LAKE.
AKAD. WISS. OSTERR., MAT.-NAT. KL., SITZ., 180:49-63.

4831

NELSON, J.S.                    1976
FISHES OF THE WORLD.
NEW YORK: WILEY, 416 PP.


4832F

AASA, R.                        1973
STUDIES OF HAGFISH TRANSFERRIN BY ELECTRON PARAMAGNETIC
RESONANCE (EPR) SPECTROSCOPY. P. 46.
IN  FANGE,R. ED. MYXINE GLUTINOSA. BIOCHEMISTRY, PHYSIOLOGY
AND STRUCTURE. REPORT FROM A SYMPOSIUM IN GOTEBORG 28-29
APRIL, 1972. GOTEBORG:KUNGL. VETENSKAPS- OCH
VITTERHETS-SAMHLLT.
*MYXINE GLUTINOSA *EPTATRETUS STOUTI


4833F

ERKELL, L.J.                    1973
IRON STORAGE OF THE HAGFISH (MYXINE GLUTINOSA). PP. 47-48.
IN  FANGE,R. ED. MYXINE GLUTINOSA. BIOCHEMISTRY, PHYSIOLOGY
AND STRUCTURE. REPORT FROM A SYMPOSIUM IN GOTEBORG 28-29
APRIL, 1972. GOTEBORG: KUNGL. VETENSKAPS- OCH VITTERHETS-
SAMHLLT.
ADULT METABOLISM CIRCULATORY SYSTEM


4834F

PALEUS, S.                      1973
LILJEQVIST, G.
BRAUNITZER, G.
A STUDY OF SOME HEMOPROTEINS OF MYXINE GLUTINOSA L., WITH
SPECIAL REFERENCE TO THE STRUCTURE OF HEMOGLOBIN III. PP.
42-45.
IN  FANGE,R. ED. MYXINE GLUTINOSA. BIOCHEMISTRY, PHYSIOLOGY
AND STRUCTURE. REPORT FROM A SYMPOSIUM IN GOTEBORG
28-29,APRIL, 1972. GOTEBORG: KUNGL. VETENSKAPS- OCH
VITTERHETS-SAMHLLET.
*LAMPETRA FLUVIATILIS *EPTATRETUS STOUTII BLOOD BIOCHEMISTRY


4835F

LAGERSTRAND, G.                 1973
NILSSON, S.
EFFECTS OF 6-OH-DOPAMINE ON THE CATECHOLAMINE LEVELS IN
THE SYSTEMATIC AND PORTAL VEINS OF MYXINE. PP. 39-41.
IN  FANGE,R. ED. MYXINE GLUTINOSA. BIOCHEMISTRY, PHYSIOLOGY
AND STRUCTURE. REPORT FROM A SYMPOSIUM IN GOTEBORG 28-29
APRIL, 1972. GOTEBORG: KUNGL. VETENSKAPS- OCH VITTERHETS-
SAMHLLET.
CIRCULATORY SYSTEM BIOCHEMISTRY PHYSIOLOGY


4836F

RADNER, S.                      1973
ARLOCK, P.
NILSSON, S.
ELECTROCARDIOGRAM (ECG) OF THE SYSTEMIC AND PORTAL HEARTS OF
MYXINE. PP. 35-36.
IN  FANGE,R. ED. MYXINE GLUTINOSA. BIOCHEMISTRY, PHYSIOLOGY
AND STRUCTURE. REPORT FROM A SYMPOSIUM IN GOTEBORG 28-29
APRIL, 1972. GOTEBORG:KUNGL. VETENSKAPS- OCH
VITTERHETS-SAMHLLET.
ANIMAL CIRCULATORY SYSTEM


4837F

FERNHOLM, B.                    1973
IS THERE ANY STEROID HORMONE FORMATION IN THE OVARY OF
MYXINE? PP. 33-34.
IN  FANGE,R. ED. MYXINE GLUTINOSA. BIOCHEMISTRY, PHYSIOLOGY
AND STRUCTURE. REPORT FROM A SYMPOSIUM IN GOTEBORG 28-29
APRIL, 1972. GOTEBORG: KUNGL. VETENSKAPS- OCH VITTERHETS-
SAMHLLET.
ADULT HISTOLOGY BIOCHEMISTRY


4838F

OSTBERG, Y.                     1973
BOQUIST, L.
EMDIN, S.O.
FALKMER, S.
HASSLER, O.
STEINER, D.F.
THOMAS, N.W.
STRUCTURE AND FUNCTION OF THE ENDOCRINE PANCREAS OF THE
HAGFISH, MYXINE GLUTINOSA. PP. 26-29.
IN  FANGE,R. ED. MYXINE GLUTINOSA. BIOCHEMISTRY, PHYSIOLOGY
AND STRUCTURE. REPORT FROM A SYMPOSIUM IN GOTEBORG 28-29,
APRIL, 1972. GOTEBORG: KUNGL. VETENSKAPS- OCH
VITTERHETS-SAMHLLET.
ENDOCRINOLOGY EVOLUTION ANATOMY HISTOLOGY BIOCHEMISTRY


4839F

HALLBACK, D.-A.                 1973
ACETYLCHOLINESTERASE CONTAINING STRUCTURES IN THE INTESTINE

OF MYXINE GLUTINOSA L. PP. 24-25.
IN FANGE,R. ED. MYXINE GLUTINOSA. BIOCHEMISTRY, PHYSIOLOGY
AND STRUCTURE. REPORT FROM A SYMPOSIUM IN GOTEBORG 28-29
APRIL, 1972. GOTEBORG:KUNGL. VETENSKAPS- OCH VITTERHETS-
SAMHLLET.
ADULT HISTOLOGY MUSCLE NERVOUS SYSTEM DIGESTION

4840F

FLOOD, P.R.                    1973
THE SKELETAL MUSCLE FIBRE TYPES OF MYXINE GLUTINOSA L.
RELATED TO THOSE OF OTHER CHORDATES. PP. 17-20..
IN FANGE,R. ED. MYXINE GLUTINOSA. BIOCHEMISTRY, PHYSIOLOGY
AND STRUCTURE. REPORT FROM A SYMPOSIUM IN GOTEBORG 28-29
APRIL, 1972. GOTEBORG:KUNGL. VETENSKAPS- OCH VITTERHETS-
SAMHLLET.
*LAMPETRA FLUVIATILIS *PETROMYZON MARINUS AMMOCOETE ADULT
MUSCLE ANATOMY HISTOLOGY

4841F

LIE, H.R.                      1973
AN INTERMEDIATE MUSCLE TYPE IN THE RIVER LAMPREY, LAMPETRA
FLUVIATILIS. PP. 21-22.
IN FANGE,R. ED. MYXINE GLUTINOSA. BIOCHEMISTRY, PHYSIOLOGY
AND STRUCTURE. REPORT FROM A SYMPOSIUM IN GOTEBORG 28-29
APRIL, 1972. GOTEBORG:KUNGL. VETENSKAPS- OCH
VITTERHETS-SAMHLLET.
ADULT ANATOMY MUSCLE HISTOCHEMISTRY MORPHOMETRY BIOCHEMISTRY

4842F

FLOOD, P.R.                    1973
THE NOTOCHORD OF MYXINE GLUTINOSA L. RELATED TO THAT OF
OTHER CHORDATES. PP. 14-16.
IN FANGE,R. ED. MYXINE GLUTINOSA. BIOCHEMISTRY, PHYSIOLOGY
AND STRUCTURE. REPORT FROM A SYMPOSIUM IN GOTEBORG 28-29
APRIL, 1972. GOTEBORG:KUNGLE. VETENSKAPS- OCH VITTERHETS-
SAMHALLET.
ADULT ANATOMY EVOLUTION CYTOLOGY

4843F

NYBELIN, O.                    1973
PHYLOGENY OF THE HAGFISH AND THE LAMPREYS. 8-13PP.
IN FANGE,R. ED. MYXINE GLUTINOSA. BIOCHEMISTRY,PHYSIOLOGY
AND STRUCTURE. REPORT FROM A SYMPOSIUM IN GOTEBORG 28-29
APRIL, 1972. GOTEBORG:KUNGL. VETENSKAPS- OCH
VITTERHETS-SAMHALLET.
*MYXINE GLUTINOSA *MAYOMYZON PIEKOENSIS ADULT ANATOMY
CIRCULATORY SYSTEM EVOLUTION RESPIRATION

4844F

TIBBLES, J.J. ED.             1976
ANNUAL REPORT 1975 OF THE SEA LAMPREY CONTROL CENTRE SAULT
STE. MARIE, ONTARIO, TO THE GREAT LAKES FISHERY COMMISSION.
GREAT LAKES FISH. COMM., ANNU. REP., 98PP.
RESTRICTED
SEA LAMPREY CANADA GREAT LAKES U.S.A. ADULT AMMOCOETE
CHEMISTRY MANAGEMENT MORTALITY OF

4845F

HOLMBERG, K.                   1973
ULTRASTRUCTURAL INVESTIGATION OF THE OPTICAL TRACT IN MYXINE
GLUTINOSA. P. 23.
IN FANGE,R. ED. MYXINE GLUTINOSA. BIOCHEMISTRY, PHYSIOLOGY
AND STRUCTURE. REPORT FROM A SYMPOSIUM IN GOTEBORG 28-29
APRIL, 1972. GOTEBORG:KUNGL. VETENSKAPS- OCH VITTERHETS-
SAMHLLET.
ADULT VISION MORPHOMETRY NERVOUS SYSTEM HISTOLOGY

4846F

NILSSON, A.                    1973
SECRETIN-LIKE AND CHOLECYSTOKININ-LIKE ACTIVITY IN MYXINE
GLUTINOSA L. PP. 30-32.
IN FANGE,R. ED. MYXINE GLUTINOSA. BIOCHEMISTRY, PHYSIOLOGY
AND STRUCTURE. REPORT FROM A SYMPOSIUM IN GOTEBORG 28-29
APRIL, 1972. GOTEBORG:KUNGL. VETENSKAPS- OCH
VITTERHETS-SAMHLLET.
DIGESTION BIOCHEMISTRY ANIMAL

4847F

HELLE, K.B.                    1973
LONNING, S.
BLASCHKO, H.
CHROMAFFIN AND MYOCARDIAL GRANULES OF MYXINE GLUTINOSA L.
PP. 37-38.
IN FANGE, R. ED. MYXINE GLUTINOSA. BIOCHEMISTRY,
PHYSIOLOGYAND STRUCTURE. REPORT FROM A SYMPOSIUM IN GOTEBORG
28-29    APRIL, 1972. GOTEBORG:KUNGL. VETENSKAPS- OCH
VITTERHETS-    SAMHLLET.
CIRCULATORY SYSTEM HISTOLOGY BIOCHEMISTRY CYTOLOGY

4848F

JOHNSON, B.G.H.               1974

SEA LAMPREY TAKEN AT DENN'S DAM, 1973. P. 88.
IN  TIBBLES,J.J. ANNUAL REPORT 1973 OF THE SEA LAMPREY
CONTROL CENTRE SAULT STE. MARIE, ONTARIO, TO THE GREAT LAKES
FISHERY COMMISSION. GREAT LAKES FISH. COMM., ANNU. REP.
RESTRICTED
*PETROMYZON MARINUS CANADA ADULT

4849F

FANGE, R.                        1973
ERICHSEN, L.
CARBONIC ANHYDRASE IN MYXINE. PP. 86-88.
IN  FANGE,R. ED. MYXINE GLUTINOSA. BIOCHEMISTRY, PHYSIOLOGY
AND STRUCTURE. REPORT FROM A SYMPOSIUM IN GOTEBORG 28-29
APRIL, 1972. GOTEBORG:KUNGL. VETENSKAPS-OCH VITTERHETS-
SAMHLLT.
ANIMAL BIOCHEMISTRY BLOOD

4850F

OLSSON, J.Y.                     1973
ACTIVITIY OF DELTA-AMINOLEVULINIC ACID DEHYDRASE (ALA-D)IN
ORGANS OF MYXINE GLUTINOSA. PP. 84-85.
IN  FANGE,R. ED. MYXINE GLUTINOSA. BIOCHEMISTRY, PHYSIOLOGY
AND STRUCTURE. REPORT FROM A SYMPOSIUM IN GOTEBORG 28-29
APRIL, 1972. GOTEBORG: KUNGL. VETENSKAPS- OCH
VITTERHETS-SAMHLLT.
BIOCHEMISTRY ENDOCRINOLOGY

4851F

HULTSJO, C.                      1973
METABOLISM OF PHENYLALANINE IN MYXINE GLUTINOSA. PP. 82-83.
IN  FANGE,R. ED. MYXINE GLUTINOSA. BIOCHEMISTRY, PHYSIOLOGY
AND STRUCTURE. REPORT FROM A SYMPOSIUM IN GOTEBORG 28-29
APRIL, 1972. GOTEBORG:KUNGL. VETENSKAPS- OCH VITTERHETS-
SAMHLLT.
ADULT BIOCHEMISTRY ENDOCRINOLOGY

4852F

FANGE, R.                        1973
MATTISSON, A.
NEOPLASMS IN MYXINE. PP. 76-79.
IN  FANGE,R. ED. MYXINE GLUTINOSA. BIOCHEMISTRY, PHYSIOLOGY
AND STRUCTURE. REPORT FROM A SYMPOSIUM IN GOTEBORG 28-29
APRIL, 1972. GOTEBORG:KUNGL. VETENSKAPS- OCH VITTERHETS-
SAMHLLT.
SWEDEN DENMARK PATHOLOGY HISTOLOGY

4853F

NYGREN, A.                       1973
JAHNKE, M.
CHROMOSOMES OF MYXINE GLUTINOSA. PP. 80-81.
IN  FANGE,R. ED. MYXINE GLUTINOSA. BIOCHEMISTRY, PHYSIOLOGY
AND STRUCTURE. REPORT FROM A SYMPOSIUM IN GOTEBORG 28-29
APRIL, 1972. GOTEBORG:KUNGL. VETENSKAPS- OCH
VITTERHETS-SAMHLLT.
SWEDEN GENETICS

4854F

OSTBERG, Y.                      1973
BOQUIST, L.
FALKMER, S.
TUMORS OR HAMARTOMAS IN ENDOCRINE PANCREAS OF THE HAGFISH.
PP. 73-75.
IN  FANGE,R. ED. MYXINE GLUTINOSA. BIOCHEMISTRY, PHYSIOLOGY
AND STRUCTURE. REPORT FROM A SYMPOSIUM IN GOTEBORG 28-29
APRIL, 1972. GOTEBORG: KUNGL. VETENSKAPS- OCH
VITTERHETS-SAMHLLT.
PATHOLOGY HISTOLOGY

4855F

FALKMER, S.                      1973
OSTBERG, Y.
EMDIN, S.O.
TUMORS OF THE LIVER OF THE HAGFISH, MYXINE GLUTINOSA. PP.
70-72.
IN  FANGE,R. ED. MYXINE GLUTINOSA. BIOCHEMISTRY, PHYSIOLOGY
AND STRUCTURE. REPORT FROM A SYMPOSIUM IN GOTEBORG 28-29
APRIL, 1972. GOTEBORG: KUNGL. VETENSKAPS- OCH VITTERHETS-
SAMHLLT.
SWEDEN PATHOLOGY

4856F

WALVIG, F.                       1973
SOME OBSERVATIONS OF EPIDERMOID CYSTS IN HAGFISH (MYXINE
GLUTINOSA L.). PP. 67-69.
IN  FANGE,R. ED. MYXINE GLUTINOSA. BIOCHEMISTRY, PHYSIOLOGY
AND STRUCTURE. REPORT FROM A SYMPOSIUM IN GOTEBORG 28-29
APRIL, 1972. GOTEBORG:KUNGL. VETENSKAPS- OCH
VITTERHETS-SAMHLLT.
NORWAY PATHOLOGY

4857F

FANGE, R.                    1973
THE LYMPHATIC SYSTEM OF MYXINE. PP. 57-64.
IN  FANGE,R. ED. MYXINE GLUTINOSA. BIOCHEMISTRY, PHYSIOLOGY
AND STRUCTURE. REPORT FROM SYMPOSIUM IN GOTEBORG 28-29
APRIL, 1972. GOTEBORG: KUNGL. VETENSKAPS- OCH VITTERHETS-
SAMHLLT.
BLOOD ENDOCRINOLOGY PHYSIOLOGY

4858F

LUNDHOLM, M.                  1973
AN ATTEMPT TO IMMUNIZE MYXINE. PP. 65-66.
IN  FANGE,R. ED. MYXINE GLUTINOSA. BIOCHEMISTRY. PHYSIOLOGY
AND STRUCTURE. REPORT FROM A SYMPOSIUM IN GOTEBORG 28-29
APRIL, 1972. GOTEBORG: KUNGL. VETENSKAPS- OCH
VITTERHTS-SAMHLLT.
IMMUNOLOGY

4859F

JOHANSSON, M.L.              1973
PEROXIDE IN BLOOD CELLS OF FISHES AND CYCLOSTOMES. PP. 53-56
IN  FANGE,R. ED. MYXINE GLUTINOSA. BIOCHEMISTRY, PHYSIOLOGY
AND STRUCTURE. REPORT FROM A SYMPOSIUM IN GOTEBORG 28-29
APRIL, 1972. GOTEBORG: KUNGL. VETENSKAPS- OCH VITTERHETS-
SAMHLLT.
*LAMPETRA PLANERI SWEDEN ANIMAL BLOOD HISTOCHEMISTRY

4860F

FANGE, R.                    1973
GIDHOLM, L.
BLOOD COAGULATION. PP. 51-52.
IN  FANGE,R. ED. MYXINE GLUTINOSA. BIOCHEMISTRY, PHYSIOLOGY
AND STRUCTURE. REPORT FROM A SYMPOSIUM GOTEBORG 28-29
APRIL,1972. GOTEBORG:KUNGL. VETENSKAPS-OCH VITTERHETS-SAMHLL

4861F

WESTMAN, R.W.                1974
SEA LAMPREY SURVEYS OF STREAMS TRIBUTARY TO THE CANADIAN
SIDE OF LAKE ONTARIO 1973. PP.15.
IN  TIBBLES,J.J. ED. ANNUAL REPORT 1973 OF THE SEA LAMPREY
CONTROL CENTRE SAULT STE. MARIE, ONTARIO, TO THE GREAT LAKES
FISHERY COMMISSION. GREAT LAKES FISH. COMM., ANNU. REP.
RESTRICTED
*PETROMYZON MARINUS CANADA AMMOCOETE BIOLOGY

4862F

SCHLEEN, L.P.                1974
SEA LAMPREY SURVEYS, ST. MARYS RIVER, 1973. PP. 12-15.
IN  TIBBLES,J.J. ED. ANNUAL REPORT 1973 OF THE SEA LAMPREY
CONTROL CENTRE SAULT STE. MARIE, ONTARIO, TO THE LAKES
FISHERY COMMISSION. GREAT LAKES FISH. COMM., ANNU. REP.
RESTRICTED
*PETROMYZON MARINUS *ICHTHYOMYZON CANADA AMMOCOETE BIOLOGY

4863F

WESTMAN, R.W.                1974
GRANULAR BAYER 73 SURVEYS OF LAKE HURON RIVER ESTUARIES,
1973. PP.12.
IN  TIBBLES, J.J., ED., ANNUAL REPORT 1973 OF THE SEA
LAMPREY CONTROL CENTRE SAULT STE. MARIE, ONTARIO, TO THE
GREAT LAKES FISHERY COMMISSION. GREAT LAKES FISH. COMM.,
ANNU. REP.
RESTRICTED
*PETROMYZON MARINUS *ICHTHYOMYZON CANADA AMMOCOETE BIOLOGY

4864F

WESTMAN, R.W.                1974
SEA LAMPREY SURVEYS OF STREAMS TRIBUTARY TO LAKE
HURON,1973.PP.7-12.
IN  TIBBLES,J.J. ED. ANNUAL REPORT 1973 OF THE SEA LAMPREY
CONTROL CENTRE SAULT STE. MARIE, ONTARIO, TO THE GREAT LAKES
FISHERY COMMISSION. GREAT LAKES FISH. COMM., ANNU. REP.
RESTRICTED
*PETROMYZON MARINUS *ICHTHYOMYZON CANADA AMMOCOETE BIOLOGY

4865F

SCHLEEN, L.P.                1974
SEA LAMPREY SURVEYS OF STREAMS TRIBUTARY TO LAKE SUPERIOR,
1973. PP.6-7.
IN  TIBBLES,J.J. ED. ANNUAL REPORT 1973 OF THE SEA LAMPREY
CONTROL CENTRE SAULT STE. MARIE, ONTARIO, TO THE GREAT LAKES
FISHERY COMMISSION GREAT LAKES FISH. COMM., ANNU. REP.
RESTRICTED
*PETROMYZON MARINUS *ICHTHYOMYZON CANADA AMMOCOETE BIOLOGY

4866F

JOHNSON, B.G.H.              1974
SEA LAMPREY ASSESSMENT BARRIER OPERATIONS: 1973. PP. 5.
IN  TIBBLES,J.J. ANNUAL REPORT 1973 OF THE SEA LAMPREY
CONTROL CENTRE SAULT STE. MARIE, ONTARIO, TO THE GREAT LAKES
FISHERY COMMISSION.GREAT LAKES FISH. COMM., ANNU. REP.
RESTRICTED

*PETROMYZON MARINUS CANADA ADULT MORTALITY OF

4867F

SCHLEEN, L.P.                    1974
SEA LAMPREY SURVEYS OF STREAMS TRIBUTARY TO THE NEW YORK
SIDE OF LAKE ONTARIO, 1973. PP.15-18.
IN  TIBBLES,J.J.ED. ANNUAL REPORT 1973 OF THE SEA LAMPREY
CONTROL CENTRE SAULT STE. MARIE, ONTARIO, TO THE GREAT LAKES
FISHERY COMMISSION. GREAT LAKES FISH. COMM., ANNU. REP.
RESTRICTED
*PETROMYZON MARINUS U.S.A. AMMOCOETE BIOLOGY


4868F

LARSEN, L.O.
POSSIBLE FACTORS INVOLVED IN INITIATION OF SEXUAL
MATURATIONIN RIVER LAMPREYS (LAMPETRA FLUVIATILIS). PP.
156-157.
IN  PETERS,H. ED. THE DEVELOPMENT AND MATURATION OF THE
OVARY AND ITS FUNCTIONS. INTERNATIONAL CONGRESS SERIES NO.
267. AMSTERDAM:EXCERPTA MEDICA AMSTERDAM.
ADULT LIFE MATURITY


4869F

RIVIERE, H.B.                    1975
COOPER, E.L.
REDDY, A.L.
HILDEMANN, W.H.
IN SEARCH OF THE HAGFISH THYMUS.
AM. ZOOL., 15:39-49.
IMMUNOLOGY ANATOMY HISTOLOGY


4870F

OSTBERG, Y.                      1976
VAN NOORDEN, S.
EVERSON OEARSE, A.G.
THOMAS, N.W.
CYTOCHEMICAL, IMMUNOFLUORESCENCE, AND ULTRASTRUCTURAL
INVESTIGATIONS ON POLYPEPTIDE HORMONE CONTAINING CELLS IN
THE INTESTINAL MUCOSA OF A CYCLOSTOME, MYXINE GLUTINOSA.
GEN. COMP. ENDOCRINOL., 28:213-227.
SWEDEN TECHNIQUES DIGESTION


4871

KERR, J.W.                       1975
CAPE STORM FORMATION A NEW SI-URIAN UNIT IN THE CANADIAN
ARCTIC.
CAN. PET. GEOL., BULL., 23:67-83.


4872

SHIBATA, Y.                      1977
YAMAMOTO, T.
GAP JUNCTIONS IN THE CARDIAC MUSCLE CELLS OF THE LAMPREY.
CELL TISSUE RES., 178:477-482.


4873

POLTAVCHUK, M.A.                 1975
FISH OF SMALL RIVERS IN RIGHT BANK POLESYE OF THE
UKRAINIAN-SSR USSR. PART 1. SPECIES COMPOSITION OF FISH
POPULATION IN THE UPPER PRIPYAT RIVER.
VESTN. ZOOL., 9(4):9-15.
(RUSS.)
ENG. SUMM.
LAMPETRA MARIAE ECOLOGY


4874

BUNDGAARD, M.                    1976
THE BLOOD BRAIN BARRIER IN THE LAMPREY.
ACTA PHYSIOL. SCAND., SUPPL., 440:83.
(ABSTR.)
*LAMPETRA FLUVIATILIS NERVOUS SYSTEM CIRCULATORY SYSTEM
PHYSIOLOGY CYTOLOGY ADULT


4875F

HEROLD, R.C.B.                   1975
SCANNING ELECTRON MICROSCOPY OF ENAMELOID AND DENTIN IN
FISH TEETH.
ARCH. ORAL BIOL., 20:635-640.
HISTOLOGY


4876F

LIE, H.R.                        1974
A QUANTITATIVE IDENTIFICATION OF 3 MUSCLE FIBER TYPES IN THE
BODY MUSCLES OF LAMPETRA FLUVIATILIS AND THEIR RELATION TO
BLOOD CAPILLARIES.
CELL TISSUE RES., 154:109-119.
*LAMPETRA FLUVIATILIS FINLAND ADULT MUSCLE HISTOCHEMISTRY
MORPHOLOGY


4877F

LEWIS, S.V.                      1976
POTTER, I.C.

GILL MORPHOMETRICS OF THE LAMPREYS LAMPETRA FLUVIATILIS AND
LAMPETRA PLANERI (BLOCH).
ACTA ZOOL., (STOCKH.), 57:103-112.
U.K. AMMOCOETE ADULT RESPIRATION

4878F

MOLNAR, B.                          1975
SZABO, S.
NEURAL AND VASCULAR CONNECTIONS IN THE HYPOTHALAMIC
HYPOPHYSIAL COMPLEX OF EUDONTOMYZON DANFORDI REGAN.
CLUJ, ROM., UNIV. BABES-BOLYAI, STUD., SER. BIOL., 20:43-47.
(ROM.)

4879

VOROB'EVA, E.I.                     1975
RELATIONSHIP BETWEEN CARTILAGE AND BONE DURING THE PHYLOGENY
OF THE EARLIEST LOWER VERTEBRATES.
ZH. OBSHCH. BIOL., 36:361-372.

4880F

YOUSON, J.H.                        1975
ABSORPTION AND TRANSPORT OF EXOGENOUS PROTEIN IN THE
ARCHINEPHRIC DUCT OF THE OPISTHONEPHRIC KIDNEY OF THE SEA
LAMPREY, PETROMYZON MARINUS L.
COMP. BIOCHEM. PHYSIOL., 52A:639-643.
EXCRETION HISTOLOGY

4881F

YOUSON, J.H.                        1975
RADIOAUTOGRAPHY OF PRESUMPTIVE INTERRENAL CELLS IN THE SEA
LAMPREY AFTER TRITIATED CHOLESTEROL INJECTION.
ACTA ZOOL., (STOCKH.), 56:219-223.
*PETROMYZON MARINUS ADULT ENDOCRINOLOGY CANADA HISTOLOGY

4882F

TSUNEKI, K.                         1974
DISTRIBUTION OF A MONOAMINE OXIDASE AND ACETYLCHOLINESTERASE
IN THE HYPOTHALAMO-HYPOPHYSIAL SYSTEM OF THE LAMPREY,
LAMPETRA JAPONICA.
CELL TISSUE RES., 154:17-27.
JAPAN ADULT NERVOUS SYSTEM ENDOCRINOLOGY

4883F

GRUZOVA, M.N.                       1975
KARIOSFERA V OOGENEZE. [THE KARYOSPHERE IN OOGENESIS.]
TSITOLOGIYA, 17:219-237.
ANIMAL EMBRYOLOGY CYTOLOGY

4884

NAKAO, T.                           1976
AN ELECTRON MICROSCOPIC STUDY OF THE NEURO MUSCULAR JUNCTION
IN THE MYOTOMES OF LARVAL LAMPREY LAMPETRA JAPONICA.
J. COMP. NEUROL., 165:1-15.
*LAMPETRA FLUVIATILIS *LAMPETRA PLANERI AMMOCOETE HISTOLOGY
MUSCLE CYTOLOGY

4885F

IDLER, D.R.                         1976
BURTON, M.P.M.
THE PRONEPHROI AS THE SITE OF PRESUMPTIVE INTERRENAL CELLS
IN THE HAGFISH MYXINE GLUTINOSA.
COMP. BIOCHEM. PHYSIOL., 53A:73-77.

4886F

HARDISTY, M.W.                      1975
ZELNIK, P.R.
MOORE, I.A.
THE EFFECTS OF SUBTOTAL AND TOTAL ISLETECTOMY IN THE RIVER
LAMPREY LAMPETRA FLUVIATILIS.
GEN. COMP. ENDOCRINOL., 27:179-192.
*LAMPETRA PLANERI ADULT BIOCHEMISTRY ENDOCRINOLOGY MIGRATION
PHYSIOLOGY

4887F

SEREBRENIKOVA, T.P.                 1975
KHLYUSTINA, T.E.
IZOFERMENTY FOSFORILAZY SKELETNYKH MYSHTS KRUGOROTIKH I
KOSTISTYKH RY'. [PHOSPHORYLASE ISOENZYMES FROM SKELETAL
MUSCLES  OF CYCLOSTOMATA AND BONY FISH.]
BIOKHIMIIA, 40:652-658.
(RUSS.)
ENGL. SUMM.
*LAMPETRA FLUVIATILIS MUSCLE METABOLISM

4888F

DAVIS, W.A.                         1974
SHERA, W.P.
WEISE, J.G.
INDIVIDUAL STREAM TREATMENT REPORTS, LAKE HURON, 1973. PP.
32-46.
IN  TIBBLES,J.J.ED. ANNUAL REPORT 1973 OF THE SEA LAMPREY

CONTROL CENTRE SAULT STE. MARIE, ONTARIO, TO THE GREAT
LAKES FISHERY COMMISSION. GREAT LAKES FISH. COMM., ANNU.REP.
RESTRICTED
*PETROMYZON MARINUS CANADA AMMOCOETE MORTALITY OF CHEMISTRY

4889F

DUSTIN, S.M.                    1974
GOOLD, R.J.
INDIVIDUAL STREAM TREATMENT REPORTS, LAKE SUPERIOR, 1973.
PP. 21-31.
IN  TIBBLES, J.J., ED., ANNUAL REPORT 1973 OF THE SEA
LAMPREY CONTROL CENTRE, SAULT STE. MARIE, ONTARIO, TO THE
GREAT LAKES FISHERY COMMISSION. GREAT LAKES FISH. COMM.,
ANNU. REP.
RESTRICTED
*PETROMYZON MARINUS CANADA AMMOCOETE CHEMISTRY MORTALITY OF

4890F

WESTMAN, R.W.                    1974
SEA LAMPREY SURVEY OF THE NIAGARA RIVER, 1973. PP. 19-20.
IN  TIBBLES,J.J. ED. ANNUAL REPORT 1973 OF THE SEA LAMPREY
CONTROL CENTRE SAULT STE. MARIE, ONTARIO, TO THE GREAT LAKES
FISHERY COMMISSION. GREAT LAKES FISH. COMM., ANNU. REP.
RESTRICTED
*LAMPETRA LAMOTTEI *PETROMYZON MARINUS CANADA AMMOCOETE

4891F

DAVIS, W.A.                     1974
SHERA, W.P.
GRANULAR BAYER 73 TREATMENT OF ST. MARYS RIVER, 1973. PP.
75-76.
IN  TIBBLES,J.J.ED. ANNUAL REPORT 1973 OF THE LAMPREY
CONTROL CENTRE SAULT STE. MARIE, ONTARIO, TO THE GREAT LAKES
FISHERY COMMISSION. GREAT LAKES FISH. COMM., ANNU. REP.
RESTRICTED
CANADA AMMOCOETE CHEMISTRY

4892F

DUSTIN, S.M.                    1974
GOOLD, R.J.
GRANULAR BAYER 73 TREATMENTS, LAKE SUPERIOR, 1973. PP 71-74.
IN  TIBBLES,J.J. ED. ANNUAL REPORT 1973 OF THE SEA LAMPREY
CONTROL CENTRE SAULT STE. MARIE, ONTARIO, TO THE GREAT LAKES
FISHERY COMMISSION. GREAT LAKES FISH. COMM., ANNU. REP.
RESTRICTED
*PETROMYZON MARINUS CANADA MORTALITY OF CHEMISTRY AMMOCOETE

4893F

DAVIS, W.A.                     1974
SHERA, W.P.
WEISE, J.G.
INDIVIDUAL STREAM TREATMENT REPORTS, CANADIAN SIDE OF LAKE
ONTARIO, 1973. PP. 47-52.
IN  TIBBLES,J.J.ED. ANNUAL REPORT 1973 OF THE SEA LAMPREY
CONTROL CENTRE SAULT STE. MARIE, ONTARIO, TO THE GREAT
LAKES FISHERY COMMISSION. GREAT LAKES FISH. COMM., ANNU.REP.
RESTRICTED
*PETROMYZON MARINUS CANADA AMMOCOETE MORTALITY OF CHEMISTRY

4894F

DUSTIN, S.M.                    1974
GOOLD, R.J.
INDIVIDUAL STREAM TREATMENT REPORTS, LAKE ONTARIO (UNITED
STATES) 1973. PP. 52-70.
IN  TIBBLES, J.J. ED. ANNUAL REPORT 1973 OF THE SEA LAMPREY
CONTROL CENTRE SAULT STE. MARIE, ONTARIO, TO THE GREAT LAKES
FISHERY COMMISSION. GREAT LAKES FISH. COMM., ANNU. REP.
RESTRICTED
*PETROMYZON MARINUS U.S.A. MORTALITY OF CHEMISTRY AMMOCOETE

4895F

KOTT, E.                        1973
EPIDERMAL PIGMENTATION IN THE SEA LAMPREY PETROMYZON
MARINUS L.
CAN. J. ZOOL., 51:101-104.
CANADA ADULT PIGMENTATION MIGRATION

4896F

FILOSOFOVA, E.M.                1969
SEREBRENIKOVA, T.P.
ISSLEDOVANIE SARKOPLAZMATICHESKIKH 'ELKOV 1 IKH
FERMENTATIVNOI AKTIVNOSTI U KRUGLOROTYKH 1 RY' SOCHETANIEM
METODOV KHROMATOGRAFII 1 ELEKTROFOREZA. [COMBINED
CHROMATOGRAPHIC AND ELECTROPHORETIC STUDIES ON SARCOPLASMIC
PROTEINS AND THEIR ENZYMIC PROPERTIES IN CYLLOSTOMES AND
FISHES.] PP. 59-66.
IN  KREPSA,E.M. ED. FERMENTY V EVOLYUTSII ZHIVOTNYKH.
LENINGRAD: NAUKA.
(RUSS.)
ENGL. SUMM.
ANIMAL BIOCHEMISTRY SYSTEMATICS

4897F

YOUSON, J.H.                    1972
STRUCTURE AND DISTRIBUTION OF INTERSTITIAL CELLS
(PRESUMPTIVE INTERRENAL CELLS) IN THE OPISTHONEPHRIC KIDNEYS
OF LARVAL AND ADULT SEA LAMPREY, PETROMYZON MARINUS L.
GEN. COMP. ENDOCRINOL., 19:56-68.
HISTOLOGY EXCRETION ADULT GREAT LAKES ENDOCRINOLOGY
EVOLUTION HISTOCHEMISTRY


4898F

YOUSON, J.H.                    1973
POTASSIUM PYROANTIMONATE AND OSMIUM-ZINC IODIDE REACTIVITY
IN THE TUBULAR EPITHELIUM OF THE OPISTHONEPHRIC KIDNEY OF
THESEA LAMPREY, PETROMYZON MARINUS L..
J. MORPHOL., 140:119-134.
CANADA AMMOCOETE ADULT EXCRETION HISTOLOGY HISTOCHEMISTRY


4899F

YOUSON, J.H.                    1973
EFFECTS OF MAMMALIAN CORTICOTROPHIN ON THE ULTRASTRUCTURE
OFPRESUMPTIVE INTERRENAL CELLS IN THE OPISTHONEPHROS OF THE
LAMPREY, PETROMYZON MARINUS L..
AM. J. ANAT., 138:235-252.
CANADA AMMOCOETE ADULT BIOCHEMISTRY HISTOLOGY HISTOCHEMISTRY


4900F

JOHNSON, B.G.H.                 1974
BIOLOGICAL DATA ON ADULT SEA  LAMPREY FROM THE HUMBER
RIVER, 1973. PP. 85-87.
IN  TIBBLES,J.J. ANNUAL REPORT 1973 OF THE SEA LAMPREY
CONTROL CENTRE SAULT STE. MARIE, ONTARIO. TO THE GREAT
LAKESFISHERY COMMISSION. GREAT LAKES FISH. COMM., ANNU. REP.
RESTRICTED
*PETROMYZON MARINUS CANADA ADULT BIOLOGY


4901F

DUSTIN, S.M.                    1974
NUMBER AND PERCENTAGE OF SEA LAMPREY AMMOCOETES TAKEN FROM
LAKE SUPERIOR STREAM TREATMENTS, 1971-1973. PP. 77.
IN  TIBBLES,J.J. ED. ANNUAL REPORT 1973 OF THE SEA LAMPREY
CONTROL CENTRE SAULT STE. MARIE, ONTARIO, TO THE GREAT LAKES
FISHERY COMMISSION. GREAT LAKES FISH. COMM., ANNU. REP.
RESTRICTED
*PETROMYZON MARINUS CANADA AMMOCOETE


4902F

JOHNSON, B.G.H.                 1974
BIOLOGICAL DATA FROM SEA LAMPREY COLLECTED BY COMMERCIAL
FISHERMEN, 1973. P. 85.
IN  TIBBLES,J.J. ANNUAL REPORT 1973 OF THE SEA LAMPREY
CONTROL CENTRE SAULT STE. MARIE, ONTARIO, TO THE GREAT
LAKESFISHERY COMMISSION. GREAT LAKES FISH. COMM., ANNU. REP.
RESTRICTED
*PETROMYZON MARINUS CANADA ADULT BIOLOGY


4903F

SHERA, W.P.                     1974
INSTALLATION OF PERMANENT STAFF GAUGES ON LAKES HURON AND
ONTARIO TRIBUTARIES. PP. 79-80.
IN  TIBBLES,J.J. ANNUAL REPORT 1973 OF THE SEA LAMPREY
CONTROL CENTRE SALUT STE. MARIE, ONTARIO, TO THE GREAT LAKES
FISHERY COMMISSION.
RESTRICTED
CANADA TECHNIQUES


4904F

DAVIS, W.A.                     1974
WEISE, J.G.
NUMBER AND PERCENTAGE OF SEA LAMPREY AMMOCOETES TAKEN FROM
LAKE HURON STREAM TREATMENTS, 1971-1973. PP. 78.
IN  TIBBLES,J.J. ANNUAL REPORT 1973 OF THE SEA LAMPREY
CONTROL CENTRE SAULT STE. MARIE, ONTARIO. TO THE GREAT
LAKESFISHERY COMMISSION. GREAT LAKES FISH. COMM., ANNU.
REP.,
RESTRICTED
*PETROMYZON MARINUS CANADA AMMOCOETE


4905F

JOHNSON, B.G.H.                 1974
BIOLOGICAL DATA ON SEA LAMPREY COLLECTED FROM ELECTRICAL
ASSESSMENT BARRIERS: LAKE HURON, 1973. PP. 81-84.
IN  TIBBLES,J.J. ANNUAL REPORT 1973 OF THE SEA LAMPREY
CONTROL CENTRE SAULT STE. MARIE, ONTARIO, TO THE GREAT
LAKES FISHERY COMMISSION. GREAT LAKES FISH. COMM., ANNU. REP
RESTRICTED
*PETROMYZON MARINUS CANADA ADULT BIOLOGY


4906F

JOHNSON, B.G.H.                 1974
SURFACE TRAWLING FOR ADULT SEA LAMPREY IN ST. MARY'S RIVER
1973. PP. 80-81.

IN TIBBLES,J.J. ANNUAL REPORT 1973 OF THE SEA LAMPREY
CONTROL CENTRE SAULT STE. MARIE, ONTARIO, TO THE GREAT
LAKES FISHERY COMMISSION. GREAT LAKES FISH. COMM., ANNU.REP.
RESTRICTED
*PETROMYZON MARINUS CANADA ADULT

4907F

DUSTIN, S.M.                     1975
ADULT SEA LAMPREY ELECTRICAL ASSESSMENT BARRIER OPERATIONS,
LAKE HURON, 1974. P. 5.
IN TIBBLES,J.J. ED. ANNUAL REPORT 1974 OF THE SEA LAMPREY
CONTROL CENTRE SAULT STE. MARIE, ONTARIO, TO THE GREAT LAKES
FISHERY COMMISSION. GREAT LAKES FISH. COMM., ANNU. REP.
RESTRICTED
*PETROMYZON MARINUS CANADA MANAGEMENT

4908F

TIBBLES, J.J.                    1975
ANNUAL REPORT TO THE DEPARTMENT OF THE ENVIRONMENT AND THE
GREAT LAKES FISHERY COMMISSION FOR 1974-1975. PP. I-VI.
IN TIBBLES,J.J. ED. ANNUAL REPORT 1974 OF THE SEA LAMPREY
CONTROL CENTRE SAULT STE. MARIE, ONTARIO, TO THE GREAT LAKES
FISHERY COMMISSION. GREAT LAKES FISH. COMM., ANNU. REP.
RESTRICTED
*PETROMYZON MARINUS CANADA ADULT AMMOCOETE BIOLOGY GROWTH
HISTORY MANAGEMENT

4909F

GRIGG, G.C.                      1974
RESPIRATORY FUNCTION OF BLOOD IN FISHES. PP. 331-368.
IN FLORKIN,M. AND BRADLEY,T.S.(ED.). CHEMICAL ZOOLOGY,
VOL.8. DEUTEROSTOMIANS, CYCLOSTOMES AND FISHES. NEW YORK;
    ACADEMIC PRESS.

4910F

TAMMAR, A.R.                     1974
BILE SALTS IN FISHES. PP. 595-612.
IN FLORKIN,M. AND BRADLEY,T.S. EDS. CHEMICAL
ZOOLOGY,VOL.8.DEUTEROSTOMIANS, CYCLOSTOMES AND FISHES. NEW
YORK: ACADEMIC PRESS.
*MYXINE GLUTINOSA *EPTATRETUS STOUTI *PETROMYZON MARINUS
BIOCHEMISTRY DIGESTION

4911

HOLTERDAHL, H.                   1975
THE GEOLOGY OF THE HARDANGER FJORD WEST NORWAY.
NOR. GEOL. UNDERS. BULL., (36):1-87.

4912F

DULMA, A.                        1973
ZUR FISCHFAUNA DER MONGOLEI. [ON THE FISH FAUNA OF
MONGOLIA.]
BERLIN. UNIV., ZOOL. MUS., MITT., 49:49-67.
(GER.)
RUSS. SUMM.
*LAMPETRA REISSNERI MONGOLIA DISTRIBUTION

4914

VINNIKOV, Y.A.                   1974
FORM CREATING FUNCTION OF BIO MEMBRANES AND PHOTO RECEPTOR
DIFFERENTIATION.
USP. SOVREM. BIOL., 77:348-359.

4915F

PICKERING, A.D.                  1974
OESTROGENIC STIMULATION OF VITELLOGENESIS IN THE RIVER
LAMPREY.
GEN. COMP. ENDOCRINOL., 22:391
(ABSTR.)
*LAMPETRA FLUVIATILIS ADULT GONADOGENESIS

4916

FUJITA, H.                       1974
COMPARATIVE STUDIES ON THE ULTRASTRUCTURE AND IODINE
METABOLISM OF THE THYROID AND ITS HOMOLOGOUS ORGAN.
GEN. COMP. ENDOCRINOL., 22:372.
(ABSTR.)
METABOLISM ADULT AMMOCOETE ANIMAL TECHNIQUES CYTOLOGY
ENDOCRINOLOGY

4918F

STEINER, D.F.                    1973
PETERSON, J.D.
TAGER, H.
EMDIN, S.O.
OSTBERG, Y.
FALKNER, S.
COMPARATIVE ASPECTS OF PROINSULIN AND INSULIN STRUCTURE AND
BIOSYNTHESIS.
AM. ZOOL., 13:591-604.
*MYXINE GLUTINOSA BIOCHEMISTRY

4919

NISHIMURA, H.                1973
OGAWA, M.
THE RENIN ANGIOTENSIN SYSTEM IN FISHES.
AM. ZOOL., 13:823-838.
*LAMPETRA ENTOSPHENUS *LAMPETRA JAPONICUS *PARAMYXINE ATAMI
PHYSIOLOGY BIOCHEMISTRY


4920F

EPPLE, A.                1973
LEWIS, T.L.
COMPARATIVE HISTO PHYSIOLOGY OF THE PANCREATIC ISLETS.
AM. ZOOL., 13:567-590.
*MYXINE ANIMAL ENDOCRINOLOGY HISTOCHEMISTRY


4921

ATZ, J.W.                1973
COMPARATIVE ENDOCRINOLOGY AND SYSTEMATICS.
AM. ZOOL., 13:933-936.
PHYSIOLOGY SYSTEMATICS


4922F

HENDERSON, N.E.                1972
ULTRASTRUCTURE OF THE NEUROHYPOPHYSIAL LOBE OF THE HAGFISH,
EPTATRETUS STOUTI (CYCLOSTOMATA).
ACTA ZOOL., 53:243-266.
CANADA HISTOLOGY ANATOMY NERVOUS SYSTEM


4923

PEARSON, R.                1976
PEARSON, L.
THE VERTEBRATE BRAIN.
NEW YORK: ACADEMIC PRESS, 746 PP.


4924

ALEKSEEVA, K.D.                1975
METODIKA OPREDELENIYA AKTIVNOGO OBMENA PRI PROTZVOL'NOM
PLAVANII MOLODI RYB. [METHODS OF DETERMINING ACTIVE
METABOLISM IN YOUNG FISH.]
VOPR. ICHTHYOL., 15:369-370.
(RUSS.)
J. ICHTHYOL., 15:334-337.
(ENG.)


4925

SUZUKI, S.                1973
STRUCTURE AND FUNCTION OF THYROID OF HAGFISH. PART 1. RADIO
IODINE UPTAKE AND IODONATION OF THYRO GLOBULIN.
ZOOL. MAG. (TOKYO), 82:269.


4926

BUCHANAN, T.M.                1973
CHECKLIST OF ARKANSAS USA FISHES.
ARKANSAS ACAD. SCI., FAYETTEVILLE, PROC., 27:27-29.


4927

STERBA, G.                1974
HOHEISEL, H.
RUEHLE, H-J.
EXTRAHYPOTHALAMIC PEPTIDERGIC NEUROSECRETION IN THE MIDBRAIN
OF LAMPREYS.
GEN. COMP. ENDOCRINOL., 22:335-336.
(ABSTR.)
NERVOUS SYSTEM ENDOCRINOLOGY METAMORPHOSIS ADULT


4928F

KOBAYASHI, H.                1973
UEMURA, H.
THE NEUROHYPOPHYSIS OF THE HAGFISH EPTATRETUS BURGERI.
GEN. COMP. ENDOCRINOL., 21:214-215.
(ABSTR.)
JAPAN NERVOUS SYSTEM CIRCULATION HISTOLOGY


4929

ROSEN, F.S.                1974
COMPLEMENT ONTOGENY AND PHYLOGENY.
TRANSPLANT PROC., 6:47-50.


4930

BUSHUYEV, V.N.                1973
VUL'FIUS, Y.A.
GAGLOYEV, V.N.
GOLOVANOV, I.B.
CHEREMISIN, A.N.
PHYSIOLOGICALLY ACTIVE COMPOUNDS CORRELATION OF FINDINGS
ON THE PHYSIOLOGICAL ACTIVITY AND FINDINGS OF MOLECULAR
SPECULAR SPECTROSCOPY. PART I  NMR SPECTRA AND PHYSIOLOGICAL
ACTIVITY OF CERTAIN...
BIOFIZIKA, 18:216-222.
(RUSS.)
BIOPHYSICS, 18:222-229.

4931

SCHMIDT-NIELSON, B.          1972
RENFRO, J.L.
BENDS, D.
ESTIMATION OF EXTRACELLULAR SPACE AND INTRA CELLULAR ION
CONCENTRATIONS IN OSMO CONFORMERS, HYPO OSMO REGULATORS AND
HYPER OSMO REGULATORS.
MT. DESERT ISL. BIOL. LAB., BULL., 12:99-104.

4932

WESSELLS, N.K.          1974
VASCULAR SYSTEM BIOLOGY INTRODUCTION. PP.82-88.
IN  WESSELLS,N.K. ED. READINGS FROM SCIENTIFIC AMERICAN.
VERTEBRATE STRUCTURES AND FUNCTIONS. SAN FRANCISCO:FREEMAN
AND COMPANY.

4933

HARDISTY, M.W.          1974
HUGGINS, R.J.
KARTAR, S.
SAINSBURY, M.
ECOLOGICAL IMPLICATIONS OF HEAVY METAL IN FISH FROM THE
SEVERN ESTUURY.
MAR. POLLUT. BULL., 5:12-18.

4934

KIRSCHNER, L.B.          1973
ELECTROLYTE TRANSPORT ACROSS THE BODY SURFACE OF FRESH
WATER FISH AND AMPHIBIA. PP. 447-463.
IN  USSING,H.H. AND N.A. THORN (EDS.) PROCEEDINGS OF THE
ALFRED BENZON SYMPOSIUM, NO. 5. TRANSPORT MECHANISMS IN
EPITHELIA. COPENHAGEN, DENMARK, SEPTEMBER 10-14, 1972. NEW
YORK: ACADEMIC PRESS.

4935

ADAMSON, R.H.          1972
GUARINO, A.M.
NATURAL LEVELS OF DDT-RELATED COMPOUNDS AND POLYCHLORINATED
BIPHENLS IN VARIOUS MARINE SPECIES.
MT. DESERT ISL. BIOL. LAB., BULL., 12:6-9.

4936

THEIL, G.B.          1972
THEIL, M.F.
THE MOLECULAR CHARACTERISTICS OF CERTIN MAMMALIAN AND
PISCINE MYOGLOBINS AND HEMOGLOBINS.
MT. DESERT ISL. BIOL. LAB., BULL., 12:105-108.

4937F

KAWATSKI, J.A.          1974
DAWSON, V.K.
REUVERS, M.L.
EFFECT OF 3-TRIFLUOROMETHYL-4-NITROPHENOL (TFM) AND BAYER 73
ON IN VIVO OXYGEN CONSUMPTION OF THE AQUATIC MIDGE
CHIRONOMUS TENTANS.
AM. FISH. SOC., TRANS., 103:551-556.
CHEMISTRY ANIMAL METABOLISM

4938

IORIYA, T.          1976
NOTES ON SOME SPECIES OF COLOURLESS EUGLENOPHYCEAE FROM
HOKKAIDO JAPAN.
JAP. SOC. PHYCOL., BULL., 24:62-67.

4939F

JANVIER, P.          1974
THE SENSORY LINE SYSTEM AND ITS INNERVATION IN THE
OSTEOSTRACI (AGNATHA, CEPHALASPIDOMORPHI).
ZOOL. SCR., 3:91-99.
ANATOMY INTEGUMENT

4940F

OCHI, J.          1974
HOSOYA, Y.
FLUORESCENCE MICROSCOPIC DIFFERENTIATION OF MONO AMINES IN
THE HYPOTHALAMUS AND SPINAL CORD OF THE LAMPREY USING A NEW
FILTER SYSTEM.
HISTOCHEMISTRY, 40:263-266.
JAPAN ADULT CYTOLOGY NERVOUS SYSTEM

4941F

KONSTANTINOVA, M.S.          1973
MONOAMINES IN THE LIQUOR-CONTACTING NERVE CELLS IN THE
HYPOTHALAMUS OF THE LAMPREY LAMPETRA FLUVIATILIS L.
Z. ZELLFORSCH., 356:1-5
ADULT ENDOCRINOLOGY HISTOCHEMISTRY

4942

THEISEN, B.                    1976
THE OLFACTORY SYSTEM IN THE PACIFIC HAGFISHES EPTATRETUS
STOUTII EPTATRETUS DEAN1 AND MYXINE CIPCIFRONS.
ACTA ZOOL., (STOCKH.), 57:167-173.
OLFACTION CYTOLOGY HISTOLOGY


                                                      4943F

KOBAYASHI, H.             1972
ICHIKAWA, T.
SUZUKI, H.
SEKIMOTO, M.
SEASONAL MIGRATION OF EPTATRETUS BURGERI.
JAP. J. ICHTHYOL., 19:191-194.
(JAP.)
ENG. SUMM.
JAPAN DISTRIBUTION GROWTH


                                                      4944F

ROBISON, H.W.            1974
NEW DISTRIBUTIONAL RECORDS OF SOME ARKANSAS FISHES WITH
ADDITION OF 3 SPECIES TO THE ICHTHYO FAUNA.
SOUTHWEST. NAT., 19:220-223.
*ICHTHYOMYZON GAGEI U.S.A. SYSTEMATICS


                                                      4945F

FANGE, R.                1974
JOHANSSON-SJOBECK, M-L.
KANJE, M.
TRANSFORMATION OF SPINDLE CELLS INTO LYMPHOCYTE-LIKE CELLS
IN THE BLOOD FROM MYXINE GLUTINOSA
ACTA PHYSIOL. SCAND., 91:13A-14A.
*MYXINE GLUTINOSA SWEDEN BLOOD HISTOLOGY BIOCHEMISTRY


                                                      4946

MURRAY, M.               1973
JONES, H.
CSERR, H.F.
THE CEREBRAL VENTRICULAR SYSTEM OF MYXINE-GLUTINOSA.
MT. DESERT ISL. BIOL. LAB., BULL., 13:84-85.


                                                      4947F

EGOROVA, V.V.            1977
IEZUITOVA, N.N.
TIMOFEEVA, N.M.
TULYAGANOVA, E.KH.
SOME TEMPERATURE CHARACTERISTICS AND TEMPERATURE ADAPTATIONS
OF ENZYMES INVOLVED IN MEMBRANE DIGESTION IN POIKLOTHERMS
AND HOMOIOTHERMS.
ZH. EVOL. BIOKHIM. FIZIOL., 10:223-231.
(RUSS.)
J. EVOL. BIOCHEM. PHYSIOL., 10:201-208.
(ENG.)
*LAMPETRA FLUVIATILIS EVOLUTION PHYSIOLOGY METABOLISM
DIGESTION


                                                      4948

KITADA, J.               1973
TAGAWA, M.
CHROMOSOMES OF CYCLOSTOMATA.
ZOOL. MAG. (TOKYO), 82:381.


                                                      4949F

MORITA, Y.               1973
DODT, E.
SLOW PHOTIC RESPONSES OF THE ISOLATED PINEAL ORGAN OF
LAMPREY, PART 1. PP. 331-339.
IN  MOTHES,K. AND JOACHIM-HERMANN,S. EDS. NOVA ACTA
LEOPOLDINA. ABHANDLUNGEN DER DEUTSCHEN AKADEMIE DER
NATURFORSCHER LEOPOLDINA, BAND 38 NO.211. EAST GERMANY:
DEUTSCHE AKADEMIE DER NATURFORSCHER LEOPOLDINA.
*PETROMYZON MARINUS *LAMPETRA FLUVIATILIS SWEDEN ADULT
NERVOUS SYSTEM SENSE RECEPTORS


                                                      4950

DODT, E.                 1973
DIFFERENCES IN DATA PROCESSING UPON LIGHT STIMULATION
BETWEEN RETINA AND PINEAL ORGAN. PP. 137-144.
IN  LUEKEN,B. AND J.-H.SCHARF EDS. NOVA ACTA LEOPOLDINA.
VOL. 37/2. NO.208. HALLE:DEUTSCHE AKADEMIE DER NATURFORSCHER
LEOPOLDINA.


                                                      4951F

RUBINSON, K.             1974
THE CENTRAL DISTRIBUTION OF VESTIBULO COCHLEAR NERVE
AFFERENTS IN LARVAL PETROMYZON MARINUS.
BRAIN BEHAV. EVOL., 10:121-129.
AMMOCOETE NERVOUS SYSTEM


                                                      4952F

FLEISHMAN, D.G.          1973
KANEVSKII, Y.P.

41

PRIMENENIE PRIRODNOGO NA22 DLYA ISSLEDOVANIYA O'MENA NATRIYA
U PROKHODNYKH KOSTISTYKH RY' 1 MINOG V ESTESTVENNYKH
USLOVIYAKH. [THE USE OF ENVIRONMENTAL NA22 IN STUDIES ON
SODIUM EXCHANGE IN MIGRATING TELEOST FISHES AND LAMPREYS
UNDER NATURAL CONDITIONS.]
ZH. EVOL. BIOKHIM. FIZIOL., 9:123-129.
(RUSS.)
J. EVOL. BIOCHEM. PHYSIOL., 9:107-112.
(ENG.)
*LAMPETRA FLUVIATILIS FINLAND ADULT IONIC REGULATION


4953

YONEYAMA, Y.                    1973
ASSOCIATION AND FUNCTION OF HEMOGLOBIN OF
ENTOSPHENUS-JAPONICUS.
ACTA HAEMATOL. JAP., 36:256-257.


4954F

DOOLITTLE, R.F.                 1974
COTTRELL, B.A.
LAMPREY FIBRINOPEPTIDE B IS A GLYCOPEPTIDE.
BIOCHEM. BIOPHYS. RES. COMMUN., 60:1090-1096.
*PETROMYZON MARINUS U.S.A. ADULT BLOOD HISTOCHEMISTRY


4955F

KHOKHOLEV, L.S.                 1974
THE HETEROGENEITY OF THE AMINO ACID COMPOSITION IN
CYTOCHROMES.
ZH. EVOL. BIOKHIM. FIZIOL., 10:300-302.
(RUSS.)
J. EVOL. BIOCHEM. PHYSIOL., 10:266-268
(ENG.)
*LAMPETRA TRIDENTATUS *ENTOSPHENUS TRIDENTATUS BIOCHEMISTRY
PHYSIOLOGY


4956

MORITA, Y.                      1973
ELECTRO RETINOGRAM OF LAMPREYS.
PHYSIOL. SOC. JAP., J., 35:528.


4957

AWAYA, K.                       1973
TOMONAGA, S.
KATO, S.
ARE THERE LYMPHOCYTES IN HAGFISH?
ACTA HAEMATOL. JAP., 36:329-330.


4958

HENDERSON, I.W.                 1974
JONES, I.C.
ACTIONS OF HORMONES ON OSMOREGULATORY SYSTEMS OF FISH. PP.
391-418.
IN  HANKE,W. AND M. LINDAUER (EDS.) FORTSCHRITTE  DER
ZOOLOGIE, BAND 22, HEFT 2/3. VERGLEICHENDE ENDOKRINOLOGIE
2.INTERNATIONALES SYMPOSIUM DER AKADEMIE DER WISSENSCHAFTEN
 UND DER LITERATURE ZU MAINZ VOM 3 BIS 5, APRIL, 1973.
 STUTTGART:GUSTAV FISHER VERLAG.


4959

JANVIER, P.                     1973
CARACTERES PRIMITIFS ET SPECIALISATIONS PRECOCES DU SYSTEME
CIRCULATOIRE SANGUIN DES OSTEOSTRACES.
COLL. INT. CENTR. NAT. RECH. SCI., PARIS.


4960F

DOOLITTLE, R.F.                 1974
WOODING, G.L.
THE SUBUNIT STRUCTURE OF LAMPREY FIBRINOGEN AND FIBRIN.
BIOCHIM. BIOPHYS. ACTA, 271:277-282.
*PETROMYZON MARINUS BLOOD HISTOCHEMISTRY


4961F

WITTENBERG, J.B.                1974
HAEDRICH, R.L.
THE CHOROID RETE MIRABILE OF THE FISH EYE. PART
2:DISTRIBUTION AND RELATION TO THE PSEUDOBRANCH AND TO THE
SWIMBLADDER RETE MIRABILE.
BIOL. BULL., 146:137-156.
*MYXINE GLUTINOSA *PETROMYZON MARINUS *LAMPETRA FLUVIATILIS
ADULT VISION EVOLUTION


4962F

JOHNSON, B.G.H.                 1975
SEA LAMPREY COLLECTED BY COMMERCIAL FISHERMEN IN LAKE
ONTARIO. PP. 68-69.
IN  TIBBLES,J.J. ED. ANNUAL REPORT 1974 OF THE SEA LAMPREY
CONTROL CENTRE SAULT STE. MARIE, ONTARIO, TO THE GREAT LAKES
FISHERY COMMISSION. GREAT LAKES FISH. COMM., ANNU. REP.
RESTRICTED
CANADA MORPHOMETRY

4963

MURRAY, M.                    1973
OSTRACH, L.
CSERR, H.F.
THE CEREBRAL CAPILLARY SYSTEM OF MYXINE GLUTINOSA.
MT. DESERT ISL. BIOL. LAB., BULL., 13:87-88


4964F

POHL, R.J.                    1973
BEND, J.R.
DEVEREUX, T.R.
FOUTS, J.R.
HEPATIC CHEMICAL AND DRUG METABOLIZING ENZYMES IN COASTAL
MAINE USA MARINE SPECIES.
MT. DESERT ISL. BIOL. LAB., BULL., 13:94-98.
METABOLISM ANIMAL BIOCHEMISTRY


4965

BARDELE, C.F.                 1974
TRANSPORT OF MATERIALS IN THE SUCTORIAN TENTACLE. PP.
191-208.
IN  SLEIGH,M.A. AND D.H.JENNINGS EDS. SYMPOSIA OF THE
SOCIETY FOR EXPERIMENTAL BIOLOGY, NO.28. TRANSPORT AT THE
CELLULAR LEVEL. LONDON, ENGLAND, AUG. 27-31, 1973. LONDON:
CAMBRIDGE UNIV. PRESS.


4966F

PETERSON  J.D.                1974
COULTER, C.L.
STEINER, D.F.
EMDIN, S.O.
FALKMER, S.
STRUCTURAL AND CRYSTALLOGRAPHIC OBSERVATIONS ON HAGFISH
INSULIN.
NATURE (LOND.), 251:239-240.
ENDOCRINOLOGY TECHNIQUES


4967

PUPPIONE, D.P.               1974
SERUM LIPO PROTEINS OF HAGFISH EPTATRETUS DEANI.
AM. OIL CHEM. SOC., CHICAGO, J., 51:521A.


4968

KAPLAN, M.A.                 1973
HAYS, L.
HAYS, R.M.
THE EVOLUTION OF A FACILITATED DIFFUSION PATHWAY FOR AMIDES
IN THE VERTEBRATE ERYTHROCYTE.
MT. DESERT ISL. BIOL. LAB., BULL., 13:62-63.


4969F

DUSTIN, S.M.                 1976
ADULT SEA LAMPREY ELECTRICAL ASSESSMENT BARRIER
OPERATIONS,LLAKE HURON: 1975. P. 5.
IN  TIBBLES, J.J. ED. ANNUAL REPORT 1975 OF THE SEA LAMPREY
CONTROL CENTRE SAULT STE. MARIE, ONTARIO, TO THE GREAT LAKES
FISHERY COMMISSION. GREAT LAKES FISH. COMM., ANNU. REP.
RESTRICTED
CANADA ADULT MANAGEMENT MORTALITY OF


4970F

HANSON, L.H.                 1974
KING, E.L. JR.
HOWELL, J.H.
SMITH, A.J.
A CULTURE METHOD FOR SEA LAMPREY LARVAE.
PROG. FISH CULT., 36:122-128.
U.S.A. AMMOCOETE CULTURE FEEDING GROWTH


4971F

MCDONALD, M.J.               1974
NOBLE, R.W.
SHARMA, V.S.
RANNEY, H.M.
EQUILIBRIUM AND KINETIC STUDIES OF HEMOGLOBIN. PART 1:
FUNCTIONALLY SILENT AMINO ACID SUBSTITUTION AT AN INVARIANT
RESIDUE.
J. MOL. BIOL., 89:245-248.
BLOOD


4972

BEND, J.R.                   1973
FOUTS, J.R.
GLUTATHIONE S ARYL TRANSFERASE DISTRIBUTION IN SEVERAL
MARINE SPECIES AND PARTIAL CHARACTERIZATION IN HEPATIC
SOLUBLE FRACTIONS FROM LITTLE SKATE RAJAERINACEA LIVER.
MT. DESERT ISL. BIOL. LAB., BULL., 13:4-8.


4973

KONSTANTINOVA, M.S.          1973
MONOAMINES IN PERIVENTRICULAR HYPOTHALAMIC NERVE CELLS OF

THE LAMPREY LAMPETRA FLUVIATILIS.
ZH. EVOL. BIOKHIM. FIZIOL., 9:307-308.
(RUSS.)
J. EVOL. BIOCHEM. PHYSIOL., 9:269-271.
(ENG.)
RUSSIA ADULT NERVOUS SYSTEM HISTOLOGY


                                                            4974F
PLETCHER, F.T.              1963
THE LIFE HISTORY AND DISTRIBUTION OF LAMPREYS IN THE SALMON
AND CERTAIN OTHER RIVERS IN BRITISH COLUMBIA,CANADA.
UNIV. B.C., DEP. ZOOL.
*LAMPETRA PLANERI *ENTOSPHENUS TRIDENTATUS CANADA ADULT
AMMOCOETE ECOLOGY FECUNDITY FEEDING GROWTH MORPHOMETRY
BEHAVIOUR PARASITISM LIFE CYCLE

                                                            4975F
O'BOYLE, R.N.               1977
BEAMISH, F.W.H.
GROWTH AND INTERMEDIARY METABOLISM OF LARVAL AND
METAMORPHOSING STAGES OF THE LANDLOCKED SEA LAMPREY,
PETROMYZON MARINUS L.
ENVIRON. BIOL. FISHES., 2:103-120.
U.S.A. ECOLOGY AMMOCOETE LIFE CYCLE BLOOD GROWTH METABOLISM
METAMORPHOSIS PHYSIOLOGY


                                                            4976
GILLY, R.                   1974
DUTRUGE, J.
MAMELLE, J.C.
USE OF APESPORINE OR AREMYXINE IN PEDIATRICS.
LYON  MED., 231:79-81.


                                                            4977
FLOOD, P.R.                 1975
DRY FRACTURING TECHNIQUES FOR THE STUDY OF SOFT INTERNAL
BIOLOGICAL TISSUES IN THE SCANNING ELECTRON MICROSCOPE.
PP. 287-294.
IN  JOHARI,O.M. AND I. CORVIN EDS. SCANNING ELECTRON
MICROSCOPY 1975. CHICAGO:IIT RESEARCH INSTITUTE.


                                                            4978
HORSTMAN, E.                1973
THE INTEGUMENT. PP. 591-613.
IN  HIRSCH,G.C., RUSKA, H. AND SITTE,P. EDS. GRUNDLAGEN DER
CYTOLOGIE. JENA: G. FISCHER.


                                                            4979
JASINSKI, A.                1969
VASCULARIZATION OF THE HYPOPHYSIAL REGION IN LOWER
VERTEBRATES (CYCLOSTOMES AND FISHES).
GEN. COMP. ENDOCRINOL., SUPPL., 2:510-521.


                                                            4980F
TOMONAGA, S.                1973
AWAYA, K.
KATO, S.
[HEMATOPOIESIS OF THE PRIMITIVE SPLEEN OF HAGFISH.]
ACTA HAEMATOL. JAP., 36:231-232.
(JAP.)


                                                            4982
HOLMQUIST, R.               1976
JUKES, T.H.
MOISE, H.
GOODMAN, M.
MOORE, G.W.
THE EVOLUTION OF THE GLOBIN FAMILY GENES: CONCORDANCE OF
STOCHASTIC AND AUGMENTED MAXIMUM PARSIMONY GENETIC DISTANCES
FOR ALPHA HEMOGLOBIN AND MYOGLOBIN PHYLOGENIES.
    PHYLOGENIES...
J. MOL. BIOL., 105:39-74.
*LAMPETRA FLUVIATILIS BIOCHEMISTRY BIOLOGY EVOLUTION


                                                            4983
CHRISTOMANOS, A.A.          1976
IS THE RELATIONSHIP BETWEEN THE STRUCTURE OF ANIMAL AND
PLANT HEMOGLOBINS BASED ON TRUE HOMOLOGY OR ON ADAPTATION
BY THE LATTER DURING THE COURSE OF EVOLUTION?
FOLIA BIOCHEM. BIOL. GRAECA., 13:25-46.


                                                            4984
KOBAYASHI, H.               1975
ABSORPTION OF CEREBRO SPINAL FLUID BY EPENDYMAL CELLS OF
THE MEDIAN EMINENCE. PP. 109-122.
IN  KNIGGE,K.M. ET AL. ED. BRAIN-ENDOCRINE INTERACTION II.
THE VENTRICULAR SYSTEM IN NEUROENDOCRINE MECHANISMS.
SYMPOSIUM. SHIZUOKA, JAPAN. OCT. 16-18. 1974. BASEL:
S. KARGER.


                                                            4985F

SILLS, J.B.                     1975
ALLEN, J.L.
ROPHENOL IN FISH MUSCLE TISSUE: LABORATORY STUDIES.
ROPHENOL IN FISH MUSCLE TISSUE LABORATORY STUDIES.
U.S. DEP. INT., FISH WILDL. SERV., INVEST. FISH CONTROL,
65:3-10.
ANIMAL BIOCHEMISTRY CHEMISTRY MANAGEMENT

                                                        4986

MORITA, Y.                      1975
DIRECT PHOTO-SENSORY ACTIVITY OF THE PINEAL. PP. 376-387.
IN  KNIGGE,K.M. ET.AL. EDS. BRAIN-ENDOCRINE INTERACTION 11.
THE VENTRICULAR SYSTEM IN NEUROENDOCRINE MECHANISMS.
SYMPOSIUM. SHIZUOKA, JAPAN. OCT. 16-18,1974. BASEL:S.KARGER.

                                                        4987

PROTASOV, V.R.                  1975
DANKOVA, A.A.
ORLOV, A.A.
THE PROBLEM OF THE EVOLUTION OF ELECTRIC ORGANS OF FISHES.
ZH. OBSHCH. BIOL., 36:483-491.

                                                        4988

HELLER, H.                      1974
HISTORY OF NEUROHYPOPHYSIAL RESEARCH. PP. 103-117.
IN  KNOBIL,E. AND W.H. SAWYER EDS. HANDBOOK OF PHYSIOLOGY,
SECTION 7. ENDOCRINOLOGY, VOL. IV. THE PITUITARY GLAND AND
NEUROENDOCRINE CONTROL, PART 1. WASHINGTON, AMERICAN
PHYSIOLOGICAL SOCIETY.

                                                        4989

SEREBRENIKOVA, T.P.             1975
KHLYUSTINA, T.E.
ISO ENZYMES OF PHOSPHORYLASE EC-2.4.1.1. FROM SKELETAL
MUSCLES OF CYCLOSTOMES AND TELEOSTS.
BIOKHIMIIA, 40:550-555.

                                                        4990F

TSUNEKI, K.                     1976
EFFECTS OF ESTRADIOL AND TESTOSTERONE IN THE HAGFISH
EPATRETUS BURGERI.
ACTA ZOOL., (STOCKH.), 57:137-146.
*EPTATRETUS BURGERI ENDOCRINOLOGY ADULT HISTOLOGY

                                                        4991

PICKERING, A.D.                 1975
MORRIS, R.
ULTASTRUCTURE OF THE PRESUMED ION TRANSPORTING CELLS IN THE
GILLS OF AMMOCOETE LAMPREYS LAMPETRA FLUVIATILIS AND
LAMPETRA PLANERI.
CELL TISSUE RES., 163:327-341.

                                                        4992

STRAHAN, R.                     1975
EPTATRETUS LONGIPINNIS NEW SPECIES A NEW HAGFISH FAMILY
EPTATRETIDAE FROM SOUTH AUSTRALIA WITH A KEY TO THE 5-7
GILLED EPTATRETIDAE.
AUST. ZOOL., 18:137-148.

                                                        4993

YOUSON, J.H.                    1976
BUTLER, D.G.
THE ADRENOCORTICAL HOMOLOG IN THE LAKE STURGEON, ACIPENSER
FULVESCENS RAFINESQUE.
AM. J. ANAT., 145:207-223.
EXCRETION ANIMAL BIOCHEMISTRY ENDOCRINOLOGY CYTOLOGY
HISTOLOGY

                                                        4994F

HOCUTT, C.H.                    1977
DENONCOURT, R.F.
STAUFFER, J.R. JR.
INVENTORY OF THE FISHES OF GAULEY RIVER WEST VIRGINIA U.S.A.
ASSOC. SOUTHEAST. BIOL., BULL., 24:58.
(ABSTR.)
*LAMPETRA AEPYPTERA PISCES

                                                        4995

LOEFFLER, E.J.                  1976
JONES, B.
AN OSTRACODERM FAUNA FROM THE LEOPOLD FORMATION SILURIAN TO
DEVONIAN OF SOMERSET ISLAND NORTHWEST TERRITORIES CANADA.
PALAEONTOLOGY (LOND.), 19:1-5.
EVOLUTION SYSTEMATICS

                                                        4996

USOVA, A.A.                     1978
OSTAPENKO, I.A.
ETINGOF, R.N.
PROTEIN INHIBITOR OF CYCLIC NUCLEOTIDE PHOSPHO-DIESTERASE IN
HEREDITARY DYSTROPHY OF THE RETINA.

4997

GORBUNOVA, M.P.                1975
STUDY OF ENDOSTYLE OF LAMPREY LARVAE ASSOCIATED WITH THE
PROBLEM OF THYROID GLAND EVOLUTION.
ZH. OBSHCH. BIOL., 36:173-188.


4998F

PAIEMENT, J.M.                1975
MCMILLAN, D.B.
THE EXTRACARDIAC CHROMAFFIN CELLS OF LARVAL LAMPREYS.
GEN. COMP. ENDOCRINOL., 27:495-508.
*PETROMYZON MARINUS *LAMPETRA (LETHENTERON) LAMOTTENII
AMMOCOETE CIRCULATORY SYSTEM HISTOCHEMISTRY MORPHOLOGY CYTOL


5000

KONSTANTINOVA, M.S.            1975
MONOAMINES IN CEREBROSPINAL FLUID - CONTACT NERVE CELLS OF
THE VERTEBRATE HYPOTHALAMUS.
ZH. EVOL. BIOKHIM. FIZIOL., 11:187-190.
(RUSS.)
J. EVOL. BIOCHEM. PHYSIOL., 11:151-155.
(ENG.)
NERVOUS SYSTEM HISTOLOGY BIOLOGY


5001F

OHMAN, P.                      1976
FINE STRUCTURE OF PHOTO RECEPTORS AND ASSOCIATED NEURONS IN
THE RETINA OF LAMPETRA FLUVIATILIS CYCLOSTOMI.
VISION RES., 16:659-662.
FINE STRUCTURE OF PHOTORECEPTORS AND ASSOCIATED NEURONS IN
THE RETINA OF LAMPETRA FLUVIATILIS (CYCLOSTOMI).


5002F

HARDISTY, M.W.                 1976
ZELNIK, P.R.
WRIGHT, V.C.
THE EFFECTS OF HYPOXIA ON BLOOD SUGAR LEVELS AND ON THE
ENDOCRINE PANCREAS, INTERRENAL, AND CHROMAFFIN TISSUES OF
THE LAMPREY, LAMPETRA FLUVIATILIS.
GEN. COMP. ENDOCRINOL., 28:184-204.
ADULT BLOOD PHYSIOLOGY RESPIRATION TECHNIQUES


5003F

PACKARD, G.C.                  1976
PACKARD, M.J.
GORBMAN, A.
SERUM THYROXINE CONCENTRATIONS IN THE PACIFIC HAGFISH AND
LAMPREY AND IN THE LEOPARD FROG.
GEN. COMP. ENDOCRINOL., 28:365-367.
*ENTOSPHENUS TRIDENTATUS ADULT ANIMALS ENDOCRINOLOGY


5004F

FOUCRIER, J.                   1976
CHALUMEAU, M.T.
BOFFA, G.A.
TRANSFERRIN OF THE NEWT PLEURODELES WALTLII MICHAH.
ISOLATION AND IMMUNOLOGICAL RELATIONSHIP WITH OTHER
VERTEBRATE TRANSFERRINS.
COMP. BIOCHEM. PHYSIOL., 53B:555-559.
*PETROMYZON MARINUS BIOCHEMISTRY IMMUNOLOGY


5005

PICKERING, A.D.                1976
IODIDE UPTAKE BY THE ISOLATED THYROID OF THE RIVER LAMPREY
LAMPETRA FLUVIATILIS.
GEN. COMP. ENDOCRINOL., 28:358-364.
ENDOCRINOLOGY PHYSIOLOGY MIGRATION TECHNIQUES ADULT


5006F

NICOLAYSEN, K.                 1976
THE SPREAD OF THE ACTION POTENTIAL THROUGH THE TRANSVERSE
TUBULAR SYSTEM IN HAGFISH TWITCH MUSCLE FIBERS.
ACTA PHYSIOL. SCAND., 96:29-49
ATLANTIC ADULT MUSCLE PHYSIOLOGY


5007

NICOLAYSEN, K.                 1976
SPREAD OF THE JUNCTION POTENTIAL IN THE TRANSVERSE TUBULAR
SYSTEM IN HAGFISH SLOW MUSCLE FIBERS.
ACTA PHYSIOL. SCAND., 96:50-57


5008

BODIN-BAUDOUIN, J.             1975
DUGUY, R.
VAN BREE, P.J.H.
CATALOGUE OF MARINE MAMMALS CONSERVED AT THE MUSEUM OF
NATURAL HISTORY OF NANTES.
SOC. SCI. NAT. OUEST FR., BULL., 73:1-6.
(FR.)

5009

HARDISTY, M.W.                    1973
HUGGINS, R.J.
LAMPREY GROWTH AND BIOLOGICAL CONDITIONS IN THE BRISTOL
CHANNEL REGION.
NATURE (LOND.), 243:229-231.
*LAMPETRA FLUVIATILIS UNITED KINGDOM ADULT ECOLOGY GROWTH
LIFE CYCLE MIGRATION


5010

ARNDT, R.G.                       1977
SERENDIPITY AND LAMPREY.
DEL. CONSERV., 20:16-20.


5011

LEWIS, M.S.                       1976
CHUNG, S.I.
GLADNER, J.A.
CHARACTERIZATION OF LAMPREY FIBRINOGEN.
FED. PROC., 35:656.


5012

RYAPOLOVA, N.I.                   1975
SPAWNING STOCK AND FISHERY FOR RIVER LAMPREY IN THE EASTERN
BALTIC 1973.
ANN. BIOL., 30:206-207.


5013

BERMAN, H.A.                      1976
TAYLOR, P.
FLUORESCENCE PROBE STUDIES OF ACETYLCHOLINESTERASE FROM
TORPEDO.
FED. PROC., 35:1131.


5014

FROEHNER, S.C.                    1976
HALL, Z.W.
PURIFICATION AND IODINATION OF EXTRAJUNCTIONAL ACETYLCHOLINE
RECEPTOR FROM RAT MUSCLE.
FED. PROC., 35:1126.


5015

LARSSON, A.                       1977
FANGE, R.
CHOLESTEROL AND FREE FATTY-ACIDS IN THE BLOOD OF MARINE
FISH.
COMP. BIOCHEM. PHYSIOL., 57B:191-196.
ANIMAL BLOOD BIOCHEMISTRY SPAWNING


5016F

ALLEGRET, P.                      1976
DENIS, C.
LE LANNIC, J.
METHOD OF FEEDING AMMOCETES OF LAMPETRA PLANERI IN THE
PRESENCE OF PHYTOPHAGOUS LARVAE OF TRICHOPTERA.
SOC. ZOOL. FR., PARIS, BULL., 101:135-136.


5017

VALTIN, H.                        1974
STEWART, J.
SOKOL, H.W.
GENETIC CONTROL OF THE PRODUCTION OF POSTERIOR PITUITARY
PRINCIPLES. PP. 131-171.
IN  KNOBIL,E. AND W.H.SAWYER EDS. HANDBOOK OF PHYSIOLOGY,
SECTION,7. ENDOCRINOLOGY, VOL.IV. THE PITUITARY GLAND AND
ITS NEUROENDOCRINE CONTROL, PART I. WASHINGTON, AMERICAN
PHYSIOLOGICAL SOCIETY.


5018F

SPITZER, R.H.                     1976
DOWNING, S.
KOCH, E.A.
KAPLAN, M.A.
HEMAGLUTININS AND PROTEINS IN THE MUCUS OF PACIFIC HAGFISH
(EPTATRETUS STOUTII): IMMUNOLOGIC SIMILARITY TO SERUM
PROTEINS.
FED. PROC., 35:321.
(ABSTR.)
BIOCHEMISTRY BLOOD IMMUNOLOGY


5019F

RAHEMTULLA, F.                    1976
HOGLUND, N-G.
LOVTRUP, S.
ACID MUCOPOLYSACCHARIDES IN THE SKIN OF SOME LOWER
VERTEBRATES (HAGFISH, LAMPREY AND CHIMAERA).
COMP. BIOCHEM. PHYSIOL., 53B:295-298.
*MYXINE GLUTINOSA *LAMPETRA FLUVIATILIS BIOCHEMISTRY
INTEGUMENT


5020F

CHRISTENSEN, B.N.            1976
MORPHOLOGICAL CORRELATES OF SYNAPTIC TRANSMISSION IN
LAMPREY SPINAL CORD.
J. NEUROPHYSIOL., 39:197-212.
*PETROMYZON MARINUS ANADROMOUS LARVA ADULT NERVOUS SYSTEM
HISTOLOGY

                                                        5021

RURAK, D.W.                 1976
PERKS, A.M.
THE NEUROHYPOPHYSIAL PRINCIPLES OF THE WESTERN BROOK
LAMPREY, LAMPETRA RICHARDSONI. STUDIES IN THE ADULT.
GEN. COMP. ENDOCRINOL., 29:301-312.

                                                        5022F

PAGEAU, Y.                  1976
PRICHONNET, G.
INTERPRETATION DE LA PALEONTOLOGIE ET DE LA SEDIMENTOLOGIE
D'UNE COUPE GEOLOGIQUE DANS LA FORMATION DE BATTERY POINT
(DEVONIEN MOYEN), GRES DE GASPE. [INTERPRETATION OF
PALEONTOLOGY AND SEDIMENTOLOGY OF A GEOLOGIC LAYER IN THE
BATTERY POINT FORMATION MIDDLE DEVONIAN GASPE SANDSTONE
CANADA.]
NAT. CAN., 103:111-118.
(FR.)
ENG. SUMM.
*ENTOSPHENUS JAPONICUS

                                                        5023F

ASAI, H.                    1976
TAKAGI, H.
TSUNODA, S.
SOME CHARACTERISTICS OF ERYTHROCYTE MEMBRANE AND ITS ATPASE
FROM LAMPREY ENTOSPHENUS JAPONICUS.
COMP. BIOCHEM. PHYSIOL., 55B:69-75.
JAPAN BLOOD EVOLUTION PHYSIOLOGY

                                                        5024F

DE JONG, W.W.               1976
TERWINDT, E.C.
GROENEWOUD, G.
SUBUNIT COMPOSITIONS OF VERTEBRAE ALPHA CRYATALLINS.
COMP. BIOCHEM. PHYSIOL., 55B:49-56.
*PETROMYZON MARINUS ANIMAL EVOLUTION PHYSIOLOGY

                                                        5025F

MONACO, F.                  1976
ANDREOLI, M.
CATAUDELLA, S.
ROCHE, J.
SUR LA BIOSYNTHESE DE LA THYROGLOBULINE CHEZ UNE LAMPROIE
ADULTE, LAMPETRA PLANERI (BLOCH). [ON THE BIOSYNTHESIS OF
THYROGLOBULIN IN AN ADULT LAMPREY, LAMPETRA PLANERI (BLOCH)]
SOC. BIOL., PARIS, C.R. HEBD. SEANCES, 170:59-64.
(FR.)
ENG. SUMM.

                                                        5026F

ROVAINEN, C.M.              1976
REGENERATION OF MUELLER AND MAUTHNER AXONS AFTER SPINAL
TRANSECTION IN LARVAL LAMPREYS.
J. COMP. NEUROL., 168:545-554.
*PETROMYZON MARINUS AMMOCOETE HISTOLOGY LOCOMOTION NERVOUS
SYSTEM TECHNIQUES

                                                        5027

EPPLE, A.                   1976
STETSON, M.H.
SOME ENDOCRINE SPECIALIZATION OF THE BIRD.
J. ORNITHOL., 117:257-278.

                                                        5028

ZYTKOVICZ, T.H.             1976
NELSESTUEN, G.L.
GAMMA CARBOXY GLUTAMIC-ACID DISTRIBUTION.
BIOCHIM. BIOPHYS. ACTA, 444:344-348.
BLOOD ANIMAL BIOCHEMISTRY BIOLOGY

                                                        5029F

TSUNEKI, K.                 1975
KOBAYASHI, H.
YANAGISAWA, M.
BANDO, T.
HISTOCHEMICAL DISTRIBUTION OF MONOAMINES IN THE
HYPOTHALAMO-HYPOPHYSIAL REGION OF THE LAMPREY, LAMPETRA
JAPONICA.
CELL TISSUE RES., 161:25-32.
*LAMPETRA JAPONICA JAPAN ADULT HISTOLOGY ENDOCRINOLOGY

                                                        5030

POTTER, I.C.                1975

CANNON, D.
MOORE, J.W.
THE ECOLOGY OF ALGAE IN THE MORUYA RIVER AUSTRALIA.
HYDROBIOLOGIA, 47:415-430.

5031

TURNER, P.                    1974
LITHO STRATIGRAPHY AND FACIES ANALYSIS OF THE RINGERIKE
GROUP OF THE OSLO REGION NORWAY.
NOR. GEOL. UNDERS. BULL., 27:101-131.

5032F

ROBISON, H.W.                 1974
THREATENED FISHES OF ARKANSAS USA.
ARKANSAS ACAD. SCI., FAYETTEVILLE, PROC., 28:59-64.
*LAMPETRA AEPYPTERA *LAMPETRA LAMOTTEI

5033

GREEN, J.F.                   1974
BEADLES, J.K.
ICTHYO FAUNAL SURVEY OF THE CURRENT RIVER WITHIN ARKANSAS
USA.
ARKANSAS ACAD. SCI., FAYETTEVILLE, PROC., 28:22-26.

5034F

KASTIN, A.J.                  1974
VIOSCA, S.
SCHALLY, A.V.
REGULATION OF MELANOCYTE-STIMULATING HORMONE RELEASE.
FIBER TYPES IN THE ATLANTIC HAGFISH MYXINE GLUTINOSA.
IN  KNOBIL,E. AND W.H.SAWYER EDS. HANDBOOK OF PHYSIOLOGY,
SECTION 7. ENDOCRINOLOGY, VOL.IV. THE PITUITARY GLAND AND
ITS NEUROENDOCRINE CONTROL, PART 2. WASHINGTON:AMERICAN
PHYSIOLOGICAL SOCIETY.
ANIMAL ENDOCRINOLOGY PHYSIOLOGY

5035F

ROBISON, H.W.                 1974
BEADLES, J.K.
FISHES OF THE STRAWBERRY RIVER SYSTEM OF NORTH CENTRAL
ARKANSAS USA.
ARKANSAS ACAD. SCI., FAYETTEVILLE, PROC., 28:65-70.
*LAMPETRA LAMOTTEI *LAMPETRA AEPYPTERA U.S.A. SYSTEMATICS
ANIMAL

5036

CODOCERO, R.M.                1974
ECHINODERMS OF EASTER ISLAND PRELIMINARY COMMUNICATION.
CHILE. MUS. NAC. HIST. NAT., BOL., 33:53-63.

5037

CORDIER, G.                   1975
THE HALLSTATTIAN TUMULUS OF SUBLAINES INDRE-ET-LOIRE FRANCE.
ANTHROPOLOGIE (PARIS), 79:451-481.

5038

WAECHTLER, K.                 1975
THE DISTRIBUTION OF ACETYLCHOLINESTERASE IN THE CYCLOSTOME
BRAIN. PART 2: MYXINE GLUTINOSA.
CELL TISSUE RES., 159:109-120.

5039

HOLMBERG, K.                  1976
OHMAN, P.
FINE STRUCTURE OF RETINAL SYNAPTIC ORGANELLES IN LAMPREY AND
HAGFISH PHOTORECEPTORS.
VISION RES., 16:237-239.
*LAMPETRA FLUVIATILIS SWEDEN VISION NERVOUS SYSTEM CYTOLOGY
SWEDEN ADULT

5040

BORISOV, I.N.                 1973
BINDING OF TOLUIDINE BLUE WITH ACID MUCO POLY SACCHARIDES
IN THE PRESENCE OF PHENOLS STAINING TECHNIQUE.
ARKH. PATOL., 35:71-72.

5041

ANNO, K.                      1975
KAWAI, Y.
MUCOPOLYSACCARIDES FROM THE CONNECTIVE TISSUE OF THE
AMPHIOXUS, BRANCHIOSTOMA BELCHERII.
COMP. BIOCHEM. PHYSIOL., 52B:547-549.
BIOCHEMISTRY ADULT

5042F

TSUNEKI, K.                   1975
GORBMAN, A.
ULTRASTRUCTURE OF PARS NERVOSA AND PARS INTERMEDIA OF THE
LAMPREY LAMPETRA TRIDENTATA.
CELL TISSUE RES., 157:165-184
JAPAN ADULT ENDOCRINOLOGY HISTOLOGY NERVOUS SYSTEM

5043

KORNELIUSSEN, H.                    1975
NICOLAYSEN, K.
DISTRIBUTION AND DIMENSION OF THE SYSTEM IN DIFFERENT MUSCLE
FIBER TYPES IN THE ATLANTIC HAGFISH MYXINE GLUTINOSA.
CELL TISSUE RES., 157:1-6.


5044F

TERAKADO, K.                        1975
OGAWA, M.
HASHIMOTO, Y.
MATSUZAKI, H.
ULTRASTRUCTURE OF THE THREAD CELLS IN THE SLIME GLAND OF
JAPANESE HAGFISHES PARAMYXINE ATAMI AND EPTATRETUS BURGERI.
CELL TISSUE RES., 159:311-323.
*PARAMYXINE ATAMI *EPTATRETUS BURGERI HISTOLOGY CYTOLOGY
ADULT


5045F

YOUSON, J.H.                        1975
ABSORPTION AND TRANSPORT OF FERRITIN AND EXOGENOUS
HORSERADISH PEROXIDASE IN THE OPISTHONEPHRIC KIDNEY OF THE
SEA LAMPREY. PART 2: THE TUBULAR NEPHRON.
CELL TISSUE RES., 157:503-516.
*PETROMYZON MARINUS CANADA GREAT LAKES ADULT EXCRETION
HISTOCHEMISTRY


5046

STORESUND, A.                       1975
HELLE, K.B.
PRACTOLOL CAFFEINE AND CALCIUM IN THE REGULATION OF
MECHANICAL ACTIVITY OF THE CARDIAC VENTRICLE IN MYXINE
GLUTINOSA.
COMP. BIOCHEM. PHYSIOL., 52:17-22.


5047F

JANVIER, P.                         1975
REMARQUES SUR L'ORIFICE NASO-HYPOPHYSAIRE DES
CEPHALASPIDOMORPHES. [REMARKS ON THE NASO-HYPOPHYSIAL
ORIFICE OF CEPHALASPIDOMORPHS.]
ANN. PALEONTOL., VERTEBR., 61:3-16.
(FR.)
ENG. SUMM.
*LAMPETRA FLUVIATILIS *PETROMYZON MARINUS *LAMPETRA PLANERI
ANATOMY OLFACTION SENSE RECEPTORS


5048

VIGNA, S.                           1975
THE EFFECTS OF GASTRIN AND CHOLECYSTOKININ ON GALLBLADDER
CONTRACTION IN SOME LOWER VERTEBRATES.
AM. ZOOL., 15:786.
(ABSTR.)
BIOCHEMISTRY ENDOCRINOLOGY


5049

SVANTE VON EULER, V.                1975
EXAMPLES OF NEUROENDOCRINE EVOLUTION. PP. 291-298.
IN   SCHARF ANLAESSLICH DER JAHRESVERSAMMLUNG. HALLE,EAST
GERMANY, OCT. 11-14,1973.
  HALLE: DEUTSCHE AKADEMIE DER NATURFORSCHER LEOPOLDINA.


5050F

PIAVIS, G.W.                        1975
HOWELL, J.H.
EFFECTS OF 3-TRIFLUOROMETHYL-4-NITROPHENOL (TFM) ON
DEVELOPMENTAL STAGES OF THE SEA LAMPREY.
U.S. DEP. INT., FISH WILDL. SERV., INVEST. FISH CONTROL,
64:3-8.
*PETROMYZON MARINUS CHEMISTRY AMMOCOETE MANAGEMENT


5051F

GILDERHUS, P.A.                     1975
SILLS, J.B.
ALLEN, J.L.
RESIDUES OF THE 3-TRIFLUOROMETHYL-4-NITROPHENOL (TFM) IN A
STREAM ECOSYSTEM AFTER TREATMENT FOR CONTROL OF SEA
LAMPREYS.
U.S. DEP. INT., FISH WILDL. SERV., INVEST. FISH CONTROL,
66:3-8.
ANIMAL CHEMISTRY ECOLOGY MANAGEMENT


5052

THORNHILL, R.A.                     1974
BIOCHEMICAL AND HISTOCHEMICAL STUDIES ON VESTIBULAR NEURO
TRANSMISSION. PP. 209-221.
IN   SCHWARTZKOPFT,J. ED. ABHANDLUNGEN DER
RHEINISCH-WESTFAELISCHEN AKADEMIE DER WISSEN-SCHAFTEN,BAND
53. SYMPOSIUM. BOCHUM, WEST GERMANY, OCT. 14-18, 1973.
OPLADEN:WESTDEUTSCHER VERLAG.


5053F

DAWSON, V.K.                1975
CUMMING, K.B.
GILDERHUS, P.A.
LABORATORY EFFICACITY OF 3-TRIFLUOROMETHYL-4-NITROPHENOL
(TFM) AS A LAMPRICIDE.
U.S. DEP. INT., FISH WILDL. SERV., INVEST. FISH CONTROL,
63:3-11.
*PETROMYZON MARINUS AMMOCOETE CHEMISTRY MANAGEMENT
TECHNIQUES


                                               5054

YAKOVLEV, V.N.              1968
THE FUNCTIONAL SIGNIFICANCE OF HETEROCERCAL TAILS. PP.
10-20.
IN OBRUCHEV, D.V., ED., [ESSAYS ON THE PHLOGENY AND
SYSTEMATICS OF FOSSIL FISH AND AGNATHA.] MOSCOW: NAUKA.
(RUSS.)


                                               5055

HARP, G.L.                 1975
MATTHEWS, W.J.
1ST ARKANSAS U.S.A. RECORDS OF LAMPETRA SPP.
(PETROMYZONIDAE).
SOUTHWEST. NAT., 20:414-416.


                                               5056F

JACKSON, I.M.D.            1974
REICHLIN, S.
THYROTROPIN RELEASING HORMONE (TRH): DISTRIBUTION IN
HYPOTHALAMIC AND EXTRAHYPOTHALAMIC BRAIN TISSUES OF
MAMMALIAN AND SUBMAMMALIAN CHORDATES.
ENDOCRINOLOGY, 95:854-862.
*PETROMYZON MARINUS ENDOCRINOLOGY


                                               5057F

EAKIN, R.M.                1973
GENERAL MORPHOLOGY. PP. 74-82.
IN EAKIN,R.M. THE THIRD EYE. BERKELEY: UNIV. OF CALIFORNIA
PRESS.
AMMOCOETE SENSE RECEPTORS HISTOLOGY


                                               5058

ACCINNI, L.                1976
NATALI, P.G.
NICOTRA, M.R.
CAPANNA, E.
CATAUDELLA, S.
DE MARTINO, C.
PHILOGENESIS OF THE GLOMERULAR CAPILLARY WALL:
MORPHOLOGICAL, HISTOCHEMICAL AND IMMUNOLOGIC STUDIES.
J. SUBMICROSC. CYTOL., 8:243
(ABSTR.)
IMMUNOLOGY HISTOCHEMISTRY


                                               5059

NATALI, P.G.               1976
DE MARTINO, C.
NICOTRA, M.R.
CAPANNA, E.
NARDONI, C.
ACCINI, L.
PHYLOGENESIS OF THE MESANGIUM MORPHOLOGICAL AND
IMMUNOLOGICAL STUDIES.
J. SUBMICROSC. CYTOL., 8:258
(ABSTR.)
HISTOLOGY


                                               5060

SAVINA, M.V.               1976
VILKOVA, V.A.
GLUCONEOGENESIS FROM GLYCEROL IN THE ORGANS OF THE LAMPREY
LMPETRA FLUVIATILIS DURING THE PRESPAWNING PERIOD.
ZH. EVOL. BIOKHIM. FIZIOL., 12:189-192.
(RUSS.)
J. EVOL. BIOCHEM. PHYSIOL., 12:174-176.
(ENG.)


                                               5061F

SWEETING, R.               1976
EXPERIMENTAL DEMONSTRTION OF THE LIFE CYCLE OF A
DIPLOSTOMULUM FROM LAMPETRA FLUVIATILIS LINNAEUS, 1758.
Z. PARASITENK., 49:233-242.
GREAT BRITAIN ADULT PARASITISM OF


                                               5062F

LARSEN, L.O.               1976
BLOOD GLUCOSE LEVELS IN INTACT AND HYPOPHYSECTOMIZED RIVER
LAMPREYS LAMPETRA FLUVIATILIS TREATED WITH INSULIN STRESS OR
GLUCOSE BEFORE AND DURING THE PERIOD OF SEXUAL MATURATION.
GEN. COMP. ENDOCRINOL., 29:1-13.
BLOOD ENDOCRINOLOGY PHYSIOLOGY

5063F

FANGE, R.                           1976
LIDMAN, U.
LARSSON, A.
COMPARATIVE STUDIES OF INORGANIC SUBSTANCES IN THE BLOOD OF
FISHES FROM THE SCAGERAC SEA.
J. FISH BIOL., 8:441-448.
*MYXINE GLUTINOSA SWEDEN BLOOD EVOLUTION


5064F

GONCHAREVSKAYA, O.A.            1976
STRUKTURNAYA ORGANIZATSIYA NEFRONOV V POCH KE I
PROKSIMAL'NAYA REABSORBCHIYA U PREDSTAVITELEI RAZLICHNYKA
KLASSOV POZVONOCHNYKH (PO DANNYM MIKRODISSEKCHII I
MIKROPUNKCHII). [NEPHRON ORGANIZATION OF THE KIDNEY AND
PROXIMAL REABSORPTION IN VERTEBRATES (MICRODISSECTION AND
MICROPUNCTURE STUDIES.)]
ZH. EVOL. BIOKHIM. FIZIOL., 12:113-119.
(RUSS.)
ENG. SUMM.
*LAMPETRA FLUVIATILIS RUSSIA ANATOMY EXCRETION EVOLUTION


5065F

PICKERING, A.D.                 1976
EFFECTS OF GONADECTOMY, OESTRADIOL AND TESTOSTERONE ON THE
MIGRATING RIVER LAMPREY, LAMPETRA FLUVIATILIS.
GEN. COMP. ENDOCRINOL., 28:473-480.
GREAT BRITAIN ADULT ENDOCRINOLOGY EVOLUTION BLOOD
GONADOGENESIS MIGRATION


5066F

EPPLE, A.                       1975
BRINN, J.E. JR.
ISLET HISTOPHYSIOLOGY: EVOLUTIONARY CORRELATIONS.
GEN. COMP. ENDOCRINOL., 27:320-349.
*EPTATRETUS BULGERI *PETROMYZON MARINUS *PETROMYZON DORSATUS
ADULT ANIMAL ENDOCRINOLOGY


5067F

ROSS, J.R.P.                    1976
BODY WALL ULTRASTRUCTURE OF LIVING CYCLOSTOME ECTOPROCTS.
J. PALEONTOL., 50:350-353.
ANATOMY HISTOLOGY ADULT


5068F

DENONCOURT, R.F.                1975
KEY TO THE FAMILIES AND GENERA OF PENNSYLVANIA FRESHWATER
FISHES AND THE SPECIES OF FRESHWATER FISHES OF THE
SUSQUEHANNA RIVER DRAINAGE ABOVE CONOWINGO DAM.
PA. ACAD. SCI., HARRISBURG, PROC., 49:82-88.
*ICHTHYOMYZON *PETROMYZONIDAE *LAMPETRA U.S.A. DISTRIBUTION
SYSTEMATICS


5069

SELZER, M.                      1976
REGENERATION OF TRANSECTED LAMPREY SPINAL CORD.
J. NEUROPATHOL. EXP. NEUROL., 35:349.
(ABSTR.)
*PETROMYZON MARINUS AMMOCOETE NERVOUS SYSTEM


5070

FEDOROV, B.A.                   1976
KROBER, R.
DAMASCHUN, G.
RUCKPAUL, K.
EXPERIMENTAL AND THEORETICAL LARGE ANGLE X-RAY DIFFUSE
SCATTERING BY GLOBINS IN SOLUTION SENSITIVITY OF THE METHOD.
FEBS LETT., 65:92-95.
BIOCHEMISTRY TECHNIQUES


5071F

BELENKII, M.A.                  1975
UL'TRASTUKTURA NEIROGIPOFIZA U MINOGI. [ULTRASTRUCTURE OF
THE NEUROHYPOPHYSIS IN THE LAMPREY, LAMPETRA FLUVIATILIS.]
ZH. EVOL. BIOKHIM. FIZIOL., 11:605-611.
(RUSS.)
ENG. SUMM.
RUSSIA ADULT ENDOCRINOLOGY HISTOLOGY NERVOUS SYSTEM


5072F

MOLNAR, B.                      1975
SZABO, S.
ACTIUNEA CLORPROMAZINEI ASUPRA SISTEMULUI NEUROSECRETOR
HIPOTALAMO-HIPOFIZAR LA EUDONTOMYZON DANFORDI REGAN. [EFFECT
OF CHLORPROMAZINE OF HYPOTHALAMO AND HYPOPHYSIAL
NEUROSECRETORY SYSTEM OF EUDONTOMYZON DANFORDI REGAN.]
STUD. CERCET. BIOL., SER. BIOL. ANIM., 27:101-104.
ROMANIA ADULT ENDOCRINOLOGY HISTOLOGY NERVOUS SYSTEM


5073

GRASGOF, V.M.                   1976

RADIOAUTOGRAPHIC AND HISTOCHEMICAL STUDY OF THE INTESTINAL
EPITHELIUM OF THE BALTIC LAMPREY.
ARKH. ANAT. GISTOL. EMBRIOL., 70:68-74.

5074

NAGANO, H.                    1976
HOSAKA, K.
SHUKUYA, R.
THE FASTEST MIGRATING COMPONENT OF LAMPREY SERUM IN ZONE
ELECTROPHORESIS.
COMP. BIOCHEM. PHYSIOL., 54B:369-374.
*ENTOSPHENUS JAPONICUS BIOCHEMISTRY BLOOD TECHNIQUES

5075

BELL, M.A.                    1978
FISHES OF THE SANTA-CLARA RIVER SYSTEM, SOUTHERN CALIFORNIA,
U.S.A.
LOS ANGELES. NAT. HIST. MUS., CONTRIB. SCI., (295):1-20.

5076F

RYE, R.P.                     1976
KING, E.L.JR.
ACUTE TOXIC EFFECTS OF 2 LAMPRICIDES TO 21 FRESH WATER
INVERTEBRATES.
AM. FISH. SOC., TRANS., 105:322-326.

5077

NIBLETT, P.D.                 1976
BALLANTYNE, J.S.
UNCOUPLING OF OXIDATIVE PHOSPORYLATION IN RAT LIVER
MITOCHONDRIA BY THE LAMPREY LARVICIDE
3-TRIFLUOROMETHYL-4-NITROPHENOL (TFM).
PEST. BIOCHEM. PHYSIOL., 6:363-366.
ANIMAL CHEMISTRY TOXICITY MANAGEMENT

5078F

TSUNEKI, K.                   1976
ADACHI, T.
ISHII, S.
OOTA, Y.
MORPHOMETRIC CLASSIFICATION OF NEUROSECRETORY GRANULES IN
THE NEUROHYPOPHYSIS OF THE HAGFISH EPTATRETUS BURGERI.
CELL TISSUE RES., 166:145-157.
JAPAN ENDOCRINOLOGY HISTOLOGY NERVOUS SYSTEM

5079F

AGARKOV, G.B.                 1976
VARICH, YU.N.
SNEZHINA, K.A.
[INNERVATION OF THE LOCOMOTOR ORGAN IN SOME FISH.]
(RUSS.)
HYDROBIOL. J., 12:57-59.
(ENG.)
(ENG.)
*LAMPETRA FLUVIATILIS ANIMAL MUSCLE NERVOUS SYSTEM HISTOLOGY
LOCOMOTION ANATOMY

5080F

OLSON, L.E.                   1973
MARKING, L.L.
TOXICITY OF TFM (LAMPRICIDE) TO SIX EARLY LIFE STAGES OF
RAINBOW TROUT (SALMO GAIRDNERI).
FISH. RES. BOARD CAN., J., 30:1047-1052.
CHEMISTRY MANAGEMENT

5081

SUTTERLIN, A.                 1975
CHEMICAL ATTRACTION OF SOME MARINE FISH. PP. 153-156.
IN  DENTON,D.A. AND J.P.COGHLAN EDS. OLFACTION AND TASTE V.
PROCEEDING OF THE FIFTH INTERNATIONAL SYMPOSIUM. MELBOURNE,
AUSTRALIA, OCT. 1974. NEW YORK: ACADEMIC PRESS.

5082

MAUGHAN, O.E.                 1976
A SURVEY OF FISHES OF THE CLEARWATER RIVER.
NORTHWEST SCI., 50:76-86.

5083F

EPPLE, A.                     1976
BRINN, J.E. JR.
A NEW LOOK AT THE EVOLUTION OF THE ISLET ORGAN.
ANAT. REC., 184:397-398.
(ABSTR.)
*PETROMYZON MARINUS SPAWNING MIGRATION ENDOCRINOLOGY
PHYSIOLOGY BLOOD

5084

REMBISZEWSKI, J.M.            1975
ROLIK, H.
CYCLOSTOMATA AND PISCES.
KAT. FAUNY. POL., 24:1-253.

5085F

YOUSON, J.H.                   1976
FREEMAN, P.A.
MORPHOLOGY OF THE GILLS OF LARVAL AND PARASITIC ADULT SEA
LAMPREY PETROMYZON MARINUS.
J. MORPHOL., 149:73-104.
ONTARIO MORPHOLOGY RESPIRATION HISTOLOGY OSMOREGULATION


5086

KUEHN, K.                     1975
STOLTE, H.
REALE, E.
THE FINE STRUCTURE OF THE KIDNEY OF THE HAGFISH MYXINE
GLUTINOSA, A THIN SECTION AND FREEZE FRACTURE STUDY.
CELL TISSUE RES., 164:201-213.


5087

OBRUCHEV, D.V.                1968
KARATAYUTE-TALIMAA, V.N.
[VERTEBRATE FAUNA AND THE CORRELATION OF THE LUDLOW AND
LOWER DEVONIAN DEPOSITS IN EASTERN EUROPE.] PP. 63-70.
IN  OBRUCHEV, D.V. ED. [ESSAYS ON THE PHYLOGENY AND
SYSTEMATICS OF FOSSIL FISH AND AGNATHA.] MOSCOW:NAUKA


5088F

JOHNSON, B.G.H.               1975
BIOLOGICAL DATA ON ADULT SEA LAMPREY COLLECTED FROM
ELECTRICAL ASSESSMENT BARRIERS, LAKE HURON, 1974. P. 68.
IN  TIBBLES,J.J. ED. ANNUAL REPORT 1974 OF THE SEA LAMPREY
CONTROL CENTRE SAULT STE. MARIE, ONTARIO, TO THE GREAT LAKES
FISHERY COMMISSION. GREAT LAKES FISH. COMM., ANNU. REP.
RESTRICTED
CANADA MORPHOMETRY


5089F

RANDOLPH, K.N.                1973
FISHES OF THE TIPPAH RIVER SYSTEM MISSISSIPPI WITH NOTES ON
HABITATS AND DISTRIBUTION.
MISS. ACAD. SCI., J., 19:128:134.
*LAMPETRA AEPYPTERA U.S.A. AMMOCOETE DISTRIBUTION


5090

GAMMELTOFT, S.                1976
OSTERGAARD, K.L.
SESTOFT, L.
INTERACTION OF INSULIN WITH ISOLATED RAT LIVER CELLS
SPECIFICITY OF RECEPTOR BINDING.
ACTA PHYSIOL. SCAND., 96:11A-12A.


5091

OBRUCHEV, D.V.                1968
[THE EVOLUTION OF HETEROSTRACI]. PP. 21-28.
IN  OBRUCHEV, D.V. ED. [ESSAYS ON THE PHYLOGENY AND
SYSTEMATICS OF FOSSIL FISH AND AGNATHA.] MOSCOW:NAUKA.


5092F

MOSKALENKO, T.A.              1968
LLANDOVERIISKIE OSTATKI AGNATHA V SIBIRI. [LLANDOVERIAN
REMAINS OF AGNATHA IN SIBERIA.] PP. 29-32.
IN  OBRUCHEV, D.V. ED. [ESSAYS ON THE PHYLOGENY AND
SYSTEMATICS OF FOSSIL FISH AND AGNATH.] MOSCOW: NAUKA
(RUSS.)


5093

LAURENCE, E.B.               1976
THORNLEY, A.L.
CHALONE TISSUE SPECIFICITY AND THE EMBRYONIC DERIVATION OF
ORGANS AN APPRAISAL OF THE PROBLEMS. PP. 273-288.
IN  HOUCK, J.C. ED. CHALONES. AMSTERDAM: NORTH-HOLLAND
PUBLISHING CO.


5094F

ITINA, N.A.                   1975
NASLEDOV, G.A.
SKOROBOVICHUK, N.F.
ATSETILKHOLINOVAYA KONTRAKTURA BYSTRYKH MYSHECHNYKH VOLOKON
MINOGI LAMPETRA FLUVIATILIS. [ACETYLCHOLINE CONTRACTURES IN
FAST MUSCLE FIBERS OF THE LAMPREY.]
ZH. EVOL. BIOKHIM. FIZIOL., 11:567-572.
(RUSS.)
ENG. SUMM.
RUSSIA MUSCLE


5095F

BILLS, T.D.                   1976
MARKING, L.L.
TOXICITY OF 3-TRIFLUOROMETHYL-4-NITROPHENOL (TFM),
2',5-DICHLORO-4'-NITROSALICYLANILIDE (BAYER 73), AND A 98:2
MIXTURE TO FINGERLINGS OF SEVEN FISH SPECIES AND TO EGGS AND
FRY OF COHO SALMON.
U.S. DEP. INT., FISH WILDL. SERV., INVEST. FISH CONTROL,

69:1-9.
MANAGEMENT

5096F

JOHN, T.M.                    1977
THOMAS, E.
GEORGE, J.C.
BEAMISH, F.W.H.
EFFECT OF VASOTOCIN ON PLASMA FREE FATTY ACID LEVEL IN THE
MIGRATING ANADROMOUS SEA LAMPREY.
ARCH. INT. PHYSIOL. BIOCHIM., 85:865-870.
CANADA NEW BRUNSWICK BIOCHEMISTRY BLOOD


5097F

DOOLITTLE, R.F.              1976
THE EVOLUTION OF VERTEBRATE FIBRINOGEN.
FED. PROC., 35:2145-2149
ANIMAL BIOCHEMISTRY EVOLUTION PHYSIOLOGY


5098

FUJITA, H.                   1975
ON THE INTRA CELLULAR DENSE GRANULES IN THE THYROID
FOLLICULAR EPITHELIAL CELL OF THE CYCLOSTOMES HAGFISH AND
LAMPREY.
J. ELECTRONMICROSC., 24:195-196.
(ABSTR.)
ADULT ENDOCRINOLOGY HISTOLOGY


5099

HAY, A.W.M.                  1975
COMPARATIVE ASPECTS OF VITAMIN D TRANSPORT. PP. 405-407.
IN  TALMAGE,R.V. OWEN,M. AND PARSONS,J.A. EDS. EXCERPTA
MEDICA INTERNATIONAL CONGRESS SERIES, NO.346. CALCIUM
REGULATING HORMONES. PROCEEDINGS OF THE FIFTH PARATHYROID
CONFERENCE, OXFORD, ENGLAND. JULY 21-26,1974. AMSTERDAM:
EXCERPTA MEDICA.


5100F

DOVING, K.B.                 1974
HOLMBERG, K.
A NOTE ON THE FUNCTION OF THE OLFACTORY ORGAN OF THE HAGFISH
MYXINE GLUTINOSA.
ACTA PHYSIOL. SCAND., 91:430-432.
SWEDEN OLFACTION BEHAVIOUR SENSE RECEPTORS


5101F

PICKERING, A.D.             1976
MORRIS, R.
FINE STRUCTURE OF THE INTER PLATELET AREA IN THE GILLS OF
THE MACROPHTHALMIA STAGE OF THE  RIVER LAMPREY LAMPETRA
FLUVIATILIS.
CELL TISSUE RES., 168:433-444.
ADULT HISTOLOGY OSMOREGULATION


5102

ORVIG, T.                    1976
PALEOHISTOLOGICAL NOTES. PART 3: THE INTERPRETATION OF
PLEROMIN (PLEROMIC HARD TISSUE) IN THE DERMAL SKELETON OF
PSAMMOSTEID HETEROSTRACHANS.
ZOOL. SCR., 5:35-47.
ANATOMY HISTOLOGY


5103F

JOHNSON, B.G.H.             1975
SURFACE TRAWLING FOR ADULT SEA LAMPREY IN ST. MARYS RIVER,
1974. PP. 66-67.
IN  TIBBLES,J.J. ED. ANNUAL REPORT 1974 OF THE SEA LAMPREY
CONTROL CENTRE SAULT STE. MARIE, ONTARIO, TO THE GREAT LAKES
FISHERY COMMISSION. GREAT LAKES FISH. COMM., ANNU. REP.
RESTRICTED
*PETROMYZON MARINUS CANADA


5104F

DAVIS, W.A.                  1975
SHERA, W.P.
WEISE, J.G.
GRANULAR BAYER 73 TREATMENTS, LAKE HURON, 1974. PP. 59-65.
IN  TIBBLES,J.J. ED. ANNUAL REPORT 1974 OF THE SEA LAMPREY
CONTROL CENTRE SAULT STE. MARIE, ONTARIO, TO THE GREAT LAKES
FISHERY COMMISSION. GREAT LAKES FISH. COMM., ANNU. REP.
RESTRICTED
*PETROMYZON MARINUS *ICHTHYOMYZON SP. *LAMPETRA LAMOTTII
CANADA AMMOCOETE GROWTH CHEMISTRY


5105F

SCHLEEN, L.P.               1975
WESTMAN, R.W.
GRANULAR BAYER 73 TREATMENTS, LAKE SUPERIOR, 1974. PP.
51-58.
IN  TIBBLES,J.J. ED. ANNUAL REPORT 1974 OF THE SEA LAMPREY
CONTROL CENTRE SAULT STE. MARIE, ONTARIO, TO THE GREAT LAKES

FISHERY COMMISSION. GREAT LAKES FISH. COMM., ANNU. REP.
RESTRICTED
*PETROMYZON MARINUS *ICHTHYOMYZON SP. *LAMPETRA LAMOTTII
CANADA AMMOCOETE GROWTH CHEMISTRY

5106F

**DAVIS, W.A.**                    1975
SHERA, W.P.
WEISE, J.G.
. PP. 39-51.
.PP.39-51.
IN   TIBBLES,J.J. ED. ANNUAL REPORT 1974 OF THE SEA LAMPREY
CONTROL CENTRE SAULT STE. MARIE, ONTARIO, TO THE GREAT LAKES
FISHERY COMMISSION. GREAT LAKES FISH. COMM., ANNU. REP.
RESTRICTED
*PETROMYZON MARINUS CANADA MANAGEMENT CHEMISTRY AMMOCOETE
ADULT

5107F

**DAVIS, W.A.**                    1975
SHERA, W.P.
WEISE, J.G.
STREAM TREATMENT REPORTS, LAKE HURON, 1974. PP. 26-39.
IN   TIBBLES,J.J. ED. ANNUAL REPORT 1974 OF THE SEA LAMPREY
CONTROL CENTRE SAULT STE. MARIE, ONTARIO, TO THE GREAT LAKES
FISHERY COMMISSION. GREAT LAKES FISH. COMM., ANNU. REP.
RESTRICTED
*PETROMYZON MARINUS CANADA MANAGEMENT ADULT AMMOCOETE
CHEMISTRY

5108F

**SCHLEEN, L.P.**                  1975
STREAM TREATMENT REPORTS, LAKE SUPERIOR, 1974. PP. 17-26.
IN   TIBBLES,J.J. ED. ANNUAL REPORT 1974 OF THE SEA LAMPREY
CONTROL CENTRE SAULT STE. MARIE, ONTARIO, TO THE GREAT LAKES
FISHERY COMMISSION. GREAT LAKES FISH. COMM., ANNU. REP.
RESTRICTED
*PETROMYZON MARINUS CANADA MANAGEMENT ADULT AMMOCOETE
CHEMISTRY

5109F

**SHERA, W.P.**                    1975
SEA LAMPREY SURVEYS OF STREAMS TRIBUTARY TO THE NEW YORK
SIDE OF LAKE ONTARIO, 1974. PP. 15-17.
IN   TIBBLES,J.J. ED. ANNUAL REPORT 1974 OF THE SEA LAMPREY
CONTROL CENTRE SAULT STE. MARIE, ONTARIO, TO THE GREAT LAKES
FISHERY COMMISSION. GREAT LAKES FISH. COMM. ANNU. REP.
RESTRICTED
*PETROMYZON MARINUS *ICHTHYOMYZON SP. *LAMPETRA LAMOTTEI
CANADA AMMOCOETE MANAGEMENT GROWTH

5110F

**SHERA, W.P.**                    1975
SEA LAMPREY SURVEYS OF STREAMS TRIBUTARY TO THE CANADIAN
SIDE OF LAKE ONTARIO. P. 15.
IN   TIBBLES,J.J. ED. ANNUAL REPORT 1974 OF THE SEA LAMPREY
CONTROL CENTRE SAULT STE. MARIE, ONTARIO, TO THE GREAT LAKES
FISHERY COMMISSION. GREAT LAKES FISH. COMM., ANNU. REP.
RESTRICTED
*PETROMYZON MARINUS *LAMPETRA LAMOTTEI *ICHTHYOMYZON SP
CANADA MANAGEMENT GROWTH AMMOCOETE

5111F

**WESTMAN, R.W.**                  1975
SCHLEEN, L.P.
SEA LAMPREY SURVEYS OF STREAMS TRIBUTARY TO LAKE SUPERIOR,
1974. PP. 6-9.
IN   TIBBLES,J.J. ED. ANNUAL REPORT 1974 OF THE SEA LAMPREY
CONTROL CENTRE SAULT STE. MARIE, ONTARIO, TO THE GREAT LAKES
FISHERY COMMISSION. GREAT LAKES FISH. COMM., ANNU. REP.
RESTRICTED
*PETROMYZON MARINUS *ICHTHYOMYZON SP. *LAMPETRA LAMOTTEI
CANADA MANAGEMENT AMMOCOETE GROWTH

5112F

**HOMMA, S.**                      1976
ROVAINEN, C.M.
GLYCINE AND GABA CONDUCTANCES IN LAMPREY INTERNEURONS.
NEUROSCI. ABSTR., 2:1124.
(ABSTR.)
NERVOUS SYSTEM TECHNIQUES

5113F

**LARSSON, A.**                    1976
JOHANSSON-SJOBECK,M-L.
FANGE, R.
COMPARATIVE STUDY OF SOME HAEMATOLOGICAL AND BIOCHEMICAL
BLOOD PARAMETERS IN FISHES FROM THE SKAGERRAK.
J. FISH BIOL., 9:425-440.
BLOOD

5114F

TOMONAGA, S.                1973
HIROKANE, T.
AWAYA, K.
LYMPHOID CELLS IN THE HAGFISH.
ZOOL. MAG. (TOKYO), 82:133-136.
BLOOD IMMUNOLOGY HISTOLOGY


5115F

MORRIS, R.                 1976
PICKERING, A.D.
CHANGES IN THE ULTRASTRUCTURE OF THE GILLS OF THE RIVER
LAMPREY, LAMPETRA FLUVIATILIS (L.), DURING THE ANADROMOUS
SPAWNING MIGRATION.
CELL TISSUE RES., 173:271-277.
MIGRATION OSMOREGULATION PHYSIOLOGY HISTOLOGY


5116F

MCDONALD, R.B.             1976
SCHLEEN, L.P.
SEA LAMPREY SURVEYS OF STREAMS TRIBUTARY TO LAKE SUPERIOR,
1975. PP. 6-7.
NEUROSCI. ABSTR., 2:786.
CONTROL CENTRE SAULT STE. MARIE, ONTARIO, TO THE GREAT
LAKES FISHERY COMMISSION. GREAT LAKES FISH. COMM., ANNU.
REP.
RESTRICTED
CANADA AMMOCOETE DISTRIBUTION MANAGEMENT


5117F

MURRAY, M.                 1975
BRAIN BARRIER SYSTEMS IN CYCLOSTOMES. PP. 123-136.
IN  CSERR,H.F. ET.AL. EDS. FLUID ENVIRONMENT OF THE BRAIN.
NEW YORK: ACADEMIC PRESS.
*PETROMYZON ANIMAL CIRCULATORY SYSTEM NERVOUS SYSTEM


5118F

MAKI, A.W.                 1975
GEISSEL, L.O.
JOHNSON, H.E.
TOXICITY OF THE LAMPRICIDE 3-TRIFLUOROMETHYL-4-NITROPHENOL
(TFM) TO 10 SPECIES OF ALGAE.
U.S. DEP. INT., FISH WILDL. SERV., INVEST. FISH CONTROL,
56:1-17.
MANAGEMENT


5119F

LIKOVSKY, Z.               1973
NUKLEOLEN IN LYMPHOZYTEN DES PERIPHEREN BLUTES BEI
WIRBELTIEREN MIT AUSNAHME VON SAUGERN.
VESTN. CESK. SPOL. ZOOL., 3:179-182.
(GER.)
*LAMPETRA PLANERI AMMOCOETE ANIMAL BLOOD


5120F

KAWATSKI, J.A.             1975
LEDVINA, M.M.
HANSEN, C.R.JR.
ACUTE TOXICITIES OF 3-TRIFLUOROMETHYL-4-NITROPHENOL (TFM)
AND 2' 5-DICHLORO-4'-NITROSALICYLANILIDE (BAYER 73) TO
LARVAE OF THE MIDGE CHIRONOMUS TENTANS.
U.S. DEP. INT., FISH WILDL. SERV., INVEST. FISH CONTROL,
57:7 PP.
MANAGEMENT


5121

JEFFERIES, R.P.S.          1978
LEWIS, D.N.
THE ENGLISH SILURIAN FOSSIL PLACOCYSTITES-FORBESIANUS AND
THE ANCESTRY OF THE VERTEBRATES.
R. SOC. LOND., PHILOS. TRANS., SER. B, BIOL. SCI.,
282-205-220.


5122F

PLISETSKAYA, E.M.          1972
SODERZHANIE GLIKOGENA V ORGANAKH KRUGLOROTYKH (CYCLOSTOMATA)
I RYB (PISCES). [GLYCOGEN CONTENT IN ORGANS OF AGNATHA
CYCLOSTOMATE (CYCLOSTOMATA) AND FISH (PISCES).]
VOPR. IKHTIOL., 12:297-306.
(RUSS.)
ENDOCRINOLOGY FEEDING MUSCLE


5123F

LEIBSON, L.                1976
PLISETSKAYA, E.
LEIBUSH, B.
THE COMPARATIVE STUDY OF MECHANISM OF INSULIN ACTION ON
MUSCLE CARBOHYDRATE METABOLISM. PP.345-362.
IN  GRILLO, T.A.I., LEIBSON, L., EPPLE, A., EDS. THE
EVOLUTION OF PANCREATIC ISLETS. OXFORD: PERGAMON PRESS.
*LAMPETRA FLUVIATILIS RUSSIA ENDOCRINOLOGY METABOLISM MUSCLE

5124F

**PLISETSKAYA, E.M.**                1976
LEIBUSH, B.N.
BONDAREVA, V.
THE SECRETION OF INSULIN AND ROLE IN CYCLOSTOMES AND FISHES.
PP. 251-269.
IN  GRILLO, T.A.I, L. LEIBSON, A. EPPLE EDS., THE EVOLUTION
OF PANCREATIC ISLETS. OXFORD: PERGAMON PRESS.
BLOOD ENDOCRINOLOGY METABOLISM TECHNIQUES


5125F

**HAEDRICH, R.L.**                1977
SEA LAMPREY FROM THE DEEP OCEAN.
COPEIA, 1977:767-768.


5126F

**LARSEN, L.O.**                1976
LE PHYSIOLOGICAL ROLE OF INSULIN AND HYPERGLYCEMIC HORMONES.
PP. 285-290.
PP.
IN  GRILLO, T.A.I. AND LIEBSON, L. AND EPPLE, A., EDS. THE
EVOLUTION OF PANCREATIC ISLETS. OXFORD: PERGAMON PRESS.
*LAMPETRA FLUVIATILIS DENMARK BLOOD ENDOCRINOLOGY GROWTH
ADULT GONADOGENESIS


5127F

**BANARESCU, P.**                1973
ZUR KENNTNIS DER FISCHAUNA DES THEISSBECKENS.
[TOWARD THE KNOWLEDGE OF THE FISH FAUNA OF THE TISZA BASIN.]
TISCIA, 7:69-77.
(GER.)
DISTRIBUTION ANIMAL SYSTEMATICS


5128F

**BEHLKE, J.**                1973
WANDT, I.
BESTIMMUNG PARTIELLER SPEZIFISCHER VOLUMINA VON HAMOGLOBINEN
UND MYOGLOBINEN. [DETERMINATION OF PARTIAL SPECIFIC VOLUMES
OF HEMOGLOBINS AND MYOGLOBINS.]
ACTA BIOL. MED. GER., 31:383-388.
(GER.)
ENGL. SUMM.
*LAMPETRA FLUVIATILIS ANIMAL BLOOD BIOCHEMISTRY


5129F

**MORRIS, S.**                1975
THE LAMPREY AT KETTLE FALLS.
BEAVER, MAG. NORTH, AUT:18-19.
ADULT MIGRATION MANAGEMENT


5130F

**SIVAK, J.G.**                1975
WOO, G.C.S.
ACCOMMODATIVE LENS MOVEMENT IN HOLOSTEANS (AMIA CALVA AND
LEPISOSTEVS OSSEVS OXYURUS) AND IN THE SEA LAMPREY
(PETROMYZON MARINUS).
CAN. J. ZOOL., 53:516-520.
ANIMAL ADULT NERVOUS SYSTEM VISION
ANIMAL ADULT NERVOUS SYSTEM VISION


5131F

**KELENYI, G.**                1976
LARSEN, L.O.
THE HAEMATOPOIETIC SUPRANEURAL ORGAN OF ADULT, SEXUALLY
IMMATURE RIVER LAMPREYS (LAMPETRA FLUVIATILIS (L.) GRAY)
WITH PARTICULAR REFERENCE TO AZUROPHIL LEUCOCYTES.
ACTA BIOL. ACAD. SCI. HUNG., 27:45-56.
BLOOD HISTOLOGY HISTOCHEMISTRY


5132F

**DUSTIN, S.M.**                1975
SUMMARY OF SURVEYS AND LAMPRICIDE TREATMENTS ON THE ST.
MARYS RIVER, 1971-74. PP. 91-93.
IN  TIBBLES, J.J. ED., ANNUAL REPORT 1974 OF THE SEA LAMPREY
CONTROL CENTRE SAULT STE. MARIE, ONTARIO, TO THE GREAT
LAKES FISHERY COMMISSION. GREAT LAKES FISH. COMM., ANNU.
REP.
RESTRICTED
CANADA MANAGEMENT AMMOCOETE


5133F

**WEISE, J.G.**                1975
SEA LAMPREY, PETROMYZON MARINUS, LARVAL (AMMOCOETE) GROWTH
RATES FROM THREE STREAMS TRIBUTARY TO LAKE ONTARIO. PP.
84-90.
IN  TIBBLES, J.J. ED., ANNUAL REPORT 1974 OF THE SEA LAMPREY
CONTROL CENTRE SAULT STE. MARIE, ONTARIO, TO THE GREAT
LAKES FISHERY COMMISSION. GREAT LAKES FISH. COMM., ANNU.
REP.
RESTRICTED
CANADA GROWTH MANAGEMENT

5134F

DUSTIN, S.M.                     1975
ADULT SEA LAMPREY BARRIER DAMS AND STREAM IMPROVEMENT,1966-
74. PP. 73-83.
IN  TIBBLES,J.J. ED. ANNUAL REPORT 1974 OF SEA LAMPREY
CONTROL CENTRE SAULT STE. MARIE, ONTARIO, TO THE GREAT LAKES
FISHERY COMMISSION. GREAT LAKES FISH. COMM., ANNU. REP.
RESTRICTED
CANADA MANAGEMENT


5135F

JOHNSON, B.G.H.                  1975
SEA LAMPREY TAG-RECAPTURE STUDIES, LAKE ONTARIO, 1974. PP.
70-73.
IN  TIBBLES,J.J. ED. ANNUAL REPORT 1974 OF SEA LAMPREY
CONTROL CENTRE SAULT STE. MARIE, ONTARIO, TO THE GREAT LAKES
FISHERY COMMISSION GREAT LAKES FISH. COMM., ANNU. REP.
RESTRICTED
CANADA ADULT MIGRATION


5136F

JOHNSON, B.G.H.                  1975
BIOLOGICAL DATA ON ADULT SEA LAMPREY FROM THE HUMBER RIVER,
1974. P. 70.
IN  TIBBLES,J.J. ED. ANNUAL REPORT 1974 OF THE SEA LAMPREY
CONTROL CENTRE SAULT STE. MARIE, ONTARIO, TO THE GREAT LAKES
FISHERY COMMISSION. GREAT LAKES FISH. COMM., ANNU. REP.
RESTRICTED
CANADA MANAGEMENT ADULT


5137F

FREMLING, C.R.                   1975
ACUTE TOXICITY OF THE LAMPRICIDE
3-TRIFLUOROMETHYL-4-NITROPHENOL TO NYMPHS OF MAYFLIES
(HEXAGENIA SP.).
U.S. DEP. INT., FISH WILDL. SERV., INVEST. FISH CONTROL,
58:7 PP.
MANAGEMENT


5138F

CLARIDGE, P.N.                   1975
POTTER, I.C.
OXYGEN CONSUMPTION, VENTILATORY FREQUENCY AND HEART RATE OF
LAMPREYS (LAMPETRA FLUVIATILIS) DURING THEIR SPAWNING RUN.
J. EXP. BIOL., 63:193-206.
*PETROMYZON MARINUS *LAMPETRA TRIDENTATA RESPIRATION ADULT
METABOLISM MIGRATION


5139F

ROHDE, F.C.                      1976
ARNDT, R.G.
WANG, C.S.
LIFE HISTORY OF THE FRESHWATER LAMPREYS, OKKELBERGIA
AEPYPTERA AND LAMPETRA LAMOTTEI  (PISCES: PETROMYZONIDAE),
ON THE DELMARVA PENINSULA (EAST COAST, U.S.)
SOUTH. CALIF. ACAD. SCI., LOS ANGELES, BULL., 75:99-111.
U.S.A. LIFE CYCLE AMMOCOETE ADULT FECUNDITY MORPHOMETRY
SPAWNING


5140F

POTTER, I.C.                     1977
BEAMISH, F.W.H.
THE FRESHWATER BIOLOGY OF ADULT ANADROMOUS SEA LAMPREYS
PETROMYZON MARINUS.
J. ZOOL., 181:113-130.
CANADA ADULT ANIMAL ECOLOGY FECUNDITY FEEDING IONIC
REGULATION PARASITISM BY EVOLUTION LIFE CYCLE


5141F

SJOBERG, K.                      1974
LEKTID, RORELSEAKTIVITET OCH LANGD/VIKT HOS FLODNEJONOGA.
[SPAWNING PERIOD, LOCOMOTOR ACTIVITY AND LENGTH/WEIGHT OF
THE RIVER LAMPREY (LAMPETRA FLUVIATILIS (L.)) IN THE RICKLED
RIVER, PROVINCE OF VASTERBOTTEN, SWEDEN.]
ZOOL. REVY., 36:41-48
(SWED.)
ENGL. SUMM.
*LAMPETRA FLUVIATILIS SWEDEN ADULT CIRCADIAN DISTRIBUTION
LOCOMOTION MIGRATION BEHAVIOUR


5142F

MOORE, J.W.                      1976
POTTER, I.C.
ASPECTS OF FEEDING AND LIPID DEPOSITION AND UTILIZATION IN
THE LAMPREYS, LAMPETRA FLUVIATILIS (L.) AND LAMPETRA PLANERI
(BLOCH).
J. ANIM. ECOL., 45:699-712.
GREAT BRITAIN ADULT DIGESTION ECOLOGY FEEDING GROWTH


5143F

APPLEGATE, V.C.                  1954

ARTIFICIAL HATCHING OF LAMPREY EGGS; LETTER TO P.J.
ROBERTSON.
U.S. DEP. INT., FISH WILDL. SERV., 2 PP.
RESTRICTED
AMMOCOETE ADULT CULTURE EMBRYOLOGY FECUNDITY MORTALITY OF
TECHNIQUES

5144F

**SANDERS, H.O.**                          1975
WALSH, D.F.
TOXICITY AND RESIDUE DYNAMICS OF THE LAMPRICIDE
3-TRIFLUOROMETHYL-4-NITROPHENOL (TFM) IN AQUATIC
INVERTEBRATES.
U.S. DEP. INT., FISH WILDL. SERV., INVEST. FISH CONTROL,
59:1-9.
MANAGEMENT

5145

**POTTER, I.C.**
BEAMISH, F.W.H.
CHANGES IN HAEMATOCRIT AND HAEMOGLOBIN CONCENTRATION DURING
THE LIFE CYCLE OF THE ANADROMOUS SEA LAMPREY, PETROMYZON
MARINUS L.
COMP. BIOCHEM. PHYSIOL., 60A:431-434.
BLOOD ADULT AMMOCOETE MIGRATION FEEDING METAMORPHOSIS

5146F

**COTTRELL, B.A.**                          1976
DOOLITTLE, R.F.
AMINO ACID SEQUENCES OF LAMPREY FIBRINO PEPTIDES A AND B AND
CHARACTERIZATION OF THE JUNCTIONS SPLIT BY LAMPREY AND
MAMMALIAN THROMBINS.
BIOCHIM. BIOPHYS. ACTA, 453:426-438.
ADULT BIOCHEMISTRY BLOOD

5147

**GHARRETT, A.J.**                          1977
SIMON, R.C.
MCINTYRE, J.D.
REASSOCIATION AND HYBRIDIZATION PROPERTIES OF DNA FROM
SEVERAL SPECIES OF FISH.
COMP. BIOCHEM. PHYSIOL., 56B:81-85.
ANIMAL BIOCHEMISTRY PHYSIOLOGY

5148F

**FOX, H.**                          1976
ANCHORING FIBRILS OF THE BASAL LAMINA AND BASEMENT LAMELLA
IN THE SKIN OF AQUATIC CHORDATES.
J. MICROSC. BIOL. CELL., 26:43-46.
ANIMAL CYTOLOGY HISTOCHEMISTRY INTEGUMENT

5149

**MARTOJA, M.**                          1975
THE KIDNEY OF POMATIAS-ELEGANS CYCLOSTOMA-ELEGANS GASTROPODA
PROSOBRANCHIA MORPHOLOGICAL AND ANALYTICAL DATA.
ANN. SCI. NAT., ZOOL. BIOL. ANIM., 17:535-558.

5150

**VLADYKOV, V.D.**                          1975
KOTT, E.
PHARAND-COAD, S.
A NEW NONPARASITIC SPECIES OF LAMPREY GENUS LETHENTERON
PETROMYZONIDAE FROM EASTERN TRIBUTARIES OF THE GULF OF
MEXICO, U.S.A.
NATL. MUS. NAT. SCI., OTTAWA, PUBL. ZOOL., 12:1-36.

5151F

**TSUNEKI, K.**                          1977
GORBMAN, A.
ULTRASTRUCTURE OF THE OVARY OF THE HAGFISH EPTATRETUS
STOUTI.
ACTA ZOOL., (STOCKH.), 58:27-40.
GONADOGENESIS HISTOLOGY PHYSIOLOGY

5152F

**TSUNEKI, K.**                          1977
GORBMAN, A.
ULTRASTRUCTURE OF THE TESTICULAR INTERSTITIAL TISSUE OF THE
HAGFISH EPTATRETUS STOUTI.
ACTA ZOOL., (STOCKH.), 58:17-25.
CANADA PACIFIC OCEAN ENDOCRINOLOGY HISTOLOGY REPRODUCTION
COMOTION MIGRATION BEHAVIOUR

5153

**SHIGIN, A.A.**                          1976
METATSERKARII RODA DIPLOSTOMUM FAUNY SSSR. [METACERCARIA OF
THE GENUS DIPLOSTOMUM IN THE FAUNA OF THE USSR.]
PARAZITOLOGIYA (LENINGR.), 10:346-351.
(RUSS.)
ENG. SUMM.
SYSTEMATICS

60

5154F

EMDIN, S.O.                    1977
GAMMELTOFT, S.
GLIEMANN, J.
DEGRADATION RECEPTOR BINDING AFFINITY AND POTENCY OF INSULIN
FROM THE ATLANTIC HAGFISH MYXINE GLUTINOSA DETERMINED IN
ISOLATED RAT FAT CELLS.
J. BIOL. CHEM., 252:602-608.
ADULT ANIMAL BIOCHEMISTRY METABOLISM PHYSIOLOGY


5155

RURAK, D.W.                    1977
PERKS, A.M.
THE NEUROHYPOPHYSIAL PRINCIPLES OF THE WESTERN BROOK
LAMPREY, LAMPETRA RICHARDSONI. STUDIES IN THE AMMOCOETE
LARVA.
GEN. COMP. ENDOCRINOL., 31:91-100.


5156F

AMOS, B.                       1977
ANDERSON, I.G.
HASLEWOOD, G.A.D.
TOKES, L.
BILE SALTS OF THE LUNGFISHES LEPIDOSIREN NEOCERATODUS AND
PROTOPTERUS AND THOSE OF THE COELACANTH LATIMERIA CHALUMNAE.
BIOCHEM. J., 161:201-204.
ANIMAL BIOCHEMISTRY SYSTEMATICS


5157F

KNAGGS, E.H.                   1975
SUNADA, J.S.
LEA, R.N.
NOTES ON SOME FISHES COLLECTED OFF THE OUTER COAST OF BAJA
CALIFORNIA.
CALIF. FISH GAME, 61:56-59.
*EPTATRETUS STOUTI DISTRIBUTION


5158

SIMANTOV, R.                   1976
GOODMAN, R.
APOSHIAN, D.
SNYDER, S.H.
PHYLOGENETIC DISTRIBUTION OF A MORPHINE-LIKE PEPTIDE
"ENKEPHALIN".
BRAIN RES., 111:204-211.
NERVOUS SYSTEM PHYSIOLOGY ANIMAL BIOCHEMISTRY


5159F

DICKHOFF, W.W.                 1977
GORBMAN, A.
IN-VITRO THYROTROPIC EFFECT OF THE PITUITARY OF THE PACIFIC
HAGFISH EPTATRETUS STOUTI.
GEN. COMP. ENDOCRINOL., 31:75-79.
ADULT ENDOCRINOLOGY TECHNIQUES BIOCHEMISTRY


5160F

KENNEDY, M.C.                  1977
RUBINSON, K.
RETINAL PROJECTIONS IN LARVAL TRANSFORMING AND ADULT SEA
LAMPREY PETROMYZON MARINUS.
J. COMP. NEUROL., 171:465-480.
VISION ADULT METAMORPHOSIS NERVOUS SYSTEM TECHNIQUES
HISTOLOGY


5161F

LAERM, J.                      1977
STRUCTURE DEVELOPMENT AND HOMOLOGY OF THE CYCLOSTOME AXIAL
SKELETON.
BIOLOGIST, 59:73-84
*CYCLOSTOMATA ANATOMY EVOLUTION


5162

CSERR, H.F.                    1977
EHLICH, B.E.
ACTIVE TRANSPORT OF IODIDE BY CHLOROID PLEXUS: A
PHYLOGENETIC SURVEY.
FED. PROC., 36:540.
*MYXINE GLUTINOSA *LAMPETRA FLUVIATILIS ANIMAL CIRCULATORY
SYSTEM NERVOUS SYSTEM PHYSIOLOGY


5163

MONACO, F.                     1976
ANDREOLI, M.
SANTOLAMAZZA, C.
DE ROS, I.
CATAUDELLA, S.
THYRO GLOBULIN BIOSYNTHESIS AND THYROID HORMONE FORMATION
INLAMPREY.
ACTA ENDOCRINOL., SUPPL., 204.


5164

MCNAMEE, M.G.                    1977
CHARACTERIZATION OF ACETYL CHOLINE RECEPTOR-RICH MEMBRANE
VESICLES FROM TORPEDO CALIFORNICA.
FED. PROC., 36:642.


                                                          5165

GONCHAREVSKAYA, O.A.             1976
NEPHRON STRUCTURE AND PROXIMAL RE ABSORPTION IN DIFFERENT
VERTEBRATE CLASSES.
J. EVOL. BIOCHEM. PHYSIOL., 12:104-109.
(ENG.)


                                                          5166

DOWNING, D.W.                    1977
SPITZER, R.H.
KAPLAN, M.A.
CALLAGHAN, O.H.
KOCH, E.A.
THE SECRETORY SYSTEM OF THE PACIFIC HAGFISH
EPATRETUS-STOUTI. UPTAKE OF RADIO LABELLED MONOSACCARIDES
AND AMINO ACIDS.
FED. PROC., 36:1239.


                                                          5167

MURTAUGH, P.A.                   1977
GLADNER, J.A.
HENSON, J.G.
BLADEN, H.A.
EFFECT OF THROMBIN-LIKE ENZYMES ON LAMPREY FIBRINOGEN
INVESTIGATION OF THE POLYMERIZATION PRODUCTS VIA ELECTRON
MICROSCOPY.
FED. PROC., 36:676.


                                                          5168

BORGESE, T.A.                    1977
HARRINGTON, J.
ROTH, E.
NAGEL, R.
ROSSETTER, S.
ORGANIC PHOSPHATES AND IRON NUCLEOTIDE COMPLEXES IN
ELASMOBRANCH AND MARINE TELEOST RED CELLS.
FED. PROC., 36:528.


                                                          5169F

PICKERING, A.D.                  1977
MORRIS, R.
SEXUAL DIMORPHISM IN THE GILLS OF THE SPAWNING RIVER
LAMPREY, LAMPETRA FLUVIATILIS L.
CELL TISSUE RES., 180:1-10.
GREAT BRITAIN ADULT HISTOLOGY IONIC REGULATION REPRODUCTION


                                                          5170F

SADO, Y.                         1976
HORI, S.H.
PROPERTIES OF HEPATIC HEXOSE 6 PHOSPHATE DEHYDROGENASE AND
GLUCOSE 6 PHOSPHATE DEHYDROGENASE FROM FISHES AND
AMPHIBIANS.
HOKKAIDO UNIV., SAPPORO, JAPAN, FAC. SCI., J., SER. VI
ZOOL., 20:277-287.
ANIMAL BIOCHEMISTRY PHYSIOLOGY


                                                          5171

KORNELIUSSEN, H.                 1976
5 HYDROXYTRYPTAMINE: AUTORADIOGRAPHIC EVIDENCE FOR UPTAKE
INTO FIBROBLAST CELL NUCLEI.
EXPERIENTIA, 32:443-445.
*MYXINE GLUTINOSA NORWAY NERVOUS SYSTEM MUSCLE


                                                          5172

SEIXAS, M.M.P.                   1976
TERRESTRIAL GASTROPODS FROM THE PORTUGUESE FAUNA.
SOC. PORT. CIENC. NAT., BOL., 16:21-46


                                                          5173

LYCHAKOV, D.V.                   1976
SVETO- I ELEKTRONNOMNKROSKOPICHESKOE ISSLEDOVANIE
FOTORETSEPTOROV MINOGI LAMPETRA FLUVIATILIS. [LIGHT
MICROSCOPIC AND ELECTRONMICROSCOPIC STUDIES OF
PHOTORECEPTORS IN THE LAMPREY LAMPETRA FLUVIATILIS.]
ZH. EVOL. BIOKHIM. FIZIOL., 12:358-361.
(RUSS.)
RUSSIA ADULT VISION HISTOCHEMISTRY


                                                          5174F

WRIGHT, G.M.                     1976
YOUSON, J.H.
TRANSFORMATION OF THE ENDOSTYLE OF THE ANADROMOUS SEA
LAMPREY PETROMYZON MARINUS DURING METAMORPHOSIS. PART 1:
LIGHT MICROSCOPY AND AUTO RADIOGRAPHY WITH IODINE-125.
GEN. COMP. ENDOCRINOL., 30:243-257.
*PETROMYZON MARINUS METAMORPHOSIS ENDOCRINOLOGY HISTOLOGY

5175F

**HOLCIK, J.** 1976
LIST OF LAMPREYS AND FISHES OF SLOVAKIA CZECHOSLOVAKIA.
BIOLOGIA (BRATISL.), 31:641-647.
*LAMPETRA PLANERI *EUDONTOMYZON DANFORDI *EUDONTOMYZON
VLADYKOV BRATISLAVA DISTRIBUTION

5176F

**PICKERING, A.D.** 1976
STIMULATION OF INTESTINAL DEGENERATION BY OESTRADIOL AND
TESTOSTERONE IMPLANTATION IN THE MIGRATING RIVER LAMPREY,
LAMPETRA FLUVIATILIS L.
GEN. COMP. ENDOCRINOL., 30:340-346.
UNITED KINGDOM ADULT ENDOCRINOLOGY MIGRATION

5177F

**DAHL, H.A.** 1976
KORNELIUSSEN, H.
LACTATE DEHYDROGENASE ISOENZYMES IN DIFFERENT TYPES OF
MUSCLE FIBERS IN THE ATLANTIC HAGFISH (MYXINE GLUTINOSA L.).
COMP. BIOCHEM. PHYSIOL., 55B:381-385.
NORWAY MUSCLE PHYSIOLOGY

5178

**MUNRO, A.L.S.** 1976
LIVERSIDGE, J.
ELSON, K.G.R.
THE DISTRIBUTION AND PREVALENCE OF INFECTIOUS PANCREATIC
NECROSIS VIRUS IN WILD FISH IN LOCH AWE.
R. SOC. EDINB., PROC., SECT. B (NAT. ENVIRON.), 75:223-232.

5179

**LYZLOVA, E.M.** 1975
SEREBRENIKOVA, T.P.
VERZHBINSKAYA, N.A.
ISOZYMES OF ASPARTATE AMINO TRANSFERASE EC-2.6.1.1. AND
GLYCOGEN PHOSPHORYLASE EC-2.4.1.1. IN MUSCLE OF LOWER
VERTEBRATES. PP.547-566.
IN  MARKERT, C.L. ED., ISOZYMES II. PHYSIOLOGICAL FUNCTION.
THIRD INTERNATIONAL CONFERENCE. NEW HAVEN, CONN., U.S.A. NEW
YORK: ACADEMIC PRESS.

5180

**YAMAGUCHI, K.** 1976
TAMECHIKA, M.
AWAYA, K.
PECULIAR SECRETORY GRANULES OF EFFERENT GILL DUCT EPITHELIAL
CELLS OF LAMPREYS LAMPETRA REISSNERI.
J. ELECTRONMICROSC., 25:119
(ABSTR.)
JAPAN AMMOCOETE ADULT HISTOLOGY

5181F

**KARAMYAN, A.I.** 1972
O FORMIROVANII STRUKTURNOI I FUNKTSIONAL'NOI ORGANIZATSII
PALEO-, ARKHI- I NEOKORTEKSH V FILOGENEZE
DOMLEKOPITAYUSHCHIKH PLZVONOCHNYKH. [ON THE FORMATION OF
STRUCTURAL AND FUNCTIONAL ORGANIZATION OF PALEO- ARCHI- AND
NEOCORTEX IN PHYLOGENESIS OF SUBMAMMALIAN VERTEBRATES.]
ZH. EVOL. BIOKHIM. FIZIOL., 8:324-332
(RUSS.)
ENG. SUMM.
*CYCLOSTOMATA EVOLUTION NERVOUS SYSTEM PHYSIOLOGY

5182F

**UMMINGER, B.L.** 1977
RELATION OF WHOLE BLOOD SUGAR CONCENTRATIONS IN VERTERATES
TO STANDARD METABOLIC RATE.
COMP. BIOCHEM. PHYSIOL., 56A:457-460.
RESPIRATION PHYSIOLOGY METABOLISM

5183F

**MENZIE, C.M.** 1976
HUNN, J.B.
CHEMICAL CONTROL OF THE SEA LAMPREY; THE ADDITION OF A
CHEMICAL TO THE ENVIRONMENT.
IN  COULSTON, F. AND F. KORTE EDS. ENVIRONMENTAL QUALITY AND
SAFETY, VOL.5.  GLOBAL ASPECTS OF CHEMISTRY TOXICOLOGY AND
TECHNOLOGY AS APPLIED TO THE ENVIRONMENT. NEW YORK: ACADEMIC
PRESS.
CHEMISTRY MANAGEMENT MORTALITY OF

5184F

**MATTY, A.J.** 1976
TSUNEKI, K.
DICKHOFF, W.W.
GORBMAN, A.
THYROID AND GONADAL FUNCTION IN HYPOPHYSECTOMIZED HAGFISH,
EPTATRETUS STOUTII.

GEN. COMP. ENDOCRINOL., 30:500-516.
U.S.A. PACIFIC OCEAN ENDOCRINOLOGY GONADOGENESIS HISTOLOGY
NERVOUS SYSTEM

5185F

**NAKAO, T.**                           1977
ELECTRON MICROSCOPIC STUDIES ON THE MYOTOMES OF LARVAL
LAMPREY LAMPETRA JAPONICA.
ANAT. REC., 187:383-403.
JAPAN AMMOCOETE ADULT MUSCLE HISTOLOGY


5186F

**FLASAR, I.**                          1975
FLASAROVA, M.
THE VERTEBRATE FAUNA OF NORTHWESTERN BOHEMIA (NORTHWESTERN
CZECHOSLOVAKIA): THE RESULTS SO FAR OF THEIR INVESTIGATION.
ZOOL. ABH. MUS. TIERK. DRESDEN, SUPPL., 33:1-150.
(GERM.)
*LAMPETRA PLANERI *PETROMYZON MARINUS *LAMPETRA FLUVIATILIS
EAST GERMANY DISTRIBUTION


5187F

**SHIODA, S.**                          1977
HONMA, Y.
YOSHIE, S.
HOSOYA, Y.
SCANNING ELECTRON MICROSCOPY OF THE 3RD VENTRICULAR WALL IN
THE LAMPREY LAMPETRA JAPONICA.
ARCH. HISTOL. JAP., 40:41-49.
ADULT HISTOLOGY


5188

**HIPPE, E.**                           1977
JORGENSEN, F.S.
OLESEN, H.
COBALAMIN BINDING PROTEINS IN STOMACH AND SERUM FROM
VARIOUSANIMAL SPECIES DATA FOR B-12 BINDING CAPACITIES AND
MOLECULAR SIZES OF THE BINDING PROTEINS.
COMP. BIOCHEM. PHYSIOL., 56:305-309.


5189F

**SILLS, J.B.**                         1976
ALLEN, J.L.
RESIDUES OF 3-TRIFLUOROMETHYL-4-NITROPHENOL (TFM) UNDETECTED
IN LAKE TROUT AND CHINOOK SALMON FROM THE UPPER GREAT
LAKES.
PROG. FISH CULT., 38:197.
ANIMAL CHEMISTRY MANAGEMENT


5190

**GURAYA, S.S.**                        1976
RECENT ADVANCES IN THE MORPHOLOGY HISTOCHEMISTRY AND
BIOCHEMISTRY OF STEROID SYNTHESIZING CELLULAR SITES IN THE
TESTES O... PP.99-136.
IN  BOURNE, G.H. AND J.F. DANIELLI EDS. INTERNATIONAL
REVIEWOF CYTOLOGY, VOL. 47. NEW YORK: ACADEMIC PRESS.


5191

**LYZLOVA, E.M.**                       1976
VERZHBINSKAYA, N.A.
[PHOSPHOENOLPHRUVATE CARBOXYKINASE IN TISSUES OF THE LAMPREY
LAMPETRA FLUVIATILIS.]
ZH. EVOL. BIOKHIM. FIZIOL., 12:75-77
(RUSS.)
J. EVOL. BIOCHEM. PHYSIOL., 12:65-67
(ENG.)
BIOCHEMISTRY METABOLISM MUSCLE


5192

**FALKMER, S.**                         1976
REVIEW OF CURRENT RESEARCH AIMING AT A FUNCTIONAL
EVOLUTIONARY ANALYSIS OF THE INSULIN MOLECULE.
GEN. COMP. ENDOCRINOL., 29:288.
(ABSTR.)


5193

**LELOUP, J.**                          1976
HARDY, A.
HORMONES THYROIDIENNES CIRCULANTES CHEZ UN CYCLOSTOME ET DES
POISSONS. [CIRCULATING THYROID HORMONES IN A CYCLOSTOME AND
IN FISHES.]
GEN. COMP. ENDOCRINOL., 29:258.
(FR.)
(ABSTR.)


5194F

**TIBBLES, J.J. ED.**                   1977
ANNUAL REPORT 1976 OF THE SEA LAMPREY CONTROL CENTRE SAULT
STE. MARIE, ONTARIO, TO THE GREAT LAKES FISHERY COMMISSION.
GREAT LAKES FISH. COMM., ANNU. REP., 101 PP.
RESTRICTED

5195

**VALE, W.**                     1976
LING, N.
RIVIER, J.
VILLARREAL, J.
RIVIER, C.
DOUGLAS, C.
ANATOMIC AND PHYLOGENETIC DISTRIBUTION OF SOMATOSTATIN.
METAB. CLIN. EXP., 25:1491-1494.


5196

**FALKMER, S.**                  1976
OSTBERG, Y.
BOQUIST, L.
ENTERO INSULAR ENDOCRINE CELLS IN MYXINE GLUTINOSA.
GEN. COMP. ENDOCRINOL., 29:287.
(ABSTR.)
*MYXINE GLUTINOSA SWEDEN BIOCHEMISTRY ENDOCRINOLOGY
TECHNIQUES HISTOCHEMISTRY CYTOLOGY


5197F

**SCHMULBACH, J.C.**             1975
GOULD, G.
GROEN, C.L.
RELATIVE ABUNDANCE AND DISTRIBUTION OF FISHES IN THE
MISSOURI RIVER GAVINS-POINT DAM TO RULO NEBRASKA.
S.D. ACAD. SCI., PROC., 54:194-222.
*ICHTHYOMYZON CASTANEUS *ICHTHYOMYZON UNICUSPIS U.S.A.


5198

**OLSSON, R.**                   1975
WHAT DO THE CIRCUM VENTRICULAR ORGANS OF THE LOWER
CHORDATESTELL US ABOUT THE EARLY EVOLUTION OF THE VERTEBRATE
BRAIN.
ZOOL. MAG. (TOKYO), 84:510.


5199F

**INANO, H.**                    1976
MORI, K.
TAMOKI, B.
GUSTAFSSON, J.A.
IN VITRO METABOLISM OF TESTOSTERONE IN HEPATIC TISSUE OF A
HAGFISH EPTATRETUS BURGERI.
GEN. COMP. ENDOCRINOL., 30:258-266.
SWEDEN ENDOCRINOLOGY EVOLUTION


5200

**ROSS, J.R.P.**                 1977
MICRO ARCHITECTURE OF BODY WALL OF EXTANT CYCLOSTOME
ECTOPROCTS.
AM. ZOOL., 17:93-105.
MICROARCHITECTURE OF BODY WALL OF CYCLOSTOME ECTOPROCTS


5201

**COUPLAND, R.E.**               1976
ENDROCRINE SYSTEM. PP. 481-503.
IN HAMILTON, W.J. ED. TEXTBOOK OF HUMAN ANATOMY. SAINT
LOUIS, MO.:MOSBY COMPANY.


5202F

**ROBISON, H.W.**                1975
NEW DISTRIBUTIONAL RECORDS OF FISHES FROM THE LOWER OUACHITA
RIVER SYSTEM IN ARKANSAS.
ARKANSAS ACAD. SCI., FAYETTEVILLE, PROC., 29:54-56.
*ICHTHYOMYZON GAGEI DISTRIBUTION PISCES


5203

**GEIHSLER, M.R.**               1975
SHORT, E.D.
KITTLE, P.D.
A PRELIMINARY CHECKLIST OF THE FISHES OF THE ILLINOIS RIVER
ARKANSAS U.S.A.
ARKANSAS ACAD. SCI., FAYETTEVILLE, PROC., 29:37-39.


5204

**LEVANOVICH, V.V.**             1976
ACTIVITY OF SODIUM. POTASSIUM, AND CALCIUM IN THE BLOOD
SERUM OF VERTEBRATES.
ZH. EVOL. BIOKHIM. FIZIOL., 12:369-372.
(RUSS.)
J. EVOL. BIOCHEM. PHYSIOL., 12:336-338.
(ENG.)
*LAMPETRA FLUVIATILIS BLOOD IONIC REGULATION


5205F

**BRINN, J.E. JR.**              1976
EPPLE, A.
NEW TYPES OF ISLET CELLS IN A CYCLOSTOME PETROMYZON MARINUS.
CELL TISSUE RES., 171:317-330.

5206F

MATTHEWS, G.                        1977
WICKELGREN, W.O.
EFFECTS OF GUANIDIDE ON TRANSMITTER RELEASE AND NEURONAL
EXITABILITY.
   J. PHYSIOL., 266:69-89.
J. PHYSIOL., 266:69-89.
*ICHTHYOMYZON FOSSOR *PETROMYZON MARINUS ADULT
HISTOCHEMISTRY


5207

MOSHIN, A.K.M.                      1977
GALLAWAY, B.J.
SEASONAL ABUNDANCE DISTRIBUTION FOOD HABITATS AND CONDITION
OF THE SOUTHERN BROOK LAMPREY ICHYHYOMYZON GAGEI IN AN EAST
TEXAS U.S.A. WATERSHED.
SOUTHWEST. NAT., 22:107-114.


5208

HORNSEY, D.J.                       1977
TRIIODOTHYRONINE AND THYROXINE LEVELS IN THE THYROID AND
SERUM OF THE SEA LAMPREY, PETROMYZON MARINUS.
GEN. COMP. ENDOCRINOL., 31:381-383.
ADULT TECHNIQUES ENDOCRINOLOGY BLOOD


5209F

JOSS, J.M.P.                        1977
HYDROXYINDOLE-O-METHYLTRANSFERASE (HIOMT) ACTIVITY AND THE
UPTAKE OF TRITATED MELATONIN IN THE LAMPREY
GEOTRIA-AUSTRALIS GRAY.
GEN. COMP. ENDOCRINOL., 31:270-275.
AUSTRALIA AMMOCOETE ENDOCRINOLOGY


5210F

SHAKHMATOVA, E.I.                   1977
[CATION EXCRETION BY THE KIDNEY OF THE LAMPREY LAMPETRA
FLUVIATILIS UNDER THE INFLUENCE OF DIURETICS.]
ZH. EVOL. BIOKHIM. FIZIOL., 13:82-83.
(RUSS.)
J. EVOL. BIOCHEM. PHYSIOL., 13:63-64.
(ENG.)
RUSSIA IONIC REGULATION


5211

PHILLIPS, J.W.                      1977
HIRD, F.J.R.
GLUCONEOGENESIS IN VERTEBRATE LIVERS.
COMP. BIOCHEM. PHYSIOL., 57B:127-132.
ANIMAL BIOCHEMISTRY MUSCLE BLOOD


5212

PHILLIPS, J.W.                      1977
HIRD, F.J.R.
KETOGENESIS IN VERTEBRATE LIVERS.
COMP. BIOCHEM. PHYSIOL., 57:133-138.


5213

GRAY, E.G.                          1976
PROBLEMS OF UNDERSTANDING THE SUBSTRUCTURE OF SYNAPSES.
PP.207-234.
IN  CORNER, M.A., AND D.F. SWAAB EDS. PROGRESS IN BRAIN
RESEARCH. VOL. 45. PERSPECTIVES IN BRAIN RESEARCH. 9TH
INTERNATIONAL SUMMER SCHOOL. AMSTERDAM. NETHERLANDS, JULY
28-AUG.1,1975. AMSTERDAM: ELSEVIER SCIENTIFIC PUBLISHING COM


5214F

OHMAN, P.                           1977
FINE STRUCTURE OF THE OPTIC NERVE OF LAMPETRA-FLUVIATILIS
CYCLOSTOMI.
VISION RES., 17:719-722.
*LAMPETRA FLUVIATILIS SWEDEN ADULT HISTOLOGY NERVOUS SYSTEM
VISION


5215

SEILER, K.                          1973
SEILER, R.
HOHEISEL, G.
ZUR CYTOLOGIE DES INTERRENAL-SYSTEMS BEIM BACHNEUNAUGE
(LAMPETRA PLANERI BLOCH).
GEGENBAURS MORPHOL. JAHRB., 119:823-856.


5216F

WESTMAN, R.W.                       1976
INDIVIDUAL STREAM TREATMENT REPORTS, LAKE HURON, 1975.
PP.31-44.
IN  TIBBLES, J.J. ED., ANNUAL REPORT 1975 OF THE SEA LAMPREY
CONTROL CENTRE SAULT STE. MARIE, ONTARIO, TO THE GREAT
LAKES FISHERY COMMISSION. GREAT LAKES FISH. COMM., ANNU.
REP.

RESTRICTED
SEA LAMPREY CANADA AMMCOETE CHEMISTRY MANAGEMENT

5218F

SAVINA, M.V.                    1977
WOJICZAK, A.B.
ENZYMES OF GLUCONEOGENESIS AND THE SYNTHESIS OF GLUCOGEN
FROM GLYCEROL IN VARIOUS ORGANS OF THE LAMPREY LAMPETRA
FLUVIATILIS.
COMP. BIOCHEM. PHYSIOL., 57B:185-190.
RUSSIA ADULT MIGRATION PHYSIOLOGY


5219F

SUTTERLIN, A.M.             1975
CHEMICAL ATTRACTION OF SOME MARINE FISH IN THEIR NATURAL
HABITAT.
FISH. RES. BOARD CAN., J., 32:729-738.
*MYXINE GLUTINOSA CANADA ATLANTIC OCEAN BEHAVIOUR OLFACTION
PHYSIOLOGY SENSE RECEPTORS


5220F

SCHLEEN, L.P.                  1976
INDIVIDUAL STREAM TREATMENT REPORTS, CANADIAN SIDE OF LAKE
ONTARIO, 1975. PP.44-55.
IN  TIBBLES, J.J. ED., ANNUAL REPORT 1975 OF THE SEA LAMPREY
CONTROL CENTRE SAULT STE. MARIE, ONTARIO, TO THE GREAT
LAKES FISHERY COMMISSION. GREAT LAKES FISH. COMM., ANNU.
REP.
RESTRICTED
SEA LAMPREY CANADA AMMOCOETE CHEMISTRY MANAGEMENT


5221F

WESTMAN, R.W.                  1976
CUDDY, D.W.
INDIVIDUAL STREAM TREATMENT REPORTS, NEW YORK STATE SIDE OF
LAKE ONTARIO, 1975. PP.55-68.
IN  TIBBLES, J.J. ED., ANNUAL REPORT 1975 OF THE SEA LAMPREY
CONTROL CENTRE SAULT STE. MARIE, ONTARIO, TO THE GREAT
LAKES FISHERY COMMISSION. GREAT LAKES FISH. COMM., ANNU.
REP.
RESTRICTED
SEA LAMPREY U.S.A. CHEMISTRY ADULTS AMMOCOETE CHEMISTRY
MANAGEMENT.


5222F

SCHLEEN, L.P.                  1976
GRANULAR BAYER 73 TREATMENTS, LAKE SUPERIOR, 1975. PP.68-75.
IN  TIBBLES, J.J. ED. ANNUAL REPORT 1975 OF THE SEA LAMPREY
CONTROL CENTRE SAULT STE. MARIE, ONTARIO, TO THE GREAT LAKES
FISHERY COMMISSION. GREAT LAKE FISH. COMM., ANNU. REP.
RESTRICTED
SEA LAMPREY CANADA AMMOCOETE CHEMISTRY MANAGEMENT


5223F

GREAT LAKES FISH. COMM.        1976
GRANULAR BAYER 73 TREATMENTS, LAKE HURON, 1975. PP.76-83.
IN  TIBBLES, J.J. ED. ANNUAL REPORT 1975 OF THE SEA LAMPREY
CONTROL CENTRE SAULT STE. MARIE, ONTARIO, TO THE GREAT
LAKESFISH COMMISSION. GREAT LAKES FISH. COMM., ANNU. REP.
RESTRICTED
*LAMPETRA LAMOTTEI *ICHTHYOMYZON SPP. SEA LAMPREY CANADA
AMMOCOETE CHEMISTRY MANAGEMENT


5224F

SCHLEEN, L.P.                  1976
SURFACE TRAWLING FOR ADULT SEA LAMPREY IN ST. MARYS RIVER,
1975. PP.83-85.
IN  TIBBLES, J.J. ED., ANNUAL REPORT 1975 OF THE SEA LAMPREY
CONTROL CENTRE SAULT STE. MARIE, ONTARIO, TO THE GREAT
LAKES FISHERY COMMISSION. GREAT LAKES FISH. COMM., ANNU.
REP.
RESTRICTED
CANADA ADULT DISTRIBUTION


5225F

JOHNSON, B.G.H.                1976
BIOLOGICAL DATA FROM SEA LAMPREY COLLECTED BY COMMERCIAL
FISHERMEN, 1975, P.86.
IN  TIBBLES, J.J. ED., ANNUAL REPORT 1975 OF THE SEA LAMPREY
CONTROL CENTRE SAULT STE. MARIE, ONTARIO, TO THE GREAT
LAKES FISHERY COMMISSION, GREAT LAKES FISH. COMM., ANNU.
REP.
RESTRICTED
*ICHTHYOMYZON SPP. ADULT BIOLOGY


5226F

JOHNSON, B.G.H.                1976
BIOLOGICAL DATA ON ADULT SEA LAMPREY COLLECTED FROM
ELECTRICAL ASSESSMENT BARRIERS, LAKE HURON, 1975. PP.86-89.
IN  TIBBLES, J.J. ED., ANNUAL REPORT 1975 OF THE SEA LAMPREY
CONTROL CENTRE SAULT STE. MARIE, ONTARIO, TO THE GREAT

LAKES FISHERY COMMISSION. GREAT LAKES FISH. COMM., ANNU.
REP.
RESTRICTED
CANADA ADULT MORPHOMETRY

5227F

**JOHNSON, B.G.H.**                     1976
BIOLOGICAL DATA ON SEA LAMPREY FROM HUMBER RIVER, 1975.
P.90.
IN  TIBBLES, J.J. ED., ANNUAL REPORT 1975 OF THE SEA LAMPREY
CONTROL CENTRE SAULT STE. MARIE, ONTARIO, TO THE GREAT
LAKES FISHERY COMMISSION. GREAT LAKES FISH. COMM. ANNU. REP.
RESTRICTED
CANADA ADULT MIGRATION

5228F

**JOHNSON, B.G.H.**                     1976
SEA LAMPREY TAG-RECAPTURE STUDIES: LAKE ONTARIO, 1975.
PP.90-93.
IN  TIBBLES, J.J. ED., ANNUAL REPORT 1975 OF THE SEA LAMPREY
CONTROL CENTRE SAULT STE. MARIE, ONTARIO, TO THE GREAT
LAKES FISHERY COMMISSION. GREAT LAKES FISH. COMM., ANNU.
REP.
RESTRICTED
CANADA ADULT BEHAVIOUR MIGRATION

5229F

**GOOLD, R.J.**                         1976
ADULT SEA LAMPREY TRAWLING AND TAGGING STUDY, LAKE ONTARIO,
1975. PP.93-95.
IN  TIBBLES, J.J. ED., ANNUAL REPORT 1975 OF THE SEA LAMPREY
CONTROL CENTRE SAULT STE. MARIE, ONTARIO, TO THE GREAT
LAKES FISHERY COMMISSION. GREAT LAKES FISH. COMM. ANNU. REP.
RESTRICTED
CANADA ADULT DISTRIBUTION

5230F

**MCDONALD, R.**                        1976
SABLE RIVER EXPERIMENTAL MECHANICAL ASSESSMENT WEIR, 1975.
PP.96-98.
IN  TIBBLES, J.J. ED., ANNUAL REPORT 1975 OF THE SEA LAMPREY
CONTROL CENTRE SAULT STE  MARIE, ONTARIO, TO THE GREAT
LAKES FISHERY COMMISSION. GREAT LAKES FISH. COMM., ANNU.
REP.
RESTRICTED
SEA LAMPREY CANADA ADULT MANAGEMENT

5231F

**TIBBLES, J.J. EDS.**                  1976
DUSTIN, S.M.
JOHNSON, B.G.H.
CANADIAN REPORT ON SEA LAMPREY CONTROL, DECEMBER 1976.
ENVIRON. CAN., FISH. MAR. SERV., SEA LAMPREY CONTROL CENT.,
SAULT STE  MARIE, REP., 14 PP.
DISTRIBUTION HISTORY MANAGEMENT MORTALITY OF

5232F

**GRAHAM, D.H.**                        1953
LAMPREY (KOROKORO). GEOTRIA AUSTRALIS GRAY. 58-62.
IN  GRAHAM, D.H.  A TREASURY OF NEW ZEALAND FISHES.
WELLINGTON: A.H. AND A.W. REED.
NEW ZEALAND LIFE CYCLE BIOLOGY

5233F

**HARDISTY, M.W.**                      1976
CYSTS AND TUMOUR-LIKE LESION IN THE ENDOCRINE PANCREAS OF
THE LAMPREY (LAMPETRA FLUVIATILIS).
J. ZOOL., 178:305-317.
ANATOMY HISTOLOGY PHYSIOLOGY

5234

**HAGELIN, L.O.**                       1974
DEVELOPMENT OF THE MEMBRANOUS LABYRINTH IN LAMPREYS.
RIVER LAMPREYS (LAMPETRA FLUVIATILIS) BEFORE AND DURING
PERIOD OF SEXUAL MATURATION.
ACTA ZOOL., SUPPL., 1974:215 PP.

5235F

**ANDERSON, W.C.**                      1977
MANION, P.J.
MORPHOLOGICAL DEVELOPMENT OF THE INTEGUMENT OF THE SEA
LAMPREY, PETROMYZON MARINUS.
FISH. RES. BOARD CAN., J. 34:159-163
U.S.A. BIOLOGY INTEGUMENT HISTOLOGY

5236F

**TIBBLES, J.J.**                       1976
ANNUAL REPORT TO THE DEPARTMENT OF THE ENVIRONMENT AND THE
GREAT LAKES FISHERY COMMISSION FOR 1975-1976. PP.1-V.
IN  TIBBLES, J.J. ED. ANNUAL REPORT 1975 OF THE SE LAMPREY
CONTROL CENTRE SAULT STE. MARIE, ONTARIO, TO GREAT LAKES

FISHERY COMMISSION. GREA LAKES FISH. COMM., ANNU. REP.
RESTRICTED

5237F

**TIBBLES, J.J.**            1974
ANNUAL REPORT TO THE DEPARTMENT OF THE ENVIRONMENT AND THE
GREAT LAKES FISHERY COMMISSION FOR 1973-1974. PP. 1-V.
IN TIBBLES, J.J., ED., ANNUAL REPORT 1973 OF THE SEA
LAMPREY CONTROL CENTRE STE. MARIE, ONTARIO, TO THE GREAT
LAKES FISHERY COMMISSION. GREAT LAKES FISH. COMM., ANNU.
REP.
RESTRICTED
*PETROMYZON MARINUS CANADA ADULT CHEMISTRY MORTALITY OF
AMMOCOETE

5238F

**PANG, P.K.T.**            1977
GRIFFITH, R.W.
ATZ, J.W.
OSMOREGULATION IN ELASMOBRANCHS.
AM. ZOOL., 17:365-377.
MYXINE BLOOD

5239

**POTTER, I.C.**            1976
THE FISH FAUNA OF THE SEVERN ESTUARY.
LINN. SOC. LOND., J. (BIOL.), 8:346.

5240

**SCHULTZE, H.P.**            1976
PALEOZOIC VERTEBRATES. PP. 217-238.
IN GRZIMEK, B., ED., GRZIMEK'S ENCYCLOPEDIA OF EVOLUTION.
NEW YORK: VAN NOSTRAND REINHOLD CO.

5241F

**MANION, P.J.**            1977
HANSON, L.H.
ABNORMAL TOOTH DEVELOPMENT IN A SEA LAMPREY
PROG. FISH CULT., 39:127-128.
U.S.A. BIOLOGY FEEDING

5242F

**MAKI, A.W.**            1976
JOHNSON, H.E.
THE FRESHWATER MUSSEL (ANODONTA SP.) AS AN INDICATOR OF
ENVIRONMENTAL LEVELS OF 3-TRIFLUOROMETHYL-4-NITROPHENOL
(TFM).
U.S. DEP. INT., FISH WILDL. SERV., INVEST. FISH CONTROL,
70:1-5.
U.S.A. MANAGEMENT ANIMAL CHEMISTRY

5243F

**TOMONAGA, S.**            1973
TAMECHIKA, M.
AWAYA, K.
FINE STRUCTURE AND PHAGOCYTIC ACTIVITY OF THE HIGH-WALLED
ENDOTHELIUM IN THE GILL OF THE HAGFISH, EPTATRETUS BURGERI.
J. ELECTRONMICROSC., 22:113.
CYTOLOGY HISTOLOGY

5244F

**BAUER, C.**            1975
ENGELS, U.
PALEUS, S.
OXYGEN BINDING TO HAEMEOGLOBINS OF THE PRIMITIVE VERTEBRATE
MYXINE GLUTINOSA L.
NATURE (LOND.), 256:66-68.
BIOCHEMISTRY BLOOD

5245F

**NORTHCUTT, R.G.**            1973
PRZYBYLSKI, R.J.
RETINAL PROJECTIONS IN THE LAMPREY PETROMYZON MARINUS L.
ANAT. REC., 175:400.
CYTOLOGY HISTOLOGY NERVOUS SYSTEM VISION

5246F

**GREAT LAKES FISH. COMM.**            1977
ADULT SEA LAMPREY ELECTRICAL ASSESSMENT BARRIER OPERATIONS,
LAKE HURON. P.5.
IN TIBBLES, J.J. ED., ANNUAL REPORT 1976 OF THE SEA LAMPREY
CONTROL CENTRE SAULT STE. MARIE, ONTARIO, TO THE GREAT
LAKES FISHERY COMMISSION. GREAT LAKES FISH. COMM., ANNU.
REP.
RESTRICTED
*PETROMYZON MARINUS CANADA GREAT LAKES CANADA MANAGEMENT
MIGRATION

5247F

**GREAT LAKES FISH. COMM.**            1977
MECHANICAL ASSESSMENT WEIR AND TRAP OPERATIONS. PP.5-7.

IN TIBBLES, J.J. ED., ANNUAL REPORT 1976 OF THE SEA LAMPREY
CONTROL CENTRE SAULT STE. MARIE, ONTARIO, TO THE GREAT
LAKES FISHERY COMMISSION. GREAT LAKES FISH. COMM., ANNU.
REP.
RESTRICTED
*PETROMYZON MARINUS CANADA GREAT LAKES LAKE SUPERIOR ADULT
MANAGEMENT MORTALITY OF

5248F

**GREAT LAKES FISH. COMM.**    1977
SEA LAMPREY SURVEYS OF STREAMS TRIBUTARY TO LAKE SUPERIOR.
PP. 7-9.
IN TIBBLES, J.J. ED., ANNUAL REPORT 1976 OF THE SEA LAMPREY
CONTROL CENTRE SAULT STE. MARIE, ONTARIO, TO THE GREAT
LAKES FISHERY COMMISSION. GREAT LAKES FISH. COMM., ANNU.
REP.
RESTRICTED
*PETROMYZON MARINUS CANADA GREAT LAKES AMMOCOETE
DISTRIBUTION MANAGEMENT

5249F

**GREAT LAKES FISH. COMM.**    1977
SEA LAMPREY SURVEYS OF STREAMS TRIBUTARY TO LAKE HURON.
PP.9-11.
IN TIBBLES, J.J. ED., ANNUAL REPORT 1976 OF THE SEA LAMPREY
CONTROL CENTRE SAULT STE. MARIE, ONTARIO, TO THE GREAT
LAKES FISHERY COMMISSION. GREAT LAKES FISH. COMM., ANNU.
REP.
RESTRICTED
*PETROMYZON MARINUS CANADA GREAT LAKES CHEMISTRY
DISTRIBUTION MANAGEMENT MORTALITY OF

5250F

**GREAT LAKES FISH. COMM.**    1977
SEA LAMPREY SURVEYS OF STREAMS TRIBUTARY TO THE CANADIAN
SIDE OF LAKE ONTARIO. PP.11-14.
IN TIBBLES, J.J. ED., ANNUAL REPORT 1976 OF THE SEA LAMPREY
CONTROL CENTRE SAULT STE. MARIE, ONTARIO, TO THE GREAT
LAKES FISHERY COMMISSION. GREAT LAKES FISH. COMM., ANNU.
REP.
RESTRICTED
*PETROMYZON MARINUS CANADA GREAT LAKES AMMOCOETE
DISTRIBUTION MANAGEMENT

5251F

**GREAT LAKES FISH. COMM.**    1977
SEA LAMPREY SURVEYS OF STREAMS TRIBUTARY TO THE UNITED
STATES (NEW YORK) SIDE OF LAKE ONTARIO. PP.14-15.
IN TIBBLES, J.J. ED., ANNUAL REPORT 1976 OF THE SEA LAMPREY
CONTROL CENTRE SAULT STE. MARIE, ONTARIO, TO THE GREAT
LAKES FISHERY COMMISSION. GREAT LAKES FISH. COMM., ANNU.
REP.
RESTRICTED
*PETROMYZON MARINUS *ICHTHYOMYZON U.S.A. GREAT LAKES
AMMOCOETE CHEMISTRY MANAGEMENT

5252F

**GREAT LAKES FISH. COMM.**    1977
INDIVIDUAL STREAM TREATMENT REPORTS, LAKE SUPERIOR.
PP.16-33.
IN TIBBLES, J.J. ED., ANNUAL REPORT 1976 OF THE SEA LAMPREY
CONTROL CENTRE SAULT STE. MARIE, ONTARIO, TO THE GREAT
LAKES FISHERY COMMISSION. GREAT LAKES FISH. COMM., ANNU.
REP.
RESTRICTED
*PETROMYZON MARINUS CANADA GREAT LAKES AMMOCOETE MANAGEMENT
MORTALITY OF

5253F

**GREAT LAKES FISH. COMM.**    1977
INDIVIDUAL STREAM TREATMENT REPORTS, LAKE HURON. PP.34-50.
IN TIBBLES, J.J. ED., ANNUAL REPORT 1976 OF THE SEA LAMPREY
CONTROL CENTRE SAULT STE. MARIE, ONTARIO, TO THE GREAT
LAKES FISHERY COMMISSION. GREAT LAKES FISH. COMM., ANNU.
REP.
RESTRICTED
*PETROMYZON MARINUS CANADA GREAT LAKES AMMOCOETE CHEMISTRY
DISTRIBUTION MANAGEMENT MORTALITY OF

5254F

**GREAT LAKES FISH. COMM.**    1977
INDIVIDUAL STREAM TREATMENT REPORTS ON STREAMS TRIBUTARY TO
THE CANADIAN (ONTARIO) WATERS OF LAKE ONTARIO. PP. 66-78.
IN TIBBLES, J.J. ED., ANNUAL REPORT 1976 OF THE SEA LAMPREY
CONTROL CENTRE SAULT STE. MARIE, ONTARIO, TO THE GREAT
LAKES FISHERY COMMISSION. GREAT LAKES FISH. COMM., ANNU.
REP.
RESTRICTED
*PETROMYZON MARINUS CANADA GREAT LAKES AMMOCOETE CHEMISTRY
DISTRIBUTION MANAGEMENT MORTALITY OF

GREAT LAKES FISH. COMM.    1977
INDIVIDUAL TREATMENT REPORTS ON STREAMS TRIBUTARY TO THE
UNITED STATES (NEW YORK) WATERS OF LAKE ONTARIO. PP.66-78.
IN  TIBBLES, J.J. ED., ANNUAL REPORT 1976 OF THE SEA LAMPREY
CONTROL CENTRE SAULT STE. MARIE, ONTARIO, TO THE GREAT
LAKES FISHERY COMMISSION. GREAT LAKES FISH. COMM., ANNU.
REP.
RESTRICTED
*PETROMYZON MARINUS U.S.A. GREAT LAKES CHEMISTRY
DISTRIBUTION MANAGEMENT MORTALITY OF

GREAT LAKES FISH. COMM.    1977
GRANULAR BAYER 73 TREATMENTS, LAKE SUPERIOR. PP.78-85.
IN  TIBBLES, J.J. ED., ANNUAL REPORT 1976 OF THE SEA LAMPREY
CONTROL CENTRE SAULT STE. MARIE, ONTARIO, TO THE GREAT
LAKES FISHERY COMMISSION. GREAT LAKES FISH. COMM., ANNU.
REP.
RESTRICTED
*PETROMYZON MARINUS CANADA GREAT LAKES AMMOCOETE CHEMISTRY
DISTRIBUTION MANAGEMENT MORTALITY OF

GREAT LAKES FISH. COMM.    1977
GRANULAR BAYER 73 TREATMENTS, LAKE HURON. PP.86-92.
IN  TIBBLES, J.J. ED., ANNUAL REPORT 1976 OF THE SEA LAMPREY
CONTROL CENTRE SAULT STE. MARIE, ONTARIO, TO THE GREAT
LAKES FISHERY COMMISSION. GREAT LAKES FISH. COMM., ANNU.
REP.
RESTRICTED
*PETROMYZON MARINUS CANADA GREAT LAKES AMMOCOETE CHEMISTRY
DISTRIBUTION MANAGEMENT MORTALITY OF

GREAT LAKES FISH. COMM.    1977
SURFACE TRAWLING FOR ADULT SEA LAMPREY IN THE ST MARYS
RIVER, LAKE HURON. P.93.
IN  TIBBLES, J.J. ED., ANNUAL REPORT 1976 OF THE SEA LAMPREY
CONTROL CENTRE SAULT STE. MARIE, ONTARIO, TO THE GREAT
LAKES FISHERY COMMISSION. GREAT LAKES FISH. COMM., ANNU.
REP.
RESTRICTED
*PETROMYZON MARINUS CANADA GREAT LAKES ADULT DISTRIBUTION

GREAT LAKES FISH. COMM.    1977
ADULT SEA LAMPREY TRAWLING AND TAGGING STUDY OFF THE CREDIT
RIVER, LAKE ONTARIO.
IN  TIBBLES, J.J. ED., ANNUAL REPORT 1976 OF THE SEA LAMPREY
CONTROL CENTRE SAULT STE. MARIE, ONTARIO, TO THE GREAT
LAKES FISHERY COMMISSION. GREAT LAKES FISH. COMM., ANNU.
REP.
RESTRICTED

GREAT LAKES FISH. COMM.    1977
BIOLOGICAL DATA ON ADULT SEA LAMPREY COLLECTED FROM
ELECTRICAL ASSESSMENT BARRIERS, LAKE HURON. PP.94-97.
IN  TIBBLES, J.J. ED., ANNUAL REPORT 1976 OF THE SEA LAMPREY
CONTROL CENTRE SAULT STE. MARIE, ONTARIO, TO THE GREAT
LAKES FISHERY COMMISSION. GREAT LAKES FISH. COMM., ANNU.
REP.
RESTRICTED
*PETROMYZON MARINUS MANAGEMENT MORPHOMETRY

GREAT LAKES FISH. COMM.    1977
BIOLOGICAL DATA FROM SEA LAMPREY COLLECTED BY COMMERCIAL
FISHERMEN. P.98.
IN  TIBBLES, J.J. ED., ANNUAL REPORT 1976 OF THE SEA LAMPREY
CONTROL CENTRE SAULT STE. MARIE, ONTARIO, TO THE GREAT
LAKES FISHERY COMMISSION. GREAT LAKES FISH. COMM., ANNU.
REP.
RESTRICTED
*PETROMYZON MARINUS GREAT LAKES ADULT MORPHOMETRY

GREAT LAKES FISH. COMM.    1977
BIOLOGICAL DATA ON SEA LAMPREY FROM THE HUMBER RIVER, LAKE
ONTARIO. PP.98-99.
IN  TIBBLES, J.J. ED., ANNUAL REPORT 1976 OF THE SEA LAMPREY
CONTROL CENTRE SAULT STE. MARIE, ONTARIO, TO THE GREAT
LAKES FISHERY COMMISSION. GREAT LAKES FISH. COMM., ANNU.
REP.
RESTRICTED
*PETROMYZON MARINUS CANADA GREAT LAKES ADULT MORPHOMETRY

GREAT LAKES FISH. COMM.    1977
RELEASE OF STERILE MALE SEA LAMPREY IN SABLE RIVER LAKE

SUPERIOR. P.100.
IN TIBBLES, J.J. ED., ANNUAL REPORT 1976 OF THE SEA LAMPREY
CONTROL CENTRE SAULT STE. MARIE, ONTARIO, TO THE GREAT
LAKES FISHERY COMMISSION. GREAT LAKES FISH. COMM., ANNU.
REP.
RESTRICTED
*PETROMYZON MARINUS CANADA U.S.A. GREAT LAKES ADULT
AMMOCOETE CHEMISTRY ENDOCRINOLOGY MANAGEMENT MIGRATION


5264F

**GREAT LAKES FISH. COMM.** 1977
IMPROVEMENT AND MODIFICATIONS TO SEA LAMPREY BARRIER DAMS.
PP.100-101.
IN TIBBLES, J.J. ED., ANNUAL REPORT 1976 OF THE SEA LAMPREY
CONTROL CENTRE SAULT STE. MARIE, ONTARIO, TO THE GREAT
LAKES FISHERY COMMISSION. GREAT LAKES FISH. COMM., ANNU.
REP.
RESTRICTED
*PETROMYZON MARINUS MANAGEMENT


5265F

**FLOOD, P.R.** 1977
KRYVI, H.
TOTLAND, G.K.
ONTO PHYLOGENETIC ASPECTS OF MUSCLE FIBER TYPES IN THE
SEGMENTAL TRUNK MUSCLE OF LOWER CHORDATES.
FOLIA MORPHOL., 25:64-67.
*LAMPETRA FLUVIATILIS *MYXINE GLUTINOSA BIOLOGY
HISTOCHEMISTRY CYTOLOGY


5266F

**SHIBATA, Y.** 1976
YAMAMOTO, T.
FINE STRUCTURE AND CYTOCHEMISTRY OF SPECIFIC GRANULES IN THE
LAMPREY ATRIUM.
CELL TISSUE RES., 172:487-502.
*ENTOSPHENUS JAPONICUS JAPAN CIRCULATORY SYSTEM CYTOLOGY
HISTOCHEMISTRY


5267F

**PLISETSKAYA, E.M.** 1974
BONDAREVA, V.M.
VLIYANIE RAZLICHNYKH INSULINOV NA POGLOSHCHENIE GLYUKOZY
MYSHTSAMI MINOGI LAMPETRA FLUVIATILIS, SKORPENY SCORPAENA
PORCUS I TSYPLYAT. [THE EFFECT OF VARIOUS INSULINS ON
GLUCOSE UPTAKE BY MUSCLES OF THE LAMPREY LAMPETRA
FLUVIATILIS, SCORPION-FISH SCORPAENA PORCUS, AND CNICKS.]
FIZIOL. BIOKHIM. NIZSHIKH POZVONO., :49-53.
(RUSS.)
ENG. SUMM.
ENDOCRINOLOGY PHYSIOLOGY


5268

**BANO, Y.** 1977
SEASONAL VARIATIONS IN THE BIOCHEMICAL COMPOSITION OF
CLARIAS-BATRACHUS.
INDIANA ACAD. SCI., PROC., 85:147-155.


5269F

**ROVAINEN, C.M.** 1977
NEURAL CONTROL OF VENTILATION IN THE LAMPREY.
FED. PROC., 36:2386-2389.
NERVOUS SYSTEM PHYSIOLOGY RESPIRATION


5270

**MORAVEC, F.** 1976
OCCURRENCE OF THE ENCYSTED LARVAE OF CUCULLANUS TRUTTAE IN
THE BROOK LAMPREY, LAMPETRA PLANERI.
SCR. FAC. SCI. NAT. UNIV. PURKYNIANAE BRUN BIOL., 6:17-20.


5271F

**FANGE. R.** 1976
DEN MYSIISKA FISKEN: ETT DJUR AV VETENSKAPLIGT INTRESSE.
[THE MYSTERIOUS FISH: AN ANIMAL OF SCIENTIFIC INTEREST.]
ZOOL. REVY., 38:113-118.
*MYXINE GLUTINOSA SWEDEN BIOLOGY


5272F

**ZIEGELS, J.** 1976
VERTEBRATE SUBCOMMISSURAL ORGAN. A STRUCTURAL AND FUNCTIONAL
REVIEW.
ARCH. BIOL., 87:429-476.


5273

**PAJOR, W.J.** 1976
AMMOCOETES ENDOSTYLE: ITS OXIDATIVE ENZYMES AS AN EVIDENCE
OF ITS HOMOLOGY WITH THE THYROID OF HIGHER CHORDATES.
FOLIA HISTOCHEM. CYTOCHEM., 14:283-308.


5274

**GOVARDOVSKII, V.I.** 1977

LYCHAKOV, D.V.
FOTORETSEPTORY I ZRITEL'NYE PIGMENTY CHERNOMORSKIKH
PLASTINOZHABERNYKH. [PHOTORECEPTORS AND VISUAL PIGMENTS IN
BLACK PIGMENTS IN BLACK SEA ELASMOBRANCHS.]
ZH. EVOL. BIOKHIM. FIZIOL., 13:162-166.
(RUSS.)
CYCLOSTOMES HISTOLOGY VISION

5275

STEINER, D.F.               1977
INSULIN TODAY.
DIABETES, 26:322-340.

5276

LEIBMAN, D.YA.              1977
ISOLATION AND SOME PROPERTIES OF GLYCERALDEHYDE 3 PHOSPHATE
DEHYDROGENASE FROM MUSCLES OF THE LAMPREY LAMPETRA
FLUVIATILIS.
ZH. EVOL. BIOKHIM. FIZIOL., 13:146-151.
(RUSS.)
J. EVOL. BIOCHEM. PHYSIOL., 13:124-129.
(ENG.)
BIOCHEMISTRY MUSCLE

5277F

HOLMBERG, K.               1977
THE HAGFISH MYXINE GLUTINOSA IN NEWSPAPERS AND IN SCIENCE.
FAUNA FLORA (STOCKH.), 72:15-17.
(SWED.)                    1977

5278

VERZHBINSKAYA, N.A.        1977
PERSHINA, L.I.
LEONT'EV, V.G.
[PROTEINS OF MITOCHONDRIAL MEMBRANES IN LOWER VERTEBRATES.]
ZH. EVOL. BIOKHIM. FIZIOL., 13:118-124.
(RUSS.)
J. EVOL. BIOCHEM. PHYSIOL., 13:99-105.
(ENG.)
*LAMPETRA FLUVIATILIS RUSSIA BIOCHEMISTRY BIOLOGY CYTOLOGY
MUSCLE

5279

KOSAREVA, A.A.            1977
VESELKIN, N.P.
ERMAKOVA, T.V.
RETINAL PROJECTIONS IN THE LAMPREY LAMPETRA FLUVIATILIS AS
REVEALED BY HORSERADISH PEROXIDASE ALONG THE OPTIC NERVE.
ZH. EVOL. BIOKHIM. FIZIOL., 13:405-407.
(RUSS.)
J. EVOL. BIOCHEM. PHYSIOL., 13:256-258.
(ENG.)
VISION HISTOLOGY

5280F

WICKELGREN, W.O.         1977
PHYSIOLOGICAL AND ANATOMICAL CHARACTERISTICS OF RETICULO
SPINAL NEURONS IN LAMPREY.
J. PHYSIOL. (LOND.), 270:89-114.
*PETROMYZON MARINUS U.S.A. GREAT LAKES ANATOMY NERVOUS
SYSTEM PHYSIOLOGY

5281F

OGORODNIKOVA, L.G.       1977
FOTINA, E.B.
GLUCOSE-6-PHOSPHATASE ACTIVITY IN THE PIAL MATTER OF SOME
VERTEBRATES.
ZH. EVOL. BIOKHIM. FIZIOL., 13:340-343.
*LAMPETRA FLUVIATILIS RUSSIA METABOLISM NERVOUS SYSTEM
PHYSIOLOGY

5282

DAWE, C.J. EDS.          1976
SCARPELLI, D.G.
WELLINGS, S.R.
PROGRESS IN EXPERIMENTAL TUMOUR RESEARCH. V. 20.
BASEL: S. KARGER.

5283F

WICKELGREN, W.O.         1977
POST-TETANIC POTENTIATION, HABITUATION AND FACILITATION OF
SYNAPTIC POTENTIALS IN RETICULOSPINAL NEURONS OF LAMPREY.
J. PHYSIOL. (LOND.), 270:115-132.
*PETROMYZON MARINUS U.S.A. GREAT LAKES ADULT NERVOUS SYSTEM
PHYSIOLOGY

5284

HOLBROOK, K.P.           1975
FECUNDITY AND SEX RATIO IN THE LAST BROOK LAMPREY LAMPETRA
AEPYPTERA IN LYNN CREEK WAYNE COUNTY WEST VIRGINIA U.S.A.
W. VA. ACAD. SCI., PROC., 47:150-153.

5285F

FALKMER, S.                    1976
EMDIN, S.O.
OSTBERG, Y.
MATTISON, A.
JOHANSSON-STOBECK, M-L.
FANGE, R.
TUMOR PATHOLOGY OF THE HAGFISH MYXINE GLUTINOSA AND THE
RIVER LAMPREY LAMPETRA FLUVIATILIS. A LIGHT MICROSCOPICAL
STUDY W... PP. 217-250.
IN  DAWE, C.J., D.G. SCARPELLI, AND S.R. WELLINGS, EDS.,
PROGRESS IN EXPERIMENTAL TUMOUR RESEARCH, VOL. 20. BASEL: S.
KARGER.
SWEDEN HISTOLOGY


5286

MUGGEO, M.                    1977
GINSBERG, B.H.
ROTH, J.
DE MEYTS, P.
FALKMER, S.
INSULIN RECEPTOR AND INSULIN OF THE ATLANTIC HAGFISH.
EXTRAORDINARY CONSERVATION OF BINDING SPECIFICITY AND
NEGATIVE COOPERATIVITY IN A VERY PRIMITIVE VERTEBRATE.
DIABETES, 26:353.


5287

YEAGER, B.E.                   1976
BEADLES, J.K.
FISHES OF THE CANE CREEK WATERSHED IN SOUTHEAST MISSOURI AND
NORTHEAST ARKANSAS U.S.A.
ARKANSAS ACAD. SCI., FAYETTEVILLE, PROC., 30:100-104.


5288F

FALKMER, S.                    1976
OSTBERG, Y.
PRODUCTION OF ISLET HORMONES IN INVERTEBRATES, CYCLOSTOMES,
AND PRIMITIVE GNATHOSTOMES, PP. 141-152.
IN  GRILLO, T.A.I., L. LEIBSON AND A.EPPLE, EDS., THE
EVOLUTION OF PANCREATIC ISLETS. OXFORD: PERGAMON PRESS.
*MYXINE GLUTINOSA *LAMPETRA FLUVIATILIS SWEDEN ENDOCRINOLOGY


5289F

EPPLE, A.                      1976
BRINN, J.E. JR.
NEW PERSPECTIVES IN COMPARATIVE ISLET RESEARCH. PP. 83-95.
IN  GRILLO, T.A.I., L. LEIBSON AND A. EPPLE, EDS., THE
EVOLUTION OF PANCREATIC ISLETS. OXFORD: PERGAMON PRESS.
*PETROMYZON MARINUS U.S.A. ANADROMOUS LANDLOCKED
ENDOCRINOLOGY


5290

GERLOVIN, E. SH.               1976
SOME PRINCIPLES OF CYTODIFFERENTATION OF PANCREATIC ISLETS
IN VERTEBRATA DURING ONTOGENY AND PHYLOGENY FROM THE
STANDP... PP. 97-112.
IN  GRILLO, T.A.I., L. LEIBSON, AND A. EPPLE, EDS., THE
EVOLUTION OF PANCREATIC ISLETS. OXFORD: PERGAMON PRESS.


5291

MULLER, J.                     1841
VERGLEICHENOE ANATOMIE DER MYXOIDEN. III. UBER DAS
GEFABSYSTEM.
AKAD. WISS., BERLIN, ABH., 1841:1-131.


5292

TRACK, N.S.                    1976
EVOLUTION OF INSULIN RELEASING MECHANISMS. PP. 311-320.
IN  GRILLO, T.A.I., L. LEIBSON AND A. EPPLE, EDS., THE
EVOLUTION OF PANCREATIC ISLETS. OXFORD: PERGAMON.


5293F

TOMONAGA, S.                   1977
YAMAGUCHI, K.
TAMECHIKA, M.
AWAYA, K.
SCANNING ELECTRON MICROSCOPE STUDY ON THE HIGH WALLED
ENDOTHELIUM OF THE HAGFISH GILLS.
J. ELECTRONMICROSC., 26:72.
(ABSTR.)
*EPTATRETUS BURGERI *PARAMYXINE ATAMI JAPAN HISTOLOGY
RESPIRATION


5294F

SHIBATA, Y.                    1977
SOME OBSERVATIONS OF THE FINE STRUCTURE OF THE CYCLOSTOME
LAMPREY HEART.
J. ELECTRONMICROSC., 26:72
HISTOLOGY CIRCULATORY SYSTEM


5295F

MARTIN, R.J.                    1977
BOWSHER, D.
ELECTROPHYSIOLOGICAL INVESTIGATION OF THE PROJECTION OF THE
INTRA MEDULLARY PRIMARY AFFERENT CELLS OF THE LAMPREY
AMMOCOETE.
NEUROSCI. LETT., 5:39-44.
ANIMAL ANATOMY HISTOCHEMISTRY


                                                    5296F
ALVESTAD-GRAEBNER, I.          1976
ADAM, H.
RELATIONSHIP BETWEEN THE CHROMATOID BODY AND THE ACROSOMAL
SYSTEM IN EARLY SPERMATIDS OF MYXINE GLUTINOSA.
CELL TISSUE RES., 174:427-430.
HISTOLOGY


                                                    5297
ANDRES, K.H.                   1975
NEW MORPHOLOGICAL FOUNDATIONS OF THE PHYSIOLOGY OF SMELL AND
TASTE.
ARCH. OTO-RHINO-LARYNGOL., 210:1-42.


                                                    5298F
MCCAULEY, R.W.                 1977
REYNOLDS, W.W.
HUGGINS, N.H.
PHOTOKINESIS AND BEHAVIOURAL THERMOREGULATION IN ADULT SEA
LAMPREYS (PETROMYZON MARINUS).
J. EXP. ZOOL., 202:431-437.
BEHAVIOUR LIGHT TEMPERATURE


                                                    5299F
HOLMBERG, K.                   1977
OHMAN, P.
DREYFERT, T.
ERG-RECORDING FROM THE RETINA THE RIVER LAMPREY (LAMPETRA
FLUVIATILIS).
VISION RES., 17:715-717.
NERVOUS SYSTEM PHYSIOLOGY VISION


                                                    5300F
LEWIS, S.V.                    1977
POTTER, I.C.
OXYGEN CONSUMPTION DURING THE METAMORPHOSIS OF THE PARASITIC
LAMPREY, LAMPETRA FLUVIATILIS (L.) AND ITS NON-PARASITIC
DERIVATIVE, LAMPETRA PLANERI (BLOCH).
J. EXP. BIOL., 69:187-198.
UNITED KINGDOM ADULT AMMOCOETE EVOLUTION METAMORPHOSIS
RESPIRATION


                                                    5301F
PERCY, R.                      1977
POTTER, I.C.
CHANGES IN HAEMOPOIETIC SITES DURING THE METAMORPHOSIS OF
THE LAMPREYS LAMPETRA FLUVIATILIS AND LAMPETRA PLANERI.
J. ZOOL., 183:111-123.
BLOOD LIFE CYCLE PHYSIOLOGY


                                                    5302F
MURTAUGH, P.A.                 1977
COMPARISON OF THE CROSS-LINKING PATTERNS OF LAMPREY
FIBRINOGREN AND FIBRIN BY THE ACTION OF THE INTRINSIC
LAMPREY FACTOR XIII AND HUMAN FACTOR XIII DURING THE PROCESS
OF BLOOD COAGULATION.
INT. SOC. THROMB. HAEMOST., J., 38:429-437.
FR. AND GER. SUMM.
BIOCHEMISTRY BLOOD EVOLUTION


                                                    5303F
WRIGHT, G.M.                   1977
YOUSON, J.H.
SERUM THYROXINE CONCENTRATIONS IN LARVAL AND METAMOPHOSING
ANADROMOUS SEA LAMPREY, PETROMYZON MARINUS L.
J. EXP. ZOOL., 202:27-32.
CANADA ATLANTIC OCEAN ADULT AMMOCOETE BLOOD ENDOCRINOLOGY
FEEDING METAMORPHOSIS


                                                    5304F
SJOBERG, K.                    1977
LOCOMOTOR ACTIVITY OF RIVER LAMPREY LAMPETRA FLUVIATILIS
(L.) DURING THE SPAWNING SEASON.
HYDROBIOLOGIA, 55:265-270.
SWEDEN ADULT ECOLOGY MIGRATION


                                                    5305
SEREBRENIKOVA, T.P.            1977
LYZLOVA, E.M.
SOME CHARACTERISTICS OF THE MOLECULAR EVOLUTION OF GYCOGEN
PHOSPHORYLASE AND AMINO TRANSFERASES IN MUSCLE TISSUE OF
VERTEBRATES.
ZH. EVOL. BIOCHEM. PHYSIOL., 13:106-113.

J. EVOL. BIOCHEM. PHYSIOL., 13:106-113.
(ENG.)
*LAMPETRA FLUVIATILIS BIOCHEMISTRY EVOLUTION MUSCLE

5306F

ZELNIK, P.R.                    1977
HORNSEY, D.J.
HARDISTY, MW.
INSULIN AND GLUCAGON-LIKE IMMUNOREACTIVITY IN THE RIVER
LAMPREY (LAMPETRA FLUVIATILIS).
GEN. COMP. ENDOCRINOL., 33:53-60.
UNITED KINGDOM ADULT ENDOCRINOLOGY FEEDING METABOLISM


5307F

HUNN, J.B.                      1974
ALLEN, J.L.
MOVEMENT OF DRUGS ACROSS THE GILLS OF FISHES.
LL.
ANNU. REV. PHARMACOL., 14:47-55.
PISCES ANIMAL PHYSIOLOGY BIOCHEMISTRY IONIC REGULATION


5308

GARCIA-CASTINEIRAS, S.         1977
WHITE, J.I.
TORO-GOYCO, E.
INHIBITION OF SODIUM AND POTASSIUM-DEPENDANT ATPASE BY
SEROTONIN.
MOL. PHARMACOL., 13:181-184.


5310F

EHINGER, B.                     1977
HOLMBERG, K.
OHMAN, P.
AMINERGIC AND INDOLEAMINE ACCUMULATING NEURONS IN THE RETINA
OF THE RIVER LAMPREY (LAMPETRA FLUVIATILIS).
ACTA ZOOL., (STOCKH.), 58:117-123.
SWEDEN ADULT HISTOCHEMISTRY HISTOLOGY VISION


5312

GOOSENS, N.                     1977
DIERICKX, K.
VANDESANDE, F.
IMMUNOCYTOCHEMICAL DEMONSTRATION OF THE
HYPOTHALAMO-HYPOPHYSIAL VASOTOCINERGIC SYSTEM OF LAMPETRA
FLUVIATILIS.
CELL TISSUE RES., 177:317-323.
BEGIUM ADULT CYTOLOGY ENDOCRINOLOGY HISTOCHEMISTRY
IMMUNOLOGY


5313

LEIBMAN, D.YA.                  1977
CONTENT AND PROPERTIES OF SULFHYDRYL GROUPS OF
GLYCERALDEHYDE 3 PHOSPHATE DEHYDROGENASE FROM MUSCLES OF THE
LAMPREY LAMPETRA FLUVIATILIS.
ZH. EVOL. BIOKHIM. FIZIOL., 13:503-505.
(RUSS.)
J. EVOL. BIOCHEM. PHYSIOL., 13:343-345.
(ENG.)
RUSSIA BIOCHEMISTRY METABOLISM


5314F

LIM, V.I.                       1977
EFIMOV, A.V.
FOLDING PATHWAY FOR GLOBINS.
FEBS LETT., 78:279-283.
BLOOD ENERGETICS


5315

JOSS, J.M.P.                    1973
THE PINEAL COMPLEX, MELATONIN, AND COLOR CHANGE IN THE
LAMPREY, LAMPETRA.
GEN. COMP. ENDOCRINOL., 21:188-195.


5316

YERGER, R.W.                    1977
FISHES OF THE APALACHICOLA RIVER.
FLA. MAR. RES. PUBL., 26:22-33.


5317

KASHAPOVA, L.A.                 1976
SAKHAROV, D.A.
DOUBLE INNERVATION OF FAST FIBERS OF THE TRUNK MUSCULATURE
OF LARVAL LAMPREY.
AKAD. NAUK SSSR, LENINGR., DOKL., SER. BIOL., 231:605-607.
(ENG.)
*LAMPETRA FLUVIATILIS RUSSIA AMMOCOETE HISTOLOGY MUSCLE
NERVOUS SYSTEM


5318

POLMORYKHINA, A.N.              1977
GROWTH AND DEVELOPMENT OF SIBERIAN LAMPREY LAMPETRA KESSLERI

IN RIVERS OF EASTERN KAZAKH-SSR U.S.S.R.
AKAD. NAUK KAZ. SSR, ALMA-ALTA, IZV., SER. BIOL., 15:40-47.

5319

ANDREASEN, J.K.          1975
THE CONSIDERATION OF ENDANGERED SPECIES IN LAND USE
DECISIONS.
OREG. ACAD. SCI., PROC., 11:47.


5320

SHIDOJI, Y.          1977
MUTO, Y.
VITAMIN A TRANSPORT IN PLASMA OF THE NONMAMMALIAN
VERTEBRATES. ISOLATION AND PARTIAL CHARACTERIZATION OF
PISCINE RETINOL BINDING PROTEIN.
J. LIPID RES., 18:679-691.
*ENTOSPHENUS JAPONICUS HOKKAIDO JAPAN BIOCHEMISTRY BLOOD
EVOLUTION PISCES ANIMAL


5321F

SVENDENIYA, O.          1949
PROMYSLOVYE RYBY SSSR OPISENIIA RYB. (TEKST K ATLASU
TSVETNYKH RISUNKOV RYB). [COMMERCIAL FISHES OF THE USSR.
DESCRIPTION. (TEXT TO THE ATLAS OF COLOUR ILLUSTRATIONS OF
THE FISHES.)] PP. 14-23.
MOSCOW: VSES. NAUCHNO-ISSLED. INST. MORSK. RYBN. (VNIRO)
(RUSS.)
*CASPIOMYZON WAGNERI (KESSLER) *PETROMYZON MARINUS *LAMPETRA
JAPONICA (MARTENS) *LAMPETRA JAPONICA (KESSLERI) LAMPETRA
FLUVIATIS *LAMPETRA FLUVIATILIS (LINNE) *LAMPETRA PLANERI
U.S.S.R. FISHERIES SYSTEMATICS


5322F

KHOR'KOV, A.D.          1971
MORFOLOGICHESKIE OSOBENNOSTI PINOTSITOZA RLADKOMYSHECHBYKH
KLETOK ARTERIAL'NYKH SOSYDOV KRUGLOROTYKH. [MORPHOLOGICAL
PECULIARITIES OF PINOCYTOSIS OF SMOOTH MUSCLE CELLS IN
ARTERIAL VESSELS OF CYCLOSTOMATA.]
NAUCHN. SOV. ELEKTRON. MIKROSK. S.S.S.R., KONFERENTSIYA,
1971I.
(RUSS.)
BIOCHEMISTRY MUSCLE


5323F

POLYAKOVA, T.I.          1977
PLISETSKAYA, E.M.
TSITOLOGICHESKIE I FUNKTSIONAL'NYE OOBENNOSTI OSTROVLOV
POLZHELYDOCHNOI ZHELEZY MINOGI V REZNYE PERIODY ZHIZNENNOGO
TSIKLA. [CYTOLOGICAL AND FUNCTIONAL PROPERTIES OF ISLET
TISSUE IN LAMPREYS (LAMPETRA FLUVIATILIS) ON DIFFERENT
STAGES OF THEIR LIFE CYCLE.]
TSITOLOGIYA, 19:1238-1244.
RUSSIA ADULT ENDOCRINOLOGY


5324F

WAINSTEIN, B.A.          1975
ONTOGENEZ I NEKOTORYE VOPROSY SISTEMATIKI AKARIFORMNYKH
KLESHCHEI (ACARIFORMES). [ONTOGENESIS AND SOME PROBLEMS OF
TAXONOMY OF ACARIFORMES.]
ZOOL. ZH., 54:526-531.
ANIMAL METAMORPHOSIS EVOLUTION EMBRYOLOGY


5325F

WEISBART, M.          1977
YOUSON, J.H.
IN VIVO FORMATION OF STEROIDS FROM
[1,2,6,7,-3H]-PROGESTERONE BY THE SEA LAMPREY, PETROMYZON
MARINUS L.
J. STEROID BIOCHEM., 8:1249-1252.
BLOOD BIOCHEMISTRY ENDOCRINOLOGY


5326F

INUI, Y.          1977
GORBMAN, A.
SENSITIVITY OF PACIFIC HAGFISH, EPTATRETUS STOUTI, TO
MAMMALIAN INSULIN.
GEN. COMP. ENDOCRINOL., 33:423-427.
CANADA BIOCHEMISTRY BLOOD ENDOCRINOLOGY


5327F

PYBUS, M.J          1978
ANDERSON, R.C.
UHAZY, L.S.
REDESCRIPTION OF TRUTTAEDACNITIS STELMIOIDES (VESSECHELLI,
1910) (NEMATODA: CUCULLANIDAE) FROM LAMPETRA LAMOTTENII
(LESUEUR, 1827).
HELMINTHOL. SOC. WASH., PROC., 45:238-245.


5328F

KERKOF, P.R.          1973
BOSCHWITZ, D.

GORBMAN, A.
RESPONSE OF HAGFISH THYROID TISSUE TO THYROID INHIBITORS AND
TO MAMMALIAN THYROID-STIMULATING HORMONE.
GEN. COMP. ENDOCRINOL., 21:231-240.
*EPTATRETUS STOUTII U.S.A. ENDOCRINOLOGY


5329

BROAD, D.S.                    1973
CONTRIBUTIONS TO CANADIAN PALEONTOLOGY: AMPHIASPIDIFORMES
(HETEROSTRACI) FROM THE SILURIAN OF THE CANADIAN ARCTIC
ARCHIPELAGO.
GEOL. SURV. CAN., BULL., 222:35-46.


5330

EMDIN, S.O.                    1973
PETERSON, J.D.
COULTER, C.L.
OSTBERG, Y.
FALKMER, S.
STEINER, D.L.
THE STRUCTURE AND BIOSYNTHESIS OF INSULIN IN A PRIMITIVE
VERTEBRATE, THE CYCLOSTOME MYXINE GLUTINOSA.
STOCKHOLM, 9TH INT. CONGR. BIOCHEM., ABSTR.


5331

SCHROLL, F.                    1959
ZUR ERNAHRUNGSBIOLOGIE DER STEIRISCHEN AMMOCOTEN LAMPETRA
PLANERI (BLOCH) UND EUDONTOMYZON DANFORDI (REGAN).
INT. REV. HYDROBIOL., 44:395-429.


5333F

ROVAINEN, C.M.                 1973
DUAL ELECTRICAL AND CHEMICAL TRANSMISSION AT LARGE SYNAPSES
IN LAMPREY SPINAL CORD.
J. GEN. PHYSIOL., 61:254.
(ABSTR.)
*PETROMYZON U.S.A. PHYSIOLOGY


5334

FLOOD, P.R.                    1962
STORM-MATHISEN, J.
A THIRD TYPE OF MUSCLE FIBRE IN THE PARIETAL MUSCLE OF THE
ATLANTIC HAGFISH MYXINE GLUTINOSA L.
Z. ZELLFORSCH., 58:638-640.


5335

KORNELIUSSEN, H.                1973
DENSE-CORE VESICLES IN MOTOR NERVE TERMINALS. MONOAMINERGIC
INNERVATION OF SLOW NON-TWITCH MUSCLES FIBRES IN ATLANTIC
HAGFISH (MYXINE GLUTINOSA, L.).
Z. ZELLFORSCH MIKROSK. ANAT., 140:425-432.


5336F

NICOLAYSEN, K.                 1966
ON THE FUNCTIONAL PROPERTIES OF THE FAST AND SLOW CRANIAL
MUSCLES OF THE ATLANTIC HAGFISH.
ACTA PHYSIOL. SCAND., 68:142.
*MYXINE NORWAY MUSCLE


5337F

CARTON, Y.                     1973
LA REPONSE IMMUNITAIRE CHEZ LES AGNATHES ET LES POISSONS.
STRUCTURE DES IMMUNOGLOBULINES.
[THE IMMUNE RESPONSE IN AGNATHS AND IN JAWED FISHES.
STRUCTURE OF THE IMMUNOGLOBINS.]
ANN. BIOL., 12:139-184.
*EPTATRETUS STOUTI *PETROMYZON MARINUS FRANCE BIOLOGY
EVOLUTION BIOCHEMISTRY


5338

TONDER, O.                     1978
LARSEN, B.
AARSKOG, D.
HANEBERG, B.
NATURAL AND IMMUNE ANTIBODIES TO RABBIT ERYTHROCYTE
ANTIGENS.
ERI.
SCAND. J. IMMUNOL., 7:245-250.


5339

PERCY, R.                      1973
LEATHERLAND, J.F.
FINE STRUCTURE OF THE PITUITARY GLAND IN LARVAL SEA
LAMPREYS(PETROMYZON MARINUS).
J. ENDOCRINOLOGY, 59.


5340F

FARMER, G.J.                   1977
BEAMISH, F.W.H.
LETT, P.F.
INFLUENCE OF WATER TEMPERATURE ON THE GROWTH RATE OF THE

LANDLOCKED SEA LAMPREY (PETROMYZON MARINUS) AND THE
ASSOCIATED RATE OF HOST MORTALITY.
FISH. RES. BOARD CAN., J., 34:1373-1378.
GREAT LAKES GROWTH MORTALITY BY

5341F

**MATTISON, A.G.M.**               1977
FANGE, R.
LIGHT- AND ELECTRONMICROSCOPIC OBSERVATIONS ON THE BLOOD
CELLS OF THE ATLANTIC HAGFISH, MYXINE GLUTINOSA (L.).
ACTA ZOOL., (STOCKH.), 58:205-221.
SWEDEN BLOOD HISTOLOGY

5342F

**MONACO, F.**               1977
ANDREOLI, M.
LA POSTA, A.
CATAUDELLA, S.
ROCHE, J.
SUR LA BIOSYNTHESE DE THYROGLOBULINE DANS L'ENDOSTYLE DES
LARVES (AMMOCOETES) D'UNE LAMPROIE D'EAU DOUCE, LAMPETRA
PLANERI BL. [ON THE BIOSYNTHESIS OF THYROGLOBULIN IN THE
LARVAE (AMMOCOETES) OF A FRESH WATER LAMPREY, LAMPETRA
PLANERI BLOCH.
SOC. BIOL., PARIS, C.R. HEBD. SEANCES, 171:308-313.
(FR.)
ENG. SUMM.
FRANCE ENDOCRINOLOGY METAMORPHOSIS

5343F

**ROOS, J.F.**               1973
GILHOUSEN, P.
KILLICK, S.R.
ZYBLUT, E.R.
PARASITISM ON JUVENILE PACIFIC SALMON (ONCORHYNCHUS) AND
PACIFIC HERRING (CLUPEA HARENGUS PALLASI) IN THE STRAIT OF
GEORGIA BY THE RIVER LAMPREY (LAMPETRA AYRESI).
FISH. RES. BOARD CAN., J., 30:565-568.
AMMOCOETE ENDOCRINOLOGY

5344F

**MATHERS, J.S.**               1974
INFLUENCE OF LIFE CYCLE AND SALINITY ON SERUM OSMOTIC AND
IONIC CONCENTRATIONS IN LANDLOCKED SEA LAMPREY, PETROMYZON
MARINUS.
UNIV. GUELPH, M.SC. THESIS: 44 PP.
RESTRICTED

5346F

**JANVIER, P.**               1977
CONTRIBUTION A LA CONNAISSANCE DE LA SYSTEMATIQUE ET DE
L'ANATOMIE DU GENRE BOREASPIS STENSIO (AGNATHA,
CEPHALASPIDOMORPHI, OSTEOSTRACI) DU DEVONIEN INFERIEUR DU
SPITSBERG.
ANN. PALEONTOL., VERTEBR., 63:1-32.
(FR.)
ADULT ANATOMY EVOLUTION MOUTH NERVOUS SYSTEM SYSTEMATICS

5347F

**PATZNER, R.A.**               1977
CYCLICAL CHANGES IN THE TESTIS OF THE HAGFISH, EPTATRETUS
BURGERI (CYCLOSTOMATA).
ACTA ZOOL., 58:223.
*LAMPETRA FLUVIATILIS *MYZINE GLUTINOSA *PARAMYXINE ATAMI
JAPAN GONADOGENESIS GROWTH HISTOLOGY REPRODUCTION

5348F

**MANION, P.J.**               1977
DENTITION THROUGHOUT THE LIFE HISTORY OF THE LANDLOCKED SEA
LAMPREY, PETROMYZON MARINUS.
COPEIA, 1977:762-766.
U.S.A. GREAT LAKES ADULT HISTOLOGY METAMORPHOSIS MOUTH

5349

**FEDOROV, B.A.**               1977
DENESYUK, A.I.
APPLICATION OF LARGE ANGLE X-RAY DIFFUSE SCATTERING TO
STUDIES OF GLOBULAR PROTEIN STRUCTURE IN SOLUTION.
ACTA CRYSTALLOGR., SECT. B, STRUCT. CRYSTALLOGR. CRYST.
CHEM., 33:3198-3204.
*LAMPETRA FLUVIATILIS *MYXINE GLUTINOSA *PARAMYXINE ATAMI JA

5350F

**OOI, E.C.**               1977
YOUSON, J.H.
MORPHOGENESIS AND GROWTH OF THE DEFINITIVE OPISTHONEPHROS
DURING METAMORPHOSIS OF ANADROMOUS SEA LAMPREY PETROMYZON
MARINUS
J. EMBRYOL. EXP. MORPHOL., 42:219-235.
CANADA ATLANTIC OCEAN ADULT EXCRETION HISTOCHEMISTRY

5351F

ALT, J.                              1976
RAGUSE-DEGENER, G.
NIERMANN, U.
WALVIG, F.
STOLTE, H.
RENAL EXCRETION OF PROTEIN AND NITROGEN END PRODUCTS IN THE
HAGFISH MYXINE GLUTINOSA.
MT. DESERT ISL. BIOL. LAB., BULL., 16:1-2.
GERMANY BLOOD IONIC REGULATION OSMOREGULATION PHYSIOLOGY


5352F

ALLEGRET, P.                         1977
DENIS, C.
LE LANNIC, J.
OBTENTION, AU LABORATOIRE, DE TAUX DE CROISSANCE ELEVES CHEZ
LES AMMOCOETES DE LAMPETRA PLANERI (BLOCH) PAR UNE NOUVELLE
METHODE D'ALIMENTATION. COMPARISON AVEC LES CROISSANCES
NATURELLES. [A NEW WAY OF LABORATORY BREEDING OF THE
AMMOCOETESOF LAMPETRA PLANERI (BLOCH). COMPARISON BETWEEN
BOTH THE EXPERIMENTAL AND NATURAL RATES OF GROWTH.]
ANN. HYDROBIOL., 17:255-262.
FRANCE AMMOCOETE CULTURE FEEDING


5353

BATUEVA, I.V.                        1977
SHAPOVALOV, A.I.
SYNAPTIC EFFECTS EVOKED IN MOTONEURONS BY DIRECT STIMULATION
OF SINGLE PRESYNAPTIC FIBERS.
NEIROFIZIOLOGIYA, 9:390-396.


5354F

GONCHAREVSKAYA, O.A.                  1977
NATOCHIN, YU.V.
TEMPERATURNAYA ZAVISIMOST' SKOROSTI REABSORBTSII V
PROKSIMAL'NOM KANAL'TSE POCHKI RECHNOI MINOGI. [THE
TEMPERATURE DEPENDENCE OF THE REABSORPTION RATE IN THE
PROXIMAL TUBULE OF THE RIVER LAMPREY KIDNEY.]
FIZIOL. ZH. SSSR IM. I.M. SECHENOVA, 63:1195-1198.
(RUSS.)
ENG. SUMM.
*LAMPETRA FLUVIATILIS RUSSIA EXCRETION


5355

NAKAO, T.                            1977
ELECTRON MICROSCOPIC STUDIES OF COATED MEMBRANES IN 2 TYPES
OF GILL EPITHELIAL CELLS OF LAMPREY.
CELL TISSUE RES., 178:385-396.


5356F

SHIN, Y.C.                           1977
SOME OBSERVATIONS ON THE FINE STRUCTURE OF LAMPREY LIVER AS
REVEALED BY ELECTRON MICROSCOPY.
OKAJIMAS FOL. ANAT. JAP., 54:25-60.
*LAMPETRA JAPONICA JAPAN AMMOCOETE ADULT HISTOLOGY


5357F

DAMAS, H.                            1950
PONTE EN AQUARIUM DES LAMPROIES FLUVIATILES ET DE PLANER.
[THE SPAWNING OF RIVER AND BROOK LAMPREYS IN AN AQUARIUM.]
SOC. R. ZOOL. BELG., ANN., 51:151-162.
*LAMPETRA FLUVIATILIS *LAMPETRA PLANERI BELGIUM ADULT
AMMOCOETE CULTURE


5358F

ZHUKOV, P.I.                         1965
RASPROSTRANENIE I EVOLYUTSIYA PRESNOVODNYKH MINOG V VODEMAKH
BSSR. [THE DISTRIBUTION AND EVOLUTION OF FRESHWATER
LAMPREYS IN WATERS OF THE BSSR.]
VOPR. IKHTIOL., 5: 7 PP.
*LAMPETRA PLANERI *LAMPETRA MARIAE U.S.S.R. UKRAINE BALTIC
SEA BLACK SEA ADULT AMMOCOETE DISTRIBUTION EVOLUTION
SYSTEMATICS


5360F

SAVINA, M.V.                         1976
PLISETSKAYA, E.M.
SINTEZ GLIKOGENA IZ GLITSERINA IZOLIROVANNYKH RKANYSKH
RECHNOI MINOGI LAMPETRA FLUVIATILIS. [SYNTHESIS OF GLYCOGEN
FROM GLYCEROL IN ISOLATED TISSUES OF THE LAMPREY LAMPETRA
FLUVIATILIS.]
ZH. EVOL. BIOKHIM. FIZIOL., 12:282-284.
(RUSS.)
ENG. SUMM.
RUSSIA BIOCHEMISTRY METABOLISM MUSCLE


5361F

PLISETSKAYA, E.M.                    1977
SOLTITSKAYA, L.
LEIBSON, L.G.
INSULIN V KROVI PROKHODMYKH MINOG I RYB V PERIOD NERESTOVOI

MIGRATSII. [INSULIN IN THE BLOOD OF DIADROMOUS LAMPREYS AND
FISHES ON VARIOUS STAGES OF SPAWNING MIGRATION.] PP.127-133.
IN PLISETSKAYA, E. AND L. LEIBSON, EDS., EVOLUTIONARY
ENDOCRINOLOGY OF PANCREAS. LENINGRAD: AKAD. NAUK SSSR.
(RUSS.)
ENG. SUMM.
*LAMPETRA FLUVIATILIS RUSSIA ADULT BLOOD ENDOCRINOLOGY
MIGRATION


5362
**NATOCHIN, YU.V.**            1977
FILTRATION REABSORPTION AND SECRETION IN EVOLUTION OF KIDNEY
FUNCTION.
ZH. EVOL. BIOKHIM. FIZIOL., 13:607-613.


5363
**BABURINA, YE. A.**          1972
DEVELOPMENT OF THE EYES OF CYCLOSTOMES AND FISHES IN
RELATION TO ECOLOGY.
MOSCOW: NAUKA PRESS.
EVOLUTION PISCES


5364F
**BEAMISH, R.J.**              1976
JORDAN, F.P.
SCARSBROOK, J.R.
PAGE, R.
INITIAL STUDY OF FISHES INHABITING THE SURFACE WATERS OF THE
STRAIT OF GEORGIA. M.V. CALIGUS JULY-AUGUST, 1974.
FISH. RES. BOARD CAN., MANUSCR. REP. SER., 1377: 37 PP.
*LAMPETRA AYRESI *LAMPETRA TRIDENTATUS PISCES PARASITISM BY


5365F
**HAVU, N.**                   1977
LUNDGREN, G.
FALKMER, S.
ZINC AND MANGANESE CONTENTS OF MICRO-DISSECTED PANCREATIC
ISETS OF SOME RODENTS. A MICROCHEMICAL STUDY IN ADULT AND
NEWBORN GUINEA PIGS, RATS, CHINESE HAMSTERS AND SPINY MICE.
ACTA ENDOCRINOL., 86:570-577.
MANAGEMENT


5366F
**HUNN, J.B.**                 1977
ANNUAL REPORT, CALENDAR YEAR 1976, OF THE HAMMOND BAY
BIOLOGICAL STATION. PP. 213-223.
IN GREAT LAKES FISH. COMM., ANNU. REP.
RESTRICTED
*PETROMYZON MARINUS U.S.A. GREAT LAKES ADULT AMMOCOETE
BEHAVIOUR CHEMISTRY ENDOCRINOLOGY INNUNOLOGY MANAGEMENT
MORTALITY BY SENSE RECEPTORS


5367F
**MEYER, F.P.**                1977
REGISTRATION-ORIENTED RESEARCH ON LAMPRICIDES IN 1976. PP.
189-199.
IN GREAT LAKES FISH. COMM., ANNU. REP., 1977.
CHEMISTRY MANAGEMENT


5368F
**BRAEM, R.A.**                1977
MOORE, H.H.
SEA LAMPREY CONTROL IN THE UNITED STATES. PP. 145-188.
IN GREAT LAKES FISH. COMM., ANNU. REP., 1977.
RESTRICTED
*PETROMYZON MARINUS U.S.A. GREAT LAKES ADULT AMMOCOETE
CHEMISTRY DISTRIBUTION MANAGEMENT MORTALITY OF


5369
**ANDREWS, S.M.**              1977
MILES, R.S.
WALKER, A.D.
LINNEAN SOCIETY SERIES NO. 4. PROBLEMS IN VERTEBRATE
EVOLUTION.
LINN. SOC. LOND., SYMP. SER., 4:411.


5370
**WEISBART, M.**               1974
STEIROIDS IN CYCLOSTOMES.
MEXICO: PROC. 4TH. INT. CONGR. HORMONAL STEROIDS.
(ABSTR.)


5371F
**HAVU, N.**                   1977
LUNDGREN, G.
FALKMER, S.
MICROCHEMICAL ASSAYS OF GLUTATHIONE ZINC COBALT AND
MANGANESE IN MICRODISSECTED AREAS OF THE ENDOCRINE PANCREAS
IN THE HAGFISH MYXINE GLUTINOSA.
ACTA ENDOCRINOL., 86:561-569.
SWEDEN ADULT HISTOCHEMISTRY ENDOCRINOLOGY

5372F

KASE, F.                          1976
STUCNY PREHLED KOAGULACNICH MECHANISMU U BEZOBRATLYCH,
OBRATLOVCU A CLOVEKA. [BRIEF REVIEW OF COAGULATION
MECHANISMS IN INVERTEBRATES, VERTEBRATES AND MAN.]
BIOL. LISTY, 41:172-185.
ANIMAL BLOOD BIOCHEMISTRY


5373

PIAVIS, G.W.                      1966
EFFECTS OF 6-METHYL MERCAPTOPURINE ON DEVELOPMENT IN SEA
LAMPREY, PETROMYZON MARINUS, EMBRYOS.
J. DENT. RES., SUPPL., 1966:75.
EMBRYOLOGY


5374F

PIAVIS, G.W.                      1965
CYTOKINESIS IN THE SEA LAMPREY, PETROMYZON MARINUS, EMBRYO.
J. DENT. RES., 1965:116.
EMBRYOLOGY


5375F

PIAVIS, G.W.                      1968
EFFECTS OF ANTIMETABOLITES ON DEVELOPMENT IN THE SEA
LAMPREY, PETROMYZON MARINUS, EMBRYOS.
MIDWEST REG. CONF. DEV. BIOL., MAY 1968.
EMBRYOLOGY CHEMISTRY


5376

BEETON, A.M.                      1974
LECH, J.J.
MICROBIAL DEGRADATION OF TFM.
UNIV. WIS.: 1P.


5377F

HUDSON, R.H.                      1971
TFM FIELD FORMULATION.
U.S. DEP. INT., FISH WILDL. SERV., DENVER WILDL. RES. CENT.,
INT. REP. SER. PHARMACOL.: 4 PP.
RESTRICTED
CHEMISTRY MANAGEMENT


5378F

HUDSON, R.H.                      1971
TFM.
U.S. DEP. INT., FISH WILDL. SERV., DENVER WILDL. RES. CENT.,
INT. REP. SER. PHARMACOL.: 4 PP.
RESTRICTED
CHEMISTRY MANAGEMENT


5379F

CHANDLER, J.H., JR.               1974
GROWTH AND REPRODUCTION OF FATHEAD MINNOWS EXPOSED TO TFM.
U.S. DEP. INT., FISH WILDL. SERV., SOUTHEAST. FISH CONTROL
LAB.:11 PP.
RESTRICTED
BIOLOGY CHEMISTRY MANAGEMENT


5380

BITTNER, M.A.                     1972
DEVELOPMENT OF METHODS FOR INVESTIGATIONS OF UPTAKE,
BIOTRANSFORMATION, AND ELIMINATION OF TFM
(3-TRIFLUOROMETHYL-4-NITROPHENOL) IN CHIRONOMID LARVAE
(CHIRONOMOUS TENTANS).
WIS., VITERBO COLL.:24 PP.


5381

BINKS, A.E.                       1972
KUB, M.
ORR, E.
STARKEY, R.
ANALYSIS OF TFM (3-TRIFLUOROMETHYL-4-NITROPHENOL) BY
LUMINESCENCE SPECTROPHOMETRY.
GEN. ELECTR., ENVIRON. SCI. LAB.:55 PP.


5382F

STATHAM, C.M.                     1976
MELANCON, M.J.
LECH, J.J.
BIOCONCENTRATION OF XENOBIOTICS IN TROUT BILE: A PROPOSED
MONITORING AID FOR SOME WATERBOURNE CHEMICALS.
SCIENCE, 193:680-681.
HISTOCHEMISTRY MANAGEMENT


5384

THINGVOLD, D.A.                   1975
ABSORPTION, DEGRADATION AND PERSISTENCE OF
3-TRIFLUOROMETHYL-4-NITROPHENOL (TFM) IN AQUATIC
ENVIRONMENTS.
UNIV. WIS., PH.D. THESIS:163 PP.

5386F

MAKI, A.W.                    1976
JOHNSON, H.E.
EVALUTION OF A TOXICANT ON THE METABOLISM OF MODEL STREAM
COMMUNITIES.
FISH. RES. BOARD CAN., J., 33:2740-2746.
CHEMISTRY MANAGEMENT ANIMAL


5387F

MAKI, A.W.                    1975
GEISSEL, L.D.
JOHNSON, H.E.
COMPARATIVE TOXICITY OF LARVAL LAMPRICIDE (TFM:
3-TRIFLUOROMETHYL-4-NITROPHENOL) TO SELETED
MACROINVERTEBRATES.
FISH. RES. BOARD CAN., J., 32:1455-1459.
CHEMISTRY MANAGEMENT ANIMAL


5388

MAKI, A.W.                    1974
EFFECTS AND FATE OF LAMPRICIDE (TFM:
3-TRIFLUOROMETHYL-4-NITROPHENOL) IN MODEL STREAM
COMMUNITIES.
MICHIGAN STATE UNIV., M.SC. THESIS:162 PP.


5389F

SAWYER, W.J.                  1977
EVOLUTION OF ACTIVE NEUROHYPOPHYSIAL PRINCIPLES AMONG THE
VERTEBRATES.
AM. ZOOL., 17:727-737.
*CYCLOSTOMATA BIOLOGY ENDOCRINOLOGY


5390F

LA POINTE, J.                 1977
COMPARATIVE PHYSIOLOGY OF NEUROHYPOPHYSIAL HORMONE ACTION ON
THE VERTEBRATE OVIDUCT-UTERUS.
AM. ZOOL., 17:763-773.
ANIMAL PHYSIOLOGY GONADOGENESIS


5391F

VLADYKOV, V.D.                1977
KOTT, E.
SATELLITE SPECIES AMONG THE HOLARCTIC LAMPREYS
PETROMYZONIDAE.
AM. ZOOL., 17:973.
*CASPIOMYZON *PETROMYZON ICHTHYOMYZON *TETRAPLEURODON
*ENTOSPENUS *LAMPETRA *EUDONTOMYZON *LETHENTERON SYSTEMATICS


5392

ELIAS, H.                     1977
FONG, B.B.
RECRUITMENT IN MAMMALIAN BILE DUCT FORMATION.
AM. ZOOL., 17:897.
(ABSTR.)
HISTOLOGY


5393

PERKS, A.M.                   1977
DEVELOPMENTAL AND EVOLUTIONARY ASPECTS OF THE
NEUROHYPOPHYSIS.
AM. ZOOL., 17:833-849.
*LAMPETRA RICHARDSONI AMMOCOETE ADULT ENDOCRINOLOGY
BIOCHEMISTRY HISTOCHEMISTRY


5395F

LARSSON, P.                   1976
RAASTAD, J.E.
NORWEGIAN ANIMAL NAMES WITH CORRESPONDING LATIN NAMES, PART
A: VERTEBRATES COMPLETE TO OCTOBER 1, 1976.
FAUNA (OSLO), 29:1-64.
(NORW.)
*PETROMYZON MARINUS *LAMPETRA FLUVIATILIS *LAMPETRA PLANERI
*LETHENTERON JAPONICUM NORWAY ANIMAL SYSTEMATICS


5396F

MARKING, L.L.                 1975
DAWSON, V.K.
METHOD FOR ASSESSMENT OF TOXICITY OF EFFICACY OF MIXTURES OF
CHEMICALS.
U.S. DEP. INT., FISH WILDL. SERV., INVEST. FISH CONTROL,
67:1-8.
PISCES CHEMISTRY MANAGEMENT


5397F

LECH, J.J.                    1972
COSTRINI, N.V.
IN VITRO AND IN VIVO METABOLISM OF
3-TRIFLUOROMETHYL-4-NITROPHENOL (TFM) IN RAINBOW TROUT.
COMP. GEN. PHARMACOL., 3:160-166.
*PETROMYZON MARINUS PISCES CHEMISTRY MANAGEMENT
PISCES CHEMISTRY MANAGEMENT TECHNIQUES

5399F

MOLNAR, B.                    1973
SZABO, S.
ASPECTELE CITOLOGICE ALE CELULELOR GONADOTROPE SI
NEUROSECRETIEI PREOPTICO-NEUROHIPOFIZARE IN DECURSUL
CICLULUI OVARIAN LA EUDONTOMYSON DANFORDI. [CYTOLOGICAL
ASPECTS OF THE GONADOTROPIC CELLS AND PREOPTIC NEUROHYPOSEAL
NEUROSECRETION DURING THE OVARIAN CYCLE IN EUDONTOMYZON
DANFORDI.]
CLUJ, ROM., UNIV. BABES-BOLYAI, STUD., SER. BIOL., 1:105-113
(ITAL.)
FR. AND RUSS. SUMM.


5400F

STEINER, D.F.                 1975
TERRIS, S.
EMDIN, S.O.
PETERSON, J.D.
FALKMER, S.
EVOLUTION AND COMPARATIVE BIOLOGY OF ISLET SECRETORY
PRODUCTS.
IN  STEINER, D.F. ET AL, EARLY DIABETES IN EARLY LIFE. NEW
YORK: ACADEMIC PRESS.
*MYXINE GLUTINOSA BIOCHEMISTRY


5401F

MAZIN, A.L.                   1975
EVOLUTION OF DNA STRUCTURE: DIRECTION, MECHANISM, RATE.
MOL. BIOL., 9:252-274.
(RUSS.)
ENG. TRANSL.
*LAMPETRA FLUVIATILIS GENETICS


5402

LECH, J.J.                    1973
PEPPLE, S.
STATHAM, C.N.
FISH BILE ANALYSIS. POSSIBLE AID IN MONITORING WATER
QUALITY.
TOXICOL. & APPL. PHARMACOL., 25:430-434.


5403F

DAWSON, V.K.                  1973
PHOTODECOMPOSITION OF THE PISCICIDES TFM
(3-TRIFLUOROMETHYL-4-NITROPHENOL) AND ANTIMYCIN.
UNIV. WIS., M.SC. THESIS.
CHEMISTRY MANAGEMENT ANIMAL ECOLOGY


5404F

KAWATSKI, J.A.                1974
MCDONALD, M.J.
EFFECT OF 3-TRIFLUOROMETHYL-4-NITROPHENOL ON IN VITRO TISSUE
RESPIRATION OF FOUR SPECIES OF FISH WITH PRELIMINARY NOTES
ON ITS IN VITRO BIOTRANSFORMATION.
COMP. GEN. PHARMACOL., 5:67-76.
MANAGEMENT


5405F

HANSEN, C.R.                  1976
KAWATSKI, J.A.
APPLICATION OF 24-HOUR POSTEXPOSURE OBSERVATION TO ACUTE
TOXICITY STUDIES WITH INVERTEBRATES.
FISH. RES. BOARD CAN., J., 33:1198-1201.
MANAGEMENT


5406

DWYER, W.P.                   1978
MAYER, F.L.
BUCKLER, D.R.
CHRONIC AND USE PATTERN EFFECTS OF
3-TRIFLUOROMETHYL-4-NITROPHENOL (TFM) TO BROOK TROUT
(LALVELINU FONTINALIS).
U.S. DEP. INT., FISH WILDL. SERV., INVEST. FISH CONTROL, IN
PROGRESS.


5407F

DAWSON, V.K.                  1976
MARKING, L.L.
BILLS, T.D.
REMOVAL OF TOXIC CHEMICALS FROM WATER WITH ACTIVATED CARBON.
AM. FISH. SOC., TRANS., 105:119-123.
CHEMISTRY MANAGEMENT


5408F

COBURN, J.A.                  1976
CHAU, A.S.Y.
GAS-LIQUID CHROMATOGRAPHIC DETERMINATION OF
3-TRIFLUOROMETHL-4-NITROPHENOL IN NATURAL WATERS.
ASSOC. OFF. ANAL. CHEM., J., 59:862-865.
CHEMISTRY MANAGEMENT TECHNIQUES

5409F

CHRISTIANSON, G.G.          1977
MICROBIAL REMOVAL OF 3-TRIFLUOROMETHYL-4-NITROPHENOL FROM
CULTURE MEDIA.
UNIV. WIS., M.SC. THESIS: 39 PP.


5410F

CAIRNS, J. JR.          1976
CALHOUN, W.F.
MCGINNISS, M.J.
STRAKA, W.
AQUATIC ORGANISMS RESPONSE TO SEVERE STRESS FOLLOWING
ACUTELY SUBLETHAL TOXICANT EXPOSURE.
WATER RESOUR. BULL., 12:1233-1243.
ANIMAL CHEMISTRY MANAGEMENT


5411F

BOTHWELL, M.L.          1973
BEETON, A.M.
LECH, J.J.
DEGRADATION OF THE LAMPRICIDE
(3-TRIFLUOROMETHYL-4-NITROPHENOL) BY BOTTOM SEDIMENTS.
FISH. RES. BOARD CAN., J., 30:1841-1846.
CHEMISTRY MANAGEMENT


5412F

LECH, J.J.          1973
PEPPLE, S.
ANDERSON, M.
EFFECTS OF NOVOBIOCIN ON THE ACUTE TOXICITY, METABOLISM AND
BILIARY EXCRETION OF 3-TRIFLUOROMETHYL-4-NITROPHENOL IN
RAINBOW TROUT.
TOXICOL. & APPL. PHARMACOL., 25:542-552.


5413F

LECH, J.J.          1972
MOYER, P.
TRITIUM LABELING OF 3-TRIFLUOROMETHYL-4-NITROPHENOL.
J. LABELLED COMPD., 8:499-504.
MANAGEMENT


5414F

LECH, J.J.          1973
PREPARATION AND PROPERITIES OF
3-TRIFLUOROMETHYL-4-AMINOPHENOL.
FISH. RES. BOARD CAN., J., 30:461-463.
MANAGEMENT


5415

LECH, J.J.          1972
ISOLATION AND IDENTIFICATION OF TFM GLUCURONIDE IN BILE OF
TFM EXPOSED RAINBOW TROUT.
FED. PROC., 31:606.


5416F

HEATH, R.G.          1972
SPANN, J.W.
HILL, E.F.
KREITZER, J.F
COMPARATIVE DIETARY TOXICITIES OF PESTICIDES TO BIRDS.
U.S. DEP. INT., FISH WILDL. SERV., SPEC. SCI. REP., 152:1-57
RESTRICTED
MANAGEMENT


5417

LIU, Y.H.          1975
LOWER DEVONIAN AGNATHANS OF YUNNAN AND SICHUAN.
VERT. PALASIAT., 13:202-216.


5418

ALVESTAD-GRAEBNER, I.          1977
ADAM, H.
ZUR FEINSTRUKTUR DER SPERMATOGENETISCHEN STADIEN VON MYXINE
GLUTINOSA L. (CYCLOSTOMATA).
ZOOL. SCR., 6:113-126.
(GER.)
ENG. SUMM.
*MYXINE GLUTINOSA REPRODUCTION


5419F

MAKI, A.W.          1977
JOHNSON, H.E.
KINETICS OF LAMPRICIDE (TFM,
3-TRIFLUOROMETHYL-4-NITROPHENOL). RESIDUES IN MODEL STREAM
COMMUNITIES.
FISH. RES. BOARD CAN., J., 34:276-281.
CHEMISTRY MANAGEMENT ANIMAL ECOLOGY


5421F

MAKI, A.W.          1977
GEISSEL, L.

JOHNSON, H.E.
INFLUENCE OF LARVAL LAMPRICIDE (TFM:
3-TRIFLUOROMETHYL-4-NITROPHENOL) ON GROWTH AND PRODUCTION OF
TWO SPECIES OF AQUATIC MACROPHYTES (ELODEA CANADENSIS
(MICHX) PLANCHON AND MYRIOPHYLLUM SPICATUM L.).
BULL. ENVIRON. CONTAM. TOXICOL., 17:57-65.
CHEMISTRY MANAGEMENT ECOLOGY

5422F

HUNN, J.B.                    1975
ALLEN, J.L.
RENAL EXCRETION IN COHO SALMON (ONCORHYNCHUS KISUTCH) AFTER
ACUTE EXPOSURE TO 3-TRIFLUOROMETHYL-4-NITROPHENOL.
FISH. RES. BOARD CAN., J., 32:1873-1876.
CHEMISTRY MANAGEMENT

5424

ALLEN, J.L.                   1977
HUNN, J.L.
RENAL EXCRETIONS IN CHANNEL CATFISH FOLLOWING INJECTION OF
QUINALDINE SULFATE OR 3-TRIFLUOROMETHYL-4-NITROPHENOL.
J. FISH BIOL., 10:473-480.
BIOCHEMISTRY CHEMISTRY MANAGEMENT

5425

GILDERHUS, P.A.               1978
ANTIMYCIN AS A CONTROL FOR SEA LAMPREY LARVAE IN LENTIC
HABITAT.
GREAT LAKES FISH. COMM., TECH. REP.

5426F

WELLBORN, T.L., JR.           1971
TOXICITY OF SOME COMPOUNDS TO STRIPED BASS FINGERLINGS.
PROG. FISH CULT., 33:32-36.
MANAGEMENT CHEMISTRY

5427

SCHULTZ, D.P.                  1975
HARMAN, P.D.
BAYER 73 - EXTRACTION AND CLEANUP.
U.S. DEP. INT., FISH WILDL. SERV., SOUTHEAST. FISH CONTROL
LAB.: 10 PP.

5428

RYE, R.P., JR.                1972
TOXICITY OF TWO LAMPREY LARVICIDES TO SELECTED
INVERTEBRATES.
U.S. DEP. INT., FISH WILDL. SERV., HAMMOND BAY BIOL. STN.: 4
PP.

5429

MARKING, L.L.                 1973
ATLANTIC SALMON: THEIR RESPONSES TO FISHERY CHEMICALS.
U.S. DEP. INT., FISH WILDL. SERV., FISH CONTROL LAB.: 5 PP.

5430F

HUDSON, R.H.                  1976
FINAL REPORT ON THE STUDY TO DETERMINE THE TOXICITY OF A
MIXTURE OF TFM AND CLONITRALID (BAYLUSCIDE, BAYER 73) TO
SEVERAL BIRD SPECIES.
U.S. DEP. INT., FISH WILDL. SERV., DENVER WILDL. RES.
CENT.:5 PP.
RESTRICTED
CHEMISTRY MANAGEMENT

5431

HARMAN, P.D.                  1976
SCHULTZ, D.P.
RAPID METHOD FOR DETERMINATION OF BAYER 73 IN NATURAL
WATERS.
U.S. DEP. INT., FISH WILDL. SERV., SOUTHEAST. FISH CONTROL
LAB: 10 PP.

5432F

GILDERHUS, P.A.               1973
STATUS OF POSSIBLE BOTTOM-RELEASE TOXICANTS FOR SEA LAMPREY
CONTROL.
U.S. DEP. INT., FISH WILDL. SERV., FISH CONTROL LAB.: 6 PP.
RESTRICTED
MANAGEMENT

5433

U.S. DEPT. INT.               1976
FIELD METHODS FOR DETERMINING CONCENTRATIONS OF BAYER 73 IN
WATER.
U.S. DEP. INT., FISH WILDL. SERV., FISH CONTROL LAB.: 5 PP.

5434

DAWSON, V.K.                  1971
PHOTOLYTIC SENSITIVITY TO BAYLUSCIDE (R).
U.S. DEP. INT., FISH WILDL. SERV., FISH CONTROL LAB.: 7 PP.

5435

CHOLETTE, C.                    1971
LA REGULATION ISOSMOTIQUE INTRACELLULAIRE CHEZ MYXINE
GLUTINOSA L.
UNIV. LAVAL, QUE., PH.D. THESIS.


5436

WINBLADH, L.
PANCREATIC ISLETS OF SOME MYXINOID CYCLOSTOMES.
IN  GRILLO, T.A.I., AND A. EPPLE, EDS., EVOLUTION OF
PANCREATIC ISLETS. OXFORD: PERGAMMON.


5437

GABE, M.
EXISTENCE, DANS L'INTESTINE MOYEN DES AGNATHES, D'ELEMENTS
COMPARABLES AUX CELLULE ENDOCRINES INTESTINALES DES
VERTEBRES GNATHOSTOMES.
ACAD. SCI. PARIS, C.R. HEBD. SEANCES, SER. D, 276:57-60.
(FR.)


5438

STATHAM, C.N.                   1974
PEPPLE, S.K.
LECH, J.J.
BILIARY EXCRETION PRODUCTS OF 1-NAPHTHL-N-METHYLCARBONATE
(CARBARYL) AND 2',5-DICHLORO-4'-NITROSALICYLANILIDE (BAYER
73) IN RAINBOW TROUT (SALMO GAIRDNERI).
PHARMACOLOGIST, 16:327.


5439

STATHAM, C.N.                   1976
LECH, J.J.
STUDIES ON THE MECHANISM OF POTENTIATION OF THE ACUTE
TOXICICITY OF 2,4-DN-BUTYL ESTER AND
2',5-DICHLORO-4'-NITROSALICYLANILIDE IN RAINBOW TROUT BY
CARBARYL.
TOXICOL. & APPL. PHARMACOL., 36:281-296.


5440

STATHAM, C.N.                   1975
LECH, J.J.
METABOLISM OF 2',5-DICHLORO-4'-NITROSALICYLANILIDE (BAYER
73) IN RAINBOW TROUT (SALMO GAIRDNERI).
FISH. RES. BOARD CAN., J., 32:515-522.


5441

SHIFF, C.J.                     1972
THE EFFECTS OF MOLLUSCICIDES ON THE MICROFLORA AND
MICROFAUNA OF AQUATIC SYSTEMS. PP. 109-115.
IN  FARVAR, M.T., AND J.P. MILTON, EDS., TECHNOLOGY,
ECOLOGY, AND INTERNATIONAL DEVELOPMENT. NEW YORK: DOUBLEDAY.


5442F

SCHULTZ, D.P.                   1977
HARMAN, P.D.
HYDROLYSIS AND PHOTOSIS OF THE LAMPRICIDE 2',
5-DICHLORO-4-NITROSALICYLANILIDE (BAYER 73)
U.S. DEP. INT., FISH WILDL. SERV., INVEST. FISH CONTROL,
85:5 PP.
CHEMISTRY MANAGEMENT


5444F

SANDERS, H.O.                   1977
TOXICITY OF THE MOLLUSCICIDE BAYER 73 AND RESIDUE DYNAMICS
OF BAYER 2353 IN AQUATIC INVERTEBRATES.
U.S. DEP. INT., FISH WILDL. SERV., INVEST. FISH CONTROL, 78:
7 PP.
MANAGEMENT


5445

SAALFELD, R.                    1974
HOWELL, J.H.
THE SEA LAMPREY AND ITS CONTROL. PP. 97-106.
IN  MICH. DEP. NAT. RESOUR., MICH. FISH. CENTEN. REP., 1873
TO 1973. FISH MANAGE. REP., #6.


5447

BARRINGTON, E.J.W.              1945
THE SUPPOSED PANCREATIC ORGANS OF PETROMYZON FLUVIATILIS AND
MYXINE GLUTINOSA.
EXPERIENTIA, 32:1537-1538.
CHEMISTRY MANAGEMENT


5448

PAFLITSCHEK, R.                 1977
INVESTIGATIONS ON THE TOXIC EFFECT OF BAYLUSCIDE AND
LABAYCID ON EGG-, JUVENILE AND ADULT STAGES OF THE
VARIEGATED PERCHES TILAPIA LEUCOSTICTA (TREWAVAS, 1933) AND
HEROTILAPIA MULTISPINOSA (GUNTNER, 1898).
UNIV. TUBINGEN, PH.D. THESIS: 70 PP.

5449

OLSON, L. E.                    1975
MARKING, L.L.
TOXICITY OF FOUR TOXICANTS TO GREEN EGGS OF SALMONIDS.
PROG. FISH CULT., 37:143-147.
CHEMISTRY MANAGEMENT


5450

LEATHERLAND, J.F.             1972
STRUCTURE AND FINE STRUCTURE OF THE PARS DISTALIS IN
CYCLOSTOME, HOLOSTEAN AND TELEOSTEAN REPRESENTATIVES.
GEN. COMP. ENDOCRINOL., 26:2-15.


5451

LEMMA, A.                      1975
AMES, B.N.
SCREENING FOR MUTAGENIC ACTIVITY OF SOME MOLLUSCICIDES.
R. SOC. TROP. MED. HYG., TRANS., 69:167-168.


5452

LECH, J.J.                     1978
STUDIES ON THE DEVELOPMENT OF A RADIOIMMUNE ASSAY FOR BAYER
73.
MED. COLL. WIS., IN PRESS.


5453F

KAWATSKI, J.A.                 1977
ZITTEL, A.E.
ACCUMULATION, ELIMINATION, AND BIOTRANSFORMATION OF THE
LAMPRICIDE 2',5-DICHLORD-4'-NITROSALICYLANILIDE BY
CHIRONOMUS TENTANS.
U.S. DEP. INT., FISH WILDL. SERV., INVEST. FISH CONTROL, 79:
8 PP.
MANAGEMENT


5454F

KAWATSKI, J.A.                 1973
ACUTE TOXICITIES OF ANTIMYCIN A, BAYER 73, AND TFM TO THE
OSTRACOD CYPRETTA KAWATAI.
AM. FISH. SOC., TRANS., 102:829-831.


5455

HUNN, J.B.                     1978
EXCRETION OF LAMPRICIDES BY FISH
ASTM, PROC., IN PROGRESS.


5456

HUDSON, R.H.                   1978
TOXICITIES OF THE LAMPRICIDES
3-TRIFLUOROMETHYL-4-NITROPHENOL (TFM) AND THE 2-AMINOETHANOL
SALT OF 2',5-DICHLORO-4'-NITROSALICYLANILIDE (BAYER 73) TO
4 BIRD SPECIES.
U.S. DEP. INT., FISH WILDL. SERV., INVEST. FISH CONTROL, IN
PROGRESS.


5457

HANSEN, C.R.                   1978
KAWATSKI, J.A.
RESIDUE DISTRIBUTION OF 2',5-DICHLORO-4'NITROSALICYLANILIDE
(BAYER 73) AND 3-TRIFLUOROMETHYL-4-NITROPHENOL (TFM) IN
SUBLETHALLY ESPOSED LARVAE OF THE AQUATIC MIDGE CHIRONOMUS
TENTANS.
TOXICOLOGY.


5458

HAMILTON, S.E.                 1974
REVIEW OF THE LITERATURE ON THE USE OF BAYLUSCIDE IN
FISHERIES.
U.S. DEP. INT., FISH WILDL. SERV., LIT. REV., 74-02: 6 PP.


5459

GILDERHUS, P.A.                1978
EFFECTS OF GRANULAR 2',5-DICHLORO-4'NITROSALICYLANILIDE
(BAYER 73) ON BENTHIC MACROINVERTEBRATES IN A LAKE
ENVIRONMENT.
GREAT LAKES FISH. COMM., TECH. REP.


5460

FARRINGER, J.E.               1972
THE DETERMINATION OF THE ACUTE TOXICITY OF ROTENONE AND
BAYER 73 TO SELECTED AQUATIC ORGANISMS.
UNIV. WIS., M.SC. THESIS: 32 PP.


5461

GRADWELL, N.                   1972
HYDROSTATIC PRESSURES AND MOVEMENTS OF THE LAMPREY,
PETROMYZON, DURING SUCTION, OLFACTION, AND GILL VENTILATION.
CAN. J. ZOOL. 50:1215-1223.


5462F

DAWSON, V.K.                   1978

HARMAN, P.D.
SCHULTZ, D.P.
ALLEN, J.L.
RAPID METHOD FOR DETERMINATION OF BAYER 73 IN WATER DURING
LAMPRICIDE TREATMENTS.
FISH. RES. BOARD CAN., J., 35:1262-1265.
CHEMISTRY MANAGEMENT TECHNIQUES


                                                              5463F

DAWSON, V.K.                    1977
CUMMING, K.B.
GILDERHUS, P.A.
EFFICACY OF 3-TRIFLUOROMETHYL-4-NITROPHENOL (TFM), 2',
5-DICHLORO-4'-NITROSALICYLANIDE (BAYER 73) AND A 98:2
MIXTURE AS LAMPRICIDES IN LABORATORY STUDIES.
U.S. DEP. INT., FISH WILDL. SERV., INVEST. FISH CONTROL,
77:11 PP.
MANAGEMENT


                                                              5464

LEMONS, D.E.                    1978
CRAWSHAW, L.I.
TEMPERATURE REGULATION IN THE PACIFIC LAMPREY, LAMPETRA
TRIDENTATA.
FED. PROC., 37:929
(ABSTR.)
BEHAVIOUR CIRCULATORY SYSTEM LOCOMOTION NERVOUS SYSTEM
RESPIRATION


                                                              5465F

SPITZER, R.H.                   1976
DOWNING, S.W.
KOCH, E.A.
KAPLAN, M.A.
HEMAGGLUTININS IN THE MUCUS OF PACIFIC HAGFISH EPTATRETUS
STOUTII.
COMP. BIOCHEM. PHYSIOL., 54B:409-411.
BLOOD BIOCHEMISTRY IMMUNOLOGY


                                                              5466F

HOMMA, S.                       1978
ORGANIZATION OF THE TRIGEMINAL MOTORNUCLEUS BEFORE AND AFTER
METAMORPHOSIS IN LAMPREYS.
BRAIN RES., 140:33-42.
*ICHTHYOMYZON UNICUSPIS *LAMPETRA AEPYPTERA *LAMPETRA
LAMOTTEI AMMOCOETE ADULT FEEDING PHYSIOLOGY


                                                              5467F

MORAVEC, F.                     1977
MALMQVIST, B.
RECORDS OF CUCULLANUS TRUTTAE (FABRICUS, 1794) (NEMATODA
CUCULLANICAE) FROM SWEDISH BROOK LAMPREYS LAMPETRA PLANERI
(BLOCH).
FOLIA PARASITOL., (PRAGUE), 24:323-329
SWEDEN AMMOCOETE PARASITISM OF ADULT


                                                              5468F

MURAT, J.C.                     1977
PLISETSKAYA, E. M.
EFFETS DU GLUCAGON SUR LA GLYCEMIE, LE GLYCOGENE ET LA
GLYCOGENE-SYNTHETASE HEPATIQUE CHEZ LA CARPE ET LA LAMPROIE.
SOC. BIOL., PARIS, C.R. HEBD. SEANCES, 171:1302-1305.
(FR.)
ENG. SUMM.
*LAMPETRA FLUVIATILIS RUSSIA ADULT ENDOCRINOLOGY


                                                              5469F

NORW. ZOOL. SOC.                1976
NORSKE DYRENAVN MED TILHORENDE LATINSKE NAVN. A. VIRVELDYR.
[NORWEGIAN ANIMALS NAMES WITH CORRESPONDING LATIN NAMES. A.
VERTEBRATES.]
FAUNA, 29:1-64.
(NORW.)
*PETROMYZON MARINUS *LAMPETRA FLUVIATILIS *LAMPETRA PLANERI
*LETHENTERON JAPONICUM NORWAY ANIMAL SYSTEMATICS


                                                              5470F

WEISBART, M.                    1978
YOUSON, J.H.
WIEBE, J.P.
BIOCHEMICAL, HISTOCHEMICAL, AND ULTRASTRUCTURAL ANALYSES OF
PRESUMED STEROID-PRODUCING TISSUES IN THE SEXUALLY MATURE
SEA LAMPREY, PETROMYZON MARINUS L.
GEN. COMP. ENDOCRINOL., 34:26-37.
*PETROMYZON MARINUS CANADA GREAT LAKES ENDOCRINOLOGY
GONADOGENESIS HISTOCHEMISTRY


                                                              5472F

HUNN, J.B.                      1973
REGISTRATION-ORIENTED RESEARCH ON TFM IN 1972. P. 310.
IN  GREAT LAKES FISH. COMM., ANNU. REP.

5473F

PURVIS, H.A.                    1973
SPECIAL REPORT ON THE STUDIES OF SEA LAMPREY EARLY LIFE
HISTORY, 1960-72. PP. 223-271.
IN  GREAT LAKES FISH. COMM., ANNU. REP., 1973.
RESTRICTED

5477F

WESTMAN, R.W.                   1973
SEA LAMPREY SURVEYS OF THE GLOUCESTER POOL DRAINAGE OF THE
SEVERN RIVER SYSTEM, 1972. PP. 101-103.
IN  TIBBLES, J.J., ED., ANNUAL REPORT 1972 OF THE SEA
LAMPREY CONTROL CENTRE SAULT STE. MARIE, ONTARIO, TO THE
GREAT LAKES FISHERY COMMISSION. GREAT LAKES FISH. COMM.,
ANNU. REP.
RESTRICTED
*PETROMYZON MARINUS CANADA AMMOCOETE

5481

JOHNSON, B.G.H.                 1973
CARTER, A.W.
SEA LAMPREY ASSESSMENT BARRIER OPERATIONS: 1972. PP. 95-96.
IN  TIBBLES, J.J., ED., ANNUAL REPORT 1972 OF THE SEA
LAMPREY CONTROL CENTRE SAULT STE. MARIE, ONTARIO, TO THE
GREAT LAKES FISHERY COMMISSION. GREAT LAKES FISH. COMM.,
ANNU. REP.
RESTRICTED
*PETROMYZON MARINUS CANADA ADULT MORTALITY OF MANAGEMENT

5485F

JOHNSON, B.G.H.                 1973
SEA LAMPREY TAKEN AT DENNY'S DAM ON THE SAUGEEN RIVER. PP.
171-172.
IN  TIBBLES, J.J., ED., ANNUAL REPORT 1972 OF THE SEA
LAMPREY CONTROL CENTRE SAULT STE. MARIE, ONTARIO, TO THE
GREAT LAKES FISHERY COMMISSION. GREAT LAKES FISH. COMM.,
ANNU. REP.
RESTRICTED
*PETROMYZON MARINUS CANADA ADULT

5486F

JOHNSON, B.G.H.                 1973
BIOLOGICAL DATA ON ADULT SEA LAMPREY COLLECTED IN THE HUMBER
RIVER, 1972. P. 171.
IN  TIBBLES, J.J., ED., ANNUAL REPORT 1972 OF THE SEA
LAMPREY CONTROL CENTRE SAULT STE. MARIE, ONTARIO, TO THE
GREAT LAKES FISHERY COMMISSION. GREAT LAKES FISH. COMM.,
ANNU. REP.
RESTRICTED
*PETROMYZON MARINUS CANADA ADULT MANAGEMENT

5487

JOHNSON, B.G.H.                 1973
BIOLOGICAL DATA ON SEA LAMPREY COLLECTED FROM CANADIAN
ELECTRICAL ASSESSMENT BARRIERS, LAKE HURON, 1972. PP.
160-170.
IN  TIBBLES, J.J., ED., ANNUAL REPORT 1972 OF THE SEA
LAMPREY CONTROL CENTRE SAULT STE. MARIE, ONTARIO, TO THE
GREAT LAKES FISHERY COMMISSION. GREAT LAKES FISH. COMM.,
ANNU. REP.
RESTRICTED
*PETROMYZON MARINUS CANADA MORPHOMETRY

5488F

JOHNSON, B.G.H.                 1973
SURFACE TRAWLING FOR ADULT SEA LAMPREY IN ST. MARYS RIVER,
1972. P. 166.
IN  TIBBLES, J.J., ED., ANNUAL REPORT 1972 OF THE SEA
LAMPREY CONTROL CENTRE SAULT STE. MARIE, ONTARIO, TO THE
GREAT LAKES FISHERY COMMISSION. GREAT LAKES FISH. COMM.,
ANNU. REP.
RESTRICTED
*PETROMYZON MARINUS CANADA ADULT

5491F

DAVIS, W.A.                     1973
GRANULAR BAYER 73 TREATMENTS, LAKE HURON, 1972. PP. 154-161.
IN  TIBBLES, J.J., ED., ANNUAL REPORT 1972 OF THE SEA
LAMPREY CONTROL CENTRE SAULT STE. MARIE, ONTARIO, TO THE
GREAT LAKES FISHERY COMMISSION. GREAT LAKES FISH. COMM.,
ANNU. REP.
RESTRICTED
*PETROMYZON MARINUS *ICHTHYOMYZON SP. *LAMPETRA LAMOTTI
CANADA AMMOCOETE CHEMISTRY GROWTH MANAGEMENT

5494F

BOUCHARD, R.P.
MODIFICATION OF THE LISTENING DEVICE OF THE SMALL, TYPE-A,

PRICE CURRENT METER. PP. 174-175.
IN  TIBBLES, J.J., ED., ANNUAL REPORT 1972 OF THE SEA
LAMPREY CONTROL CENTRE SAULT STE. MARIE, ONTARIO, TO THE
GREAT LAKES FISHERY COMMISSION. GREAT LAKES FISH. COMM.,
ANNU. REP.
RESTRICTED

5497F

JOHNSON, B.G.H.              1973
CARTER  A.W.
TRANSFER OF STORAGE AREA FROM ROOT RIVER TO ST. MARYS
ISLAND. P. 174.
IN  TIBBLES, J.J., ED., ANNUAL REPORT 1972 OF THE SEA
LAMPREY CONTROL CENTRE SAULT STE. MARIE, ONTARIO, TO THE
GREAT LAKES FISHERY COMMISSION. GREAT LAKES FISH. COMM.,
ANNU. REP.
RESTRICTED

5498F

HILL, E.F.                   1975
HEATH, R.G.
SPANN, J.W.
WILLIAMS, J.D.
LETHAL DIETARY TOXICITIES OF ENVIRONMENTAL POLLUTANTS TO
BIRDS.
U.S. DEP. INT., FISH WILDL. SERV., SPEC. SCI. REP., 191:1-61
CHEMISTRY MANAGEMENT

5499F

GREAT LAKES FISH. COMM.      1976
REPORT OF THE ANNUAL MEETING.
GREAT LAKES FISH. COMM., ANNU. REP., 361 PP.

5500F

MEYER, F.P.                  1976
REGISTRATION-ORIENTED RESEARCH ON LAMPRICIDES IN 1975. PP.
189-195.
IN  GREAT LAKES FISH. COMM., ANNU. REP., 1976.
RESTRICTED
CHEMISTRY MANAGEMENT

5501F

EDSALL, T.A.                 1976
EARCH. CHEMICAL SENSING IN SEA LAMPREYS. PP. 211-216.
IN  GREAT LAKES FISH. COMM., ANNU. REP., 1976.
RESTRICTED
*PETROMYZON MARINUS U.S.A. GREAT LAKES MANAGEMENT

5502F

SMITH, B.R.                  1976
BRAEM, R.A.
LAMPREY CONTROL IN THE UNITED STATES. PP. 153-181.
IN  GREAT LAKES FISH. COMM., ANNU. REP., 1976.
RESTRICTED
*PETROMYZON MARINUS U.S.A. GREAT LAKES ADULT AMMOCOETE
CHEMISTRY DISTRIBUTION MANAGEMENT MORTALITY OF

5503

JENSEN, D.                   1965
THE ANEURAL HEART OF THE HAGFISH.
N.Y. ACAD. SCI., ANN., 127:443-458.
CIRCULATORY SYSTEM PHYSIOLOGY

5504

GOVYRIN, V.A.                1977
DEVELOPMENT OF VASOMOTOR ADRENERGIC INNERVATION IN
ONTOGENESIS AND PHYLOGENESIS.
ZH. EVOL. BIOKHIM. FIZIOL., 13:614-620.
(RUSS.)
J. EVOL. BIOCHEM. PHYSIOL., 13:429-433.
(ENG.)
*LAMPETRA FLUVIATILIS RUSSIA BIOLOGY CIRCULATORY SYSTEM
HISTOLOGY NERVOUS SYSTEM

5505

SHAPOVALOV, A.I.             1977
INTERNEURONAL SYNAPSES WITH ELECTRICAL AND CHEMICAL MODES OF
TRANSMISSION AND EVOLUTION OF THE CENTRAL NERVOUS SYSTEM.
ZH. EVOL. BIOKHIM. FIZIOL., 13:621-632.
(RUSS.)
J. EVOL. BIOCHEM. PHYSIOL., 13:434-443.
(ENG.)
*LAMPETRA FLUVIATILIS RUSSIA BIOLOGY HISTOLOGY NERVOUS
SYSTEM PHYSIOLOGY

5506F

WILKENS, H.                  1977
KOEHLER, A.
DIE FISCHFAUNA DER UNTEREN UND MITTLEREN ELBE: DIE GENUTZTEN
ARTEN, 1950-1975. [THE FISH FAUNA OF THE LOWER AND MIDDLE
ELBE RIVER, WEST GERMANY. THE SPECIES USED 1950-1975.]

NATURWISS. VER., HAMBURG, ABH. VERH., 20:185-222.
(GERM.)
*PETROMYZON MARINUS *LAMPETRA FLUVIATILIS ADULT BIOLOGY
DISTRIBUTION MIGRATION


5507

**BEAMISH, F.W.H.**                    1979
MIGRATION AND SPAWNING ENERGETICS OF THE ANADROMOUS SEA
LAMPREY, PETROMYZON MARINUS.
ENV. BIOL. FISH., 4.


5507 **

**BEAMISH, F.W.H.**                    1979
MIGRATION AND SPAWNING ENERGETICS OF THE ANADROMOUS SEA
LAMPREY, PETROMYZON MARINUS.
ENV. BIOL. FISH., 4.
MIKROPUNKTSIONNOE IZUCHENIE REABSORBTSII KHLORIDA I
BIKARBONATA NATRIYA V PROKSIMAL'NOM KANAL'TSE POCHKI MINOGI
LAMPETRAFLUVIATILIS. [MICROPUNCTURE STUDY ON SODIUM CHLORIDE
AND SODIUM BICARBONATE REABSORPTION IN PROXIMAL TUBULES OF
THE KIDNIN THE LAMPREY.]
(RUSS.)
ENG. SUMM.
RUSSIA EXCRETION PHYSIOLOGY


5509

**LARSEN, L.O.**                    1978
SUBTOTAL HEPATECTOMY IN INTACT OR HYPOSECTOMIZED RIVER
LAMPREYS (LAMPETRA FLUVIATILIS L.): EFFECTS ON REGENERATION,
BLOOD GLUCOSE REGULATION, AND VITELLOGENESIS.
GEN. COMP. ENDOCRINOL., 35:197-204.
DENMARK ADULT ENDOCRINOLOGY METABOLISM SPAWNING REPRODUCTION


5510

**WHITING, H.P.**                    1977
CRANIAL NERVES IN LAMPREYS AND CEPHALASPIDS.
IN  ANDREWS, S., R. MAHALA, S. MILES AND A.D. WALKER, EDS.,
LINNEAN SOCIETY SYMPOSIUM SERIES, NO. 4. PROBLEMS IN
VERTEBRATE EVOLUTION. LONDON: ACADEMIC PRESS.


5511

**BOIVINET, P.**                    1977
CALORIMETRIC INVESTIGATIONS ON ANIMAL SUBORGANISMS,
ORGANITES, TISSUES AND ISOLATED ORGANS.
IN  LAMPRECHT, I. AND B. SCHAARSCHMEDT, EDS., APPLICATION OF
CALORIMETRY IN LIFE SCIENCES. BERLIN: WALTER DE GRUYTER.


5512F

**KREPS, E.M.**                    1977
AVROVA, N.F.
ZABELINSKII, S.A.
KRUGLOVA, E.E.
BIOCHEMICAL EVOLUTION OF THE VERTEBRATE BRAIN.
ZH. EVOL. BIOKHIM. FIZIOL., 13:556-569
BIOCHEMISTRY


5513F

**EHRLICH, B.E.**                    1978
CSERR, H.F.
COMPARATIVE ASPECTS OF BRAIN BARRIER SYSTEMS FOR IODIDE.
AM. J. PHYSIOL., 234:R61-R65.
PHYSIOLOGY ANIMAL NERVOUS SYSTEM


5514F

**SHELTON, R.G.J.**                    1978
ON THE FEEDING OF THE HAGFISH MYXINE GLUTINOSA IN THE NORTH
SEA.
MAR. BIOL. ASSOC. U.K., J., 58:81-86
SCOTLAND FEEDING


5515

**CSERR, H.F.**                    1978
FENSTERMACHER, J.D.
RALL, D.P.
COMPARATIVE ASPECTS OF BRAIN BARRIER SYSTEMS FOR
NONELECTROLTES.
AM. J. PHYSIOL., 234:R51-R60.
BLOOD ENDOCRINOLOGY NERVOUS SYSTEM PHYSIOLOGY


5516F

**HANSEN, S.J.**                    1978
YOUSEN, J.H.
CELL RENEWAL IN THE EPITHELIUM OF THE ALIMENTARY TRACT OF
THE LARVAL LAMPREY PETROMYZON MARINUS.
J. MORPHOL., 155:219-236.
CANADA NEW BRUNSWICK ANADROMOUS HISTOLOGY CYTOLOGY DIGESTION


5517

**VAN NOORDEN, S.**                    1977
OSTBERG, Y.
PEARSE, A.G.E.

LOCALIZATION OF SOMATOSTATIN-LIKE IMMUNOREACTIVITY IN THE
PANCREATIC ISLETS OF THE HAGFISH MYXINE GLUTINOSA AND THE
LAMPREY LAMPETRA FLUVIATILIS.
CELL TISSUE RES., 177:281-286.
ENDOCRINOLOGY HISTOLOGY PHYSIOLOGY

                                                      5518F

HANSEN, S.J.                  1978
YOUSEN, J.H.
MORPHOLOGY OF THE EPITHELIUM IN THE ALIMENTARY TRACT OF THE
LARVAL LAMPREY PETROMYZON MARINUS.
J. MORPHOL., 155:193-218.
CANADA NEW BRUNSWICK AMMOCOETE ANADROMOUS CYTOLOGY DIGESTION
HISTOLOGY

                                                      5519

NAKAO, T.                     1978
UCHINOMIYA, K.
A STUDY ON THE BLOOD VASCULAR SYSTEM OF THE LAMPREY GILL
FILAMENT.
AM. J. ANAT., 151:239-264.
*LAMPETRA JAPONICA CIRCULATORY SYSTEM ADULT

                                                      5520F

PATZNER, R.A.                 1978
CYCLICAL CHANGES IN THE OVARY OF THE HAGFISH, EPTATRETUS
BURGERI (CYCLOSTOMATA).
ACTA ZOOL., (STOCKH.), 59:57-62.

                                                      5521

BATUEVA, I.V.                 1977
SHAPOVALOV, A.I.
ELECTRICAL AND CHEMICAL EXCITATORY POSTSYNAPTIC POTENTIALS
EVOKED IN LAMPREY MOTONEURONS BY STIMULATION OF DESCENDING
TRACT AND DORSAL ROOT AFFERENTS.
NEIROFIZIOLOGIYA, 9:512-517.

                                                      5522

ELFIMOVA, L.I.                1977
LEIBMAN, D.YA.
THE EFFECT OF COENZYME ON CONFORMATIONAL STABILITY OF
GLYCERALDEHYDE 3 PHOSPHATE DEHYDROGENASE FROM THE MUSCLES OF
ECTOTHERMIC AND ENDOTHERMIC ANIMALS.
BIOKHIMIIA, 42:1960-1964.

                                                      5523F

ROHDE, F.C.                   1976
FIRST RECORD OF THE LEAST BROOK LAMPREY, OKKELBERGIA
AEPYPTERA, (PISCES: PETROMYZONIDAE) FROM ILLINOIS.
ILL. STATE ACAD. SCI., TRANS., 69:313-314.
U.S.A. ADULT DISTRIBUTION

                                                      5524F

KORNELIUSSEN, H.              1977
TUBULES INVAGINATING FROM THE SARCOLEMMA IN THE SUBNEURAL
REGION OF MUSCLE FIBERS.
CELL TISSUE RES., 181:73-79.
*MYXINE GLUTINOSA NORWAY MUSCLE NERVOUS SYSTEM HISTOLOGY

                                                      5525F

ICHIKAWA, T.                  1977
EFFECTS OF HYPOPHYSECTOMY ON THE TESTIS OF THE HAGFISH,
EPTATRETUS BURGERI GIRARD (CYCLOSTOMATA).
ZOOL. ANZ., 199:371-380.
JAPAN ENDOCRINOLOGY HISTOLOGY GONADOGENESIS

                                                      5526F

PATZNER, R.A.                 1976
ADAM, H.
THE GONADAL DEVELOPMENT OF THE FEMALE ATLANTIC HAGFISH,
MYXINE GLUTINOSA L.
JAP. SOC. COMP. ENDOCRINOL., GIFU, PROC., 1976:1 P.
(ABSTR.)
AUSTRIA GONADOGENESIS FEEDING

                                                      5527

OLE, B.                       1975
LARSEN, L.O.
ABSENCE OF KNOWN CORTICOSTEROIDS IN BLOOD OF RIVER LAMPREYS
(LAMPETRA FLUVIATILIS) AFTER TREATMENT WITH MAMMALIAN
CORTICOTROPIN.
GEN. COMP. ENDOCRINOL., 26:96-99.

                                                      5528

HEINTZ, A.                    1962
LES ORGANES OLFACTIFS DES HETEROSTRACI.
IN  COLLOQUE INT. CENT. NAT. RECH. SCI., PARIS, 104:13-29.
PROBLEMES ACTUELS DE PALEONTOLOGIE - EVOLUTION DES
VERTEBRES.PARIS: INT. CENT. NAT. RECH. SCI.

                                                      5529F

NAKAO, T.                          1966
DESMOSOMES FOUND IN THE SKELETAL MUSCLE OF THE LAMPREY. PP.
405-406.
IN  UYEDA, R., ELECTRON MICROSCOPY. VOL. II. TOKYO: MARUZEN.
*ENTOSPHENUS JAPAN AMMOCOETE HISTOLOGY MUSCLE


                                                                5530
KRAUSE, C.                         1886
DIE RETINA - II: DIE RETINA DER FISCHE. CYCLOSTOMATA.
INT. MSCHR. ANAT. HISTOL., 3:8-21.


                                                                5531
SHIBATA, Y.                        1976
YAMAMOTO, T.
SPECIFIC GRANULES IN THE LAMPREY ATRIUM.
J. ELECTRONMICROSC., 25:217.


                                                                5532
COHEN, M.J.                        1976
CELLULAR EVENTS IN THE EVOLUTION OF BEHAVIOUR. PP. 39-51.
IN  FENTRESS, J.C., ED., SIMPLER NETWORKS AND BEHAVIOUR.
SUNDERLAND: SINAUER.


                                                                5533
CLEGG, M.                          1977
NOTES ON THE FISHES OF THE YORKSHIRE RIVER DERWENT.
NATURALIST (LEEDS), 102:105-108.


                                                                5534
GOOS, H.J.T.                       1977
TERLOU, M.
HYPOTHALAMIC CONTROL OF MELANOCYTE STIMULATING HORMONE
SECRETION IN LOWER VERTEBRATES. PP. 51-62.
IN  TILDERS, F.J.H., D.F. SWAAB AND T.B. VAN WILMERSMA
GREIDANUS, EDS., FRONTIERS OF HORMONE RESEARCH, VOL. 4.
MELANOCYTE STIMULATING HORMONE: CONTROL, CHEMISTRY AND
EFFECTS. BASEL: S. KARGER.


                                                                5535
HOFFMAN, R.A.                      1975
PINEAL GLAND AND BEHAVIOUR. PP. 697-721.
IN  ELEFTHERIOU, B.E. AND R.L. SPROTT, EDS., HORMONAL
CORRELATES OF BEHAVIOUR, VOL. 1. A LIFESPAN VIEW. NEW YORK:


                                                                5536F
RIEGEL, J.A.                       1978
FACTORS AFFECTING GLOMERULAR FUNCTIONS IN THE PACIFIC
HAGFISH EPTATRETUS STOUTI (LOCKINGTON).
J. EXP. BIOL., 73:261.
ENGLAND EXCRETION


                                                                5537F
POTTER, I.C.                       1978
WRIGHT, G.M.
YOUSON, J.H.
METAMORPHOSIS IN THE ANADROMOUS SEA LAMPREY, PETROMYZON
MARINUS L.
CAN. J. ZOOL., 56:561-570.
METAMORPHOSIS GROWTH MORPHOMETRY


                                                                5538F
WRIGHT, G.M.                       1978
FILOSA, M.F.
YOUSON, J.H.
IMMUNOCHEMICAL LOCALIZATION OF THYROGLOBULIN IN THE
ENDOSTYLE OF THE ANADROMOUS SEA LAMPREY, PETROMYZON MARINUS
L.
AM. J. ANAT., 152:263-268.
IMMUNOLOGY ENDOCRINOLOGY CYTOLOGY BIOCHEMISTRY


                                                                5539F
WRIGHT, G.M.                       1978
FILOSA, M.F.
YOUSON, J.H.
LIGHT AND ELECTRON MICROSCOPIC IMMUNOCYTOCHEMICAL
LOCALIZATION OF THYROGLOBULIN IN THE THYROID GLAND OF THE
ANADROMOUS SEA LAMPREY, PETROMYZON MARINUS L., DURING ITS
UPSTREAM MIGRATION.
CELL TISSUE RES., 187:473-478.
CANADA ATLANTIC OCEAN ADULT HISTOLOGY ENDOCRINOLOGY
IMMUNOLOGY


                                                                5540F
NAKAO, T.                          1978
AN ELECTRON MICROSCOPIC STUDY OF THE CAVERNOUS BODIES IN
THELAMPREY GILL FILAMENTS.
AM. J. ANAT., 151:319-336.
*LAMPETRA JAPONICA JAPAN ADULT RESPIRATION HISTOLOGY


                                                                5541F
PYBUS, M.J.                        1978

UHAZY, L.S.
ANDERSON, B.C.
LIFE CYCLE OF TRUTTAEDACNITIS STELMIOIDES (VESSICHELLI,
1910) (NEMATODA: CUCULLANIDAE) IN AMERICAN BROOK LAMPREY
(LAMPETRA LAMOTTENII).
CAN. J. ZOOL., 56:1420-1429.
CANADA ADULT AMMOCOETE PARASITISM OF

5542F

VLADYKOV, V.D.          1978
KOTT, E.
A NEW NONPARASITIC SPECIES OF THE HOLARCTIC LAMPREY GENUS
LETHENTERON CREASER AND HUBBS, 1922, (PETROMYZONIDAE) FROM
NORTHWESTERN NORTH AMERICA WITH NOTES ON OTHER SPECIES OF
THE SAME GENUS.
UNIV. ALASKA  BIOL. PAP., 19:74 PP.
AMMOCOETE SYSTEMATICS ADULT DISTRIBUTION

5543F

PATZNER, R.A.          1977
GEORGIEVA, V.
ADAM, H.
SINNESZELLEN AN DEN TENTAKELN DER SCHLEIMAALE MYXINE
GLUTINOSA UND EPTATRETUS BURGERI (CYCLOSTOMATA). EINE
RASTERELEKTRONENOPTISCHE UNTERSUCHUNG. [SENSORY CELLS ON THE
TENTACLES OF THE HAGFISH MYXINE GLUTINOSA AND EPTATRETUS
BURGERI. A RASTERELECTRONMICROSCOPIC STUDY.]
AKAD. WISS. OSTERR., MAT.-NAT. KL., SITZ., 21:3 PP.
(GER.)
NORWAY JAPAN SENSE RECEPTORS HISTOLOGY

5544

LECH, J.J.          1973
ISOLATION AND IDENTIFICATION OF
3-TRIFLUOROMETHYL-4-NITROPHEVOL GLUCURONIDE FROM BILE OF
RAINBOW TROUT EXPOSED TO 3-TRIFLUOROMETHYL-4-NITROPHENOL.
OREGON STATE UNIV., PH.D. THESIS.

5545F

MACEY, D.J.          1978
POTTER, I.C.
LETHAL TEMPERATURES OF AMMOCOETES OF THE SOUTHERN HEMISPERE
LAMPREY, GEOTRIA AUSTRALIS GRAY.
ENV. BIOL. FISH., 3:241-243.
AUSTRALIA PHYSIOLOGY MORTALITY TEMPERATURE

5546

FALKMER, S.          1974
PRIMARY LIVER CARCINOMA AND OTHER TUMORS IN A LARGE
POPULATION OF A PRIMITIVE VERTEBRATE, THE ATLANTIC HAGFISH,
MYXINE GLUTINOSA.
FLORENCE, 11TH INT. CANCER CONGR., ABSTR., 2:176.

5547

JANVIER, P.          1975
SPECIALISATIONS PRECOCES ET CARATERES PRIMITIFS DU SYSTEME
CIRCULATOIRE DES OSTEOSTRACES.
COLL. INT. CENTR. NAT. RECH. SCI., PARIS, 218:15-31.

5549

TRETJAKOFF, D.          1915
DIE PARIETALORGANE VON PETROMYZON FLUVIATILIS.
Z. WISS. ZOOL., 11:1-112.

5550

STRUFE, R.          1962
GONNERT, R.
COMPARATIVE STUDIES ON THE INFLUENCE OF ENVIRONMENTAL
FACTORS ON THE EFFICIENCY OF BAYLUSCIDE.
PFLANZENSCHUTS-NACHR., 15:50-70.

5551

MEYLING, A.H.          1962
SCHUTTE, C.H.J.
PITCHFORD, R.J.
SOME LABORATORY INVESTIGATIONS ON BAYER 73 AND ICI 24223 AS
MOLLUSCICIDES.
W.H.O., BULL., 27:95-98.

5552

HOWELL, J.H.          1964
KING, E.L.
SMITH, A.J.
HANSON, L.H.
SYNERGISM OF 5,2'-DICHLORO-4'-NITROSALICYLANILIDE AND
3-TRIFLUOROMETHL-4-NITROPHENOL AS A SELECTIVE LAMPREY
LARVICIDE.
GREAT LAKES FISH. COMM., TECH. REP., 8:21 PP.

5553

KOTT, E.          1974

A MORPHOMETRIC AND MERISTIC STUDY OF A POPULATION OF THE
AMERICAN BROOK LAMPREY, LETHENTERON LAMOTTEI (LE SEUR), FROM
ONTARIO.
CAN. J. ZOOL., 52:1047-1055.
*LETHENTERON JAPONICUM MORPHOMETRY

5554

KAN, T.T.                          1975
SYSTEMATICS, VARIATION, DISTRIBUTION AND BIOLOGY OF LAMPREYS
OF THE GENUS LAMPETRA IN OREGON.
UNIV. OREGON STATE, PH.D. THESIS.


5555

BOND, C.E.                         1973
KAN, T.T.
LAMPETRA (ENTOSPHENUS) MINIMA, A DWARFED PARASITIC LAMPREY
FROM OREGON.
COPEIA, 1973:568-574.
*LAMPETRA TRIDENTATA *LAMPETRA LETHOPHAGA U.S.A. SYSTEMATICS
MORPHOMETRY MUSCLE


5556

PIETSCHMANN, V.                    1962
CYCLOSTOMA. PP. 369-372.
IN  KEKENTHAL, W., ED., HANDBUCH DER ZOOLOGIE. VOL. 6.
BERLIN: WALTER DE GRUYTER & CO.


5558

VAN NOORDEN, S.                    1974
PEARSE, A.G.E.
LES CELLULES HORMONALES DU SYSTEME DIGESTIF DES CYCLOSTOMES.
RECH. BIOL. CONTEMP., 4:155-162.
(FR.)
ENG. SUMM.
*MYXINE GLUTINOSA *LAMPETRA FLUVIATILIS HISTOLOGY
ENDOCRINOLOGY DIGESTION PHYSIOLOGY


5559

SUZUKI, S.                         1973
KONDO, Y.
THYROIDAL MORPHOGENESIS AND BIOSYNTHESIS OF THYROGLOBULIN
BEFORE AND AFTER METAMORPHOSIS IN THE LAMPREY, LAMPETRA
REISSNERI.
GEN. COMP. ENDOCRINOL., 21:451-460.
ADULT ENDOCRINOLOGY BIOCHEMISTRY METABOLISM TECHNIQUES


5560

POTTER, I.C.                       1973
ROBINSON, E.S.
THE CHROMOSOMES OF THE CYCLOSTOMES.
IN  CHIARELLI, S.B. AND E. CAPANNA, EDS., CYTOTAXONOMY AND
VERTEBRATE EVOLUTION. LONDON: ACADEMIC PRESS.


5561F

VAN NOORDEN, S.                    1973
OSTBERG, Y.
PEARSE, A.G.E.
INTESTINAL HORMONES IN CYCLOSTOMES.
IRCS MED. SCI., 13-1-5.
*MYXINE GLUTINOSA *LAMPETRA PLANERI *LAMPETRA FLUVIATILIS
SWEDEN ADULT LARVA ENDOCRINOLOGY DIGESTION


5562F

VAN NOORDEN, S.                    1973
PEARSE, A.G.E.
INTESTINAL HORMONES IN LAMPREYS: IMMUNOFLUORESCENCE REACTION
WITH ANTISERUM TO CAERULEIN.
IRCS MED. SCI., 13-1-1.
*LAMPETRA FLUVIATILIS *LAMPETRA PLANERI UNITED KINGDOM LARVA
DIGESTION ENDOCRINOLOGY


5563

SEILER, K.                         1973
SEILER, R.
ZUR TOPOGRAPHIE DIS INTERRENAL-UND ADRENAL-SYSTEMS DES
BACHNEUNAUGES (LAMPETRA PLANERI BLOCH).
GEGENBAURS MORPHOL. JAHRB., 119:796-808.


5564

TIMMONS, T.J.                      1978
SHELTON, W.L.
DAVIES, W.D.
FISH POPULATION CHANGES FOLLOWING IMPOUNDMENT OF WEST POINT
RESERVOIR CATTAHOOCHEE RIVER, ALABAMA GEORGIA.
ASSOC. SOUTHEAST. BIOL., BULL., 25:55.


5565

ROBBINS, D.L.                      1978
GERSHWIN, M.E.
IDENTIFICATION AND CHARACTERIZATION OF LYMPHOCYTE
SUBPOPULATIONS.

SEMIN. ARTHRITIS RHEUM., 7:245-277.

5566

FREDGA, K.                     1977
CHROMOSOMAL CHANGES IN VERTEBRATE EVOLUTION.
R. SOC. EDINB., PROC., SECT. B (BIOL. SCI.), 199:377-397.

5567

SHOLDICE, J.A.                 1976
MCMILLAN, D.B.
HISTOCHEMICAL IDENTIFICATION OF ADENOHYPOPHYSIAL CELLS OF
THE SEA LAMPREY, PETROMYZON MARINUS.
CAN. FED. BIOL. SOC., PROC., 19:24.

5568

COHEN, N.                      1977
PHYLOGENETIC EMERGENCE OF LYMPHOID TISSUES AND CELLS. PP.
149-202.
IN MARCHALONIS, J.J., ED., IMMUNOLOGY SERIES, V.5. THE
LYMPHOCYTE: STRUCTURE AND FUNCTION. PART 1. NEW YORK: BASEL.

5569

GRAY, R.H.                     1977
DAUBLE, D.D.
CHECKLIST AND RELATIVE ABUNDANCE OF FISH SPECIES FROM THE
HANFORD REACH OF THE COLUMBIA RIVER.
NORTHWEST SCI., 51:208-215.

5570

DAVIDSON, W.S.                 1976
FLYNN, T.G.
SPECIES AND TISSUE DISTRIBUTION OF NADPH-DEPENDANT ALDEHYDE
REDUCTASE.
CAN. FED. BIOL. SOC., PROC., 19:150.
ANIMAL BIOCHEMISTRY TECHNIQUES EXCRETION

5571

SIELFELD, K.W.H.               1976
PRESENCE OF EXOMEGAS MACROSTOMUS NEW RECORD MYXINE
PETROMYZONIDAE IN WATERS OF THE MAGALLANES CHILE REGION.
ANN. INST. PATAGONIA, 7:211-214.

5572F

HONMA, S.                      1973
ORGANIZATION OF SYNAPTIC CONNECTIONS AMONG THE GIANT
INTERNEURONS OF THE LAMPREY, LAMPETRA JAPONICA.
ACTA MED. & BIOL., 21:33-43.
ADULT NERVOUS SYSTEM TECHNIQUES

5573F

DYK, V.V.                      1960
MATERIALY PO EKOLOGII KHARIUSA V REKAKH ZAKARPATSKOI OBLASTI
FLORA & FAUNA (MOSCOW), 1960:171-178.

5574F

SADO, Y.                       1978
HORI, S.H.
IMMUNOLOGICAL RELATEDNESS OF GLUCOSE 6-PHOSPHATE
DEHYDROGENASES FROM VERTEBRATE AND INVERTEBRATE SPECIES.
JAP. J. GENET., 53:91-102
*ENTOSPHENUS JAPONICUS ANIMAL BIOCHEMISTRY IMMUNOLOGY

5575F

POHLA, H.                      1977
LEMETSCHWANDTNER, A.
ADAM, H.
DIE VASKULARISATION DER KIEMEN VON MYXINE GLUTINOSA L.
(CYCLOSTOMATA).
ZOOL. SCR., 6:331-341.
(GER.)
ENG. SUMM.
HISTOLOGY RESPIRATION

5577F

DWYER, W.P.                    1977
MAYER, F.L.
ALLEN, J.L.
BUCKLER, D.R.
CHRONIC AND SIMULATED USE-PATTERN EXPOSURES OF BROOK TROUT
(SALVELINUS FONTINALIS) TO 3-TRIFLUOROMETHYL-4-NITROPHENOL
(TFM).
U.S. DEP. INT., FISH WILDL. SERV., INVEST. FISH CONTROL, 84:
6 PP.
CHEMISTRY MANAGEMENT

5578F

MATTHEWS, G.                   1978
WICKELGREN, W.O.
TRIGEMINAL SENSORY NEURONS OF THE LAMPREY.
J. COMP. PHYSIOL., 123A:329-334.
*PETROMYZON MARINUS U.S.A. GREAT LAKES NERVOUS SYSTEM

5579F

**SELZER, M.E.**                    1978
MECHANISMS OF FUNCTIONAL RECOVERY AND REGENERATION AFTER
SPINAL CORD TRANSECTION IN LARVAL SEA LAMPREY.
J. PHYSIOL., 277:395-408.
NERVOUS SYSTEM TECHNIQUES LOCOMOTION


5580F

**OESER, R.**                       1957
FANG EINES MEERNEUNAGES (PETROMYZON MARINUS L.) IM
GOTTIN-SEE BEI POTSDAM.
WISS. Z., MATH-NATURWISS. ZEIHE, (POTSDAM), 3:141-144.


5581F

**KUX, Z.**                         1972
**STEINER, H.M.**
LAMPETRA LANCEOLATA, EINE NEUE NEUNAGENART AUS DEM
EINZUGSGEBIET DES SCHWARZEN MEERES IN DER NORDOSTLICHEN
TURKEI.
CAS. MORAV. MUS., 56-57:375-384.


5582

**HOLCIK, J.**                      1970
NUMBER AND VARIATION OF TRUNK MYOMERES IN LAMPETRA PLANERI
WITH REGARD TO POPULATION FROM THE POPRAD AND HORNAD RIVER
BASINS.
BIOLOGIA, (BRATISL.), 25:123-128.


5583

**DEAN, B.**                        1904
NOTES ON JAPANESE MYXINOIDS.
J. COLL. SCI., IMP. UNIV. TOKYO, 19:1-23.


5584

**CONEL, J.L.**                     1931
THE GENITAL SYSTEM OF THE MYXINOIDEA: A STUDY BASED ON NOTES
AND DRAWINGS OF THESE ORGANS IN BDELLOSTOMA MADE BY
BASHFORD DEAN. PP.67-102.
IN  GUDGER, E.W., ED., THE BASHFORD DEAN MEMORIAL VOLUME
ARCHAIC FISHES. ART III. NEW YORK:AMERICAN MUSEUM NATURAL
HISTORY.


5585

**MORITA, Y.**                      1971
**DODT, E.**
PHOTOSENSORY RESPONSES FROM THE PINEAL EYE OF THE LAMPREY
(PETROMYZON FLUVIATILIS).


5586

**COLLIN, J.P.**
CELLULES GANGLIONNAIRES ET TRACTUS DE L'ORGANE PINEAL DE
LAMPETRA PLANERI.
J. NEURO-VISC. REL. 31:308-333.
(FR.)
ENG. SUMM.
NERVOUS SYSTEM VISION TECHNIQUES CYTOLOGY BIOLOGY OLFACTION


5587

**LEWIS, S.V.**                     1976
**POTTER, I.C.**
A SCANNING ELECTRON MICROSCOPIC STUDY OF THE GILLS OF THE
LAMPREY, LAMPETRA FLUVIATILIS (L.)
MICRON. 7:205-211.


5588

**RAUTHER, M.**                     1937
KIEMEN DER ANAMNIER-KIEMENDARMDERIVATE DER CYCLOSTOMEN  UND
FISCHE. P. 223.
IN  HANDBUCH DER VERGLEICHENDEN ANATOMIE DER WIRBELTIERE.
BD. III.


5590

**TOMONAGA, S.**                    1977
**YAMAGUCHI, K.**
**AWAYA, K.**
FATE OF INJECTED HORSERADISH PEROXIDASE IN HEPATIC
PARENCHYMAL CELLS OF HAGFISH EPTATRETUS BURGERI.
J. ELECTRONMICROSC., 26:228-229.
(ABSTR.)
CYTOLOGY HISTOLOGY BIOCHEMISTRY PHYSIOLOGY


5591

**AWAYA, K.**                       1977
**TOMONAGA, S.**
**YAMAGUCHI, K.**
MONONUCLEAR PHAGOCYTES KUPFFER CELLS IN LIVER SINUSOIDS OF
HAGFISH.
J. ELECTRONMICROSC., 26:228.
(ABSTR.)
CYTOLOGY

5592

NIEWENHUYS, R.                    1977
THE BRAIN OF THE LAMPREY IN A COMPARATIVE PERSPECTIVE. PP.
97-145.
IN  DIMOND, S.J. AND D.A. BLIZZARD, EDS., ANNALS OF THE NEW
YORK ACADEMY OF SCIENCES, V.299. EVOLUTION AND
LATERALIZATION OF THE BRAIN. NEW YORK: N.Y. ACAD. SCI.


5593

MIYOSHI, M.                       1977
NAKANO, M.
SOEJIMA, K.
MACROPHAGES IN THE RENAL URINARY SPACE.
J. ELECTRONMICROSC., 26:275.
(ABSTR.)
ANIMAL EXCRETION TECHNIQUES CYTOLOGY


5594

STAHL, B.J.                       1977
EARLY AND RECENT PRIMITIVE BRAIN FORMS. PP. 87-96.
IN  DIMOND, S.J. AND D.A. BLIZZARD, EDS., ANNALS OF THE NEW
YORK ACADEMY OF SCIENCES, V. 299. EVOLUTION AND
LATERALIZATION OF THE BRAIN. NEW YORK: N.Y. ACAD. SCI.


5595

FALKMER, S.
MARKLUND, S.
MATTSON, P.E.
RAPPE, C.
HEPATOMAS AND OTHER NEOPLASMS IN THE ATLANTIC HAGFISH MYXINE
GLUTINOSA. A HISTOPATHIC AND CHEMICAL STUDY. PP. 342-355.
IN  KRAYBILL, H.F. ET AL, EDS., ANNALS OF THE NEW YORK
ACADEMY OF SCIENCES, V. 298. AQUATIC POLLUTANTS AND
BIOLOGIC EFFECTS WITH EMPHASIS ON NEOPLASIS. NEW YORK: N.Y.
ACAD. SCI.


5596

NAKAI, Y.                         1977
SHIODA, S.
HONMA, Y.
ULTRASTRUCTURE OF THE CEREBROSPINAL FLUID CONTACTING NEURONS
IN THE HYPOTHALAMUS OF THE LAMPREY, LAMPETRA JAPONICA.
J. ELECTRONMICROSC., 26:264.
(ABSTR.)
ADULT CYTOLOGY NERVOUS SYSTEM HISTOLOGY


5597

PERMITIN, YU.E.                   1977
SPECIES COMPOSITION AND ZOOGEOGRAPHIC ANALYSIS OF BENTHIC
FISH FAUNA OF THE SCOTIA SEA.
UOPR. IKHTIOL., 17:843-861.


5598

PAJOR, W.J.                       1977
COMPARATIVE HISTOENZYMOLOGICAL STUDIES ON THE ACTIVITY OF
SOME OXIDATIVE ENZYMES IN THYROIDS OF POST METAMORPHIC
LAMPREYS OF LAMPETR LAMPETRA FLUVIATILIS.
FOLIA BIOL., 25:409-414.


5599

REPETSKI, J.E.                    1978
A FISH FROM THE UPPER CAMBRIAN OF NORTH AMERICA.
SCIENCE, 200:529-531.
ANIMAL ANATOMY EVOLUTION


5600

DICKHOFF, W.W.                    1978
CRIM, J.W.
GORBMAN, A.
LACK OF EFFECT OF SYNTHETIC THYROTROPIN RELEASING HORMONE OF
PACIFIC HAGFISH EPTATRETUS STOUTI PITUITARY THYROID TISSUES
IN VITRO.
GEN. COMP. ENDOCRINOL., 35:96-98.


5601

DE MEYTS, P.                      1978
VAN OBBERGHEN, E.
ROTH, J.
WOLLMER, A.
BRANDENBURG, D.
MAPPING OF THE RESIDUES RESPONSIBLE FOR THE NEGATIVE
COOPERATIVITY OF THE RECEPTOR BINDING REGION OF INSULIN.
NATURE (LOND.), 273:504-509.
ANIMAL ENDOCRINOLOGY BIOCHEMISTRY


5602

SHIBATA, Y.                       1977
COMPARATIVE ULTRASTRUCTURE OF CELL REMBRANE SPECIALIZATIONS
IN VERTEBRATE CARDIAC MUSCLES.
ARCH. HISTOL. JAP., 40:391-406.
CIRCULATORY SYSTEM MUSCLE ANIMAL TECHNIQUES PHYSIOLOGY

5603

MONACO, F.
ANDREOLI, M.
LA POSTA, A.
ROCHE, J.
THYROGLOBULIN BIOSYNTHESIS IN A LARVAL AMMOCOETE AND ADULT
FRESH WATER LAMPREY LAMPETRA PLANERI.
COMP. BIOCHEM. PHYSIOL., 60B:87-92.


5604F

BEAMISH, F.W.H.               1978
STRACHAN, P.D.
THOMAS, E.
OSMOTIC AND IONIC PERFORMANCE OF THE ANADROMOUS SEA LAMPREY,
PETROMYZON MARINUS.
COMP. BIOCHEM. PHYSIOL., 60A:435-443.
NEW BRUNSWICK AMMOCOETE OSMOREGULATION ENDOCRINOLOGY


5605F

FALKMER, S.                   1978
EMDIN, S.O.
CHEMICAL ASPECTS OF THE FINE STRUCTURE OF THE SECRETORY
GRANULES OF THE INSULIN-PRODUCING B-CELLS OF THE MYXINE
ISLET PARENCHYMA.
GEN. COMP. ENDOCRINOL., 34.
(ABSTR.)
SWEDEN ENDOCRINOLOGY BIOCHEMISTRY IMMUNOLOGY


5606F

HOMMA, S.                     1978
ROVAINEN, C.M.
CONDUCTANCE INCREASES FOUND BY GLYCINE AND Y-AMINOBUTYRIC
ACID IN LAMPREY INTERNEURONES.
J. PHYSIOL., 279:231-252.
*ICHTHYOMYZON UNICUSPIS *LAMPETRA AEPYPTERA *LAMPETRA
LAMOTTEI U.S.A. PHYSIOLOGY NERVOUS SYSTEM


5607F

LARSEN, L.O.                  1974
EFFECTS OF OESTRADIOL-17B AND TESTOSTERONE ON INTACT MALE
AND FEMALE RIVER LAMPREYS (LAMPETRA FLUVIATILIS).
GEN. COMP. ENDOCRINOL., 22:384.
(ABSTR.)


5608F

MATTHEWS, G.                  1978
WICKELGREN, W.O.
EVOKED DEPOLARING AND HYPERPOLARIZING POTENTIALS IN
RETICULOSPINAL AXONS OF AXONS OF LAMPREY.
J. PHYSIOL., 279:551-567.
J. PHYSIOL., 279:551-567.
*PETROMYZON MARINUS U.S.A. GREAT LAKES ADULT NERVOUS SYSTEM
PHYSIOLOGY


5610F

FARMER, G.J.                  1974
FOOD CONSUMPTION, GROWTH AND HOST PREFERENCES OF THE SEA
LAMPREY, PETROMYZON MARINUS.
UNIV. GUELPH, PH.D. THESIS: 76 PP.
RESTRICTED


5611F

STRACHAN, P.                  1977
SERUM OSMOTIC AND IONIC CONCENTRATIONS OF ANADROMOUS SEA
LAMPREY (PETROMYZON MARINUS) IN RELATION TO AMBIENT
SALINITY.
UNIV. GUELPH, M. SC. THESIS: 67 PP.
RESTRICTED


5612

ELFIMOVA, L.I.                1977
LEIBMAN, D. YA.
INFLUENCE OF COENZYME ON THE CONFORMATIONAL STABILITY OF
GYCERALDEHYDE-3-PHOSPHATE DEHYDROGENASE FROM MUSCLES OF
ECTOTHERMIC AND ENDOTHERMIC ANIMALS.
BIOKHIMIIA, 42:1960-1964.
(RUSS.)
BIOCHEMISTRY, 42:1545-1548.
(ENG.)
ADULT BIOCHEMISTRY PHYSIOLOGY


5613

PIAVIS, G.W.                  1978
PIAVIS. M.G.
IMPLANTATION OF AN ABDOMINAL WINDOW IN THE SEA LAMPREY
PETROMYZON MARINUS.
COPEIA, 1978:349-352.


5614

**LEE, K.Y.**                1977
KIMM, S.W.
TCHAI, B.S.
SEO, J.S.
STUDIES ON GENOME ANALYSIS OF SEVERAL FISHES AND
DISSOCIATION PROFILE OF VARYING EUKARYOTIC DAM.
KOREAN J. BIOCHEM., 9:58.


                                                5615
**KIER, E.L.**                1976
PHYLOGENETIC AND ONTOGENETIC CHANGES OF THE BRAIN RELEVANT
TO THE EVOLUTION OF THE SKULL. PP. 468-499.
IN  BOSMA, J.F., ED., SYMPOSIUM ON DEVELOPMENT OF THE
BASICRANIUM. BETHESDA, MD., JUNE 23-25, 1975. BETHESDA: DEP.
HEALTH EDUC. WELFARE.


                                                5616F
**O'BOYLE, R.N.**              1975
GROWTH AND INTERMEDIARY METABOLISM OF LARVAL AND
METAMORPHOSING STAGES OF THE LANDLOCKED SEA LAMPREY,
PETROMYZON MARINUSL.
UNIV. GUELPH, M.SC. THESIS: 85 PP.
RESTRICTED
U.S.A. AMMOCOETE ECOLOGY LIFE CYCLE BLOOD GROWTH METABOLISM
METAMORPHOSIS PHYSIOLOGY

# Author Index

| Name | Year | No. | * | Name | Year | No. | * |
|---|---|---|---|---|---|---|---|
| AARSKOG, D. | 1978 | 5338 | | BARRINGTON, E.J.W. | 1945 | 5447 | * |
| AASA, R. | 1973 | 4832F | * | BATUEVA, I.V. | 1974 | 4601F | * |
| ACCINI, L. | 1976 | 5059 | | BATUEVA, I.V. | 1977 | 5353 | * |
| ACCINNI, L. | 1976 | 5058 | * | BATUEVA, I.V. | 1977 | 5521 | * |
| ADACHI, T. | 1975 | 4538F | | BAUER, C. | 1975 | 5244F | * |
| ADACHI, T. | 1976 | 5078F | | BAUMGARTEN, H.G. | 1973 | 4735F | * |
| ADAM, H. | 1975 | 4809F | | BEADLES, J.K. | 1974 | 5033 | |
| ADAM, H. | 1976 | 5296F | | BEADLES, J.K. | 1974 | 5035F | |
| ADAM, H. | 1976 | 5526F | | BEADLES, J.K. | 1976 | 5287 | |
| ADAM, H. | 1977 | 5418 | | BEAMISH, F.W.H. | | 5145 | |
| ADAM, H. | 1977 | 5543F | | BEAMISH, F.W.H. | 1973 | 4754F | |
| ADAM, H. | 1977 | 5575F | | BEAMISH, F.W.H. | 1973 | 4755F | * |
| ADAMSON, R.H. | 1972 | 4935 | * | BEAMISH, F.W.H. | 1974 | 4593F | * |
| AGARKOV, G.B. | 1976 | 5079F | * | BEAMISH, F.W.H. | 1974 | 4599F | |
| AGAYAN, A.L. | 1973 | 4612 | | BEAMISH, F.W.H. | 1974 | 4743F | |
| ALEKSEEVA, K.D. | 1975 | 4924 | * | BEAMISH, F.W.H. | 1974 | 4759F | |
| ALLEGRET, P. | 1976 | 5016F | * | BEAMISH, F.W.H. | 1975 | 4565F | |
| ALLEGRET, P. | 1977 | 5352F | * | BEAMISH, F.W.H. | 1975 | 4673F | |
| ALLEN, J.L. | | 5422F | | BEAMISH, F.W.H. | 1975 | 4710F | * |
| ALLEN, J.L. | 1974 | 5307F | | BEAMISH, F.W.H. | 1975 | 4758F | |
| ALLEN, J.L. | 1975 | 4660F | | BEAMISH, F.W.H. | 1975 | 4769F | |
| ALLEN, J.L. | 1975 | 4985F | | BEAMISH, F.W.H. | 1977 | 4975F | |
| ALLEN, J.L. | 1975 | 5051F | | BEAMISH, F.W.H. | 1977 | 5096F | |
| ALLEN, J.L. | 1976 | 5189F | | BEAMISH, F.W.H. | 1977 | 5140F | |
| ALLEN, J.L. | 1977 | 5424 | * | BEAMISH, F.W.H. | 1977 | 5340F | |
| ALLEN, J.L. | 1977 | 5577F | | BEAMISH, F.W.H. | 1978 | 5604F | * |
| ALLEN, J.L. | 1978 | 5462F | | BEAMISH, F.W.H. | 1979 | 5507 | * |
| ALT, J. | 1976 | 5351F | * | BEAMISH, R.J. | 1976 | 4633F | * |
| ALVESTAD-GRAEBNER, I. | 1976 | 5296F | * | BEAMISH, R.J. | 1976 | 5364F | * |
| ALVESTAD-GRAEBNER, I. | 1977 | 5418 | * | BEETON, A.M. | 1973 | 5411F | |
| AMES, B.N. | 1975 | 5451 | | BEETON, A.M. | 1974 | 5376 | * |
| AMOS, B. | 1977 | 5156F | * | BEHLKE, J. | 1973 | 4760F | |
| ANDERSON, B.C. | 1978 | 5541F | | BEHLKE, J. | 1973 | 5128F | * |
| ANDERSON, I.G. | 1977 | 5156F | | BELEKHOVA, M.G. | 1975 | 4540F | |
| ANDERSON, J.B. | 1972 | 4794F | | BELEN'KII, M.A. | 1973 | 4556F | |
| ANDERSON, M. | 1973 | 5412F | | BELENKII, M.A. | 1975 | 5071F | * |
| ANDERSON, R.C. | 1978 | 5327F | | BELENKY, M.A. | 1974 | 4586 | |
| ANDERSON, W.C. | 1977 | 5235F | * | BELL, M.A. | 1978 | 5075 | * |
| ANDREASEN, J.K. | 1975 | 5319 | * | BELOZERSKII, A.N. | 1973 | 4548F | |
| ANDREOLI, M. | | 5603 | | BEND, J.R. | 1973 | 4964F | |
| ANDREOLI, M. | 1976 | 5025F | | BEND, J.R. | 1973 | 4972 | * |
| ANDREOLI, M. | 1976 | 5163 | | BEND, J.R. | 1974 | 4604F | |
| ANDREOLI, M. | 1977 | 5342F | | BENDS, D. | 1972 | 4931 | |
| ANDRES, K.H. | 1975 | 5297 | * | BERMAN, H.A. | 1976 | 5013 | * |
| ANDREWS, S.M. | 1977 | 5369 | * | BERMAN, M. | 1973 | 4571 | |
| ANNO, K. | 1975 | 5041 | * | BEST, J.B. | 1974 | 4536 | |
| ANTONOV, A.S. | 1973 | 4802F | | BIGNAMI, A. | 1973 | 4570F | |
| ANTONOV, A.S. | 1974 | 4567F | | BILLS, T.D. | 1975 | 4691F | |
| APOSHIAN, D. | 1976 | 5158 | | BILLS, T.D. | 1976 | 5095F | * |
| APPLEGATE, V.C. | 1954 | 5143F | * | BILLS, T.D. | 1976 | 5407F | |
| ARLOCK, P. | 1973 | 4836F | | BINKS, A.E. | 1972 | 5381 | * |
| ARLOCK, P. | 1975 | 4573F | * | BIRD, D.J. | 1976 | 4678F | * |
| ARNDT, R.G. | 1974 | 4676F | | BIRNBERGER, K.L. | 1974 | 4661 | * |
| ARNDT, R.G. | 1975 | 4720F | | BITTNER, M.A. | 1972 | 5380 | * |
| ARNDT, R.G. | 1976 | 5139F | | BITTNER, M.A. | 1975 | 4647F | |
| ARNDT, R.G. | 1977 | 5010 | * | BIUW, L.W. | 1974 | 4725F | |
| ASAI, H. | 1976 | 5023F | * | BJORKLUND, A. | 1973 | 4735F | |
| ATKIN, N.B. | 1975 | 4696 | | BLACKER, R.W. | 1974 | 4816 | * |
| ATZ, J.W. | 1973 | 4921 | * | BLADEN, H.A. | 1977 | 5167 | |
| ATZ, J.W. | 1977 | 5238F | | BLANCK, J. | 1972 | 4775F | * |
| AUSTIN, J.C. | 1976 | 4827 | | BLANCK, J. | 1973 | 4760F | |
| AVROVA, N.F. | 1977 | 5512F | | BLASCHKO, H. | 1972 | 4789F | |
| AWAYA, K. | 1973 | 4795F | | BLASCHKO, H. | 1973 | 4847F | |
| AWAYA, K. | 1973 | 4796F | | BLOCH, W.W. | 1976 | 4827 | * |
| AWAYA, K. | 1973 | 4957 | * | BODIN-BAUDOUIN, J. | 1975 | 5008 | * |
| AWAYA, K. | 1973 | 4980F | | BOFFA, G.A. | 1976 | 5004F | |
| AWAYA, K. | 1973 | 5114F | | BOIVINET, P. | 1977 | 5511 | * |
| AWAYA, K. | 1973 | 5243F | | BOND, C.E. | 1973 | 5555 | * |
| AWAYA, K. | 1975 | 4618F | | BONDAREVA, V. | 1976 | 5124F | |
| AWAYA, K. | 1976 | 5180 | | BONDAREVA, V.M. | 1974 | 5267F | |
| AWAYA, K. | 1977 | 5293F | | BOQUIST, L. | 1973 | 4838F | |
| AWAYA, K. | 1977 | 5590 | | BOQUIST, L. | 1973 | 4854F | |
| AWAYA, K. | 1977 | 5591 | * | BOQUIST, L. | 1974 | 4706F | |
| BABURINA, YE. A. | 1972 | 5363 | * | BOQUIST, L. | 1975 | 4790F | * |
| BAGE, G. | 1975 | 4646F | * | BOQUIST, L. | 1976 | 4786F | |
| BALL, J.N. | 197 | 4732 | | BOQUIST, L. | 1976 | 4806F | |
| BALLANTYNE, J.S. | 1976 | 5077 | | BOQUIST, L. | 1976 | 5196 | |
| BANARESCU, P. | 1973 | 5127F | * | BORGESE, T.A. | 1977 | 5168 | * |
| BANDO, T. | 1975 | 5029F | | BORISOV, I.N. | 1973 | 5040 | * |
| BANO, Y. | 1977 | 5268 | * | BOSCHWITZ, D. | 1973 | 5328F | |
| BARDELE, C.F. | 1974 | 4965 | * | BOTHWELL, M.L. | 1973 | 5411F | * |

| Name | Year | No. | |
|---|---|---|---|
| BOUCHARD, R.P. | | 5494F | * |
| BOWSHER, D. | 1977 | 5295F | |
| BOYLAN, J.W. | 1971 | 4773F | |
| BRAEM, R.A. | 1976 | 5502F | |
| BRAEM, R.A. | 1977 | 5368F | * |
| BRANDENBURG, D. | 1978 | 5601 | |
| BRAUNITZER, G. | 1973 | 4834F | |
| BRIGHAM, W.U. | 1973 | 4653F | * |
| BRILEY, P.M. | 1972 | 4794F | |
| BRINN, J.E. JR. | 1975 | 5066F | |
| BRINN, J.E. JR. | 1976 | 5083F | |
| BRINN, J.E. JR. | 1976 | 5205F | * |
| BRINN, J.E. JR. | 1976 | 5289F | |
| BROAD, D.S. | 1973 | 5329 | * |
| BRONSHTEIN, A.A. | 1974 | 4634F | * |
| BROWN, I.D. | 1974 | 4649F | |
| BROWN, I.D. | 1975 | 4584F | |
| BRUKMOZER, P. | 1972 | 4739 | * |
| BUCHANAN, T.M. | 1973 | 4926 | * |
| BUCHWALD, D. | 1972 | 4761F | |
| BUCKLER, D.R. | 1977 | 5577F | |
| BUCKLER, D.R. | 1978 | 5406 | |
| BUNDGAARD, M. | 1976 | 4874 | * |
| BURTON, M.P.M. | 1976 | 4885F | |
| BUSHUYEV, V.N. | 1973 | 4930 | * |
| BUSTOS-VALDES, S.E. | 1974 | 4579 | * |
| BUTLER, D.G. | 1976 | 4993 | |
| BUUS, O. | 1975 | 4574F | * |
| CAIRNS, J. JR. | 1976 | 5410F | * |
| CALHOUN, W.F. | 1976 | 5410F | |
| CALLAGHAN, O.H. | 1977 | 5166 | |
| CAMERON, B.F. | 1975 | 4818 | |
| CAMPBELL, I.M. | 1974 | 4726 | |
| CANNON, D. | 1975 | 5030 | |
| CAPANNA, E. | 1976 | 5058 | |
| CAPANNA, E. | 1976 | 5059 | |
| CARTER A.W. | 1973 | 5497F | |
| CARTER, A.W. | 1973 | 5481 | |
| CARTON, Y. | 1973 | 5337F | * |
| CARTON, Y. | 1974 | 4602F | * |
| CASLEY-SMITH, J.R. | 1975 | 4640F | * |
| CATAUDELLA, S. | 1976 | 5025F | |
| CATAUDELLA, S. | 1976 | 5058 | |
| CATAUDELLA, S. | 1976 | 5163 | |
| CATAUDELLA, S. | 1977 | 5342F | |
| CHALUMEAU, M.T. | 1976 | 5004F | |
| CHANDA, S.K. | 1974 | 4579 | |
| CHANDLER, J.H. | 1975 | 4557F | * |
| CHANDLER, J.H. | 1975 | 4691F | |
| CHANDLER, J.H., JR. | 1974 | 5379F | * |
| CHAU, A.S.Y. | 1976 | 5408F | |
| CHEBOTAREVA, M.A. | 1974 | 4741F | |
| CHEREMISIN, A.N. | 1973 | 4930 | |
| CHEZE, G. | 1973 | 4665 | * |
| CHOLETTE, C. | 1971 | 5435 | * |
| CHOLETTE, C. | 1973 | 4767F | * |
| CHRISTENSEN, B.N. | 1976 | 5020F | * |
| CHRISTIANSON, G.G. | 1977 | 5409F | * |
| CHRISTOMANOS, A.A. | 1976 | 4983 | * |
| CHUNG, S.I. | 1976 | 5011 | |
| CIACCIO, G. | 1972 | 4619F | * |
| CIACCIO, G. | 1973 | 4807F | |
| CLARIDGE, P.N. | 1973 | 4719F | * |
| CLARIDGE, P.N. | 1974 | 4651F | * |
| CLARIDGE, P.N. | 1975 | 5138F | * |
| CLEGG, M. | 1977 | 5533 | * |
| COBURN, J.A. | 1976 | 5408F | * |
| CODOCERO, R.M. | 1974 | 5036 | * |
| COHEN, M.J. | 1976 | 5532 | * |
| COHEN, N. | 1977 | 5568 | * |
| COLLIN, J.P. | | 5586 | * |
| CONEL, J.L. | 1931 | 5584 | * |
| COOPER, E.L. | 1975 | 4869F | |
| CORBEL, M.J. | 1975 | 4708F | * |
| CORDIER, G. | 1975 | 5037 | * |
| COSTRINI, N.V. | 1972 | 5397F | |
| COTTRELL, B.A. | 1974 | 4954F | |
| COTTRELL, B.A. | 1975 | 4683 | |
| COTTRELL, B.A. | 1976 | 4715F | |
| COTTRELL, B.A. | 1976 | 5146F | * |
| COULTER, C.L. | 1973 | 5330 | |
| COULTER, C.L. | 1974 | 4966F | |
| COUPLAND, R.E. | 1976 | 5201 | * |
| CRAWFORD, R.R. JR. | 1974 | 4724 | * |
| CRAWSHAW, L.I. | 1978 | 5464 | |
| CRIM, J.W. | 1978 | 5600 | |
| CROWLEY, T.E. | 1973 | 4544 | * |
| CSERR, H.F. | 1973 | 4946 | |
| CSERR, H.F. | 1973 | 4963 | |
| CSERR, H.F. | 1974 | 4627F | |
| CSERR, H.F. | 1975 | 4805F | |
| CSERR, H.F. | 1977 | 5162 | * |
| CSERR, H.F. | 1978 | 5513F | |
| CSERR, H.F. | 1978 | 5515 | * |
| CUDDY, D.W. | 1976 | 5221F | |
| CUMMING, K.B. | 1975 | 5053F | |
| CUMMING, K.B. | 1977 | 5463F | |
| CUTFIELD, J.F. | 1974 | 4629 | * |
| CUTFIELD, J.F. | 1974 | 4725F | |
| CUTFIELD, J.F. | 1975 | 4744F | |
| CUTFIELD, S.M. | 1974 | 4629 | |
| CUTFIELD, S.M. ET.AL. | 1975 | 4744F | |
| CZECZUGA, B. | 1973 | 4777F | * |
| DAHL, D. | 1973 | 4570F | * |
| DAHL, F.H. | 1974 | 4746F | |
| DAHL, H.A. | 1976 | 5177F | * |
| DAMAS, H. | 1950 | 5357F | * |
| DAMASCHUN, G. | 1976 | 5070 | |
| DANKOVA, A.A. | 1975 | 4987 | |
| DANNER, H. | 1974 | 4700F | |
| DATHE, H. | 1975 | 4822 | * |
| DAUBLE, D.D. | 1977 | 5569 | |
| DAVIDSON, W.S. | 1976 | 5570 | * |
| DAVIES, W.D. | 1978 | 5564 | |
| DAVIS, W.A. | 1973 | 5491F | * |
| DAVIS, W.A. | 1974 | 4888F | * |
| DAVIS, W.A. | 1974 | 4891F | * |
| DAVIS, W.A. | 1974 | 4893F | * |
| DAVIS, W.A. | 1974 | 4904F | * |
| DAVIS, W.A. | 1975 | 5104F | * |
| DAVIS, W.A. | 1975 | 5106F | * |
| DAVIS, W.A. | 1975 | 5107F | * |
| DAVISON, P.F. | 1973 | 4571 | * |
| DAWE, C.J. EDS. | 1976 | 5282 | * |
| DAWSON, V.K. | 1971 | 5434 | * |
| DAWSON, V.K. | 1973 | 5403F | * |
| DAWSON, V.K. | 1974 | 4937F | |
| DAWSON, V.K. | 1975 | 5053F | |
| DAWSON, V.K. | 1975 | 5396F | |
| DAWSON, V.K. | 1976 | 5407F | * |
| DAWSON, V.K. | 1977 | 5463F | * |
| DAWSON, V.K. | 1978 | 5462F | * |
| DE GROOT, S.J. | 1975 | 4825 | |
| DE JONG, W.W. | 1976 | 5024F | * |
| DE MARTINO, C. | 1976 | 5058 | |
| DE MARTINO, C. | 1976 | 5059 | |
| DE MEYTS, P. | 1977 | 5286 | |
| DE MEYTS, P. | 1978 | 5601 | * |
| DE ROS, I. | 1976 | 5163 | |
| DE VOS, R. | 1973 | 4734F | * |
| DE WOLF-PEETERS, C. | 1973 | 4734F | |
| DEAN, B. | 1904 | 5583 | * |
| DELL'AGATA, M. | 1972 | 4619F | |
| DELL'AGATA, M. | 1973 | 4807F | * |
| DENESYUK, A.I. | 1977 | 5349 | |
| DENIS, C. | 1976 | 5016F | |
| DENIS, C. | 1977 | 5352F | |
| DENONCOURT, R.F. | 1975 | 5068F | * |
| DENONCOURT, R.F. | 1977 | 4994F | |
| DESMET, V. | 1973 | 4734F | |
| DEVEREUX, T.R. | 1973 | 4964F | |
| DICKHOFF, W.W. | 1976 | 5184F | |
| DICKHOFF, W.W. | 1977 | 5159F | * |
| DICKHOFF, W.W. | 1978 | 5600 | * |
| DIERICKX, K. | 1977 | 5312 | |
| DOBRYLKO, A.K. | 1972 | 4739 | |
| DODD, J.M. | 1971 | 4657 | * |
| DODSON, E.J. | 1974 | 4629 | |
| DODSON, G.G. | 1974 | 4629 | |
| DODT, E. | 1971 | 5585 | |
| DODT, E. | 1973 | 4949F | |
| DODT, E. | 1973 | 4950 | * |

| Name | Year | No. | |
|---|---|---|---|
| DONNER, P.J. | 1972 | 4830 | * |
| DOOLITTLE, R.F. | 1974 | 4954F | * |
| DOOLITTLE, R.F. | 1974 | 4960F | * |
| DOOLITTLE, R.F. | 1975 | 4683 | * |
| DOOLITTLE, R.F. | 1976 | 4715F | * |
| DOOLITTLE, R.F. | 1976 | 5097F | * |
| DOOLITTLE, R.F. | 1976 | 5146F | |
| DOUGLAS, C. | 1976 | 5195 | |
| DOVING, K.B. | 1974 | 5100F | * |
| DOWNING, D.W. | 1977 | 5166 | * |
| DOWNING, S. | 1976 | 5018F | |
| DOWNING, S.W. | 1976 | 5465F | |
| DREYFERT, T. | 1977 | 5299F | |
| DUBIN, N.H. | 1975 | 4780F | |
| DUGUY, R. | 1975 | 5008 | |
| DULMA, A. | 1973 | 4912F | * |
| DUSTIN, S.M. | 1974 | 4889F | * |
| DUSTIN, S.M. | 1974 | 4892F | * |
| DUSTIN, S.M. | 1974 | 4894F | * |
| DUSTIN, S.M. | 1974 | 4901F | * |
| DUSTIN, S.M. | 1975 | 4907F | * |
| DUSTIN, S.M. | 1975 | 5132F | * |
| DUSTIN, S.M. | 1975 | 5134F | * |
| DUSTIN, S.M. | 1976 | 4969F | * |
| DUSTIN, S.M. | 1976 | 5231F | |
| DUTRUGE, J. | 1974 | 4976 | |
| DWYER, W.P. | 1977 | 5577F | * |
| DWYER, W.P. | 1978 | 5406 | * |
| DYK, V.V. | 1960 | 5573F | * |
| EAKIN, R.M. | 1973 | 4731F | * |
| EAKIN, R.M. | 1973 | 5057F | * |
| EDMIN, S.O. | 1974 | 4725F | |
| EDSALL, T.A. | 1976 | 5501F | * |
| EDSTROM, A. | 1973 | 4804F | |
| EFIMOV, A.V. | 1977 | 5314F | |
| EGOROVA, V.V. | 1977 | 4947F | * |
| EHINGER, B. | 1977 | 5310F | * |
| EHLICH, B.E. | 1977 | 5162 | |
| EHRLICH, B.E. | 1978 | 5513F | * |
| EISENBACH, G.M. | 1971 | 4773F | * |
| EISNBACH, G.M. | 1973 | 4799F | |
| ELFIMOVA, L.I. | 1977 | 5522 | * |
| ELFIMOVA, L.I. | 1977 | 5612 | * |
| ELIAS, H. | 1977 | 5392 | * |
| ELSON, K.G.R. | 1976 | 5178 | |
| EMDIN, S.O. | 1973 | 4765F | |
| EMDIN, S.O. | 1973 | 4838F | |
| EMDIN, S.O. | 1973 | 4855F | |
| EMDIN, S.O. | 1973 | 4918F | |
| EMDIN, S.O. | 1973 | 5330 | * |
| EMDIN, S.O. | 1974 | 4966F | |
| EMDIN, S.O. | 1975 | 4654F | |
| EMDIN, S.O. | 1975 | 5400F | |
| EMDIN, S.O. | 1976 | 5285F | |
| EMDIN, S.O. | 1977 | 5154F | * |
| EMDIN, S.O. | 1978 | 5605F | |
| ENGELS, U. | 1975 | 5244F | |
| ENGLEMANN, W.E. | 1973 | 4747F | |
| EPPLE, A. | 1973 | 4920F | * |
| EPPLE, A. | 1975 | 5066F | * |
| EPPLE, A. | 1976 | 5027 | * |
| EPPLE, A. | 1976 | 5083F | * |
| EPPLE, A. | 1976 | 5205F | |
| EPPLE, A. | 1976 | 5289F | * |
| ERICHSEN, L. | 1973 | 4849F | |
| ERKELL, L.J. | 1973 | 4833F | * |
| ERMAKOVA, T.V. | 1977 | 5279 | |
| ETINGOF, R.N. | 1978 | 4996 | |
| ETNIER, D.A. | 1974 | 4590F | |
| EVERSON OEARSE, A.G. | 1976 | 4870F | |
| FAHRENBACH, W.H. | 1975 | 4693F | * |
| FALKMER, S. | | 5595 | * |
| FALKMER, S. | 1973 | 4596F | |
| FALKMER, S. | 1973 | 4765F | |
| FALKMER, S. | 1973 | 4838F | |
| FALKMER, S. | 1973 | 4854F | |
| FALKMER, S. | 1973 | 4855F | * |
| FALKMER, S. | 1973 | 5330 | |
| FALKMER, S. | 1974 | 4706F | |
| FALKMER, S. | 1974 | 4725F | * |
| FALKMER, S. | 1974 | 4966F | |
| FALKMER, S. | 1974 | 5546 | * |
| FALKMER, S. | 1975 | 4654F | |
| FALKMER, S. | 1975 | 4744F | * |
| FALKMER, S. | 1975 | 5400F | |
| FALKMER, S. | 1976 | 5192 | * |
| FALKMER, S. | 1976 | 5196 | * |
| FALKMER, S. | 1976 | 5285F | * |
| FALKMER, S. | 1976 | 5288F | * |
| FALKMER, S. | 1977 | 5286 | |
| FALKMER, S. | 1977 | 5365F | |
| FALKMER, S. | 1977 | 5371F | |
| FALKMER, S. | 1978 | 5605F | * |
| FALKNER, S. | 1973 | 4918F | |
| FANGE. R. | 1976 | 5271F | * |
| FANGE, R. | 1973 | 4582F | |
| FANGE, R. | 1973 | 4804F | * |
| FANGE, R. | 1973 | 4849F | * |
| FANGE, R. | 1973 | 4852F | * |
| FANGE, R. | 1973 | 4857F | * |
| FANGE, R. | 1973 | 4860F | * |
| FANGE, R. | 1974 | 4945F | * |
| FANGE, R. | 1976 | 4800F | |
| FANGE, R. | 1976 | 5063F | * |
| FANGE, R. | 1976 | 5113F | |
| FANGE, R. | 1976 | 5285F | |
| FANGE, R. | 1977 | 5015 | |
| FANGE, R. | 1977 | 5341F | |
| FANGE, R. ED. | 1973 | 4793F | * |
| FARMER, G.J. | 1973 | 4754F | * |
| FARMER, G.J. | 1974 | 5610F | * |
| FARMER, G.J. | 1975 | 4673F | * |
| FARMER, G.J. | 1975 | 4769F | |
| FARMER, G.J. | 1977 | 5340F | * |
| FARRINGER, J.E. | 1972 | 5460 | * |
| FEDOROV, B.A. | 1976 | 4813 | * |
| FEDOROV, B.A. | 1976 | 5070 | |
| FEDOROV, B.A. | 1977 | 5349 | * |
| FENSTERMACHER, J.D. | 1978 | 5515 | |
| FERNHOLM, B. | 1972 | 4791F | * |
| FERNHOLM, B. | 1973 | 4837F | * |
| FERNHOLM, B. | 1974 | 4699F | * |
| FERNHOLM, B. | 1975 | 4609F | |
| FERNHOLM, B. | 1975 | 4631F | * |
| FERNHOLM, B. | 1975 | 4646F | |
| FERNHOLM, B. | 1975 | 4669F | |
| FERNHOLM, B. | 1975 | 4703F | * |
| FERNHOLM, B. | 1975 | 4782F | |
| FETTEROLF, C.M. ED. | 1975 | 4756F | * |
| FILOSA, M.F. | 1978 | 5538F | |
| FILOSA, M.F. | 1978 | 5539F | |
| FILOSOFOVA, E.M. | 1969 | 4803F | |
| FILOSOFOVA, E.M. | 1969 | 4896F | * |
| FINSTAD, J. | 1975 | 4550F | |
| FLASAR, I. | 1975 | 5186F | * |
| FLASAROVA, M. | 1975 | 5186F | |
| FLEISHMAN, D.G. | 1973 | 4952F | * |
| FLOOD, P.R. | 1962 | 5334 | * |
| FLOOD, P.R. | 1973 | 4840F | * |
| FLOOD, P.R. | 1973 | 4842F | * |
| FLOOD, P.R. | 1975 | 4977 | * |
| FLOOD, P.R. | 1977 | 5265F | * |
| FLORKIN, M. EDS. | 1974 | 4707F | * |
| FLYNN, T.G. | 1976 | 5570 | |
| FOLLETT, B.K. | 1971 | 4657 | |
| FONG, B.B. | 1977 | 5392 | |
| FOTINA, E.B. | 1977 | 5281F | |
| FOUCRIER, J. | 1976 | 5004F | * |
| FOUTS, J.R. | 1973 | 4964F | * |
| FOUTS, J.R. | 1973 | 4972 | |
| FOUTS, J.R. | 1974 | 4604F | |
| FOX, H. | 1976 | 5148F | * |
| FRANKENBERG, R. | 1974 | 4680 | * |
| FREDGA, K. | 1977 | 5566 | * |
| FREEMAN, P.A. | 1976 | 5085F | |
| FREMLING, C.R. | 1975 | 5137F | * |
| FRIDAY, A.E. | 1973 | 4737 | |
| FROEHNER, S.C. | 1976 | 5014 | * |
| FUJITA, H. | 1974 | 4916 | * |
| FUJITA, H. | 1975 | 4587F | * |
| FUJITA, H. | 1975 | 4690F | * |
| FUJITA, H. | 1975 | 5098 | * |

| Name | Year | Number | |
|---|---|---|---|
| GABE, M. | | 5437 | * |
| GAGLOYEV, V.N. | 1973 | 4930 | |
| GAGNON, A. | 1973 | 4767F | |
| GALLAWAY, B.J. | 1977 | 5207 | |
| GAMMELTOFT, S. | 1976 | 5090 | * |
| GAMMELTOFT, S. | 1977 | 5154F | |
| GARCIA-CASTINEIRAS, S. | 1977 | 5308 | * |
| GARTMAN, D.K. | 1974 | 4637F | * |
| GAS, N. | 1973 | 4665 | |
| GEIHSLER, M.R. | 1975 | 5203 | * |
| GEISSEL, L. | 1977 | 5421F | |
| GEISSEL, L.D. | 1975 | 5387F | |
| GEISSEL, L.O. | 1975 | 5118F | |
| GEORGE, J.C. | 1974 | 4743F | * |
| GEORGE, J.C. | 1977 | 5096F | |
| GEORGIEVA, V. | 1977 | 5543F | |
| GERLOVIN, E. SH. | 1976 | 5290 | * |
| GERSHWIN, M.E. | 1978 | 5565 | |
| GETSEVICHYUTE, S. | 1974 | 4664 | * |
| GHARRETT, A.J. | 1977 | 5147 | |
| GIDHOLM, L. | 1973 | 4860F | |
| GILDERHUS, P.A. | 1973 | 5432F | * |
| GILDERHUS, P.A. | 1975 | 5051F | * |
| GILDERHUS, P.A. | 1975 | 5053F | * |
| GILDERHUS, P.A. | 1977 | 5463F | |
| GILDERHUS, P.A. | 1978 | 5425 | * |
| GILDERHUS, P.A. | 1978 | 5459 | * |
| GILHOUSEN, P. | 1973 | 5343F | |
| GILLY, R. | 1974 | 4976 | * |
| GINSBERG, B.H. | 1977 | 5286 | |
| GLADNER, J.A. | 1973 | 4738 | |
| GLADNER, J.A. | 1974 | 4594F | |
| GLADNER, J.A. | 1975 | 4682 | |
| GLADNER, J.A. | 1976 | 5011 | |
| GLADNER, J.A. | 1977 | 5167 | |
| GLIEMANN, J. | 1977 | 5154F | |
| GLOVER, C.J.M. | 1976 | 4718F | * |
| GOLOVANOV, I.B. | 1973 | 4930 | |
| GONCHAREVSKAYA, O.A. | 1976 | 5064F | * |
| GONCHAREVSKAYA, O.A. | 1976 | 5165 | * |
| GONCHAREVSKAYA, O.A. | 1977 | 4569 | * |
| GONCHAREVSKAYA, O.A. | 1977 | 5354F | * |
| GONCHAROVSKAYA, O.A. | 1975 | 4694F | * |
| GONNERT, R. | 1962 | 5550 | |
| GOOD, R.A. | 1975 | 4550F | |
| GOODMAN, M. | 1976 | 4982 | |
| GOODMAN, R. | 1976 | 5158 | |
| GOOLD, R.J. | 1974 | 4889F | |
| GOOLD, R.J. | 1974 | 4892F | |
| GOOLD, R.J. | 1974 | 4894F | |
| GOOLD, R.J. | 1976 | 4547F | * |
| GOOLD, R.J. | 1976 | 4605F | * |
| GOOLD, R.J. | 1976 | 5229F | * |
| GOOS, H.J.T. | 1977 | 5534 | * |
| GOOSENS, N. | 1977 | 5312 | * |
| GORBMAN, A. | 1973 | 5328F | |
| GORBMAN, A. | 1974 | 4564F | |
| GORBMAN, A. | 1975 | 4577F | |
| GORBMAN, A. | 1975 | 4585F | * |
| GORBMAN, A. | 1975 | 4729F | |
| GORBMAN, A. | 1975 | 5042F | |
| GORBMAN, A. | 1976 | 5003F | |
| GORBMAN, A. | 1976 | 5184F | |
| GORBMAN, A. | 1977 | 5151F | |
| GORBMAN, A. | 1977 | 5152F | |
| GORBMAN, A. | 1977 | 5159F | |
| GORBMAN, A. | 1977 | 5326F | |
| GORBMAN, A. | 1978 | 5600 | |
| GORBUNOVA, M.P. | 1975 | 4997 | * |
| GORDIENKO, V.M. | 1974 | 4635 | |
| GOROKHOV, YU.A. | 1974 | 4650F | |
| GOULD, G. | 1975 | 5197F | |
| GOVARDOVSKII, V.I. | 1975 | 4817F | * |
| GOVARDOVSKII, V.I. | 1977 | 5274 | * |
| GOVYRIN, V.A. | 1977 | 5504 | * |
| GRABDA, J. | 1971 | 4688F | * |
| GRADWELL, N. | 1972 | 5461 | * |
| GRAHAM, D.H. | 1953 | 5232F | * |
| GRASGOF, V.M. | 1976 | 5073 | |
| GRAY, E.G. | 1976 | 5213 | * |
| GRAY, R.H. | 1977 | 5569 | * |
| GREAT LAKES FISH. COMM. | 1976 | 5223F | * |
| GREAT LAKES FISH. COMM. | 1976 | 5499F | * |
| GREAT LAKES FISH. COMM. | 1977 | 5246F | * |
| GREAT LAKES FISH. COMM. | 1977 | 5247F | * |
| GREAT LAKES FISH. COMM. | 1977 | 5248F | * |
| GREAT LAKES FISH. COMM. | 1977 | 5249F | * |
| GREAT LAKES FISH. COMM. | 1977 | 5250F | * |
| GREAT LAKES FISH. COMM. | 1977 | 5251F | * |
| GREAT LAKES FISH. COMM. | 1977 | 5252F | * |
| GREAT LAKES FISH. COMM. | 1977 | 5253F | * |
| GREAT LAKES FISH. COMM. | 1977 | 5254F | * |
| GREAT LAKES FISH. COMM. | 1977 | 5255F | * |
| GREAT LAKES FISH. COMM. | 1977 | 5256F | * |
| GREAT LAKES FISH. COMM. | 1977 | 5257F | * |
| GREAT LAKES FISH. COMM. | 1977 | 5258F | * |
| GREAT LAKES FISH. COMM. | 1977 | 5259F | * |
| GREAT LAKES FISH. COMM. | 1977 | 5260F | * |
| GREAT LAKES FISH. COMM. | 1977 | 5261F | * |
| GREAT LAKES FISH. COMM. | 1977 | 5262F | * |
| GREAT LAKES FISH. COMM. | 1977 | 5263F | * |
| GREAT LAKES FISH. COMM. | 1977 | 5264F | * |
| GREEN, J.F. | 1974 | 5033 | * |
| GREEN, J.R. | 1973 | 4544 | |
| GRIFFITH, R.W. | 1977 | 5238F | |
| GRIGG, G.C. | 1974 | 4909F | * |
| GRIGOR'EV, N.I. | 1975 | 4701F | * |
| GRODZINSKI, Z. | 1974 | 4778F | * |
| GROEN, C.L. | 1975 | 5197F | |
| GROENEWOUD, G. | 1976 | 5024F | |
| GRUZOVA, M.N. | 1975 | 4883F | * |
| GUARINO, A.M. | 1972 | 4794F | * |
| GUARINO, A.M. | 1972 | 4935 | |
| GUARINO, A.M. | 1974 | 4604F | |
| GURAYA, S.S. | 1976 | 5190 | * |
| GUSTAFSSON, J.A. | 1976 | 5199F | |
| HAEDRICH, R.L. | 1974 | 4961F | |
| HAEDRICH, R.L. | 1977 | 5125F | |
| HAGELIN, L.O. | 1974 | 5234 | * |
| HALL, Z.W. | 1976 | 5014 | |
| HALLBACK, D.-A. | 1973 | 4839F | * |
| HALSTEAD, L.B. | 1973 | 4611 | * |
| HALVER, J.E. | 1973 | 4738 | |
| HALVER, J.E. | 1974 | 4594F | |
| HALVER, J.E. | 1975 | 4682 | |
| HAMILTON, S.E. | 1974 | 5458 | * |
| HANEBERG, B. | 1978 | 5338 | |
| HANKE, K. | 1971 | 4773F | |
| HANSEN, C.R. | 1976 | 5405F | * |
| HANSEN, C.R. | 1978 | 5457 | * |
| HANSEN, C.R.JR. | 1975 | 5120F | |
| HANSEN, S.J. | 1974 | 4726 | |
| HANSEN, S.J. | 1978 | 5516F | * |
| HANSEN, S.J. | 1978 | 5518F | * |
| HANSON, L.H. | 1964 | 5552 | |
| HANSON, L.H. | 1974 | 4970F | * |
| HANSON, L.H. | 1977 | 5241F | |
| HARDISTY, M.W. | 1973 | 5009 | * |
| HARDISTY, M.W. | 1974 | 4933 | * |
| HARDISTY, M.W. | 1975 | 4886F | * |
| HARDISTY, M.W. | 1976 | 5002F | * |
| HARDISTY, M.W. | 1976 | 5233F | * |
| HARDISTY, MW. | 1977 | 5306F | |
| HARDY, A. | 1976 | 5193 | |
| HARMAN, P.D. | 1975 | 5427 | |
| HARMAN, P.D. | 1976 | 5431 | * |
| HARMAN, P.D. | 1977 | 5442F | |
| HARMAN, P.D. | 1978 | 5462F | |
| HARP, G.L. | 1975 | 5055 | * |
| HARRINGTON, J. | 1977 | 5168 | |
| HASHIMOTO, Y. | 1975 | 5044F | |
| HASLEWOOD, G.A.D. | 1977 | 5156F | |
| HASSLER, O. | 1973 | 4838F | |
| HAVU, N. | 1973 | 4765F | |
| HAVU, N. | 1974 | 4725F | |
| HAVU, N. | 1977 | 5365F | * |
| HAVU, N. | 1977 | 5371F | * |
| HAY, A.W.M. | 1975 | 5099 | * |
| HAYS, L. | 1973 | 4968 | |
| HAYS, R.M. | 1973 | 4968 | |
| HAZLETT, C.A. | 1975 | 4687F | |
| HEATH-EVES, M.J. | 1974 | 4568 | * |

| | | | | | | | | |
|---|---|---|---|---|---|---|---|---|
| HEATH, R.G. | 1972 | 5416F | * | HUDSON, R.H. | 1976 | 5430F | * |
| HEATH, R.G. | 1975 | 5498F | | HUDSON, R.H. | 1978 | 5456 | * |
| HEINTZ, A. | 1962 | 5528 | * | HUGGINS, N.H. | 1977 | 5298F | |
| HELLE, K.B. | 1972 | 4789F | * | HUGGINS, R.J. | 1973 | 5009 | |
| HELLE, K.B. | 1973 | 4597F | * | HUGGINS, R.J. | 1974 | 4933 | |
| HELLE, K.B. | 1973 | 4847F | * | HUGHES, G.M. | 1973 | 4719F | |
| HELLE, K.B. | 1975 | 4792F | * | HULTSJO, C. | 1973 | 4851F | * |
| HELLE, K.B. | 1975 | 5046 | | HUNN, J.B. | 1973 | 5472F | * |
| HELLER, H. | 1974 | 4988 | * | HUNN, J.B. | 1974 | 5307F | * |
| HENDERSON, I.W. | 1972 | 4624 | * | HUNN, J.B. | 1975 | 4660F | * |
| HENDERSON, I.W. | 1974 | 4958 | * | HUNN, J.B. | 1975 | 5422F | * |
| HENDERSON, N.E. | 1972 | 4922F | * | HUNN, J.B. | 1976 | 5183F | * |
| HENDERSON, N.E. | 1975 | 4656F | * | HUNN, J.B. | 1977 | 5366F | * |
| HENDERSON, N.E. | 1976 | 4723F | * | HUNN, J.B. | 1978 | 5455 | * |
| HENSON, J.G. | 1977 | 5167 | | HUNN, J.L. | 1977 | 5424 | |
| HEROLD, R.C.B. | 1975 | 4875F | * | ICHIKAWA, T. | 1972 | 4943F | |
| HERZFELD, J. | 1974 | 4552 | * | ICHIKAWA, T. | 1977 | 5525F | * |
| HESSLER, R.R. | 1974 | 4575F | | IDLER, D.R. | 1976 | 4885F | * |
| HILDEMANN, W.H. | 1975 | 4869F | | IEZUITOVA, N.N. | 1974 | 4947F | |
| HILL, E.F. | 1972 | 5416F | | INANO, H. | 1976 | 5199F | * |
| HILL, E.F. | 1975 | 5498F | * | INUI, Y. | 1977 | 5326F | * |
| HILL, R.L. | 1973 | 4549 | | IORIYA, T. | 1976 | 4938 | * |
| HIPPE, E. | 1977 | 5188 | * | ISHII, S. | 1975 | 4538F | |
| HIRABAYASHI, T. | 1974 | 4628F | * | ISHII, S. | 1976 | 5078F | |
| HIRABAYASHI, T. | 1974 | 4671F | * | ITINA, N.A. | 1975 | 5094F | * |
| HIRD, F.J.R. | 1977 | 5211 | | JACKSON, I.M.D. | 1974 | 5056F | * |
| HIRD, F.J.R. | 1977 | 5212 | | JAHNKE, M. | 1973 | 4853F | |
| HIROKANE, T. | 1973 | 4795F | | JANVIER, P. | 1971 | 4692F | * |
| HIROKANE, T. | 1973 | 5114F | | JANVIER, P. | 1971 | 4810F | * |
| HIROSE, K. | 1975 | 4609F | * | JANVIER, P. | 1973 | 4578 | * |
| HITCH, R.K. | 1974 | 4590F | * | JANVIER, P. | 1973 | 4959 | * |
| HOCUTT, C.H. | 1977 | 4994F | * | JANVIER, P. | 1974 | 4675F | * |
| HOFFMAN, G.L. | 1973 | 4662 | * | JANVIER, P. | 1974 | 4939F | * |
| HOFFMAN, P. | 1974 | 4572 | | JANVIER, P. | 1975 | 4648F | * |
| HOFFMAN, R.A. | 1975 | 5535 | * | JANVIER, P. | 1975 | 4709F | * |
| HOGLUND, N-G. | 1976 | 5019F | | JANVIER, P. | 1975 | 5047F | * |
| HOHEISEL, G. | 1973 | 4747F | | JANVIER, P. | 1975 | 5547 | * |
| HOHEISEL, G. | 1973 | 5215 | | JANVIER, P. | 1977 | 5346F | * |
| HOHEISEL, H. | 1974 | 4927 | | JARLFORS, U. | 1975 | 4818 | |
| HOLBROOK, K.P. | 1975 | 5284 | * | JASINSKI, A. | 1969 | 4979 | * |
| HOLCIK, J. | 1970 | 5582 | * | JEFFERIES, R.P.S. | 1978 | 5121 | * |
| HOLCIK, J. | 1974 | 4541F | * | JENSEN, D. | 1965 | 5503 | * |
| HOLCIK, J. | 1976 | 5175F | * | JESPERSEN, A. | 1975 | 4704F | * |
| HOLMBERG, K. | 1973 | 4543F | * | JOHANSSON-SJOBECK, M-L. | 1974 | 4945F | |
| HOLMBERG, K. | 1973 | 4845F | * | JOHANSSON-SJOBECK,M-L. | 1976 | 5113F | |
| HOLMBERG, K. | 1974 | 5100F | | JOHANSSON-STOBECK, M-L. | 1976 | 5285F | |
| HOLMBERG, K. | 1975 | 4631F | | JOHANSSON, M.L. | 1973 | 4859F | * |
| HOLMBERG, K. | 1976 | 5039 | * | JOHN, T.M. | 1977 | 5096F | * |
| HOLMBERG, K. | 1977 | 5277F | * | JOHNSON, B.G.H. | 1973 | 5481 | * |
| HOLMBERG, K. | 1977 | 5299F | * | JOHNSON, B.G.H. | 1973 | 5485F | * |
| HOLMBERG, K. | 1977 | 5310F | | JOHNSON, B.G.H. | 1973 | 5486F | * |
| HOLMES, R.L. | 197 | 4732 | * | JOHNSON, B.G.H. | 1973 | 5487 | * |
| HOLMQUIST, R. | 1976 | 4982 | * | JOHNSON, B.G.H. | 1973 | 5488F | * |
| HOLTERDAHL, H. | 1975 | 4911 | * | JOHNSON, B.G.H. | 1973 | 5497F | * |
| HOMMA, S. | 1973 | 4560 | * | JOHNSON, B.G.H. | 1974 | 4630F | |
| HOMMA, S. | 1975 | 4788F | * | JOHNSON, B.G.H. | 1974 | 4759F | |
| HOMMA, S. | 1976 | 5112F | * | JOHNSON, B.G.H. | 1974 | 4826F | |
| HOMMA, S. | 1978 | 5466F | * | JOHNSON, B.G.H. | 1974 | 4848F | * |
| HOMMA, S. | 1978 | 5606F | * | JOHNSON, B.G.H. | 1974 | 4866F | * |
| HONMA, S. | 1973 | 5572F | * | JOHNSON, B.G.H. | 1974 | 4900F | * |
| HONMA, Y. | 1977 | 5187F | | JOHNSON, B.G.H. | 1974 | 4902F | * |
| HONMA, Y. | 1977 | 5596 | | JOHNSON, B.G.H. | 1974 | 4905F | * |
| HOPSON, J.A. | 1974 | 4537 | * | JOHNSON, B.G.H. | 1974 | 4906F | * |
| HORI, S.H. | 1976 | 5170F | | JOHNSON, B.G.H. | 1975 | 4962F | * |
| HORI, S.H. | 1977 | 4717 | | JOHNSON, B.G.H. | 1975 | 5088F | * |
| HORI, S.H. | 1978 | 5574F | | JOHNSON, B.G.H. | 1975 | 5103F | * |
| HORNSEY, D.J. | 1977 | 5208 | * | JOHNSON, B.G.H. | 1975 | 5135F | * |
| HORNSEY, D.J. | 1977 | 5306F | | JOHNSON, B.G.H. | 1975 | 5136F | * |
| HORSTEDT, P. | 1975 | 4762F | | JOHNSON, B.G.H. | 1976 | 5225F | * |
| HORSTMAN, E. | 1973 | 4978 | * | JOHNSON, B.G.H. | 1976 | 5226F | * |
| HOSAKA, K. | 1976 | 5074 | | JOHNSON, B.G.H. | 1976 | 5227F | * |
| HOSHINO, T. | 1975 | 4663F | * | JOHNSON, B.G.H. | 1976 | 5228F | * |
| HOSOYA, Y. | 1974 | 4940F | | JOHNSON, B.G.H. | 1976 | 5231F | |
| HOSOYA, Y. | 1977 | 5187F | | JOHNSON, H.E. | 1975 | 5118F | |
| HOWELL, J.H. | 1964 | 5552 | * | JOHNSON, H.E. | 1975 | 5387F | |
| HOWELL, J.H. | 1974 | 4970F | | JOHNSON, H.E. | 1976 | 5242F | |
| HOWELL, J.H. | 1974 | 5445 | | JOHNSON, H.E. | 1976 | 5386F | |
| HOWELL, J.H. | 1975 | 5050F | | JOHNSON, H.E. | 1977 | 5419F | |
| HSU, D-S. | 1974 | 4572 | * | JOHNSON, H.E. | 1977 | 5421F | |
| HUDSON, R.H. | 1971 | 5377F | * | JOHNSON, J.A. | 1975 | 4685 | |
| HUDSON, R.H. | 1971 | 5378F | * | JONES, A.N. | 1975 | 4638 | * |

| Name | Year | Number | |
|---|---|---|---|
| JONES, B. | 1976 | 4995 | |
| JONES, H. | 1973 | 4946 | |
| JONES, H. | 1975 | 4805F | |
| JONES, I.C. | 1972 | 4624 | |
| JONES, I.C. | 1974 | 4958 | |
| JORDAN, F.P. | 1976 | 5364F | |
| JORGENSEN, F.S. | 1977 | 5188 | |
| JOSS, J.M.P. | 1973 | 5315 | * |
| JOSS, J.M.P. | 1977 | 5209F | * |
| JOYSEY, K.A. | 1973 | 4737 | |
| JUKES, T.H. | 1976 | 4982 | |
| KABAT, E.A. | | 4626F | |
| KAMYSHNAYA, M.S. | 1973 | 4622F | * |
| KAN, T.T. | 1973 | 5555 | |
| KAN, T.T. | 1975 | 5554 | * |
| KANEVSKII, Y.P. | 1973 | 4952F | |
| KANJE, M. | 1974 | 4945F | |
| KAPLAN, M.A. | 1973 | 4968 | * |
| KAPLAN, M.A. | 1976 | 5018F | |
| KAPLAN, M.A. | 1976 | 5465F | |
| KAPLAN, M.A. | 1977 | 5166 | |
| KARAMYAN, A.I. | 1972 | 5181F | * |
| KARAMYAN, A.I. | 1973 | 4612 | * |
| KARAMYAN, A.I. | 1975 | 4540F | * |
| KARAMYAN, A.I. | 1975 | 4642F | * |
| KARATAYUTE-TALIMAA, V.N. | 1968 | 5087 | |
| KARTAR, S. | 1974 | 4933 | |
| KASE, F. | 1976 | 5372F | * |
| KASHAPOVA, L.A. | 1976 | 5317 | * |
| KASTIN, A.J. | 1974 | 5034F | * |
| KATO, S. | 1973 | 4957 | |
| KATO, S. | 1973 | 4980F | |
| KAWABATA, I. | 1975 | 4581 | |
| KAWAI, Y. | 1975 | 5041 | |
| KAWATSKI, J.A. | 1973 | 5454F | * |
| KAWATSKI, J.A. | 1974 | 4937F | * |
| KAWATSKI, J.A. | 1974 | 5404F | * |
| KAWATSKI, J.A. | 1975 | 4647F | * |
| KAWATSKI, J.A. | 1975 | 5120F | * |
| KAWATSKI, J.A. | 1976 | 5405F | |
| KAWATSKI, J.A. | 1977 | 5453F | * |
| KAWATSKI, J.A. | 1978 | 5457 | |
| KELENYI, G. | 1976 | 5131F | * |
| KEMPE, L.L. | 1973 | 4736F | * |
| KENNEDY, M.C. | 1975 | 4785F | |
| KENNEDY, M.C. | 1977 | 5160F | * |
| KERKOF, P.R. | 1973 | 5328F | * |
| KERR, J.W. | 1975 | 4871 | * |
| KEVERN, N.R. | 1975 | 4614F | |
| KHLYUSTINA, T.E. | 1975 | 4887F | |
| KHLYUSTINA, T.E. | 1975 | 4989 | |
| KHOKHOLEV, L.S. | 1974 | 4955F | * |
| KHOR'KOV, A.D. | 1971 | 5322F | * |
| KIER, E.L. | 1976 | 5615 | * |
| KIER, P.M. | 1975 | 4819F | * |
| KILLICK, S.R. | 1973 | 5343F | |
| KIMM, S.W. | 1977 | 5614 | |
| KING, E.L. | 1964 | 5552 | |
| KING, E.L. JR. | 1974 | 4970F | |
| KING, E.L.JR. | 1976 | 5076F | |
| KING, G.R. | 1975 | 4711 | |
| KINTER, M.A. | 1972 | 4794F | |
| KIRSCHNER, L.B. | 1973 | 4934 | * |
| KITADA, J. | 1973 | 4948 | * |
| KITTLE, P.D. | 1975 | 5203 | |
| KNAGGS, E.H. | 1975 | 5157F | * |
| KNUTSON, D.D. | 1975 | 4693F | |
| KOBAYASHI, H. | | 4609F | |
| KOBAYASHI, H. | 1972 | 4943F | * |
| KOBAYASHI, H. | 1973 | 4928F | * |
| KOBAYASHI, H. | 1974 | 4591F | |
| KOBAYASHI, H. | 1975 | 4782F | |
| KOBAYASHI, H. | 1975 | 4984 | * |
| KOBAYASHI, H. | 1975 | 5029F | |
| KOCH, E.A. | 1976 | 5018F | |
| KOCH, E.A. | 1976 | 5465F | |
| KOCH, E.A. | 1977 | 5166 | |
| KOEHLER, A. | 1977 | 5506F | |
| KONDO, Y. | 1973 | 5559 | |
| KONSTANTINOVA, M.S. | 1973 | 4941F | * |
| KONSTANTINOVA, M.S. | 1973 | 4973 | * |
| KONSTANTINOVA, M.S. | 1974 | 4586 | |
| KONSTANTINOVA, M.S. | 1975 | 4606F | * |
| KONSTANTINOVA, M.S. | 1975 | 5000 | * |
| KORNELIUSSEN, H. | 1973 | 4588 | * |
| KORNELIUSSEN, H. | 1973 | 4610F | * |
| KORNELIUSSEN, H. | 1973 | 5335 | * |
| KORNELIUSSEN, H. | 1975 | 5043 | * |
| KORNELIUSSEN, H. | 1976 | 5171 | * |
| KORNELIUSSEN, H. | 1976 | 5177F | |
| KORNELIUSSEN, H. | 1977 | 5524F | * |
| KORNELUISSEN, H. | 1973 | 4563F | * |
| KOSAREVA, A.A. | | 4540F | |
| KOSAREVA, A.A. | 1977 | 5279 | * |
| KOSMACH, P.I. | 1974 | 4635 | * |
| KOTT, E. | 1973 | 4895F | * |
| KOTT, E. | 1974 | 5553 | * |
| KOTT, E. | 1975 | 5150 | |
| KOTT, E. | 1976 | 4750F | |
| KOTT, E. | 1976 | 4768F | |
| KOTT, E. | 1976 | 4821 | |
| KOTT, E. | 1977 | 5391F | |
| KOTT, E. | 1978 | 5542F | |
| KOZITSYN, S.A. | 1974 | 4632F | * |
| KRAUSE, C. | 1886 | 5530 | * |
| KRAYUSHKINA, L.S. | 1974 | 4616F | * |
| KREITZER, J.F | 1972 | 5416F | |
| KREPS, E.M. | 1977 | 5512F | * |
| KRIPPNER, M. | 1973 | 4828 | * |
| KROBER, R. | 1976 | 5070 | |
| KRUGLOVA, E.E. | 1974 | 4741F | |
| KRUGLOVA, E.E. | 1977 | 5512F | |
| KRYVI, H. | 1977 | 5265F | |
| KUB, M. | 1972 | 5381 | |
| KUEHN, K. | 1975 | 5086 | * |
| KUHLENBECK, H. | 1975 | 4730 | * |
| KUX, Z. | 1972 | 5581F | * |
| KUZ'MINA, V.V. | 1971 | 4681F | |
| KUZNETSOV, N.V. | 1974 | 4650F | * |
| LA POINTE, J. | 1977 | 5390F | * |
| LA POSTA, A. | | 5603 | |
| LA POSTA, A. | 1977 | 5342F | |
| LACHENMAYER, L. | 1973 | 4735F | |
| LAERM, J. | 1977 | 5161F | * |
| LAGERSTRAND, G. | 1973 | 4835F | * |
| LAKOSKI, J.M. | 1975 | 4685 | |
| LAMPE, J. | 1972 | 4775F | |
| LAMPE, J. | 1973 | 4760F | * |
| LAMSA, A.K. | 1974 | 4746F | |
| LARSEN, B. | 1978 | 5338 | |
| LARSEN, L.O. | | 4868F | * |
| LARSEN, L.O. | 1974 | 4592F | * |
| LARSEN, L.O. | 1974 | 5607F | * |
| LARSEN, L.O. | 1975 | 4574F | |
| LARSEN, L.O. | 1975 | 5527 | |
| LARSEN, L.O. | 1976 | 5062F | * |
| LARSEN, L.O. | 1976 | 5126F | * |
| LARSEN, L.O. | 1976 | 5131F | |
| LARSEN, L.O. | 1978 | 5509 | * |
| LARSSON, A. | 1973 | 4582F | |
| LARSSON, A. | 1976 | 5063F | |
| LARSSON, A. | 1976 | 5113F | * |
| LARSSON, A. | 1977 | 5015 | * |
| LARSSON, P. | 1976 | 5395F | * |
| LAURENCE, E.B. | 1976 | 5093 | * |
| LAWRENCE, A.L. | 1974 | 4724 | |
| LE LANNIC, J. | 1976 | 5016F | |
| LE LANNIC, J. | 1977 | 5352F | |
| LEA, R.N. | 1975 | 5157F | |
| LEAKE, L.D. | 1975 | 4829 | * |
| LEATHERLAND, J.F. | 1972 | 5450 | * |
| LEATHERLAND, J.F. | 1973 | 5339 | |
| LEATHERLAND, J.F. | 1975 | 4758F | |
| LEATHERLAND, J.F. | 1976 | 4779F | * |
| LECH, J.J. | 1972 | 5397F | * |
| LECH, J.J. | 1972 | 5413F | * |
| LECH, J.J. | 1972 | 5415 | * |
| LECH, J.J. | 1973 | 5402 | * |
| LECH, J.J. | 1973 | 5411F | |
| LECH, J.J. | 1973 | 5412F | * |
| LECH, J.J. | 1973 | 5414F | * |
| LECH, J.J. | 1973 | 5544 | * |

| Name | Year | Ref | | Name | Year | Ref | |
|---|---|---|---|---|---|---|---|
| LECH, J.J. | 1974 | 4652F | * | MAKI, A.W. | 1977 | 5419F | * |
| LECH, J.J. | 1974 | 5376 | | MAKI, A.W. | 1977 | 5421F | * |
| LECH, J.J. | 1974 | 5438 | | MALMQVIST, B. | 1977 | 5467F | |
| LECH, J.J. | 1975 | 4751F | * | MAMELLE, J.C. | 1974 | 4976 | |
| LECH, J.J. | 1975 | 5440 | | MANION, P.J. | 1977 | 5235F | |
| LECH, J.J. | 1976 | 5382F | | MANION, P.J. | 1977 | 5241F | * |
| LECH, J.J. | 1976 | 5439 | | MANION, P.J. | 1977 | 5348F | * |
| LECH, J.J. | 1978 | 5452 | * | MANUSOVA, N.B. | 1973 | 4742F | * |
| LEDVINA, M.M. | 1975 | 5120F | | MARKING, L.L. | 1973 | 5080F | |
| LEE, K.Y. | 1977 | 5614 | * | MARKING, L.L. | 1973 | 5429 | * |
| LEHMANN, H. | 1973 | 4737 | | MARKING, L.L. | 1975 | 4557F | |
| LEHTOLA, K.A. | 1973 | 4615 | * | MARKING, L.L. | 1975 | 4691F | * |
| LEIBMAN, D. YA. | 1977 | 5612 | | MARKING, L.L. | 1975 | 4712F | * |
| LEIBMAN, D.YA. | 1977 | 5276 | * | MARKING, L.L. | 1975 | 5396F | * |
| LEIBMAN, D.YA. | 1977 | 5313 | | MARKING, L.L. | 1975 | 5449 | |
| LEIBMAN, D.YA. | 1977 | 5522 | | MARKING, L.L. | 1976 | 5095F | |
| LEIBSON, L. | 1976 | 5123F | * | MARKING, L.L. | 1976 | 5407F | |
| LEIBSON, L.G. | 1973 | 4598F | | MARKLUND, S. | | 5595 | |
| LEIBSON, L.G. | 1977 | 5361F | | MARTIN, A.R. | 1974 | 4714 | * |
| LEIBUSH, B. | 1976 | 5123F | | MARTIN, A.R. | 1975 | 4580F | * |
| LEIBUSH, B.N. | 1972 | 4797F | | MARTIN, R.J. | 1977 | 5295F | * |
| LEIBUSH, B.N. | 1974 | 4721F | | MARTOJA, M. | 1975 | 5149 | * |
| LEIBUSH, B.N. | 1976 | 5124F | | MASHANSKII, V.F. | 1973 | 4583 | * |
| LELEK, A. | 1973 | 4733F | * | MASHBURN, T.A. JR. | 1974 | 4572 | |
| LELOUP, J. | 1976 | 5193 | * | MASLOVA, M.N. | 1975 | 4670F | |
| LEMETSCHWANDTNER, A. | 1977 | 5575F | | MATHERS, J.S. | 1974 | 4599F | * |
| LEMMA, A. | 1975 | 5451 | * | MATHERS, J.S. | 1974 | 5344F | * |
| LEMONS, D.E. | 1978 | 5464 | * | MATHIS, B.J. | 1975 | 4614F | * |
| LEONT'EV, V.G. | 1975 | 4670F | | MATSUZAKI, H. | 1975 | 5044F | |
| LEONT'EV, V.G. | 1977 | 5278 | | MATTHEWS, G. | 1977 | 5206F | * |
| LETT, P.F. | 1975 | 4769F | * | MATTHEWS, G. | 1978 | 5578F | * |
| LETT, P.F. | 1977 | 5340F | | MATTHEWS, G. | 1978 | 5608F | * |
| LEVANOVICH, V.V. | 1976 | 5204 | * | MATTHEWS, W.J. | 1975 | 5055 | |
| LEWIS, D.N. | 1978 | 5121 | | MATTISON, A. | 1976 | 5285F | |
| LEWIS, J.H. | 1973 | 4576 | * | MATTISON, A.G.M. | 1977 | 5341F | * |
| LEWIS, M.S. | 1974 | 4594F | | MATTISSON, A. | 1973 | 4852F | |
| LEWIS, M.S. | 1976 | 5011 | | MATTISSON, A. | 1976 | 4800F | |
| LEWIS, S.V. | 1976 | 4877F | * | MATTSON, P.E. | | 5595 | |
| LEWIS, S.V. | 1976 | 5587 | * | MATTY, A.J. | 1976 | 5184F | * |
| LEWIS, S.V. | 1977 | 5300F | * | MAUGHAN, O.E. | 1976 | 5082 | * |
| LEWIS, T.L. | 1973 | 4920F | | MAYER, F.L. | 1977 | 5577F | |
| LIDMAN, U. | 1973 | 4582F | | MAYER, F.L. | 1978 | 5406 | |
| LIDMAN, U. | 1976 | 5063F | | MAZIN, A.L. | 1973 | 4548F | * |
| LIE, H.R. | 1973 | 4841F | * | MAZIN, A.L. | 1975 | 5401F | * |
| LIE, H.R. | 1974 | 4876F | * | MCCAULEY, R.W. | 1977 | 5298F | * |
| LIKOVSKY, Z. | 1973 | 5119F | * | MCDONALD, M.J. | 1974 | 4971F | * |
| LILJEQVIST, G. | 1973 | 4834F | | MCDONALD, M.J. | 1974 | 5404F | |
| LIM, V.I. | 1977 | 5314F | * | MCDONALD, R. | 1976 | 5230F | * |
| LING, N. | 1976 | 5195 | | MCDONALD, R.B. | 1976 | 5116F | * |
| LINNA, T.J. | 1975 | 4550F | * | MCGINNISS, M.J. | 1976 | 5410F | |
| LISSITZKY, S. | 1975 | 4729F | | MCINERNEY, J.E. | 1974 | 4621 | * |
| LIU, Y.H. | 1975 | 5417 | * | MCINTYRE, J.D. | 1977 | 5147 | |
| LIVERSIDGE, J. | 1976 | 5178 | | MCKEE, P.A. | 1973 | 4549 | |
| LOEFFLER, E.J. | 1976 | 4995 | * | MCKEOWN, B.A. | 1975 | 4687F | * |
| LOHNISKY, K. | 1975 | 4764F | * | MCMILIAN, N.F. | 1973 | 4544 | |
| LONNING, S. | 1972 | 4789F | | MCMILLAN, D.B. | 1974 | 4568 | |
| LONNING, S. | 1973 | 4597F | | MCMILLAN, D.B. | 1975 | 4998F | |
| LONNING, S. | 1973 | 4847F | | MCMILLAN, D.B. | 1976 | 5567 | |
| LORSCHEIDER, F.L. | 1975 | 4656F | | MCNAMEE, M.G. | 1977 | 5164 | * |
| LOVTRUP, S. | 1976 | 5019F | | MEDNIKOV, B.M. | 1973 | 4802F | * |
| LOWENSTEIN, O.ED. | 1971 | 4689 | * | MEDNIKOV, B.M. | 1974 | 4567F | * |
| LUKOMSKAYA, N.YA. | 1974 | 4636F | * | MELANCON, M.J. | 1976 | 5382F | |
| LUNDGREN, G. | 1977 | 5365F | | MENZIE, C.M. | 1976 | 5183F | * |
| LUNDGREN, G. | 1977 | 5371F | | MEYER, F.P. | 1976 | 5500F | * |
| LUNDGREN, G. ET.AL. | 1973 | 4765F | | MEYER, F.P. | 1977 | 5367F | * |
| LUNDHOLM, M. | 1973 | 4858F | * | MEYLING, A.H. | 1962 | 5551 | * |
| LUNDIN, L-G. | 1973 | 4562F | * | MIKHEL'SON, M. YA. | 1975 | 4668 | * |
| LUNDIN, V. | 1973 | 4543F | | MILES, R.S. | 1977 | 5369 | |
| LUTZ, P.L. | 1976 | 4678F | | MIYOSHI, M. | 1977 | 5593 | * |
| LYCHAKOV, D.V. | 1976 | 5173 | * | MOHNIKE, W. | 1973 | 4760F | |
| LYCHAKOV, D.V. | 1977 | 5274 | | MOISE, H. | 1976 | 4982 | |
| LYZLOVA, E.M. | 1975 | 5179 | * | MOLNAR, B. | 1973 | 5399F | * |
| LYZLOVA, E.M. | 1976 | 5191 | * | MOLNAR, B. | 1975 | 4878F | * |
| LYZLOVA, E.M. | 1977 | 5305 | | MOLNAR, B. | 1975 | 5072F | * |
| MACEY, D.J. | 1978 | 5545F | * | MONACO, F. | | 5603 | * |
| MAGAZANIK, L.G. | 1974 | 4636F | | MONACO, F. | 1976 | 5025F | * |
| MAKI, A.W. | 1974 | 5388 | * | MONACO, F. | 1976 | 5163 | * |
| MAKI, A.W. | 1975 | 5118F | * | MONACO, F. | 1977 | 5342F | * |
| MAKI, A.W. | 1975 | 5387F | * | MONTFORT, M-F. | 1975 | 4729F | |
| MAKI, A.W. | 1976 | 5242F | * | MOORE, G.W. | 1976 | 4982 | |
| MAKI, A.W. | 1976 | 5386F | * | MOORE, H.H | 1974 | 4746F | * |

| | | | | | | | |
|---|---|---|---|---|---|---|---|
| MOORE, H.H. | 1977 | 5368F | | NORTHCUTT, R.G. | 1973 | 5245F | * |
| MOORE, I.A. | 1975 | 4886F | | NORW. ZOOL. SOC. | 1976 | 5469F | * |
| MOORE, J.W. | 1975 | 5030 | | NOVITSKAYA, L.I. | 1975 | 4771F | * |
| MOORE, J.W. | 1976 | 4776F | * | NOZAKI, M. | 1975 | 4782F | * |
| MOORE, J.W. | 1976 | 5142F | * | NURSALL, J.R. | 1972 | 4761F | * |
| MORAVEC, F. | 1976 | 5270 | * | NYBELIN, O. | 1973 | 4843F | * |
| MORAVEC, F. | 1977 | 5467F | * | NYGREN, A. | 1973 | 4853F | * |
| MORI, K. | 1976 | 5199F | | O'BOYLE, R.N. | 1975 | 5616F | * |
| MORITA, M. | 1974 | 4536 | | O'BOYLE, R.N. | 1977 | 4975F | * |
| MORITA, Y. | 1971 | 5585 | * | OBRUCHEV, D.V. | 1968 | 4823 | * |
| MORITA, Y. | 1973 | 4949F | * | OBRUCHEV, D.V. | 1968 | 5087 | * |
| MORITA, Y. | 1973 | 4956 | * | OBRUCHEV, D.V. | 1968 | 5091 | * |
| MORITA, Y. | 1975 | 4986 | * | OCHI, J. | 1974 | 4940F | * |
| MORRIS, R. | 1975 | 4991 | | OESER, R. | 1957 | 5580F | * |
| MORRIS, R. | 1976 | 5101F | | OGAWA, M. | 1973 | 4919 | |
| MORRIS, R. | 1976 | 5115F | * | OGAWA, M. | 1975 | 5044F | |
| MORRIS, R. | 1977 | 5169F | | OGORODNIKOVA, L.G. | 1977 | 5281F | * |
| MORRIS, S. | 1975 | 5129F | * | OHMAN, P. | 1974 | 4655F | * |
| MOSHIN, A.K.M. | 1977 | 5207 | * | OHMAN, P. | 1976 | 5001F | * |
| MOSKALENKO, T.A. | 1968 | 5092F | * | OHMAN, P. | 1976 | 5039 | |
| MOYER, P. | 1972 | 5413F | | OHMAN, P. | 1977 | 5214F | * |
| MUGGEO, M. | 1977 | 5286 | * | OHMAN, P. | 1977 | 5299F | |
| MULLER, J. | 1841 | 5291 | * | OHMAN, P. | 1977 | 5310F | |
| MUNRO, A.L.S. | 1976 | 5178 | * | OHTA, Y. | 1975 | 4538F | |
| MURAT, J.C. | 1977 | 5468F | * | OLE, B. | 1975 | 5527 | * |
| MURRAY, M. | 1973 | 4946 | * | OLESEN, H. | 1977 | 5188 | |
| MURRAY, M. | 1973 | 4963 | * | OLSON, L. E. | 1975 | 5449 | * |
| MURRAY, M. | 1975 | 4805F | * | OLSON, L.E. | 1973 | 5080F | * |
| MURRAY, M. | 1975 | 5117F | * | OLSON, L.E. | 1975 | 4712F | |
| MURTAUGH, P.A. | 1974 | 4594F | * | OLSSON, J.Y. | 1973 | 4850F | * |
| MURTAUGH, P.A. | 1974 | 4738 | * | OLSSON, R. | 1975 | 5198 | * |
| MURTAUGH, P.A. | 1975 | 4682 | * | OOI, E.C. | 1976 | 4539 | * |
| MURTAUGH, P.A. | 1977 | 5167 | * | OOI, E.C. | 1977 | 4695 | |
| MURTAUGH, P.A. | 1977 | 5302F | * | OOI, E.C. | 1977 | 4698 | * |
| MUTO, Y. | 1977 | 5320 | | OOI, E.C. | 1977 | 5350F | * |
| NAGANO, H. | 1976 | 5074 | * | OOTA, Y. | 1974 | 4608 | * |
| NAGEL, R. | 1977 | 5168 | | OOTA, Y. | 1976 | 5078F | |
| NAKAI, Y. | 1977 | 5596 | * | ORLOV, A.A. | 1975 | 4987 | |
| NAKANO, M. | 1977 | 5593 | | ORR, E. | 1972 | 5381 | |
| NAKAO, T. | 1966 | 5529F | * | ORVIG, T. | 1976 | 5102 | * |
| NAKAO, T. | 1973 | 4722 | * | OSBORNE, T.S. | 1975 | 4713 | |
| NAKAO, T. | 1974 | 4603F | * | OSTAPENKO, I.A. | 1978 | 4996 | |
| NAKAO, T. | 1975 | 4644F | * | OSTBERG, Y. | 1973 | 4596F | |
| NAKAO, T. | 1975 | 4757F | * | OSTBERG, Y. | 1973 | 4838F | * |
| NAKAO, T. | 1976 | 4884 | * | OSTBERG, Y. | 1973 | 4854F | * |
| NAKAO, T. | 1977 | 5185F | * | OSTBERG, Y. | 1973 | 4855F | |
| NAKAO, T. | 1977 | 5355 | * | OSTBERG, Y. | 1973 | 4918F | |
| NAKAO, T. | 1978 | 5519 | * | OSTBERG, Y. | 1973 | 5330 | |
| NAKAO, T. | 1978 | 5540F | * | OSTBERG, Y. | 1973 | 5561F | |
| NAKAZAWA, T. | 1974 | 4658 | * | OSTBERG, Y. | 1975 | 4617F | * |
| NALBANT, T. | 1974 | 4541F | | OSTBERG, Y. | 1975 | 4790F | |
| NARDELL, B. | 1975 | 4780F | | OSTBERG, Y. | 1976 | 4774F | * |
| NARDONI, C. | 1976 | 5059 | | OSTBERG, Y. | 1976 | 4786F | * |
| NASLEDOV, G.A. | 1975 | 5094F | | OSTBERG, Y. | 1976 | 4800F | * |
| NATALI, P.G. | 1976 | 5058 | | OSTBERG, Y. | 1976 | 4806F | * |
| NATALI, P.G. | 1976 | 5059 | * | OSTBERG, Y. | 1976 | 4870F | * |
| NATOCHIN, YU.V. | 1974 | 4684 | * | OSTBERG, Y. | 1976 | 5196 | |
| NATOCHIN, YU.V. | 1975 | 4670F | * | OSTBERG, Y. | 1976 | 5285F | |
| NATOCHIN, YU.V. | 1977 | 5354F | | OSTBERG, Y. | 1976 | 5288F | |
| NATOCHIN, YU.V. | 1977 | 5362 | * | OSTBERG, Y. | 1977 | 5517 | |
| NELSESTUEN, G.L. | 1976 | 5028 | | OSTERGAARD, K.L. | 1976 | 5090 | |
| NELSON, J.S. | 1976 | 4831 | * | OSTRACH, L. | 1973 | 4963 | |
| NIBLETT, P.D. | 1976 | 5077 | * | OSTRACH, L.H. | 1974 | 4627F | |
| NICOLAYSEN, K. | 1966 | 5336F | * | PACKARD, G.C. | 1976 | 5003F | * |
| NICOLAYSEN, K. | 1973 | 4588 | | PACKARD, M.J. | 1976 | 5003F | |
| NICOLAYSEN, K. | 1975 | 5043 | | PAFLITSCHEK, R. | 1977 | 5448 | * |
| NICOLAYSEN, K. | 1976 | 5006F | * | PAGE, R. | 1976 | 5364F | |
| NICOLAYSEN, K. | 1976 | 5007 | * | PAGEAU, Y. | 1976 | 5022F | * |
| NICOTRA, M.R. | 1976 | 5058 | | PAIEMENT, J.M. | 1975 | 4998F | * |
| NICOTRA, M.R. | 1976 | 5059 | | PAJOR, W.J. | 1976 | 5273 | * |
| NIERMANN, U. | 1976 | 5351F | | PAJOR, W.J. | 1977 | 5598 | * |
| NIEUWENHUYS, R. | 1974 | 4697F | * | PALEUS, S. | 1973 | 4834F | * |
| NIEWENHUYS, R. | 1977 | 5592 | * | PALEUS, S. | 1975 | 5244F | |
| NIIMI, A.J. | 1974 | 4727F | * | PANG, P.K.T. | 1977 | 5238F | * |
| NILSSEN, H. | 1975 | 4825 | * | PATZNER, R.A. | 1974 | 4589F | * |
| NILSSON, A. | 1973 | 4846F | * | PATZNER, R.A. | 1975 | 4667F | * |
| NILSSON, S. | 1973 | 4835F | | PATZNER, R.A. | 1976 | 5526F | * |
| NILSSON, S. | 1973 | 4836F | | PATZNER, R.A. | 1977 | 5347F | * |
| NISHIMURA, H. | 1973 | 4919 | * | PATZNER, R.A. | 1977 | 5543F | * |
| NOBIN, A. | 1973 | 4735F | | PATZNER, R.A. | 1978 | 5520F | * |
| NOBLE, R.W. | 1974 | 4971F | | PEARSE, A.G.E. | 1973 | 5561F | |

| Author | Year | Ref | |
|---|---|---|---|
| PEARSE, A.G.E. | 1973 | 5562F | |
| PEARSE, A.G.E. | 1974 | 5558 | |
| PEARSE, A.G.E. | 1975 | 4617F | |
| PEARSE, A.G.E. | 1976 | 4806F | |
| PEARSE, A.G.E. | 1977 | 5517 | |
| PEARSE, G.E. | 1974 | 4748F | |
| PEARSON, L. | 1976 | 4923 | |
| PEARSON, R. | 1976 | 4923 | * |
| PEPPLE, S. | 1973 | 5402 | |
| PEPPLE, S. | 1973 | 5412F | |
| PEPPLE, S.K. | 1974 | 5438 | |
| PERCY, R. | 1973 | 5339 | * |
| PERCY, R. | 1975 | 4758F | * |
| PERCY, R. | 1976 | 4716F | * |
| PERCY, R. | 1976 | 4779F | |
| PERCY, R. | 1977 | 5301F | * |
| PERKS, A.M. | 1974 | 4558F | |
| PERKS, A.M. | 1976 | 5021 | |
| PERKS, A.M. | 1977 | 5155 | |
| PERKS, A.M. | 1977 | 5393 | * |
| PERMITIN, YU.E. | 1977 | 5597 | * |
| PERRY, S.V. | 1974 | 4628F | |
| PERRY, S.V. | 1974 | 4671F | |
| PERSHINA, L.I. | 1977 | 5278 | |
| PETERSON J.D. | 1974 | 4966F | * |
| PETERSON, J.D. | 1973 | 4918F | |
| PETERSON, J.D. | 1973 | 5330 | |
| PETERSON, J.D. | 1975 | 4654F | * |
| PETERSON, J.D. | 1975 | 5400F | |
| PETTER, A.J. | 1974 | 4546 | * |
| PFEIL, W. | 1973 | 4760F | |
| PFENNINGER, K.H. | 1974 | 4595F | * |
| PFISTER, C. | 1974 | 4700F | * |
| PHARAND-COAD, S. | 1975 | 5150 | |
| PHARAND, S. | 1972 | 4752F | |
| PHILLIPS, J.W. | 1977 | 5211 | * |
| PHILLIPS, J.W. | 1977 | 5212 | * |
| PIAVIS. M.G. | 1978 | 5613 | |
| PIAVIS, G.W. | 1965 | 5374F | * |
| PIAVIS, G.W. | 1966 | 5373 | * |
| PIAVIS, G.W. | 1968 | 5375F | * |
| PIAVIS, G.W. | 1975 | 4780F | * |
| PIAVIS, G.W. | 1975 | 5050F | * |
| PIAVIS, G.W. | 1978 | 5613 | * |
| PICKERING, A.D. | 1974 | 4915F | * |
| PICKERING, A.D. | 1975 | 4991 | * |
| PICKERING, A.D. | 1976 | 5005 | * |
| PICKERING, A.D. | 1976 | 5065F | * |
| PICKERING, A.D. | 1976 | 5101F | * |
| PICKERING, A.D. | 1976 | 5115F | * |
| PICKERING, A.D. | 1976 | 5176F | * |
| PICKERING, A.D. | 1977 | 5169F | * |
| PIETSCHMANN, V. | 1962 | 5556 | * |
| PIGON, A. | 1974 | 4536 | * |
| PITCHFORD, R.J. | 1962 | 5551 | |
| PIZZO, S.V. | 1973 | 4549 | |
| PLETCHER, F.T. | 1963 | 4974F | * |
| PLISETSKAYA, E. | 1976 | 5123F | |
| PLISETSKAYA, E. M. | 1977 | 5468F | |
| PLISETSKAYA, E.M. | 1971 | 4681F | * |
| PLISETSKAYA, E.M. | 1972 | 4797F | * |
| PLISETSKAYA, E.M. | 1972 | 5122F | * |
| PLISETSKAYA, E.M. | 1973 | 4561F | * |
| PLISETSKAYA, E.M. | 1973 | 4598F | * |
| PLISETSKAYA, E.M. | 1974 | 4721F | * |
| PLISETSKAYA, E.M. | 1974 | 5267F | * |
| PLISETSKAYA, E.M. | 1975 | 4772F | * |
| PLISETSKAYA, E.M. | 1976 | 5124F | * |
| PLISETSKAYA, E.M. | 1976 | 5360F | |
| PLISETSKAYA, E.M. | 1977 | 5323F | |
| PLISETSKAYA, E.M. | 1977 | 5361F | * |
| POHL, R.J. | 1973 | 4964F | * |
| POHL, R.J. | 1974 | 4604F | * |
| POHLA, H. | 1977 | 5575F | * |
| POLENOV, A.L. | 1973 | 4556F | * |
| POLENOV, A.L. | 1974 | 4586 | * |
| POLMORYKHINA, A.N. | 1974 | 4623F | * |
| POLMORYKHINA, A.N. | 1977 | 5318 | * |
| POLTAVCHUK, M.A. | 1975 | 4873 | * |
| POLYAKOVA, T.I. | 1977 | 5323F | * |
| POMAZANSKAYA, L.F. | 1975 | 4670F | |
| POPOV, L.S. | 1973 | 4802F | |
| POSMNOV, I.E. | 1974 | 4650F | |
| POTTER, I.C. | | 5145 | * |
| POTTER, I.C. | 1973 | 4719F | |
| POTTER, I.C. | 1973 | 5560 | * |
| POTTER, I.C. | 1974 | 4649F | * |
| POTTER, I.C. | 1974 | 4651F | |
| POTTER, I.C. | 1974 | 4759F | * |
| POTTER, I.C. | 1975 | 4565F | * |
| POTTER, I.C. | 1975 | 4584F | * |
| POTTER, I.C. | 1975 | 4696 | |
| POTTER, I.C. | 1975 | 4710F | |
| POTTER, I.C. | 1975 | 4713 | * |
| POTTER, I.C. | 1975 | 5030 | * |
| POTTER, I.C. | 1975 | 5138F | |
| POTTER, I.C. | 1976 | 4678F | |
| POTTER, I.C. | 1976 | 4716F | |
| POTTER, I.C. | 1976 | 4776F | |
| POTTER, I.C. | 1976 | 4877F | |
| POTTER, I.C. | 1976 | 5142F | |
| POTTER, I.C. | 1976 | 5239 | * |
| POTTER, I.C. | 1976 | 5587 | |
| POTTER, I.C. | 1977 | 5140F | * |
| POTTER, I.C. | 1977 | 5300F | |
| POTTER, I.C. | 1977 | 5301F | |
| POTTER, I.C. | 1978 | 5537F | * |
| POTTER, I.C. | 1978 | 5545F | |
| PRAVDINA, N.I. | 1974 | 4741F | * |
| PRICHONNET, G. | 1976 | 5022F | |
| PRITCHARD, J.B. | 1972 | 4794F | |
| PROTASOV, V.R. | 1975 | 4987 | * |
| PRZYBYLSKI, R.J. | 1973 | 5245F | |
| PTITSYN, O.B. | 1974 | 4632F | |
| PUPPIONE, D.P. | 1974 | 4967 | * |
| PURVIS, H.A. | 1973 | 5473F | * |
| PYBUS, M.J | 1978 | 5327F | * |
| PYBUS, M.J. | 1978 | 5541F | * |
| PYCHA, R.L. | 1975 | 4711 | * |
| QUERO, J.C. | 1974 | 4815 | * |
| RAASTAD, J.E. | 1976 | 5395F | |
| RADNER, S. | 1973 | 4836F | * |
| RAGUSE-DEGENER, G. | 1976 | 5351F | |
| RAHEMTULLA, F. | 1976 | 5019F | * |
| RALL, D.P. | 1972 | 4794F | |
| RALL, D.P. | 1975 | 4805F | |
| RALL, D.P. | 1978 | 5515 | |
| RANDOLPH, K.N. | 1973 | 5089F | * |
| RANNEY, H.M. | 1974 | 4971F | |
| RAPPE, C. | | 5595 | |
| RAUTHER, M. | 1937 | 5588 | * |
| READ, L.J. | 1975 | 4620 | * |
| REALE, E. | 1975 | 5086 | |
| REDDY, A.L. | 1975 | 4869F | |
| REICHLIN, S. | 1974 | 5056F | |
| REMBISZEWSKI, J.M. | 1975 | 5084 | * |
| RENFRO, J.L. | 1972 | 4931 | |
| REPETSKI, J.E. | 1978 | 5599 | * |
| REUVERS, M.L. | 1974 | 4937F | |
| REYNOLDS, W.W. | 1977 | 5298F | |
| RIEGEL, J.A. | 1978 | 5536F | * |
| RILEY, M. | 1976 | 4715F | |
| RINGHAM, G. | 1974 | 4714 | |
| RINGHAM, G.L. | 1975 | 4580F | |
| RINGHAM, G.L. | 1975 | 4643F | * |
| RIVIER, C. | 1976 | 5195 | |
| RIVIER, J. | 1976 | 5195 | |
| RIVIERE, H.B. | 1975 | 4869F | * |
| RIZHAMADZE, N.A. | 1973 | 4583 | |
| ROBBINS, D.L. | 1978 | 5565 | * |
| ROBERTSON, J.D. | 1974 | 4705F | * |
| ROBERTSON, J.D. | 1976 | 4811F | * |
| ROBINSON, E.S. | 1973 | 5560 | |
| ROBINSON, E.S. | 1974 | 4649F | |
| ROBINSON, E.S. | 1975 | 4696 | * |
| ROBINSON, G.A. | 1975 | 4673F | |
| ROBISON, H.W. | 1974 | 4944F | * |
| ROBISON, H.W. | 1974 | 5032F | * |
| ROBISON, H.W. | 1974 | 5035F | * |
| ROBISON, H.W. | 1975 | 5202F | * |
| ROCHE, J. | | 5603 | |
| ROCHE, J. | 1976 | 5025F | |

| Name | Year | Ref | | Name | Year | Ref | |
|---|---|---|---|---|---|---|---|
| ROCHE, J. | 1977 | 5342F | | SCHULTZ, D.P. | 1977 | 5442F | * |
| ROHDE, F.C. | 1974 | 4676F | * | SCHULTZ, D.P. | 1978 | 5462F | |
| ROHDE, F.C. | 1975 | 4720F | * | SCHULTZ, H.J. | 1975 | 4809F | * |
| ROHDE, F.C. | 1976 | 5139F | * | SCHULTZE, H.P. | 1976 | 5240 | * |
| ROHDE, F.C. | 1976 | 5523F | * | SCHUTTE, C.H.J. | 1962 | 5551 | |
| ROLIK, H. | 1975 | 5084 | | SCHWAB, M.E. | 1973 | 4559F | * |
| ROLLAND, M. | 1975 | 4729F | | SCHWARTZ, M.L. | 1973 | 4549 | * |
| ROMERO-HERRERA, A. | 1973 | 4737 | * | SEILER, K. | 1973 | 5215 | |
| ROOS, J.F. | 1973 | 5343F | * | SEILER, K. | 1973 | 5563 | * |
| ROSEN, F.S. | 1974 | 4929 | * | SEILER, R. | 1973 | 5215 | |
| ROSENGREN, E. | 1973 | 4735F | | SEILER, R. | 1973 | 5563 | |
| ROSS, J.R.P. | 1976 | 5067F | * | SEIXAS, M.M.P. | 1976 | 5172 | * |
| ROSS, J.R.P. | 1977 | 5200 | * | SEKIMOTO, M. | 1972 | 4943F | |
| ROSSETTER, S. | 1977 | 5168 | | SELZER, M. | 1976 | 5069 | * |
| ROTH, E. | 1977 | 5168 | | SELZER, M.E. | 1978 | 5579F | * |
| ROTH, J. | 1977 | 5286 | | SEO, J.S. | 1977 | 5614 | |
| ROTH, J. | 1978 | 5601 | | SEREBRENIKOVA, T.P. | 1969 | 4803F | * |
| ROVAINEN, C.M. | 1973 | 5333F | * | SEREBRENIKOVA, T.P. | 1969 | 4896F | |
| ROVAINEN, C.M. | 1974 | 4542F | * | SEREBRENIKOVA, T.P. | 1975 | 4887F | * |
| ROVAINEN, C.M. | 1974 | 4595F | | SEREBRENIKOVA, T.P. | 1975 | 4989 | |
| ROVAINEN, C.M. | 1974 | 4625F | * | SEREBRENIKOVA, T.P. | 1975 | 5179 | |
| ROVAINEN, C.M. | 1974 | 4674F | * | SEREBRENIKOVA, T.P. | 1977 | 5305 | * |
| ROVAINEN, C.M. | 1975 | 4787F | * | SESTOFT, L. | 1976 | 5090 | |
| ROVAINEN, C.M. | 1976 | 4798F | * | SHAKHMATOVA, E.I. | 1977 | 5210F | * |
| ROVAINEN, C.M. | 1976 | 5026F | * | SHAPOVALOV, A.I. | 1974 | 4601F | |
| ROVAINEN, C.M. | 1976 | 5112F | * | SHAPOVALOV, A.I. | 1977 | 5353 | |
| ROVAINEN, C.M. | 1977 | 5269F | * | SHAPOVALOV, A.I. | 1977 | 5505 | * |
| ROVAINEN, C.M. | 1978 | 5606F | * | SHAPOVALOV, A.I. | 1977 | 5521 | |
| RUBINSON, K. | 1974 | 4951F | * | SHARMA, V.S. | 1974 | 4971F | |
| RUBINSON, K. | 1975 | 4785F | | SHARP, P.J. | 1971 | 4657 | |
| RUBINSON, K. | 1977 | 5160F | | SHELTON, R.G.J. | 1978 | 5514F | * |
| RUCKPAUL, K. | 1976 | 5070 | | SHELTON, W.L. | 1978 | 5564 | |
| RUDBACH, J.A. | 1974 | 4566F | | SHERA, W.P. | 1974 | 4888F | |
| RUEHLE, H-J. | 1974 | 4927 | | SHERA, W.P. | 1974 | 4891F | |
| RUHLE, H.J. | 1973 | 4747F | | SHERA, W.P. | 1974 | 4893F | |
| RURAK, D.W. | 1974 | 4558F | * | SHERA, W.P. | 1974 | 4903F | * |
| RURAK, D.W. | 1976 | 5021 | * | SHERA, W.P. | 1975 | 4645F | * |
| RURAK, D.W. | 1977 | 5155 | * | SHERA, W.P. | 1975 | 5104F | |
| RYAPOLOVA, N.I. | 1974 | 4814 | * | SHERA, W.P. | 1975 | 5106F | |
| RYAPOLOVA, N.I. | 1975 | 5012 | * | SHERA, W.P. | 1975 | 5107F | |
| RYBAK, B. | 1974 | 4740 | | SHERA, W.P. | 1975 | 5109F | * |
| RYE, R.P. | 1976 | 5076F | * | SHERA, W.P. | 1975 | 5110F | * |
| RYE, R.P., JR. | 1972 | 5428 | * | SHIBATA, Y. | 1976 | 5266F | * |
| SAALFELD, R. | 1974 | 5445 | * | SHIBATA, Y. | 1976 | 5531 | * |
| SABESAN, M.N. | 1974 | 4629 | | SHIBATA, Y. | 1977 | 4872 | * |
| SADO, Y. | 1976 | 5170F | * | SHIBATA, Y. | 1977 | 5294F | * |
| SADO, Y. | 1978 | 5574F | * | SHIBATA, Y. | 1977 | 5602 | * |
| SAINSBURY, M. | 1974 | 4933 | | SHIDOJI, Y. | 1977 | 5320 | * |
| SAKAI, K. | 1975 | 4618F | | SHIFF, C.J. | 1972 | 5441 | * |
| SAKHAROV, D.A. | 1976 | 5317 | | SHIGIN, A.A. | 1976 | 5153 | * |
| SANDERS, H.O. | 1975 | 5144F | * | SHIN, Y.C. | 1977 | 5356F | * |
| SANDERS, H.O. | 1977 | 5444F | * | SHINKAWA, Y. | 1975 | 4690F | |
| SANTOLAMAZZA, C. | 1976 | 5163 | | SHINOHARA, H. | 1973 | 4795F | |
| SAVINA, M.V. | 1975 | 4613F | * | SHINOHARA, H. | 1973 | 4796F | |
| SAVINA, M.V. | 1976 | 5060 | * | SHIODA, S. | 1977 | 5187F | * |
| SAVINA, M.V. | 1976 | 5360F | * | SHIODA, S. | 1977 | 5596 | |
| SAVINA, M.V. | 1977 | 5218F | * | SHOLDICE, J.A. | 1976 | 5567 | * |
| SAWYER, W.J. | 1977 | 5389F | * | SHORT, E.D. | 1975 | 5203 | |
| SCARPELLI, D.G. | 1976 | 5282 | | SHUKUYA, R. | 1976 | 5074 | |
| SCARSBROOK, J.R. | 1976 | 5364F | | SIELFELD, K.W.H. | 1976 | 5571 | * |
| SCHALLY, A.V. | 1974 | 5034F | | SILLS, J.B. | 1975 | 4985F | * |
| SCHEER, B.T. | 1974 | 4707F | | SILLS, J.B. | 1975 | 5051F | |
| SCHELER, W. | 1972 | 4775F | | SILLS, J.B. | 1976 | 5189F | * |
| SCHELER, W. | 1973 | 4760F | | SIMANTOV, R. | 1976 | 5158 | * |
| SCHIEBER, M.H. | 1975 | 4787F | | SIMON, M. | 1974 | 4740 | |
| SCHLEEN, L.P. | 1974 | 4862F | * | SIMON, R.C. | 1977 | 5147 | |
| SCHLEEN, L.P. | 1974 | 4865F | * | SIVAK, J.G. | 1974 | 4555F | * |
| SCHLEEN, L.P. | 1974 | 4867F | * | SIVAK, J.G. | 1975 | 5130F | * |
| SCHLEEN, L.P. | 1975 | 5105F | * | SJOBERG, K. | 1973 | 4672F | * |
| SCHLEEN, L.P. | 1975 | 5108F | * | SJOBERG, K. | 1974 | 5141F | * |
| SCHLEEN, L.P. | 1975 | 5111F | | SJOBERG, K. | 1977 | 5304F | * |
| SCHLEEN, L.P. | 1976 | 4554F | * | SKOROBOVICHUK, N.F. | 1975 | 5094F | |
| SCHLEEN, L.P. | 1976 | 5116F | | SMITH, A.J. | 1964 | 5552 | |
| SCHLEEN, L.P. | 1976 | 5220F | * | SMITH, A.J. | 1974 | 4970F | |
| SCHLEEN, L.P. | 1976 | 5222F | * | SMITH, B.R. | 1974 | 4630F | * |
| SCHLEEN, L.P. | 1976 | 5224F | * | SMITH, B.R. | 1976 | 5502F | * |
| SCHMIDT-NIELSON, B. | 1972 | 4931 | * | SMITH, C.G. | 1975 | 4685 | * |
| SCHMULBACH, J.C. | 1975 | 5197F | * | SMITH, D.S. | 1975 | 4818 | * |
| SCHROLL, F. | 1959 | 5331 | * | SMITH, K.L. JR. | 1974 | 4575F | * |
| SCHULTZ, D.P. | 1975 | 5427 | * | SNEZHINA, K.A. | 1976 | 5079F | |
| SCHULTZ, D.P. | 1976 | 5431 | | SNYDER, S.H. | 1976 | 5158 | |

| Name | Year | Number | |
|---|---|---|---|
| SOEJIMA, K. | 1977 | 5593 | |
| SOKOL, H.W. | 1974 | 5017 | |
| SOLTITSKAYA, L. | 1977 | 5361F | |
| SPANN, J.W. | 1972 | 5416F | |
| SPANN, J.W. | 1975 | 5498F | |
| SPITZER, R.H. | 1976 | 5018F | * |
| SPITZER, R.H. | 1976 | 5465F | * |
| SPITZER, R.H. | 1977 | 5166 | |
| STAHL, B.J. | 1977 | 5594 | * |
| STANLEY, H.E. | 1974 | 4552 | |
| STARKEY, R. | 1972 | 5381 | |
| STATHAM, C.M. | 1976 | 5382F | * |
| STATHAM, C.N. | 1973 | 5402 | |
| STATHAM, C.N. | 1974 | 5438 | * |
| STATHAM, C.N. | 1975 | 4751F | |
| STATHAM, C.N. | 1975 | 5440 | * |
| STATHAM, C.N. | 1976 | 5439 | * |
| STAUFFER, J.R. JR. | 1977 | 4994F | |
| STEINER, D.F. | 1973 | 4838F | |
| STEINER, D.F. | 1973 | 4918F | * |
| STEINER, D.F. | 1974 | 4966F | |
| STEINER, D.F. | 1975 | 4654F | |
| STEINER, D.F. | 1975 | 5400F | * |
| STEINER, D.F. | 1977 | 5275 | * |
| STEINER, D.L. | 1973 | 5330 | |
| STEINER, H.M. | 1972 | 5581F | |
| STERBA, G. | 1973 | 4747F | * |
| STERBA, G. | 1974 | 4927 | * |
| STETSON, M.H. | 1976 | 5027 | |
| STEWART, J. | 1974 | 5017 | |
| STOLTE, H. | 1971 | 4773F | |
| STOLTE, H. | 1973 | 4799F | * |
| STOLTE, H. | 1975 | 5086 | |
| STOLTE, H. | 1976 | 5351F | |
| STONE, D.J. | 1975 | 4785F | * |
| STORESUND, A. | 1975 | 4792F | |
| STORESUND, A. | 1975 | 5046 | * |
| STORM-MATHISEN, J. | 1962 | 5334 | |
| STRACHAN, P. | 1977 | 5611F | * |
| STRACHAN, P.D. | 1978 | 5604F | |
| STRAHAN, R. | 1975 | 4992 | * |
| STRAKA, W. | 1976 | 5410F | |
| STRUFE, R. | 1962 | 5550 | * |
| SULLIVAN, J.B. | 1973 | 4549 | |
| SUNADA, J.S. | 1975 | 5157F | |
| SUNDBY, F. | 1974 | 4725F | |
| SUTTERLIN, A. | 1975 | 5081 | * |
| SUTTERLIN, A.M. | 1975 | 5219F | * |
| SUZUKI, H. | 1972 | 4943F | |
| SUZUKI, N. | 1975 | 4728 | * |
| SUZUKI, S. | 1973 | 4925 | * |
| SUZUKI, S. | 1973 | 5559 | * |
| SUZUKI, S. | 1974 | 4564F | * |
| SUZUKI, S. | 1975 | 4581 | * |
| SUZUKI, S. | 1975 | 4729F | * |
| SVANTE VON EULER, V. | 1975 | 5049 | * |
| SVENDENIYA, O. | 1949 | 5321F | * |
| SWEETING, R. | 1976 | 5061F | * |
| SZABO, S. | 1973 | 5399F | |
| SZABO, S. | 1975 | 4878F | |
| SZABO, S. | 1975 | 5072F | |
| TAGAWA, M. | 1973 | 4948 | |
| TAGER, H. | 1973 | 4918F | |
| TAKAGI, H. | 1976 | 5023F | |
| TAMAOKI, B-I. | 1975 | 4609F | |
| TAMECHIKA, M. | 1973 | 5243F | |
| TAMECHIKA, M. | 1976 | 5180 | |
| TAMECHIKA, M. | 1977 | 5293F | |
| TAMMAR, A.R. | 1974 | 4910F | * |
| TAMOKI, B. | 1976 | 5199F | |
| TASHIRO, J. | 1975 | 4618F | |
| TAYLOR, P. | 1976 | 5013 | |
| TCHAI, B.S. | 1977 | 5614 | |
| TEGELSTROM, H. | 1973 | 4562F | |
| TELFORD, M. | 1974 | 4545 | * |
| TERAKADO, K. | 1975 | 5044F | * |
| TERLOU, M. | 1977 | 5534 | |
| TERRIS, S. | 1975 | 5400F | |
| TERWINDT, E.C. | 1976 | 5024F | |
| THEIL, G.B. | 1972 | 4936 | * |
| THEIL, M.F. | 1972 | 4936 | |
| THEISEN, B. | 1973 | 4808F | * |
| THEISEN, B. | 1976 | 4942 | * |
| THINGVOLD, D.A. | 1975 | 5384 | * |
| THOENES, G.H. | 1972 | 4801F | * |
| THOMAS, E. | 1977 | 5096F | |
| THOMAS, E. | 1978 | 5604F | |
| THOMAS, N.W. | 1973 | 4596F | * |
| THOMAS, N.W. | 1973 | 4838F | |
| THOMAS, N.W. | 1974 | 4706F | |
| THOMAS, N.W. | 1976 | 4800F | |
| THOMAS, N.W. | 1976 | 4870F | |
| THORNHILL, R.A. | 1974 | 5052 | * |
| THORNLEY, A.L. | 1976 | 5093 | |
| TIBBLES, J.J. | 1974 | 4630F | |
| TIBBLES, J.J. | 1974 | 5237F | * |
| TIBBLES, J.J. | 1975 | 4908F | * |
| TIBBLES, J.J. | 1976 | 5236F | * |
| TIBBLES, J.J. ED. | 1974 | 4783F | * |
| TIBBLES, J.J. ED. | 1975 | 4784F | * |
| TIBBLES, J.J. ED. | 1976 | 4844F | * |
| TIBBLES, J.J. ED. | 1977 | 5194F | * |
| TIBBLES, J.J. EDS. | 1976 | 5231F | * |
| TIJSKENS, J. | 1974 | 4702F | * |
| TIMMONS, T.J. | 1976 | 4812F | * |
| TIMMONS, T.J. | 1978 | 5564 | * |
| TIMOFEEVA, N.M. | 1974 | 4947F | |
| TOKES, L. | 1977 | 5156F | |
| TOMONAGA, S. | 1973 | 4795F | * |
| TOMONAGA, S. | 1973 | 4796F | * |
| TOMONAGA, S. | 1973 | 4957 | |
| TOMONAGA, S. | 1973 | 4980F | * |
| TOMONAGA, S. | 1973 | 5114F | * |
| TOMONAGA, S. | 1973 | 5243F | * |
| TOMONAGA, S. | 1975 | 4618F | * |
| TOMONAGA, S. | 1977 | 5293F | * |
| TOMONAGA, S. | 1977 | 5590 | * |
| TOMONAGA, S. | 1977 | 5591 | |
| TONDER, O. | 1978 | 5338 | * |
| TORO-GOYCO, E. | 1977 | 5308 | |
| TOTLAND, G.K. | 1977 | 5265F | |
| TRACK, N.S. | 1976 | 5292 | * |
| TREBILCOCK, M.A. | 1974 | 4579 | |
| TRETJAKOFF, D. | 1915 | 5549 | * |
| TSEPKIN, YE. A. | 1973 | 4622F | |
| TSUNEKI, K. | 1974 | 4591F | * |
| TSUNEKI, K. | 1974 | 4882F | * |
| TSUNEKI, K. | 1975 | 4538F | * |
| TSUNEKI, K. | 1975 | 4577F | * |
| TSUNEKI, K. | 1975 | 4585F | |
| TSUNEKI, K. | 1975 | 4669F | * |
| TSUNEKI, K. | 1975 | 5029F | * |
| TSUNEKI, K. | 1975 | 5042F | * |
| TSUNEKI, K. | 1976 | 4990F | * |
| TSUNEKI, K. | 1976 | 5078F | * |
| TSUNEKI, K. | 1976 | 5184F | |
| TSUNEKI, K. | 1977 | 5151F | * |
| TSUNEKI, K. | 1977 | 5152F | * |
| TSUNODA, S. | 1976 | 5023F | |
| TULYAGANOVA, E.KH. | 1974 | 4947F | |
| TURNER, P. | 1974 | 4679 | * |
| TURNER, P. | 1974 | 5031 | * |
| U.S. DEPT. INT. | 1976 | 5433 | * |
| UCHINOMIYA, K. | 1974 | 4603F | |
| UCHINOMIYA, K. | 1978 | 5519 | |
| UEMURA, H. | 1973 | 4928F | |
| UHAZY, L.S. | 1978 | 5327F | |
| UHAZY, L.S. | 1978 | 5541F | |
| UMMINGER, B.L. | 1977 | 5182F | * |
| URANO, A. | 1974 | 4591F | |
| USOVA, A.A. | 1978 | 4996 | * |
| VALE, W. | 1976 | 5195 | * |
| VALTIN, H. | 1974 | 5017 | * |
| VAN BREE, P.J.H. | 1975 | 5008 | |
| VAN NOORDEN, S. | 1973 | 5561F | * |
| VAN NOORDEN, S. | 1973 | 5562F | * |
| VAN NOORDEN, S. | 1974 | 4748F | * |
| VAN NOORDEN, S. | 1974 | 5558 | * |
| VAN NOORDEN, S. | 1975 | 4617F | |
| VAN NOORDEN, S. | 1976 | 4806F | |
| VAN NOORDEN, S. | 1976 | 4870F | |
| VAN NOORDEN, S. | 1977 | 5517 | * |

| | | | | | | |
|---|---|---|---|---|---|---|
| VAN OBBERGHEN, E. | 1978 | 5601 | | WILLIAMS, N.E. | 1976 | 4633F |
| VANDESANDE, F. | 1977 | 5312 | | WILSON, J.H. | 1973 | 4576 |
| VANYUSHIN, B.F. | 1973 | 4548F | | WINBLADH, L. | | 5436 * |
| VARICH, YU.N. | 1976 | 5079F | | WINBLADH, L. | 1975 | 4762F * |
| VERZHBINSKAYA, N.A. | 1975 | 5179 | | WINBLADH, L. | 1976 | 4763F * |
| VERZHBINSKAYA, N.A. | 1976 | 5191 | | WITTENBERG, J.B. | 1974 | 4961F * |
| VERZHBINSKAYA, N.A. | 1977 | 5278 * | | WOJICZAK, A.B. | 1977 | 5218F |
| VESELKIN, N.P. | | 4540F | | WOJTCZAK, L. | 1975 | 4613F |
| VESELKIN, N.P. | 1973 | 4612 | | WOLLMER, A. | 1978 | 5601 |
| VESELKIN, N.P. | 1977 | 5279 | | WOO, G.C.S. | 1975 | 5130F |
| VIGNA, S. | 1975 | 5048 * | | WOODING, G.L. | 1974 | 4960F |
| VILKOVA, V.A. | 1976 | 5060 | | WRIGHT, G.M. | 1976 | 5174F * |
| VILLARREAL, J. | 1976 | 5195 | | WRIGHT, G.M. | 1977 | 4695 |
| VINNIKOV, Y.A. | 1974 | 4914 * | | WRIGHT, G.M. | 1977 | 5303F * |
| VIOSCA, S. | 1974 | 5034F | | WRIGHT, G.M. | 1978 | 5537F |
| VLADYKOV, V.D. | 1972 | 4752F * | | WRIGHT, G.M. | 1978 | 5538F * |
| VLADYKOV, V.D. | 1972 | 4753F * | | WRIGHT, G.M. | 1978 | 5539F * |
| VLADYKOV, V.D. | 1973 | 4600F * | | WRIGHT, V.C. | 1976 | 5002F |
| VLADYKOV, V.D. | 1973 | 4770F * | | WRONISZEWSKA, A. | 1975 | 4613F |
| VLADYKOV, V.D. | 1974 | 4766F * | | WU, T.T. | 1973 | 4626F * |
| VLADYKOV, V.D. | 1975 | 5150 * | | YAKOVLEV, V.N. | 1968 | 5054 * |
| VLADYKOV, V.D. | 1976 | 4750F * | | YAMADA, Y. | 1973 | 4745F * |
| VLADYKOV, V.D. | 1976 | 4768F * | | YAMAGUCHI, K. | 1976 | 5180 * |
| VLADYKOV, V.D. | 1976 | 4821 * | | YAMAGUCHI, K. | 1977 | 5293F |
| VLADYKOV, V.D. | 1977 | 5391F * | | YAMAGUCHI, K. | 1977 | 5590 |
| VLADYKOV, V.D. | 1978 | 5542F * | | YAMAGUCHI, K. | 1977 | 5591 |
| VON ESCHEN, K.B. | 1974 | 4566F * | | YAMAMOTO, T. | 1976 | 5266F |
| VOROB'EVA, E.I. | 1975 | 4879 * | | YAMAMOTO, T. | 1976 | 5531 |
| VUL'FIUS, Y.A. | 1973 | 4930 | | YAMAMOTO, T. | 1977 | 4872 |
| WAECHTLER, K. | 1974 | 4641 * | | YANAGISAWA, M. | 1975 | 5029F |
| WAECHTLER, K. | 1975 | 5038 * | | YEAGER, B.E. | 1976 | 5287 * |
| WAHREN, H. | 1973 | 4562F | | YERGER, R.W. | 1977 | 5316 * |
| WAINSTEIN, B.A. | 1975 | 5324F * | | YONEYAMA, Y. | 1973 | 4953 * |
| WALKER, A.D. | 1977 | 5369 | | YONEZAWA, S. | 1977 | 4717 * |
| WALSH, D.F. | 1975 | 5144F | | YOSHIE, S. | 1977 | 5187F |
| WALVIG, F. | 1973 | 4856F * | | YOUSEN, J.H. | 1978 | 5516F |
| WALVIG, F. | 1976 | 5351F | | YOUSEN, J.H. | 1978 | 5518F |
| WANDT, I. | 1973 | 5128F | | YOUSON, J.H. | 1972 | 4897F * |
| WANG, C.S. | 1976 | 5139F | | YOUSON, J.H. | 1973 | 4639F * |
| WANG, J.C.S. | 1974 | 4676F | | YOUSON, J.H. | 1973 | 4898F * |
| WANG, J.C.S. | 1975 | 4720F | | YOUSON, J.H. | 1973 | 4899F * |
| WARD, P.H. | 1974 | 4579 | | YOUSON, J.H. | 1974 | 4659F |
| WEISBART, M. | 1974 | 4659F * | | YOUSON, J.H. | 1974 | 4726 * |
| WEISBART, M. | 1974 | 5370 * | | YOUSON, J.H. | 1975 | 4781F |
| WEISBART, M. | 1975 | 4607F * | | YOUSON, J.H. | 1975 | 4880F * |
| WEISBART, M. | 1975 | 4781F * | | YOUSON, J.H. | 1975 | 4881F * |
| WEISBART, M. | 1977 | 5325F * | | YOUSON, J.H. | 1975 | 5045F * |
| WEISBART, M. | 1978 | 5470F * | | YOUSON, J.H. | 1976 | 4539 |
| WEISE, J.G. | 1974 | 4888F | | YOUSON, J.H. | 1976 | 4749F * |
| WEISE, J.G. | 1974 | 4893F | | YOUSON, J.H. | 1976 | 4993 * |
| WEISE, J.G. | 1974 | 4904F | | YOUSON, J.H. | 1976 | 5085F * |
| WEISE, J.G. | 1975 | 5104F | | YOUSON, J.H. | 1976 | 5174F * |
| WEISE, J.G. | 1975 | 5106F | | YOUSON, J.H. | 1977 | 4695 * |
| WEISE, J.G. | 1975 | 5107F | | YOUSON, J.H. | 1977 | 5303F |
| WEISE, J.G. | 1975 | 5133F * | | YOUSON, J.H. | 1977 | 5325F |
| WEISE, J.G. | 1976 | 4553F * | | YOUSON, J.H. | 1977 | 5350F |
| WEISE, M. | 1971 | 4773F | | YOUSON, J.H. | 1978 | 5470F |
| WEISE, R. | 1971 | 4773F | | YOUSON, J.H. | 1978 | 5537F |
| WELLBORN, T.L., JR. | 1971 | 5426F * | | YOUSON, J.H. | 1978 | 5538F |
| WELLINGS, S.R. | 1976 | 5282 | | YOUSON, J.H. | 1978 | 5539F |
| WENT, A.E.J. | 1974 | 4686F * | | ZABELINSKII, S.A. | 1977 | 5512F |
| WESSELLS, N.K. | 1974 | 4932 * | | ZAGORULKO, T.M. | 1975 | 4540F |
| WESTMAN, R.W. | 1973 | 5477F * | | ZELNIK, P.R. | 1975 | 4886F |
| WESTMAN, R.W. | 1974 | 4861F * | | ZELNIK, P.R. | 1976 | 5002F |
| WESTMAN, R.W. | 1974 | 4863F * | | ZELNIK, P.R. | 1977 | 5306F * |
| WESTMAN, R.W. | 1974 | 4864F * | | ZHELUDKOVA, Z.P. | 1973 | 4561F |
| WESTMAN, R.W. | 1974 | 4890F * | | ZHUKOV, P.I. | 1965 | 5358F * |
| WESTMAN, R.W. | 1975 | 5105F | | ZIEGELS, J. | 1976 | 5272F * |
| WESTMAN, R.W. | 1975 | 5111F * | | ZITTEL, A.E. | 1977 | 5453F |
| WESTMAN, R.W. | 1976 | 5216F * | | ZYBLUT, E.R. | 1973 | 5343F |
| WESTMAN, R.W. | 1976 | 5221F * | | ZYTKOVICZ, T.H. | 1976 | 5028 * |
| WHITE, J.I. | 1977 | 5308 | | | | |
| WHITING, H.P. | 1977 | 5510 * | | | | |
| WICKELGREN, W.O. | 1977 | 5206F | | | | |
| WICKELGREN, W.O. | 1977 | 5280F * | | | | |
| WICKELGREN, W.O. | 1977 | 5283F * | | | | |
| WICKELGREN, W.O. | 1978 | 5578F | | | | |
| WICKELGREN, W.O. | 1978 | 5608F | | | | |
| WIEBE, J.P. | 1978 | 5470F | | | | |
| WILKENS, H. | 1977 | 5506F * | | | | |
| WILLIAMS, J.D. | 1975 | 5498F | | | | |

# Subject Index

. + ADULT ANIMAL NERVOUS SY   .I. THE NEUROMAST OF LAMPREY,ENTOSPHENUS JAPONICUS   4745F
LUVIATILIS. + ADULT AMMOCOE   /BLOOD CELL FORMATION IN RIVER LAMPREY, LAMPETRA F   4717
RATES PURIFICATION AND PROP   /STUDIES ON PHOSPHORYLASE ISOZYMES IN LOWER VERTEB   4717
ONDING LATIN NAMES. A.VERTE   A. VIRVELDYR.[NORWEGIAN ANIMALS NAMES WITH CORRESP   5469F
THE CENTRAL NERVOUS SYSTEM   A.A. UKHTOMSKII ONTHE HIERARCHICAL ORGANIZATION OF   4642F
TSENTRL'NOI NERVNOI SISTEMY   A.A. UKHTOMSKOGO NA IERARKHICHESKUYU ORGANIZATSIYU   4642F
UVIATILIS *LAMPETRA PLANERI   A.VERTEBRATES.] + *PETROMYZON MARINUS *LAMPETRA FL   5469F
AI. /STRACOD CYPRETTA KAWAT   A, BAYER 73, AND TFM TO THE OSTRACOD CYPRETTA KAWA   5454F
N MARINUS U.S.A. AMMOCOETEA   ABBOTT), FROM THE DELMARVA PENINSULA. + *PETROMYZO   4676F
NUS. /MPREY PETROMYZON MARI   ABDOMINAL WINDOW IN THE SEA LAMPREY PETROMYZON MAR   5613
CCI. / /VARIATIONS IN   ABILITIES OF ANIMAL FIBRINOGENS TO CLUMPSTAPHYLOCO   4576
Y ELECTRON MICROGRAFIC AND   ABLATION STUDIES. /±NTIFICATION OF THE RECEPTORS B   4536
A. BIOLOGY FEEDING /   ABNORMAL TOOTH DEVELOPMENT IN A SEA LAMPREY + U.S.   5241F
/ OF THE VERTEBRATE BRAIN.   ABOUT THE EARLY EVOLUTION OF THE VERTEBRATE BRAIN.   5198
L.): A CYTOARCHITECTURAL AN   ABOUT THE PROSENCEPHALON OF LAMPETRA FLUVIATILIS (   4559F
AE *LAMPETRA U.S.A. DISTRIB   ABOVE CONOWINGO DAM. + *ICHTHYOMYZON *PETROMYZONID   5068F
SPECIMENS AT DIFFERENT STAG   ABRUZZO. OBSERVATION ON THE PROLONGED SURVIVAL OF   4807F
ZA DI ESEMPLARI A DIVERSO S   ABRUZZO. OSSERVAZIONI SULLA PROLUNGATA SOPRAVVIVEN   4807F
LAMPREYSLAMPETRA FLUVIATIL   ABSENCE OF KNOWN CORTICOSTEROIDS IN BLOOD OF RIVER   4574F
LAMPREYS (LAMPETRA FLUVIAT   ABSENCE OF KNOWN CORTICOSTEROIDS IN BLOOD OF RIVER   5527
ISH BDELLOSTOMA CIRRHATUM.   ABSENCE OF UREOGENIC PATHWAYS IN LIVER OF THE HAGF   4620
HE ARCHINEPHRIC DUCT OF THE   ABSORPTION AND TRANSPORT OF EXOGENOUS PROTEIN IN T   4880F
HORSERADISH PEROXIDASE IN   ABSORPTION AND TRANSPORT OF FERRITIN AND EXOGENOUS   5045F
STRUCTURE AND PROXIMAL RE   ABSORPTION IN DIFFERENT VERTEBRATE CLASSES. / HRON   5165
HPOLISTOTREME STOUTII. + EV   ABSORPTION OF AMINO ACIDS BY THE GUT OF THE HAGFIS   4724
LLS OFTHE MEDIAN EMINENCE.   ABSORPTION OF CEREBRO SPINAL FLUID BY EPENDYMAL CE   4984
F THE HAGFISH EPTATRETUS BU   ABSORPTION OF PEROXIDASE INTO THE THIRD VENTRICLEO   4782F
UOROMETHYL-4-NITROPHENOL (T   ABSORPTION, DEGRADATION AND PERSISTENCE OF 3-TRIFL   5384
RI RIVER GAVINS-POINT DAM T   ABUNDANCE AND DISTRIBUTION OF FISHES IN THE MISSOU   5197F
OF THE SOUTHERN BROOK LAMPR   ABUNDANCE DISTRIBUTION FOOD HABITATS AND CONDITION   5207
THE COLUMBIA RIVER. /CH OF   ABUNDANCE OF FISH SPECIES FROM THE HANFORD REACH O   5569
CHEMISTRY. VOL.4.NEW YORK:   ACADEMIC PRESS. /BN COMPARATIVE PHYSIOLOGY AND BIO   4689
TOTHE PERTIN....NEW YORK;   ACADEMIC PRESS, 3V. /TANATOMY WITH AN INTRODUCTION   4730
YOLOGY /OSIS EVOLUTION EMBR   ACARIFORMES.] + ANIMAL METAMORPHOSIS EVOLUTION EMB   5324F
XONOMY OF ACARIFORMES.] + A   ACARIFORMES). [ONTOGENESIS AND SOME PROBLEMS OF TA   5324F
LVA AND LEPISOSTEVS OSSEVS   ACCOMMODATIVE LENS MOVEMENT IN HOLOSTEANS (AMIA CA   5130F
TRUCTURE AND FUNCTION 4.).   ACCOUNT.CAMBRIDGE: UNIV. PRESS. 397P.(BIOLOGICAL S   4732
MPREY (LAMPETRA FLUVIATILIS   ACCUMULATING NEURONS IN THE RETINA OF THE RIVER LA   5310F
F THE LAMPRICIDE 2',5-DICHL   ACCUMULATION, ELIMINATION, AND BIOTRANSFORMATION O   5453F
TORPEDO CALIFORNICA. /FROM   ACETYL CHOLINE RECEPTOR-RICH MEMBRANE VESICLES FRO   5164
THE LAMPREY.] + RUSSIA MUS   ACETYLCHOLINE CONTRACTURES INFAST MUSCLE FIBERS OF   5094F
INATION OF EXTRAJUNCTIONAL   ACETYLCHOLINE RECEPTOR FROM RAT MUSCLE. /I AND IOD   5014
INTESTINEOF MYXINE GLUTINOS   ACETYLCHOLINESTERASE CONTAINING STRUCTURES IN THE   4839F
ORESCENCE PROBE STUDIES OF   ACETYLCHOLINESTERASE FROM TORPEDO. / /FLU   5013
2: MYXINE GLUTINOSA. /PART   ACETYLCHOLINESTERASE IN THE CYCLOSTOME BRAIN. PART   5038
: LAMPETRA PLANERI. /PART I   ACETYLCHOLINESTERASE IN THE CYCLOSTOMEBRAIN. PART   4641
L SYSTEM OF THE LAMPREY, LA   ACETYLCHOLINESTERASE IN THE HYPOTHALAMO-HYPOPHYSIA   4882F
HAGFISH EPTATRETUS BURGERI   ACETYLCHOLINESTERASE IN THE NEUROHYPOPHYSIS OF THE   4591F
L NERVOUS SYSTEMAND MODERN   ACHIEVEMENTS OF EVOLUTIONARY NEUROPHYSIOLOGY.] / A   4642F
NTATUS *ENTOSPHENUS TRIDENT   ACID COMPOSITION IN CYTOCHROMES. + *LAMPETRA TRIDE   4955F
FIBRINOGEN. EVOLUTIONARY S   ACID COMPOSITIONS OF THE SUBUNIT CHAINS OF LAMPREY   4715F
A. PP. 84-85. + BIOCHEMISTR   ACID DEHYDRASE (ALA-D)IN ORGANS OF MYXINE GLUTINOS   4850F
CUSPIS *LAMPETRA AEPYPTERA   ACID IN LAMPREY INTERNEURONES. + *ICHTHYOMYZON UNI   5606F
. + CANADA NEW BRUNSWICK BI   ACID LEVEL IN THE MIGRATING ANADROMOUS SEA LAMPREY   5096F
S STAINING TECHNIQUE. /ENOL   ACID MUCO POLY SACCHARIDESIN THE PRESENCE OF PHENO   5040
VERTEBRATES (HAGFISH, LAMP   ACID MUCOPOLYSACCHARIDES IN THE SKIN OF SOME LOWER   5019F
LIS. + SWEDEN ADULT BIOCHEM   ACID PHOSPHATASEIN THE LAMPREY PETROMYZON FLUVIATI   4562F
EBRATE THE ATLANTIC HAGFISH   ACID SEQUENCE OF THE INSULIN FROM A PRIMITIVE VERT   4654F
ND FIBRINO PEPTIDES. /BRINA   ACID SEQUENCE STUDIES ON LAMPREY FIBRINOGEN FIBRIN   4683
AND CHARACTERIZATION OF TH   ACID SEQUENCES OF LAMPREY FIBRINO PEPTIDES A AND B   5146F
/NVARIANT RESIDUE. + BLOOD   ACID SUBSTITUTION AT AN INVARIANT RESIDUE. + BLOOD   4971F
TA BIOCHEMISTRY /A TRIDENTA   ACIDIC PROTEIN IN VERTEBRATES. + *LAMPETRA TRIDENT   4570F
OF AMINO ACIDS IN PROTEINS   ACIDS (N-1) AND (N+1) ON THE BACKBONE CONFORMATION   4626F
I. + EVOLUTION PHYSIOLOGY M   ACIDS BY THE GUT OF THE HAGFISHPOLISTOTREME STOUTI   4724
TES.] + *LAMPETRA FLUVIATIL   ACIDS IN BRAIN MYELIN AND MITOCHONDRIA IN VERTEBRA   4741F
ONAL STRUCTUREOF THE POLY P   ACIDS IN PROTEINS. USE IN PREDICTING THE 3-DIMENSI   4626F
D MONOSACCARIDES AND AMINO   ACIDS. /EPATRETUS-STOUTI. UPTAKE OF RADIO LABELLE   5166
OOD AND MUSCLE CELLS. + PHY   ACIDS, TRIMETHYLAMINE OXIDE AND POTASSIUMOF THE BL   4767F
L BIOCHEMISTRY ENDOCRINOLOG   ACIPENSER FULVESCENS RAFINESQUE. + EXCRETION ANIMA   4993
INOSA. + HISTOLOGY /NE GLUT   ACROSOMAL SYSTEM IN EARLY SPERMATIDS OF MYXINE GLU   5296F
IBIA. PP. 447-463. /ND AMPH   ACROSS THE BODY SURFACE OF FRESHWATER FISH AND AMP   4934
SIOLOGY BIOCHEMISTRY IONIC   ACROSS THE GILLS OF FISHES.LL. + PISCES ANIMAL PHY   5307F
OFMOLECULAR WEIGHT BY POLY   ACRYLAMIDE GEL ELECTROPHORESIS. /ND DETERMINATION   4572
LT AMMOCOETE ANIMAL ENDOCRI   ACTINOPTERYGIANS. + *PETROMYZON MARINUS CANADA ADU   4779F
MAN FACTOR XIII DURING THE   ACTION OF THE INTRINSIC LAMPREY FACTOR XIII AND HU   5302F
62. + *LAMPETRA FLUVIATILIS   ACTION ON MUSCLE CARBOHYDRATE METABOLISM. PP.345-3   5123F
HYSIOLOGY GONADOGENESIS / P   ACTION ON THE VERTEBRATE OVIDUCT-UTERUS. + ANIMAL   5390F
STEM IN HAGFISH TWITCH MUSC   ACTION POTENTIAL THROUGH THE TRANSVERSE TUBULAR SY   5006F
AL ANDINTRA SPINAL STIMULAT   ACTIONS EVOKED IN LAMPREY MOTONEURONS BY SUPRASPIN   4601F
ISH. PP.391-418. / /   ACTIONS OF HORMONES ON OSMOREGULATORY SYSTEMS OF F   4958
RETOR HIPOTALAMO-HIPOFIZAR   ACTIUNEA CLORPROMAZINEI ASUPRA SISTEMULUI NEUROSEC   5072F

CHEMICALS FROM WATER WITH  
SIOLOGICAL ACTIVITY AND FIN  
B. [METHODS OF DETERMINING  
BRATES. + *CYCLOSTOMATA BIO  
HYLOGENETIC SURVEY. + *MYXI  
] + *LAMPETRA FLUVIATILIS B  
ES.] + *LAMPETRA FLUVIATILI  
LAKE TROUT SALVELINUS NAMAY  
ALA-D)IN ORGANS OF MYXINE G  
SCOPY. PART I  NMR SPECTRA  
F THE LAMPREY LAMPETRA FLUV  
MPETRA FLUVIATILIS (L.)) IN  
ART AND THE PORTAL VEIN HEA  
HE LAMPREY GEOTRIA-AUSTRALI  
ION BIOCHEMISTRY ANIMAL /ST  
LAMPETRA FLUVIATILIS.] + U.  
*LAMPETRA FLUVIATILIS RUSS  
SPECTRA AND PHYSIOLOGICAL  
) DURING THE SPAWNING SEASO  
INE USA. + ANIMAL ADULT BIO  
BLOOD SERUM OF VERTEBRATES.  
/SCREENING FOR MUTAGENIC  
POST METAMORPHIC LAMPREYS O  
SA. /ICLE IN MYXINE GLUTINO  
L OF THE HAGFISH, EPTATRETU  
ND SCORPIONFISH.] + *LAMPET  
/DIRECT PHOTO-SENSORY  
E ADULT RIVER LAMPREY LAMPE  
ATER INVERTEBRATES. /     /  
CHLORO-4'-NITROSALICYLANILI  
TO THE OSTRACOD CYPRETTA K  
L (TFM) AND 2' 5-DICHLORO-4  
UOROMETHYL-4-NITROPHENOL. +  
AQUATIC ORGANISMS. /LECTED  
-4-NITROPHENOL TO NYMPHS OF  
MENT /VERTEBRATES. + MANAGE  
3-TRIFLUOROMETHYL-4-NITROP  
STRY MANAGEMENT /IMAL CHEMI  
TION? / THE COURSE OF EVOLU  
THE FREE AMINO ACIDS, TRIME  
ION IN POIKLOTHERMS AND HOM  
N. + *ENTOSPENUS TRIDENTATU  
STRY MANAGEMENT MORTALITY O  
THYOMYZON GAGEI U.S.A. SYST  
LBERGIA AEPYPTERA (ABBOTT),  
TA.) [ELECTRON OPTICAL INDI  
ON MARINUS. /PREY, PETROMYZ  
A.] + NORWAY NERVOUS SYSTEM  
GFISH EPTATRETUS BURGERI. /  
BLOCH). /(LAMPETRA PLANERI  
TS MINOGI LAMPETRA FLUVIATI  
ESIS. + *LAMPETRA FLUVIATIL  
FROM THE SEA LAMPREY PETRO  
E ADULT SEA LAMPREY PETROMY  
NSER FULVESCENS RAFINESQUE.  
THE THYROID GLAND. + JAPAN  
PHOSIS LIFE CYCLE SPAWNING  
*PETROMYZON MARINUS CANADA  
*PETROMYZON MARINUS CANADA  
ROCTS. + ANATOMY HISTOLOGY  
U.S.A. ANIMAL BIOCHEMISTRY  
APONICA CIRCULATORY SYSTEM  
M HISTOCHEMISTRY AMMOCOETE  
PETROMYZON MARINUS IRELAND  
BELCHERII. + BIOCHEMISTRY  
CULATORY SYSTEM PHYSIOLOGY  
NDOCRINOLOGY METAMORPHOSIS  
NERVOUS SYSTEM PHYSIOLOGY  
PYPTERA. + U.S.A. SPAWNING  
P. 70. + CANADA MANAGEMENT  
BURGERI HISTOLOGY CYTOLOGY  
VOUS SYSTEM CYTOLOGYSWEDEN  
OLOGY MIGRATION TECHNIQUES  
GEMENT CHEMISTRY AMMOCOETE  
RINUS GERMANY DISTRIBUTION  
SYSTEM PHYSIOLOGY CYTOLOGY  
+ SWEDEN VISION HISTOLOGY  
*PETROMYZON MARINUS CANADA  
*PETROMYZON MARINUS CANADA  
EN AMMOCOETE PARASITISM OF  
STOCHEMISTRY GONADOGENESIS  

ACTIVATED CARBON. + CHEMISTRY MANAGEMENT /UF TOXIC  5407F  
ACTIVE COMPOUNDS CORRELATION OF FINDINGSON THE PHY  4930  
ACTIVE METABOLISM IN YOUNG FISH.] /MANII MOLODI RY  4924  
ACTIVE NEUROHYPOPHYSIAL PRINCIPLES AMONG THE VERTE  5389F  
ACTIVE TRANSPORT OF IODIDE BY CHLOROID PLEXUS: A P  5162  
ACTIVITIES IN CELLS OF THE NEPHRON IN VERTEBRATES.  4742F  
ACTIVITIES IN MUSCLE SARCOPLASM IN LOWER VERTEBRAT  4803F  
ACTIVITIES OF SEA LAMPREYS, PETROMYZON MARINUS ON  4769F  
ACTIVITIY OF DELTA-AMINOLEVULINIC ACID DEHYDRASE (  4850F  
ACTIVITY AND FINDINGS OF MOLECULARSPECULAR SPECTRO  4930  
ACTIVITY AND IMMUNOREACTIVE INSULIN IN THE BLOOD O  4797F  
ACTIVITY AND LENGTH/WEIGHT OF THERIVER LAMPREY (LA  5141F  
ACTIVITY AND MECHANICAL RESPONSE IN THE SYSTEMICHE  4573F  
ACTIVITY AND THE UPTAKE OF TRITATED MELATONIN IN T  5209F  
ACTIVITY IN MYXINE GLUTINOSA L. PP. 30-32. + DIGES  4846F  
ACTIVITY IN THE LIVER AND MUSCLES OF THE LAMPREY,  4561F  
ACTIVITY IN THE PIAL MATTER OF SOME VERTEBRATES. +  5281F  
ACTIVITY OF CERTAIN... + BIOCHEMISTRY /YART I  NMR  4930  
ACTIVITY OF RIVER LAMPREY LAMPETRA FLUVIATILIS (L.  5304F  
ACTIVITY OF SEVERAL MARINE SPECIES FROM COASTAL MA  4604F  
ACTIVITY OF SODIUM. POTASSIUM, AND CALCIUM IN THE  5204  
ACTIVITY OF SOME MOLLUSCICIDES. /S  5451  
ACTIVITY OF SOME OXIDATIVE ENZYMES IN THYROIDS OF  5598  
ACTIVITY OF THE CARDIAC VENTRICLE IN MYXINE GLUTIN  5046  
ACTIVITY OF THE HIGH-WALLED ENDOTHELIUM IN THE GIL  5243F  
ACTIVITY OF THE LIVER AND MUSCLES OF THE LAMPREY A  4598F  
ACTIVITY OF THE PINEAL. PP. 376-387. /  4986  
ACTIVITY VENTILATORY FREQUENCY ANDHEART RATE IN TH  4719F  
ACUTE TOXIC EFFECTS OF 2 LAMPRICIDES TO 21 FRESH W  5076F  
ACUTE TOXICICITY OF 2,4-DN-BUTYL ESTER AND 2',5-DI  5439  
ACUTE TOXICITIES OF ANTIMYCIN A, BAYER 73, AND TFM  5454F  
ACUTE TOXICITIES OF 3-TRIFLUOROMETHYL-4-NITROPHENO  5120F  
ACUTE TOXICITY CONJUGATION AND EXCRETIONOF 3-TRIFL  4652F  
ACUTE TOXICITY OF ROTENONE AND BAYER 73 TO SELECTE  5460  
ACUTE TOXICITY OF THE LAMPRICIDE 3-TRIFLUOROMETHYL  5137F  
ACUTE TOXICITY STUDIES WITH INVERTEBRATES. + MANAG  5405F  
ACUTE TOXICITY, METABOLISM ANDBILIARY EXCRETION OF  5412F  
ACUTELY SUBLETHAL TOXICANT EXPOSURE. + ANIMAL CHEM  5410F  
ADAPTATION BY THE LATTER DURING THE COURSE OF EVOL  4983  
ADAPTATION IN MYXINE GLUTINOSA L.-II.VARIATIONSOF  4767F  
ADAPTATIONS OF ENZYMES INVOLVED IN MEMBRANE DIGEST  4947F  
ADAPTATIONS OF THE VERTEBRATE EPIDERMIS TO FRICTIO  4693F  
ADDITION OF A CHEMICAL TO THE ENVIRONMENT. + CHEMI  5183F  
ADDITION OF 3 SPECIES TO THE ICHTHYO FAUNA. + *ICH  4944F  
ADDITIONAL RECORDS OF THE LEAST BROOK LAMPREY OKKE  4676F  
ADENOHYPOPHYSE BEI MYXINE GLUTINOSA L. (CYCLOSTOMA  4809F  
ADENOHYPOPHYSIAL CELLS OF THE SEA LAMPREY, PETROMY  5567  
ADENOHYPOPHYSIS IN MYXINE GLUTINOSA L. CYCLOSTOMAT  4809F  
ADENOHYPOPHYSIS OF THEHAGFISH EPTATRETUS BURGERI.  4608  
ADRENAL-SYSTEMS DES BACHNEUNAUGES (LAMPETRA PLANER  5563  
ADRENALINA NA AMNLAZNUYU AKTIVNOST' PECHENI I MYSH  4561F  
ADRENERGIC INNERVATION IN ONTOGENESIS AND PHYLOGEN  5504  
ADRENOCORTICAL CELLS ANDTESTICULAR TISSUE OBTAINED  4659F  
ADRENOCORTICAL CELLS FROM THE OPISTHONEPHROS OF TH  4607F  
ADRENOCORTICAL HOMOLOG IN THE LAKE STURGEON, ACIPE  4993  
ADULT  HISTOLOGY CYTOLOGY ENDOCRINOLOGY /AUDIES OF  4690F  
ADULT /±THE LARVA INTO THE ADULT.] + ITALY METAMOR  4807F  
ADULT / DAM ON THE SAUGEEN RIVER. PP. 171-172. +  5485F  
ADULT /AMPREY IN ST. MARYS RIVER,1972. P. 166. +  5488F  
ADULT /A ULTRASTRUCTURE OF LIVING CYCLOSTOME ECTOP  5067F  
ADULT /C ELASMOBRANCH. + *ENTOSPHENUS TRIDENTATUS  4729F  
ADULT /CF THE LAMPREY GILL FILAMENT. + *LAMPETRA J  5519  
ADULT /CNLAMPREY LIVER. + *LAMPETRA PLANERI BELGIU  4734F  
ADULT /EISHES TAKEN FROM IRISH WATERS IN 1973. + *  4686F  
ADULT /EIVE TISSUE OF THE AMPHIOXUS, BRANCHIOSTOMA  5041  
ADULT /EMATICITY STUDY ON MYXINE GLUTINOSA.] + CIR  4740  
ADULT /GE MIDBRAIN OF LAMPREYS. + NERVOUS SYSTEM E  4927  
ADULT /GHTHYOMYZON CASTANEUS U.S.A AMMOCOETEMUSCLE  4674F  
ADULT /IT CONSTRUCTION OF THE LAMPREY, LAMPETRA AE  4653F  
ADULT /KT SEA LAMPREY FROM THE HUMBER RIVER,1974.  5136F  
ADULT /N BURGERI. + *PARAMYXINE ATAMI *EPTATRETUS  5044F  
ADULT /O + *LAMPETRA FLUVIATILIS SWEDEN VISION NER  5039  
ADULT /OAMPETRA FLUVIATILIS. + ENDOCRINOLOGY PHYSI  5005  
ADULT /OP.39-51. + *PETROMYZON MARINUS CANADA MANA  5106F  
ADULT /OY IN MIDWATER OFF EUROPE. + *PETROMYZON MA  4733F  
ADULT /PRA FLUVIATILIS NERVOUS SYSTEM CIRCULATORY  4874  
ADULT /RR LAMPREY LAMPETRA FLUVIATILIS CYCLOSTOMI.  4655F  
ADULT /SMPREY TAKEN AT DENN'S DAM, 1973. P. 88. +  4848F  
ADULT /TREY IN ST. MARY'S RIVER1973. PP. 80-81. +  4906F  
ADULT /U LAMPREYS LAMPETRA PLANERI (BLOCH). + SWED  5467F  
ADULT /VNUS DURING THEIR PARASTIC LIFE STAGE. + HI  4659F

| | | |
|---|---|---|
| RI(BLOCH). + GREAT BRITAIN | **ADULT** DIGESTION ECOLOGY FEEDING GROWTH /RTRA PLANE | 5142F |
| O,1975. PP.93-95. + CANADA | **ADULT** DISTRIBUTION /.ND TAGGING STUDY, LAKE ONTARI | 5229F |
| MARINUS CANADA GREAT LAKES | **ADULT** DISTRIBUTION /AE HURON. P.93. + *PETROMYZON | 5258F |
| E) FROM ILLINOIS. + U.S.A. | **ADULT** DISTRIBUTION /DPTERA, (PISCES: PETROMYZONIDA | 5523F |
| S. + AMMOCOETE SYSTEMATICS | **ADULT** DISTRIBUTION /LTHER SPECIES OF THE SAME GENU | 5542F |
| , 1975. PP.83-85. + CANADA | **ADULT** DISTRIBUTION /MEA LAMPREY IN ST. MARYS RIVER | 5224F |
| T COAST, UNITED STATES). + | **ADULT** DISTRIBUTION /NM THE DELMARVA PENINSULA (EAS | 4720F |
| LMON AND HERRING STOCKS. + | **ADULT** DISTRIBUTION FEEDING GROWTH / REDATION ON SA | 4633F |
| MANAGEMENT /DING MIGRATION | **ADULT** DISTRIBUTION PARASITISM BY FEEDING MIGRATION | 4746F |
| FLUVIATILIS UNITED KINGDOM | **ADULT** ECOLOGY GROWTH LIFE CYCLE MIGRATION /CPETRA | 5009 |
| SPAWNING SEASON. + SWEDEN | **ADULT** ECOLOGY MIGRATION /CVIATILIS (L.) DURING THE | 5304F |
| PTATRETUS BURGERI. + JAPAN | **ADULT** EMBRYOLOGY ENDOCRINOLOGY FECUNDITY /UGFISH E | 4703F |
| AMPETRA FLUVIATILIS RUSSIA | **ADULT** ENDOCRINOLOGY / A CARPE ET LA LAMPROIE. + *L | 5468F |
| HEIR LIFE CYCLE.] + RUSSIA | **ADULT** ENDOCRINOLOGY /JIS) ON DIFFERENT STAGES OF T | 5323F |
| ETRA FLUVIATILIS. + EUROPE | **ADULT** ENDOCRINOLOGY /NHE BLOOD OF THE LAMPREY LAMP | 4797F |
| ETRA RICHARDSONI AMMOCOETE | **ADULT** ENDOCRINOLOGY BIOCHEMISTRY HISTOCHEMISTRY /N | 5393 |
| UES /TRY METABOLISM TECHNIQ | **ADULT** ENDOCRINOLOGY BIOCHEMISTRY METABOLISM TECHNI | 5559 |
| ION. + *PETROMYZON MARINUS | **ADULT** ENDOCRINOLOGY CANADA HISTOLOGY /AEROL INJECT | 4881F |
| IGRATION /D GONADOGENESIS M | **ADULT** ENDOCRINOLOGY EVOLUTION BLOOD GONADOGENESIS | 5065F |
| IATILIS). + UNITED KINGDOM | **ADULT** ENDOCRINOLOGY FEEDING METABOLISM /IETRA FLUV | 5306F |
| LAMPETRA FLUVIATILIS L. + | **ADULT** ENDOCRINOLOGY HISTOCHEMISTRY /LF THE LAMPREY | 4941F |
| LAMPETRA JAPONICA. + JAPAN | **ADULT** ENDOCRINOLOGY HISTOLOGY / RGERI AND LAMPREY | 4587F |
| MES HAGFISH AND LAMPREY. + | **ADULT** ENDOCRINOLOGY HISTOLOGY /ILL OF THE CYCLOSTO | 5098 |
| CTRON MICROSCOPY. + SWEDEN | **ADULT** ENDOCRINOLOGY HISTOLOGY BIOCHEMISTRY /RG ELE | 4762F |
| MYXINE GLUTINOSA. + SWEDEN | **ADULT** ENDOCRINOLOGY HISTOLOGY CYTOLOGY /A HAGFISH | 4596F |
| CTION /ONADOGENESIS REPRODU | **ADULT** ENDOCRINOLOGY HISTOLOGY GONADOGENESIS REPROD | 4592F |
| TRA FLUVIATILIS.] + RUSSIA | **ADULT** ENDOCRINOLOGY HISTOLOGY NERVOUS SYSTEM /*MPE | 5071F |
| MPETRA TRIDENTATA. + JAPAN | **ADULT** ENDOCRINOLOGY HISTOLOGY NERVOUS SYSTEM /O LA | 5042F |
| DANFORDI REGAN.] + ROMANIA | **ADULT** ENDOCRINOLOGY HISTOLOGY NERVOUS SYSTEM /PON | 5072F |
| TROMYZON MARINUS. + U.S.A. | **ADULT** ENDOCRINOLOGY METABOLISM REPRODUCTION /IY PE | 4607F |
| N /ISM SPAWNING REPRODUCTIO | **ADULT** ENDOCRINOLOGY METABOLISM SPAWNING REPRODUCTI | 5509 |
| ATILIS L. + UNITED KINGDOM | **ADULT** ENDOCRINOLOGY MIGRATION / EY, LAMPETRA FLUVI | 5176F |
| TRA FLUVIATILIS BALTIC SEA | **ADULT** ENDOCRINOLOGY TECHNIQUES /IBRATES.] + *LAMPE | 4721F |
| GFISH EPTATRETUS STOUTI. + | **ADULT** ENDOCRINOLOGY TECHNIQUES BIOCHEMISTRY /EICHA | 5159F |
| EMISTRY /RVOUS SYSTEM BIOCH | **ADULT** ENDOCRINOLOGY TECHNIQUES NERVOUS SYSTEM BIOC | 4558F |
| NOSA *LAMPETRA FLUVIATILIS | **ADULT** EVOLUTION ENDOCRINOLOGY BIOCHEMISTRY /CGLUTI | 4725F |
| XINE GLUTINOSA. + ATLANTIC | **ADULT** EXCRETION / DNEY OF THE ATLANTIC HAGFISH, MY | 4568 |
| US + CANADA ATLANTIC OCEAN | **ADULT** EXCRETION HISTOCHEMISTRY /E PETROMYZON MARIN | 5350F |
| MARINUS CANADA GREAT LAKES | **ADULT** EXCRETION HISTOCHEMISTRY /EN. + *PETROMYZON | 5045F |
| NUS L.. + CANADA AMMOCOETE | **ADULT** EXCRETION HISTOLOGY HISTOCHEMISTRY / ON MARI | 4898F |
| GFISH, MYXINE GLUTINOSA. + | **ADULT** EXCRETION HISTOLOGY IONIC REGULATION /THE HA | 4773F |
| .S.A. LIFE CYCLE AMMOCOETE | **ADULT** FECUNDITY MORPHOMETRY SPAWNING /A, U.S.) + U | 5139F |
| AMPETRA LAMOTTEI AMMOCOETE | **ADULT** FEEDING PHYSIOLOGY /( *LAMPETRA AEPYPTERA *L | 5466F |
| IN A LARVAL AMMOCOETE AND | **ADULT** FRESH WATER LAMPREY LAMPETRA PLANERI. /AESIS | 5603 |
| Y. + *LAMPETRA FLUVIATILIS | **ADULT** GONADOGENESIS /GOGENESIS IN THE RIVER LAMPRE | 4915F |
| OLOGY PHYSIOLOGY CYTOLOGY / | **ADULT** GONADOGENESIS HISTOLOGY PHYSIOLOGY CYTOLOGY | 4646F |
| ISTRY / EVOLUTION HISTOCHEM | **ADULT** GREAT LAKES ENDOCRINOLOGY EVOLUTION HISTOCHE | 4897F |
| Y ENTOSPHENUS JAPONICUS. + | **ADULT** HEARING ANATOMY HISTOLOGY CYTOLOGY /E LAMPRE | 4663F |
| FOSSOR *PETROMYZON MARINUS | **ADULT** HISTOCHEMISTR /A 266:69-89. + *ICHTHYOMYZON | 5206F |
| RA FLUVIATILIS L. + POLAND | **ADULT** HISTOCHEMISTRY /MF THE RIVER LAMPREY, LAMPET | 4778F |
| ETRA FLUVIATILIS. + SWEDEN | **ADULT** HISTOCHEMISTRY BIOCHEMISTRY NERVOUS SYSTEM / | 4735F |
| MYXINE GLUTINOSA. + SWEDEN | **ADULT** HISTOCHEMISTRY ENDOCRINOLOGY /A THE HAGFISH | 5371F |
| TROMYZON MARINUS AMMOCOETE | **ADULT** HISTOCHEMISTRY EVOLUTION BLOOD HISTOLOGY /TE | 4743F |
| TRA FLUVIATILIS). + SWEDEN | **ADULT** HISTOCHEMISTRY HISTOLOGY VISION /IREY (LAMPE | 5310F |
| ISSNERI. + JAPAN AMMOCOETE | **ADULT** HISTOLOGY / AL CELLS OF LAMPREYS LAMPETRA RE | 5180 |
| TOME PETROMYZON MARINUS. + | **ADULT** HISTOLOGY /*TYPES OF ISLET CELLS IN A CYCLOS | 5205F |
| ETUS BURGERI ENDOCRINOLOGY | **ADULT** HISTOLOGY /HISH EPATRETUS BURGERI. + *EPTATR | 4990F |
| A JAPONICA JAPAN AMMOCOETE | **ADULT** HISTOLOGY /R ELECTRON MICROSCOPY. + *LAMPETR | 5356F |
| MPREY LAMPETRA JAPONICA. + | **ADULT** HISTOLOGY /THE 3RD VENTRICULAR WALL INTHE LA | 5187F |
| ARY OFMYXINE? PP. 33-34. + | **ADULT** HISTOLOGY BIOCHEMISTRY /EFORMATION IN THE OV | 4837F |
| SYSTEM PHYSIOLOGY /NERVOUS | **ADULT** HISTOLOGY BIOCHEMISTRY ENDOCRINOLOGY NERVOUS | 4591F |
| SA. + SWEDEN ENDOCRINOLOGY | **ADULT** HISTOLOGY CYTOLOGY /ILOSTOME, MYXINE GLUTINO | 4774F |
| + *LAMPETRA JAPONICA JAPAN | **ADULT** HISTOLOGY ENDOCRINOLOGY /CAMPETRA JAPONICA. | 5029F |
| URGERI SWEDEN CANADA JAPAN | **ADULT** HISTOLOGY ENDOCRINOLOGY /YOUTI *EPTATRETUS B | 4763F |
| MYXINE GLUTINOSA. + SWEDEN | **ADULT** HISTOLOGY ENDOCRINOLOGY EVOLUTION /IOSTOME, | 4790F |
| N. + CANADA ATLANTIC OCEAN | **ADULT** HISTOLOGY ENDOCRINOLOGY IMMUNOLOGY /EIGRATIO | 5539F |
| IATILIS L. + GREAT BRITAIN | **ADULT** HISTOLOGY IONIC REGULATION REPRODUCTION /TUV | 5169F |
| INUS. + U.S.A. GREAT LAKES | **ADULT** HISTOLOGY METAMORPHOSIS MOUTH /NTROMYZON MAR | 5348F |
| HAGFISH MYXINE GLUTINOSA + | **ADULT** HISTOLOGY MUSCLE /DE FIBERS IN THE ATLANTIC | 4610F |
| GLUTINOSA L. PP. 24-25. + | **ADULT** HISTOLOGY MUSCLE NERVOUS SYSTEM DIGESTION /M | 4839F |
| PETRA TRIDENTATA. + U.S.A. | **ADULT** HISTOLOGY NERVOUS SYSTEM ENDOCRINOLOGY /ELAM | 4577F |
| AMPETRA FLUVIATILIS SWEDEN | **ADULT** HISTOLOGY NERVOUS SYSTEMVISION /ASTOMI. + *L | 5214F |
| EY LAMPETRA FLUVIATILIS. + | **ADULT** HISTOLOGY OSMOREGULATION /O THE RIVER LAMPR | 5101F |
| MPETRA FLUVIATILIS FINLAND | **ADULT** IONIC REGULATION /SATURAL CONDITIONS.] + *LA | 4952F |
| SIS OF THYROGLOBULIN IN AN | **ADULT** LAMPREY, LAMPETRA PLANERI (BLOCH)] /OOSYNTHE | 5025F |
| AN AMMOCOETE ADULT BLOOD CY | **ADULT** LAMPREYS LAMPETRA FLUVIATILIS. + GREAT BRITI | 4649F |
| CANADA DIGESTION EXCRETION | **ADULT** LANDLOCKED SEA LAMPREYPETROMYZON MARINUS. + | 4673F |
| AMPETRA FLUVIATILIS SWEDEN | **ADULT** LARVA ENDOCRINOLOGY DIGESTION /NA PLANERI *L | 5561F |
| UVIATILIS). PP. 156-157. + | **ADULT** LIFE MATURITY /O RIVER LAMPREYS (LAMPETRA FL | 4868F |
| M OF BIOLOGY /M BYPARASITIS | **ADULT** LIFECYCLE DISTRIBUTION PARASITISM BYPARASITI | 4761F |
| AND TEMPERATURE. + CANADA | **ADULT** LOCOMOTION MIGRATION /KIN RELATION TO WEIGHT | 4593F |

THE SEA LAMPREY (PETROMYZON
Y /CULATORY SYSTEM HISTOLOG
ATED DIFFUSION PATHWAY FOR
THE RETINA OF THE RIVER LAM
LAMPREY USING A NEW FILTER
    TRIDENTATUS *ENTOSPHENUS T
AMPREY FIBRINOGEN. EVOLUTIO
E VERTEBRATE THE ATLANTIC H
FIBRINAND FIBRINO PEPTIDES.
  AND B AND CHARACTERIZATION
BLOOD /NVARIANT RESIDUE. +
MATION OF AMINO ACIDS IN PR
STOUTII. + EVOLUTION PHYSIO
IMENSIONAL STRUCTUREOF THE
ABELLED MONOSACCARIDES AND
THE BLOOD AND MUSCLE CELLS.
RYLASE EC-2.4.1.1. IN MUSCL
. + *LAMPETRA FLUVIATILIS B
1-44. + SEA LAMPREY CANADA
HYTOPHAGOUS LARVAE OF TRICH
*PETROMYZON MARINUS CANADA
*PETROMYZON MARINUS CANADA
*PETROMYZON MARINUS CANADA
P CANADA MANAGEMENT GROWTH
*PETROMYZON MARINUS CANADA
91-93. + CANADA MANAGEMENT
ULT CHEMISTRY MORTALITY OF
ADA MORTALITY OF CHEMISTRY
ETROMYZON MARINUS. + ADULT
.A. MORTALITY OF CHEMISTRY
ERI BELGIUM HISTOCHEMISTRY
ANADA MANAGEMENT CHEMISTRY
L AND COMPARATIVE STUDY. +
EEDING MANAGEMENT MORTALITY
ITY OF /BIOCHEMISTRY MORTAL
RY / HISTOLOGY HISTOCHEMIST
RA PLANERI (BLOCH,1784). +
UVIATILIS. + GREAT BRITIAN
MPETRA FLUVIATILIS ENGLAND
GYHISTOCHEMISTRY /GY CYTOLO
ITY OF TECHNIQUES /Y MORTAL
MISTRY /OCHEMISTRY HISTOCHE
/ HISTOLOGY HISTOCHEMISTRY
U.S.) + U.S.A. LIFE CYCLE
PYPTERA *LAMPETRA LAMOTTEI
OLOGY /EVOLUTION BLOOD HIST
AMPETRA REISSNERI. + JAPAN
+ *LAMPETRA JAPONICA JAPAN
YPARASITISM OF BIOLOGY /M B
TALITY BY /MORTALITY OF MOR
*PETROMYZON MARINUS U.S.A.
LAMPETRA JAPONICA. + JAPAN
RA PLANERI (BLOCH) . + U.K.
AMORPHOSIS IONIC REGULATION
ATILIS *PETROMYZON MARINUS
OLOGY DIGESTION HISTOLOGY /
HENUS MINIMUS U.S.A. ADULT
5-7. + *PETROMYZON MARINUS
ANERI. /LAMPREY LAMPETRA PL
UGERN. + *LAMPETRA PLANERI
NOLOGY HISTOLOGY CYTOLOGY /
  SKELETAL MUSCLES. + ADULT
/ES CYTOLOGY ENDOCRINOLOGY
LOGY MANAGEMENT MORTALITY B
SIOLOGY RESPIRATION /IS PHY
*PETROMYZON MARINUS CANADA
*PETROMYZON MARINUS U.S.A.
RINUS *ICHTHYOMYZON CANADA
RINUS *ICHTHYOMYZON CANADA
RINUS *ICHTHYOMYZON CANADA
RINUS *ICHTHYOMYZON CANADA
*PETROMYZON MARINUS CANADA
LOG /LOGY LIFE CYCLE PHYSIO
OLOG /STUDIES ON PHOSPHORYL
LE METAMORPHOSIS MIGRATION
MYZON MARINUS CANADA ADULT
  /OGY FEEDING METAMORPHOSIS
US CANADA MANAGEMENT ADULT
, 1973. PP.75-76. + CANADA
US CANADA MANAGEMENT ADULT
TY OF /N MANAGEMENT MORTALI

| | |
|---|---|
| AMIA CALVA AND LEPISOSTEVS OSSEVS OXYURUS) AND IN | 5130F |
| AMIA CALVA L.. + ANIMAL CIRCULATORY SYSTEM HISTOLO | 4749F |
| AMIDESIN THE VERTEBRATE ERYTHROCYTE. / F A FACILIT | 4968 |
| AMINERGIC AND INDOLEAMINE ACCUMULATING NEURONS IN | 5310F |
| AMINES IN THE HYPOTHALAMUS AND SPINAL CORD OF THE | 4940F |
| AMINO ACID COMPOSITION IN CYTOCHROMES. + *LAMPETRA | 4955F |
| AMINO ACID COMPOSITIONS OF THE SUBUNIT CHAINS OF L | 4715F |
| AMINO ACID SEQUENCE OF THE INSULIN FROM A PRIMITIV | 4654F |
| AMINO ACID SEQUENCE STUDIES ON LAMPREY FIBRINOGEN | 4683 |
| AMINO ACID SEQUENCES OF LAMPREY FIBRINO PEPTIDES A | 5146F |
| AMINO ACID SUBSTITUTION AT AN INVARIANT RESIDUE. + | 4971F |
| AMINO ACIDS (N-1) AND (N+1) ON THE BACKBONE CONFOR | 4626F |
| AMINO ACIDS BY THE GUT OF THE HAGFISHPOLISTOTREME | 4724 |
| AMINO ACIDS IN PROTEINS. USE IN PREDICTING THE 3-D | 4626F |
| AMINO ACIDS. /IEPATRETUS-STOUTI. UPTAKE OF RADIO L | 5166 |
| AMINO ACIDS, TRIMETHYLAMINE OXIDE AND POTASSIUMOF | 4767F |
| AMINO TRANSFERASE EC-2.6.1.1. AND GLYCOGEN PHOSPHO | 5179 |
| AMINO TRANSFERASES IN MUSCLE TISSUE OF VERTEBRATES | 5305 |
| AMMCOETE CHEMISTRY MANAGEMENT /N HURON, 1975. PP.3 | 5216F |
| AMMOCETES OF LAMPETRA PLANERI IN THE PRESENCE OF P | 5016F |
| AMMOCOETE /+REAM TREATMENTS, 1971-1973. PP. 77. + | 4901F |
| AMMOCOETE /EREAM TREATMENTS, 1971-1973. PP. 78. + | 4904F |
| AMMOCOETE /M1973. PP. 19-20. + *LAMPETRA LAMOTTEI | 4890F |
| AMMOCOETE /NNUS *LAMPETRA LAMOTTEI *ICHTHYOMYZON S | 5110F |
| AMMOCOETE /ORN RIVER SYSTEM, 1972. PP. 101-103. + | 5477F |
| AMMOCOETE /R ON THE ST. MARYS RIVER, 1971-74. PP. | 5132F |
| AMMOCOETE /T. 1-V. + *PETROMYZON MARINUS CANADA AD | 5237F |
| AMMOCOETE /T3. PP 71-74. + *PETROMYZON MARINUS CAN | 4892F |
| AMMOCOETE /Y ON THE ORAL DISC OF THE SEA LAMPREY P | 4702F |
| AMMOCOETE /Y. PP. 52-70. + *PETROMYZON MARINUS U.S | 4894F |
| AMMOCOETE ADULT /CNLAMPREY LIVER. + *LAMPETRA PLAN | 4734F |
| AMMOCOETE ADULT /OP.39-51. + *PETROMYZON MARINUS C | 5106F |
| AMMOCOETE ADULT ANATOMY NERVOUS SYSTEM EVOLUTION | 4559F |
| AMMOCOETE ADULT ANIMAL BIOLOGY CHEMISTRY ECOLOGY F | 4600F |
| AMMOCOETE ADULT ANIMAL CHEMISTRYBIOCHEMISTRY MORTA | 4751F |
| AMMOCOETE ADULT BIOCHEMISTRY HISTOLOGY HISTOCHEMIS | 4899F |
| AMMOCOETE ADULT BIOLOGY GROWTH SPAWNING /A, LAMPET | 4764F |
| AMMOCOETE ADULT BLOOD CYTOLOGY /MPREYS LAMPETRA FL | 4649F |
| AMMOCOETE ADULT BLOOD EVOLUTION LIFE CYCLE /UI *LA | 4584F |
| AMMOCOETE ADULT CIRCULATORY SYSTEM HISTOLOGY CYTOL | 4619F |
| AMMOCOETE ADULT CULTURE EMBRYOLOGY FECUNDITY MORTA | 5143F |
| AMMOCOETE ADULT ENDOCRINOLOGY BIOCHEMISTRY HISTOCH | 5393 |
| AMMOCOETE ADULT EXCRETION HISTOLOGY HISTOCHEMISTRY | 4898F |
| AMMOCOETE ADULT FECUNDITY MORPHOMETRY SPAWNING /A, | 5139F |
| AMMOCOETE ADULT FEEDING PHYSIOLOGY /( *LAMPETRA AE | 5466F |
| AMMOCOETE ADULT HISTOCHEMISTRY EVOLUTION BLOOD HIS | 4743F |
| AMMOCOETE ADULT HISTOLOGY / AL CELLS OF LAMPREYS L | 5180 |
| AMMOCOETE ADULT HISTOLOGY /R ELECTRON MICROSCOPY. | 5356F |
| AMMOCOETE ADULT LIFECYCLE DISTRIBUTION PARASITISM | 4761F |
| AMMOCOETE ADULT MANAGEMENT HISTORY MORTALITY OF MO | 4630F |
| AMMOCOETE ADULT MORPHOLOGY NERVOUS SYSTEM /GRD. + | 4595F |
| AMMOCOETE ADULT MUSCLE HISTOLOGY /NLARVAL LAMPREY | 5185F |
| AMMOCOETE ADULT RESPIRATION /OUVIATILIS AND LAMPET | 4877F |
| AMMOCOETE ADULT SPAWNING LIFE CYCLE PHYSIOLOGY MET | 4599F |
| AMMOCOETE ADULTMUSCLE ANATOMY HISTOLOGY /ERA FLUVI | 4840F |
| AMMOCOETE ANADROMOUS CYTOLOGY DIGESTION HISTOLOGY | 5518F |
| AMMOCOETE ANATOMY MORPHOMETRY /VRIDENTATUS *ENTOSP | 4750F |
| AMMOCOETE ANATOMY SENSE RECEPTORS /PENTATUS). PP. | 4731F |
| AMMOCOETE AND ADULT FRESH WATER LAMPREY LAMPETRA P | 5603 |
| AMMOCOETE ANIMAL BLOOD /+IEREN MIT AUSNAHME VON SA | 5119F |
| AMMOCOETE ANIMAL ENDOCRINOLOGY HISTOLOGY CYTOLOGY | 4779F |
| AMMOCOETE ANIMAL HISTOLOGY /OS" IN SOME VERTEBRATE | 4757F |
| AMMOCOETE ANIMAL TECHNIQUES CYTOLOGY ENDOCRINOLOGY | 4916 |
| AMMOCOETE BEHAVIOUR CHEMISTRY ENDOCRINOLOGY INNUNO | 5366F |
| AMMOCOETE BIOCHEMISTRY METABOLISM METAMORPHOSIS PH | 4678F |
| AMMOCOETE BIOLOGY / F LAKE ONTARIO 1973. PP.15. + | 4861F |
| AMMOCOETE BIOLOGY /AKE ONTARIO, 1973. PP.15-18. + | 4867F |
| AMMOCOETE BIOLOGY /D3. PP. 12-15. + *PETROMYZON MA | 4862F |
| AMMOCOETE BIOLOGY /I,1973. PP.12. + *PETROMYZON MA | 4863F |
| AMMOCOETE BIOLOGY /I1973.PP.7-12. + *PETROMYZON MA | 4864F |
| AMMOCOETE BIOLOGY /O1973. PP.6-7. + *PETROMYZON MA | 4865F |
| AMMOCOETE BIOLOGY /TST. MARY'S RIVER.PP. 88-91. + | 4826F |
| AMMOCOETE BIOLOGY BLOOD HISTOLOGY LIFE CYCLE PHYSI | 4716F |
| AMMOCOETE BIOLOGY BLOOD HISTOLOGY LIFE CYCLE PHYSI | 4717 |
| AMMOCOETE BIOLOGY FECUNDITY FEEDING GROWTH LIFECYC | 4710F |
| AMMOCOETE BIOLOGY GROWTH HISTORY MANAGEMENT /GETRO | 4908F |
| AMMOCOETE BLOOD ENDOCRINOLOGY FEEDING METAMORPHOSI | 5303F |
| AMMOCOETE CHEMISTRY /A. 17-26. + *PETROMYZON MARIN | 5108F |
| AMMOCOETE CHEMISTRY /ITREATMENT OF ST. MARYS RIVER | 4891F |
| AMMOCOETE CHEMISTRY /N. 26-39. + *PETROMYZON MARIN | 5107F |
| AMMOCOETE CHEMISTRY DISTRIBUTION MANAGEMENT MORTAL | 5253F |

| | | |
|---|---|---|
| TY OF /N MANAGEMENT MORTALI | AMMOCOETE CHEMISTRY DISTRIBUTION MANAGEMENT MORTAL | 5254F |
| TY OF /N MANAGEMENT MORTALI | AMMOCOETE CHEMISTRY DISTRIBUTION MANAGEMENT MORTAL | 5256F |
| TY OF /N MANAGEMENT MORTALI | AMMOCOETE CHEMISTRY DISTRIBUTION MANAGEMENT MORTAL | 5257F |
| TY OF /N MANAGEMENT MORTALI | AMMOCOETE CHEMISTRY DISTRIBUTION MANAGEMENT MORTAL | 5368F |
| TY OF /N MANAGEMENT MORTALI | AMMOCOETE CHEMISTRY DISTRIBUTION MANAGEMENT MORTAL | 5502F |
| ION /LOGY MANAGEMENT MIGRAT | AMMOCOETE CHEMISTRY ENDOCRINOLOGY MANAGEMENT MIGRA | 5263F |
| P. *LAMPETRA LAMOTTICANADA | AMMOCOETE CHEMISTRY GROWTH MANAGEMENT /NHYOMYZON S | 5491F |
| YOMYZON U.S.A. GREAT LAKES | AMMOCOETE CHEMISTRY MANAGEMENT / ON MARINUS *ICHTH | 5251F |
| 4-55. + SEA LAMPREY CANADA | AMMOCOETE CHEMISTRY MANAGEMENT / TARIO, 1975. PP.4 | 5220F |
| , 1975. PP.16-30. + CANADA | AMMOCOETE CHEMISTRY MANAGEMENT /DTS, LAKE SUPERIOR | 4554F |
| 8-75. + SEA LAMPREY CANADA | AMMOCOETE CHEMISTRY MANAGEMENT /IERIOR, 1975. PP.6 | 5222F |
| ON SPP. SEA LAMPREY CANADA | AMMOCOETE CHEMISTRY MANAGEMENT /NOTTEI *ICHTHYOMYZ | 5223F |
| A GREAT LAKES U.S.A. ADULT | AMMOCOETE CHEMISTRY MANAGEMENT MORTALITY OF /RANAD | 4844F |
| IDE. + *PETROMYZON MARINUS | AMMOCOETE CHEMISTRY MANAGEMENT TECHNIQUE /OLAMPRIC | 5053F |
| EY U.S.A. CHEMISTRY ADULTS | AMMOCOETE CHEMISTRY MANAGEMENT. /Y-68. + SEA LAMPR | 5221F |
| *PETROMYZON MARINUS CANADA | AMMOCOETE CHEMISTRY MORTALITY OF /M3.PP. 21-31. + | 4889F |
| GREAT BRITIAN CANADA ADULT | AMMOCOETE CIRCULATORY SYSTEM /MS *LAMPETRA PLANERI | 4651F |
| OGY CYTOLOGY /ISTRY MORPHOL | AMMOCOETE CIRCULATORY SYSTEM HISTOCHEMISTRY MORPHO | 4998F |
| ETRA PLANERI BELGIUM ADULT | AMMOCOETE CULTURE /E + *LAMPETRA FLUVIATILIS *LAMP | 5357F |
| RATES OF GROWTH.] + FRANCE | AMMOCOETE CULTURE FEEDING /.ERIMENTAL AND NATURAL | 5352F |
| A LAMPREY LARVAE. + U.S.A. | AMMOCOETE CULTURE FEEDING GROWTH /+E METHOD FOR SE | 4970F |
| *LAMPETRA AEPYPTERA U.S.A. | AMMOCOETE DISTRIBUTION /TTATS AND DISTRIBUTION. + | 5089F |
| ZON CASTANEUS U.S.A. ADULT | AMMOCOETE DISTRIBUTION ECOLOGY / ELLIUM *ICHTHYOMY | 4590F |
| BALTIC SEA BLACK SEA ADULT | AMMOCOETE DISTRIBUTION EVOLUTION SYSTEMATICS /ANE | 5358F |
| *LAMPETRA LAMOTTEI CANADA | AMMOCOETE DISTRIBUTION GROWTH // *ICHTHYOMYZON SP. | 4645F |
| MARINUS CANADA GREAT LAKES | AMMOCOETE DISTRIBUTION MANAGEMENT / + *PETROMYZON | 5248F |
| , 1975. PP.14-16. + U.S.A. | AMMOCOETE DISTRIBUTION MANAGEMENT / NEW YORK STATE | 4553F |
| 1975. PP. 12-14. + CANADA | AMMOCOETE DISTRIBUTION MANAGEMENT // LAKE ONTARIO, | 4547F |
| , 1975. PP. 7-12. + CANADA | AMMOCOETE DISTRIBUTION MANAGEMENT /B TO LAKE HURON | 4605F |
| MARINUS CANADA GREAT LAKES | AMMOCOETE DISTRIBUTION MANAGEMENT /I+ *PETROMYZON | 5250F |
| R, 1975. PP. 6-7. + CANADA | AMMOCOETE DISTRIBUTION MANAGEMENT /NO LAKE SUPERIO | 5116F |
| ETRY BEHAVIOUR PARASITISM L | AMMOCOETE ECOLOGY FECUNDITY FEEDING GROWTH MORPHOM | 4974F |
| SMMETAMORPHOSIS PHYSIOLOGY | AMMOCOETE ECOLOGY LIFE CYCLE BLOOD GROWTH METABOLI | 5616F |
| MPREY (LAMPETRA AYRESI). + | AMMOCOETE ENDOCRINOLOGY / GEORGIA BY THE RIVER LA | 5343F |
| USTRALIS GRAY. + AUSTRALIA | AMMOCOETE ENDOCRINOLOGY /EIN THE LAMPREY GEOTRIA-A | 5209F |
| NUS. + CANADA LAKE ONTARIO | AMMOCOETE ENDOCRINOLOGY EXCRETION /TETROMYZON MARI | 4639F |
| CLE. + CANADA U.S.A. ADULT | AMMOCOETE ENDOCRINOLOGY HISTOLOGY METAMORPHOSIS /R | 4758F |
| H). + UNITED KINGDOM ADULT | AMMOCOETE EVOLUTION METAMORPHOSIS RESPIRATION /IOC | 5300F |
| LOGY / ENDOCRINOLOGY PHYSIO | AMMOCOETE EXCRETION INTEGUMENT ENDOCRINOLOGY PHYSI | 4705F |
| LAMOTTEI CANADA MANAGEMENT | AMMOCOETE GROWTH / US *ICHTHYOMYZON SP. *LAMPETRA | 5111F |
| *LAMPETRA LAMOTTII CANADA | AMMOCOETE GROWTH CHEMISTRY /INUS *ICHTHYOMYZON SP. | 5104F |
| *LAMPETRA LAMOTTII CANADA | AMMOCOETE GROWTH CHEMISTRY /TNUS *ICHTHYOMYZON SP. | 5105F |
| IQUES /NERVOUS SYSTEM TECHN | AMMOCOETE HISTOLOGY LOCOMOTION NERVOUS SYSTEM TECH | 5026F |
| -406. + *ENTOSPHENUS JAPAN | AMMOCOETE HISTOLOGY MUSCLE /TF THE LAMPREY. PP.405 | 5529F |
| VIATILIS *LAMPETRA PLANERI | AMMOCOETE HISTOLOGY MUSCLE CYTOLOGY /SLAMPETRA FLU | 4884 |
| AMPETRA FLUVIATILIS RUSSIA | AMMOCOETE HISTOLOGY MUSCLE NERVOUS SYSTEM /E. + *L | 5317 |
| *PETROMYZON MARINUS CANADA | AMMOCOETE HISTORY. /CIN SEDIMENT-WATER SYSTEMS. + | 4736F |
| A PLANERI. /LIS AND LAMPETR | AMMOCOETE LAMPREYS LAMPETRA FLUVIATILIS AND LAMPET | 4991 |
| ICHARDSONI. STUDIES IN THE | AMMOCOETE LARVA. / STERN BROOK LAMPREY, LAMPETRA R | 5155 |
| PHOSIS PHYSIOLOGY /MMETAMOR | AMMOCOETE LIFE CYCLE BLOOD GROWTH METABOLISMMETAMO | 4975F |
| TROMYZON MARINUS CHEMISTRY | AMMOCOETE MANAGEMENT /VS OF THE SEA LAMPREY. + *PE | 5050F |
| *LAMPETRA LAMOTTEI CANADA | AMMOCOETE MANAGEMENT GROWTH / US *ICHTHYOMYZON SP. | 5109F |
| MARINUS CANADA GREAT LAKES | AMMOCOETE MANAGEMENTMORTALITY OF /Y + *PETROMYZON | 5252F |
| N MARINUS L. + BLOOD ADULT | AMMOCOETE MIGRATION FEEDING METAMORPHOSIS /RROMYZO | 5145 |
| *PETROMYZON MARINUS CANADA | AMMOCOETE MORTALITY OF CHEMISTRY /H. PP. 47-52. + | 4893F |
| *PETROMYZON MARINUS CANADA | AMMOCOETE MORTALITY OF CHEMISTRY /S3. PP.32-46. + | 4888F |
| RVAL PETROMYZON MARINUS. + | AMMOCOETE NERVOUS SYSTEM /AR NERVE AFFERENTS IN LA | 4951F |
| ORD. + *PETROMYZON MARINUS | AMMOCOETE NERVOUS SYSTEM /SSECTED LAMPREY SPINAL C | 5069 |
| MYZON MARINUS U.S.A. ADULT | AMMOCOETE NERVOUS SYSTEM CYTOLOGY /OPREY. + *PETRO | 4625F |
| N MARINUS. + NEW BRUNSWICK | AMMOCOETE OSMOREGULATION ENDOCRINOLOGY /HPETROMYZO | 5604F |
| AMOTTENII). + CANADA ADULT | AMMOCOETE PARASITISM OF /IROOK LAMPREY (LAMPETRA L | 5541F |
| PLANERI (BLOCH). + SWEDEN | AMMOCOETE PARASITISM OF ADULT /U LAMPREYS LAMPETRA | 5467F |
| TENII *ICHTHYOMYZON FOSSOR | AMMOCOETE PHYSIOLOGY /YAMPETRA (LETHENTERON) LAMOT | 4565F |
| L MORPHOLOGY. PP. 74-82. + | AMMOCOETE SENSE RECEPTORS HISTOLOGY / /GENERA | 5057F |
| ECIES OF THE SAME GENUS. + | AMMOCOETE SYSTEMATICS ADULT DISTRIBUTION /LTHER SP | 5542F |
| LE /METRY PIGMENTATION MUSC | AMMOCOETE SYSTEMATICS MORPHOMETRY PIGMENTATION MUS | 4713 |
| ERENT CELLS OF THE LAMPREY | AMMOCOETE. + ANIMAL ANATOMY HISTOCHEMISTRY /OY AFF | 5295F |
| RY TO LAKE ONTARIO. PP. 84- | AMMOCOETE) GROWTH RATES FROM THREE STREAMS TRIBUTA | 5133F |
| YSTEMATICS MORPHOMETRY /Y S | AMMOCOETEADULT METAMORPHOSIS DISTRIBUTION ECOLOGY | 4676F |
| HTHYOMYZON CASTANEUS U.S.A | AMMOCOETEMUSCLE NERVOUS SYSTEM PHYSIOLOGY ADULT /G | 4674F |
| VELLE METHODE D'ALIMENTATIO | AMMOCOETES DE LAMPETRA PLANERI (BLOCH) PAR UNE NOU | 5352F |
| EVIDENCE OF ITS HOMOLOGY WI | AMMOCOETES ENDOSTYLE: ITS OXIDATIVE ENZYMES AS AN | 5273 |
| IA AUSTRALIS GRAY. + AUSTRA | AMMOCOETES OF THE SOUTHERN HEMISPERELAMPREY, GEOTR | 5545F |
| LANERI *PETROMYZON MARINUS | AMMOCOETES OF 4 SPECIES OF LAMPREYS. + *LAMPETRA P | 4565F |
| 1971-1973. PP. 78. + *PETR | AMMOCOETES TAKEN FROMLAKE HURON STREAM TREATMENTS, | 4904F |
| TS, 1971-1973. PP. 77. + *P | AMMOCOETES TAKEN FROMLAKE SUPERIOR STREAM TREATMEN | 4901F |
| LANERI BL. [ON THE BIOSYNTH | AMMOCOETES) D'UNE LAMPROIE D'EAU DOUCE, LAMPETRA P | 5342F |
| NERI BLOCH + FRANCE ENDOCRI | AMMOCOETES) OF A FRESH WATER LAMPREY, LAMPETRA PLA | 5342F |
| BETWEEN BOTH THE EXPERIMENT | AMMOCOETESOF LAMPETRA PLANERI (BLOCH). COMPARISON | 5352F |
| DANFORDI (REGAN). /TOMYZON | AMMOCOTEN LAMPETRA PLANERI (BLOCH) UND EUDONTOMYZO | 5331 |
| ETRA FLUVIATILIS. [THE EFFE | AMNLAZNUYU AKTIVNOST' PECHENI I MYSHTS MINOGI LAMP | 4561F |

TRA JAPONICA. + ADULT NERVO
SPIOMYZON *PETROMYZON ICHTH
CRINOLOGY /ATA BIOLOGY ENDO
RY EVOLUTION NERVOUS SYSTEM
  OF THE CANADIAN ARCTIC ARC
ACE OF FRESHWATER FISH AND
4, VERTEBRATES: 1: PISCES,
YDROGENASE FROM FISHES AND
ADULT /RII. + BIOCHEMISTRY
DA NEW BRUNSWICK AMMOCOETE
US. + CANADA NEW BRUNSWICK
*PETROMYZON MARINUS U.S.A.
ORD. + *PETROMYZON MARINUS
URING METAMORPHOSIS OF THE
ATION TO AMBIENT SALINITY.
MARINUS L. /IC, PETROMYZON
  ATLANTIC OCEAN ADULT EXCRE
ETAMORPHOSIS. PART 1: LIGHT
UNSWICK. + CANADA ADULT AMM
HEMISTRY BLOOD /NSWICK BIOC
OOD ADULT AMMOCOETE MIGRATI
NADA ATLANTIC OCEAN ADULT A
MUNOLOGY ENDOCRINOLOGY CYTO
TAMORPHOSIS GROWTH MORPHOME
  DURING LARVAL LIFE IN THE
ING ITS UPSTREAM MIGRATION.
BRUNSWICK AMMOCOETE OSMOREG
SPAWNING ENERGETICS OF THE
DA ADULT ANIMAL ECOLOGY FEC
ULATION PHYSIOLOGY HISTOLOG
PLANATA. /AND ASPATHRIA COM
HE SEXUALLY MATURE SEA LAMP
  FAUNA OF THE SCOTIA SEA. /
E OF VARYING EUKARYOTIC DAM
BY LUMINESCENCE SPECTROPHOM
. + *LAMPETRA FLUVIATILIS A
A FUNCTIONAL EVOLUTIONARY
NORWAY. /F THE OSLO REGION
  /      /FISH BILE
BRANCHIA MORPHOLOGICAL AND
STEMATICS. /OGENESIS AND SY
CHE. P. 223. /TOMEN UNDFIS
TAIL ANDLATERALFIN FOLD OF
ATIN. /           /
ONS IN LAMPREY. + *PETROMYZ
          /VERGLEICHENOE
(VERTEBRATA, CYCLOSTOMATA)
STOMI, OSTEOSTRACI), CEPHAL
YSTEM HISTOLOGY LOCOMOTION
VERTEBRATA CYCLOSTOMATA CEP
SYSTEM / OLFACTION NERVOUS
N. + *LAMPETRA FLUVIATILIS
MAYOMYZON PIEKOENSIS ADULT
LUTINOSA L. ATLANTIC ADULT
IAE *EUDONTOMYZON VLADYKOV
S /Y MORPHOMETRY SYSTEMATIC
OF NORTH AMERICA. + ANIMAL
  SKELETON. + *CYCLOSTOMATA
RDATES. PP. 14-16. + ADULT
NAMPHIASPIFORMS. + *MYXINE
/ERVOUS SYSTEM SYSTEMATICS
VONIAN OF YUNNAN CHINA.] +
A FLUVIATILIS.] + U.S.S.R.
AMPETRA FLUVIATILIS RUSSIA
AMPREY AMMOCOETE. + ANIMAL
GFISH THYMUS. + IMMUNOLOGY
MOSTEID HETEROSTRACHANS. +
INUS AMMOCOETE ADULTMUSCLE
G CYCLOSTOME ECTOPROCTS. +
  + ENDOCRINOLOGY EVOLUTION
JAPONICUS. + ADULT HEARING
(LAMPETRA FLUVIATILIS). +
HA, CEPHALASPIDOMORPHI). +
MUS U.S.A. ADULT AMMOCOETE
TRY / MORPHOMETRY BIOCHEMIS
OMATA). + CANADA HISTOLOGY
E STUDY. + AMMOCOETE ADULT
MARINUS U.S.A. GREAT LAKES
STRACI. + *PETROMYZON SPP.
TROMYZON MARINUS AMMOCOETE
EUS *LAMPETRA *LETHENTERON

AMONG THE GIANT INTERNEURONS OF THE LAMPREY, LAMPE   5572F
AMONG THE HOLARCTIC LAMPREYS PETROMYZONIDAE. + *CA   5391F
AMONG THE VERTEBRATES. + *CYCLOSTOMATA BIOLOGY END   5389F
AMONG THE VERTEBRATES. + ANIMAL BIOLOGY BIOCHEMIST   4685
AMPHIASPIDIFORMES (HETEROSTRACI) FROM THE SILURIAN   5329
AMPHIBIA. PP. 447-463. /YPORT ACROSS THE BODY SURF   4934
AMPHIBIA, REPTILIA. 244 PP. /S BOOK OF ZOOLOGY, V.   4822
AMPHIBIANS. + ANIMAL BIOCHEMISTRY PHYSIOLOGY /YDEH   5170F
AMPHIOXUS, BRANCHIOSTOMA BELCHERII. + BIOCHEMISTRY   5041
ANADROMOUS CYTOLOGY DIGESTION HISTOLOGY /O. + CANA   5518F
ANADROMOUS HISTOLOGY CYTOLOGY DIGESTION / ON MARIN   5516F
ANADROMOUS LANDLOCKED ENDOCRINOLOGY /EP. 83-95. +   5289F
ANADROMOUS LARVA ADULT NERVOUS SYSTEM HISTOLOGY /M   5020F
ANADROMOUS LARVAL LAMPREY, PETROMYZON MARINUS L. /   4695
ANADROMOUS SEA LAMPREY (PETROMYZON MARINUS) IN REL   5611F
ANADROMOUS SEA LAMPREY OF THE ATLANTIC, PETROMYZON   4726
ANADROMOUS SEA LAMPREY PETROMYZON MARINUS + CANADA   5350F
ANADROMOUS SEA LAMPREY PETROMYZON MARINUS DURING M   5174F
ANADROMOUS SEA LAMPREY PETROMYZON MARINUSIN NEW BR   4710F
ANADROMOUS SEA LAMPREY. + CANADA NEW BRUNSWICK BIO   5096F
ANADROMOUS SEA LAMPREY, PETROMYZON MARINUS L. + BL   5145
ANADROMOUS SEA LAMPREY, PETROMYZON MARINUS L. + CA   5303F
ANADROMOUS SEA LAMPREY, PETROMYZON MARINUS L. + IM   5538F
ANADROMOUS SEA LAMPREY, PETROMYZON MARINUS L. + ME   5537F
ANADROMOUS SEA LAMPREY, PETROMYZON MARINUS L. /TEY   4539
ANADROMOUS SEA LAMPREY, PETROMYZON MARINUS L., DUR   5539F
ANADROMOUS SEA LAMPREY, PETROMYZON MARINUS. + NEW   5604F
ANADROMOUS SEA LAMPREY, PETROMYZON MARINUS. /SAND   5507
ANADROMOUS SEA LAMPREYS PETROMYZON MARINUS. + CANA   5140F
ANADROMOUS SPAWNING MIGRATION. + MIGRATION OSMOREG   5115F
ANALYSES OF 2 SPECIES PILA WERNEI AND ASPATHRIA CO   4544
ANALYSES OFPRESUMED STEROID-PRODUCING TISSUES IN T   5470F
ANALYSIS OF BENTHIC FISH FAUNA OF THE SCOTIA SEA.   5597
ANALYSIS OF SEVERAL FISHES AND DISSOCIATION PROFIL   5614
ANALYSIS OF TFM (3-TRIFLUOROMETHYL-4-NITROPHENOL)   5381
ANALYSIS OF THE BRAIN STEM, A GENERAL INTRODUCTION   4697F
ANALYSIS OF THE INSULIN MOLECULE. / ARCH AIMING AT   5192
ANALYSIS OF THE RINGERIKE GROUP OF THE OSLO REGION   5031
ANALYSIS. POSSIBLE AID IN MONITORING WATER QUALITY   5402
ANALYTICAL DATA. /HOSTOMA-ELEGANS GASTROPODA PROSO   5149
ANALYZING PROBLEMS IN VERTEBRATEPHYLOGENESIS AND S   4701F
ANAMNIER-KIEMENDARMDERIVATE DER CYCLOSTOMEN  UNDFI   5588
ANASPID OSTRACODERMS. /IFICANCE OF THE HYPOCERCAL   4537
ANATOMIC AND PHYLOGENETIC DISTRIBUTION OF SOMATOST   5195
ANATOMICAL CHARACTERISTICS OF RETICULO SPINAL NEUR   5280F
ANATOMIE DER MYXOIDEN. III. UBER DAS GEFABSYSTEM.   5291
ANATOMIE ET POSITION SYSTEMATIQUE DES GALEASPIDES   4709F
ANATOMIE ET SYSTEMATIQUE DU GENRE BOREASPIS (CYCLO   4578
ANATOMY / ETRA FLUVIATILIS ANIMAL MUSCLE NERVOUS S   5079F
ANATOMY AND TAXONOMIC POSITION OF THE GALEASPIDES   4709F
ANATOMY ANIMAL BIOLOGY EVOLUTION OLFACTION NERVOUS   4771F
ANATOMY ANIMAL NERVOUS SYSTEM /RENERAL INTRODUCTIO   4697F
ANATOMY CIRCULATORY SYSTEM EVOLUTION RESPIRATION /   4843F
ANATOMY CYTOLOGY MUSCLE /UAMINA DENSA. + *MYXINE G   4563F
ANATOMY DISTRIBUTION /AN DANFORD *EUDONTOMYZON MAR   4541F
ANATOMY DISTRIBUTION ECOLOGY MORPHOMETRY SYSTEMATI   4623F
ANATOMY EVOLUTION /M FISH FROM THE UPPER CAMBRIAN   5599
ANATOMY EVOLUTION /MMOLOGY OF THE CYCLOSTOME AXIAL   5161F
ANATOMY EVOLUTION CYTOLOGY /  TO THAT OF OTHER CHO   4842F
ANATOMY EVOLUTION EMBRYOLOGY /LOMES WITH REMARKS O   4675F
ANATOMY EVOLUTION MOUTH NERVOUS SYSTEM SYSTEMATICS   5346F
ANATOMY EVOLUTION NERVOUS SYSTEM EVOLUTION / ER DE   4709F
ANATOMY EXCRETION /ONEPHRON IN THE LAMPREY LAMPETR   4694F
ANATOMY EXCRETION EVOLUTION / TURE STUDIES.)] + *L   5064F
ANATOMY HISTOCHEMISTRY /OY AFFERENT CELLS OF THE L   5295F
ANATOMY HISTOLOGY /         /IN SEARCH OF THE HA   4869F
ANATOMY HISTOLOGY /EIN THE DERMAL SKELETON OF PSAM   5102
ANATOMY HISTOLOGY /ERA FLUVIATILIS *PETROMYZON MAR   4840F
ANATOMY HISTOLOGY ADULT /A ULTRASTRUCTURE OF LIVIN   5067F
ANATOMY HISTOLOGY BIOCHEMISTRY /YINOSA. PP. 26-29.   4838F
ANATOMY HISTOLOGY CYTOLOGY /E LAMPREY ENTOSPHENUS   4663F
ANATOMY HISTOLOGY PHYSIOLOGY / REAS OF THE LAMPREY   5233F
ANATOMY INTEGUMENT /OION IN THE OSTEOSTRACI (AGNAT   4939F
ANATOMY MORPHOMETRY /VRIDENTATUS *ENTOSPHENUS MINI   4750F
ANATOMY MUSCLE HISTOCHEMISTRY MORPHOMETRY BIOCHEMI   4841F
ANATOMY NERVOUS SYSTEM /MPTATRETUS STOUTI (CYCLOST   4922F
ANATOMY NERVOUS SYSTEM EVOLUTION /A AND COMPARATIV   4559F
ANATOMY NERVOUS SYSTEM PHYSIOLOGY /L+ *PETROMYZON   5280F
ANATOMY OLFACTION /MME DU SAC NASAL CHEZ LES OSTEO   4692F
ANATOMY SENSE RECEPTORS /PENTATUS). PP. 5-7. + *PE   4731F
ANATOMY SYSTEMATICS /MTATUS *TETRAPLEURODON SPADIC   4752F

RK; ACADEMIC PRESS, 3V. /YO          ANATOMY WITH AN INTRODUCTION TOTHE PERTIN....NEW Y      4730
YSTITES-FORBESIANUS ANDTHE          ANCESTRY OF THE VERTEBRATES. /AURIAN FOSSIL PLACOC       5121
  LAMELLA IN THE SKIN OF AQU        ANCHORING FIBRILS OF THE BASAL LAMINA AND BASEMENT       5148F
ENOL IN RAINBOW TROUT. /OPH         ANDBILIARY EXCRETION OF 3-TRIFLUOROMETHYL-4-NITROP       5412F
AMPETRA FLUVIATILIS ADULT E         ANDBIOSYNTHESIS OF INSULIN. + *MYXINE GLUTINOSA *L       4725F
NE GLUTINOSA BIOCHEMISTRY /         ANDBIOSYNTHESIS. + *MYXINE GLUTINOSA BIOCHEMISTRY       4918F
] PP.127-133. + *LAMPETRA F         ANDFISHES ON VARIOUS STAGES OF SPAWNING MIGRATION.       5361F
FLUVIATILIS. + ADULT BEHAVI         ANDHEART RATE IN THE ADULT RIVER LAMPREY LAMPETRA        4719F
CLOSTOMATA). + CANADA BLOOD         ANDHYPOPHYSECTOMIZED HAGFISH, MYXINE GLUTINOSA (CY       4723F
LIS U.S.S.R. NERVOUS SYSTEM         ANDINTRA SPINAL STIMULATION.] + *LAMPETRA FLUVIATI       4601F
NCE OF THE HYPOCERCAL TAIL          ANDLATERALFIN FOLD OF ANASPID OSTRACODERMS. /IFICA       4537
AMPETRA FLUVIATILIS. + SWED         ANDNORADRENALINE-CONTAINING NEURONS IN THE GUT OFL       4735F
ULA (EAST COAST, UNITED STA         ANDOKKELBERGIA AEPYPTERA, FROM THE DELMARVA PENINS       4720F
ADAPTATION BY THE LATTER DU         ANDPLANT HEMOGLOBINS BASED ON TRUE HOMOLOGY OR ON        4983
  PETROMYZON MARINUS DURING         ANDTESTICULAR TISSUE OBTAINED FROM THE SEA LAMPREY       4659F
PLACOCYSTITES-FORBESIANUS           ANDTHE ANCESTRY OF THE VERTEBRATES. /AURIAN FOSSIL       5121
NORWAY CIRCULATORY SYSTEM           ANDVENTRICLE OF THE CYCLOSTOME MYXINE GLUTINOSA. +       4597F
MENTARY RETINAL EPITHELIUM          ANDVISUALCELLS IN A TELEOST CYPRINUS CARPIO. / PIG       4665
  PHYSIOLOGY /         /THE         ANEURAL HEART OF THE HAGFISH. + CIRCULATORY SYSTEM       5503
NUS *LAMPETRA JAPONICUS *PA         ANGIOTENSIN SYSTEM IN FISHES. + *LAMPETRA ENTOSPHE       4919
ION SENSITIVITY OF THE METH         ANGLE X-RAY DIFFUSE SCATTERING BY GLOBINS IN SOLUT       5070
LAR PROTEIN STRUCTURE IN SO         ANGLE X-RAY DIFFUSE SCATTERING TO STUDIES OF GLOBU       5349
TRY BLOOD /         /CARBONIC       ANHYDRASE IN MYXINE. PP. 86-88. + ANIMAL BIOCHEMIS       4849F
TOMY DISTRIBUTION ECOLOGY M         ANIKIN) OF UPPER IRTYSH BASIN.] + RUSSIA ADULT ANA       4623F
  PECULIARITIES AND CHANGEAB        ANIKIN) VODOEMOV VERKHNEGO IRTYSHA. [MORPHOLOGICAL       4623F
F CARRIE-BOW CAY BELIZE. +          ANIMAL /         /THE ECHINOIDS O       4819F
LIS BIOCHEMISTRY EXCRETION          ANIMAL /ERON IN VERTEBRATES.] + *LAMPETRA FLUVIATI       4742F
OLY PEPT... + BIOCHEMISTRY          ANIMAL /ICTING THE 3-DIMENSIONAL STRUCTUREOF THE P       4626F
ES. + CHEMISTRY MANAGEMENT          ANIMAL /L-NITROPHENOL) TO SELETED MACROINVERTEBRAT       5387F
PYPTERA U.S.A. SYSTEMATICS          ANIMAL /LAS USA. + *LAMPETRA LAMOTTEI *LAMPETRA AE       5035F
TRY BLOOD EVOLUTION PISCES          ANIMAL /OPHENUS JAPONICUS HOKKAIDO JAPAN BIOCHEMIS       5320
  NERVOUS SYSTEM PHYSIOLOGY         ANIMAL /RTHE LAMPREY LAMPETRA FLUVIATILIS.] + USSR       4636F
ES. + CHEMISTRY MANAGEMENT          ANIMAL /T THE METABOLISM OF MODEL STREAM COMMUNITI       5386F
. + DIGESTION BIOCHEMISTRY          ANIMAL /YACTIVITY IN MYXINE GLUTINOSA L. PP. 30-32       4846F
ETION /LOGY METABOLISM EXCR         ANIMAL ADULT BIOCHEMISTRY HISTOLOGY METABOLISM EXC       4604F
EY (PETROMYZON MARINUS). +          ANIMAL ADULT NERVOUS SYSTEM VISION /ITHE SEA LAMPR       5130F
MBRIAN OF NORTH AMERICA. +          ANIMAL ANATOMY EVOLUTION /M FISH FROM THE UPPER CA       5599
F THE LAMPREY AMMOCOETE. +          ANIMAL ANATOMY HISTOCHEMISTRY /OY AFFERENT CELLS O       5295F
  OR ON ADAPTATION BY THE LA        ANIMAL ANDPLANT HEMOGLOBINS BASED ON TRUE HOMOLOGY       4983
F TROPONIN-C. PP. 65-67. +          ANIMAL BIOCHEMISTRY /         /IMMUNOCHEMICAL STUDIES O     4671F
RINE SPECIES. + METABOLISM          ANIMAL BIOCHEMISTRY / YMES IN COASTAL MAINE USA MA       4964F
  NERVOUS SYSTEM PHYSIOLOGY         ANIMAL BIOCHEMISTRY /ELIKE PEPTIDE "ENKEPHALIN". +       5158
SPHENUS TRIDENTATUS U.S.A.          ANIMAL BIOCHEMISTRY ADULT /C ELASMOBRANCH. + *ENTO       4729F
ACID DISTRIBUTION. + BLOOD          ANIMAL BIOCHEMISTRY BIOLOGY /AMA CARBOXY GLUTAMIC-       5028
SE IN MYXINE. PP. 86-88. +          ANIMAL BIOCHEMISTRY BLOOD /         /CARBONIC ANHYDRA       4849F
D ENDOCRINOLOGYIMMUNOLOGY /         ANIMAL BIOCHEMISTRY BLOOD ENDOCRINOLOGYIMMUNOLOGY       4708F
SSUE LABORATORY STUDIES. +          ANIMAL BIOCHEMISTRY CHEMISTRY MANAGEMENT /FSCLE TI       4985F
Y /NOLOGY CYTOLOGY HISTOLOG         ANIMAL BIOCHEMISTRY ENDOCRINOLOGY CYTOLOGY HISTOLO       4993
F VERTEBRATE FIBRINOGEN. +          ANIMAL BIOCHEMISTRY EVOLUTION PHYSIOLOGY /LUTION O       5097F
PETRA FLUVIATILIS U.S.S.R.          ANIMAL BIOCHEMISTRY HISTOCHEMISTRY PISCES /M+ *LAM       4803F
. + *ENTOSPHENUS JAPONICUS          ANIMAL BIOCHEMISTRY IMMUNOLOGY /EERTEBRATE SPECIES       5574F
TED RAT FAT CELLS. + ADULT          ANIMAL BIOCHEMISTRY METABOLISM PHYSIOLOGY /F ISOLA       5154F
IS IN VERTEBRATE LIVERS. +          ANIMAL BIOCHEMISTRY MUSCLE BLOOD /         /GLUCONEOGENES       5211
SEVERAL SPECIES OF FISH. +          ANIMAL BIOCHEMISTRY PHYSIOLOGY /DTIES OF DNA FROM        5147
M FISHES AND AMPHIBIANS. +          ANIMAL BIOCHEMISTRY PHYSIOLOGY /YDEHYDROGENASE FRO       5170F
NTH LATIMERIA CHALUMNAE. +          ANIMAL BIOCHEMISTRY SYSTEMATICS /AE OF THE COELACA       5156F
  AND FISHES.] PP. 59-66. +        ANIMAL BIOCHEMISTRY SYSTEMATICS /MS IN CYLLOSTOMES       4896F
  AND RAT LIVER HISTONES. +        ANIMAL BIOCHEMISTRY TECHNIQUES /V STUDY OF HAGFISH       4579
NDANT ALDEHYDEREDUCTASE. +          ANIMAL BIOCHEMISTRY TECHNIQUES EXCRETION /*PH-DEPE       5570
  AND ALPHA-1 3 COLLAGENS. +       ANIMAL BIOCHEMISTRY TECHNIQUES VISION /I2,ALPHA-2        4571
.] + *LAMPETRA FLUVIATILIS          ANIMAL BIOCHEMISTRY VISION /RN CERTAIN VERTEBRATES       4817F
M / EVOLUTION NERVOUS SYSTE         ANIMAL BIOLOGY BIOCHEMISTRY EVOLUTION NERVOUS SYST       4685
  MORTALITY OF /G MANAGEMENT       ANIMAL BIOLOGY CHEMISTRY ECOLOGY FEEDING MANAGEMEN       4600F
  OLFACTION NERVOUS SYSTEM /       ANIMAL BIOLOGY EVOLUTION OLFACTION NERVOUS SYSTEM        4771F
LAMPETRA PLANERI AMMOCOETE          ANIMAL BLOOD /+IEREN MIT AUSNAHME VON SAUGERN. + *       5119F
.] + *LAMPETRA FLUVIATILIS          ANIMAL BLOOD BIOCHEMISTRY / GLOBINS AND MYOGLOBINS       5128F
  CLASSES OF VERTEBRATES. +        ANIMAL BLOOD BIOCHEMISTRY /I FIBRIN FROM THE MAJOR       4549
S, VERTEBRATES AND MAN.] +          ANIMAL BLOOD BIOCHEMISTRY /OHANISMS ININVERTEBRATE       5372F
ED BY ART1FICIAL STRESS. +          ANIMAL BLOOD BIOCHEMISTRY PHYSIOLOGY /ATIONS INDUC       4545
THE BLOOD OF MARINE FISH +          ANIMAL BLOOD BIOCHEMISTRY SPAWNING /TTTY-ACIDS IN       5015
+ *LAMPETRA PLANERI SWEDEN          ANIMAL BLOOD HISTOCHEMISTRY /CLOSTOMES. PP. 53-56       4859F
(TFM). + U.S.A. MANAGEMENT          ANIMAL CHEMISTRY /Y-TRIFLUOROMETHYL-4-NITROPHENOL       5242F
CONTROL OF SEA LAMPREYS. +          ANIMAL CHEMISTRY ECOLOGY MANAGEMENT /REATMENT FOR       5051F
OM THE UPPER GREAT LAKES +          ANIMAL CHEMISTRY MANAGEMENT / ND CHINOOK SALMON FR       5189F
ETHAL TOXICANT EXPOSURE. +          ANIMAL CHEMISTRY MANAGEMENT /ILLOWING ACUTELY SUBL       5410F
HYL-4-NITROPHENOL (TFM). +          ANIMAL CHEMISTRY TOXICITY MANAGEMENT /ORIFLUOROMET       5077
NUS U.S.A. AMMOCOETE ADULT          ANIMAL CHEMISTRYBIOCHEMISTRY MORTALITY OF /TN MARI       4751F
TS OF MYXINE. PP. 35-36. +          ANIMAL CIRCULATORY SYSTEM /+STEMIC AND PORTAL HEAR       4836F
  VEINS OF AMIA CALVA L.. +        ANIMAL CIRCULATORY SYSTEM HISTOLOGY /H CARDINALAND       4749F
  PP. 123-136. + *PETROMYZON      ANIMAL CIRCULATORY SYSTEM NERVOUS SYSTEM /ATOMES.        5117F
  /NERVOUS SYSTEM PHYSIOLOGY       ANIMAL CIRCULATORY SYSTEM NERVOUS SYSTEM PHYSIOLOG       5162

TEBRATES.] + *CYCLOSTOMATA        **ANIMAL** CYTOLOGY BIOCHEMISTRY /YACTORY ORGAN IN VER     4634F
IN OF AQUATIC CHORDATES. +        **ANIMAL** CYTOLOGY HISTOCHEMISTRY INTEGUMENT / THE SK     5148F
.S.A. MANAGEMENT CHEMISTRY        **ANIMAL** DISTRIBUTION CULTURE /PPETROMYZON MARINUS U     4756F
ZON CASTANEUS U.S.A. ADULT        **ANIMAL** DISTRIBUTION MORTALITY BY /UNT + *ICHTHYOMY     4812F
ES. + CHEMISTRY MANAGEMENT        **ANIMAL** ECOLOGY /AESIDUES IN MODEL STREAM COMMUNITI    5419F
IN. + CHEMISTRY MANAGEMENT        **ANIMAL** ECOLOGY /MMETHYL-4-NITROPHENOL) AND ANTIMYC    5403F
PARASITISM BY EVOLUTION LIF       **ANIMAL** ECOLOGY FECUNDITY FEEDING IONIC REGULATION     5140F
RYOSPHERE IN OOGENESIS.] +        **ANIMAL** EMBRYOLOGY CYTOLOGY /HA V OOGENEZE. [THE KA    4883F
  VERTEBRATES.] + EVOLUTION       **ANIMAL** EMBRYOLOGY ENDOCRINOLOGY NERVOUS SYSTEM /±F    4556F
*PETROMYZON DORSATUS ADULT        **ANIMAL** ENDOCRINOLOGY / ULGERI *PETROMYZON MARINUS     5066F
NDING REGION OF INSULIN. +        **ANIMAL** ENDOCRINOLOGY BIOCHEMISTRY /AHE RECEPTOR BI    5601
NCREATIC ISLETS. + *MYXINE        **ANIMAL** ENDOCRINOLOGY HISTOCHEMISTRY /HGY OF THE PA    4920F
NUS CANADA ADULT AMMOCOETE        **ANIMAL** ENDOCRINOLOGY HISTOLOGY CYTOLOGY /DZON MARI    4779F
TUS BURGERI. + JAPAN ADULT        **ANIMAL** ENDOCRINOLOGY METABOLISM /M HAGFISH EPTATRE    4669F
AGFISH MYXINE GLUTINOSA. +        **ANIMAL** ENDOCRINOLOGY PHYSIOLOGY /UN THE ATLANTIC H    5034F
E VERTEBRATE MEDIAN EYE. +        **ANIMAL** EVOLUTION CYTOLOGY SENSE RECEPTORS /ETE: TH    4555F
INS. + *PETROMYZON MARINUS        **ANIMAL** EVOLUTION PHYSIOLOGY /LEBRAE ALPHA CRYATALL    5024F
THE RENAL URINARY SPACE. +        **ANIMAL** EXCRETION TECHNIQUES CYTOLOGY /EOPHAGES IN     5593
VARIATIONS IN ABILITIES OF        **ANIMAL** FIBRINOGENS TO CLUMPSTAPHYLOCOCCI. /    /      4576
 FISH GENOMES.] + U.S.S.R.        **ANIMAL** GENETICS /Y[GENOSYTEMATICS AND EVOLUTION OF     4802F
ADA PISCES MIGRATION ADULT        **ANIMAL** GROWTH /N BASS MICROPTERUS SASMOIDES. + CAN    4727F
MUSCLES. + ADULT AMMOCOETE        **ANIMAL** HISTOLOGY /OS" IN SOME VERTEBRATE SKELETAL     4757F
PONSE. PP. 69-74. + FRANCE        **ANIMAL** IMMUNOLOGY /RTURE TOWARDS IMMUNOLOGICAL RES    4801F
 + *ENTOSPENUS TRIDENTATUS        **ANIMAL** INTEGUMENT /CTEBRATE EPIDERMIS TO FRICTION.    4693F
.] + *LAMPETRA FLUVIATILIS        **ANIMAL** IONIC REGULATION /AIN THE VERTEBRATE KIDNEY    4670F
Y MUSCLE EVOLUTION /HEMISTR       **ANIMAL** KINGDOM. + *LAMPETRA FLUVIATILIS BIOCHEMIST    4668
NOMUS TENTANS. + CHEMISTRY        **ANIMAL** METABOLISM /OION OF THE AQUATIC MIDGE CHIRO    4937F
AXONOMY OF ACARIFORMES.] +        **ANIMAL** METAMORPHOSIS EVOLUTION EMBRYOLOGY /RS OF T    5324F
NATOMY /TOLOGY LOCOMOTION A       **ANIMAL** MUSCLE NERVOUS SYSTEM HISTOLOGY LOCOMOTION     5079F
: VERTEBRATES COMPLETE TO O       **ANIMAL** NAMES WITH CORRESPONDING LATIN NAMES, PARTA    5395F
S FOR IODIDE. + PHYSIOLOGY        **ANIMAL** NERVOUS SYSTEM /FTS OF BRAIN BARRIER SYSTEM    5513F
MPETRA FLUVIATILIS ANATOMY        **ANIMAL** NERVOUS SYSTEM /RENERAL INTRODUCTION. + *LA    4697F
SPHENUS JAPONICUS. + ADULT        **ANIMAL** NERVOUS SYSTEM SENSE RECEPTORS HISTOLOGY /T    4745F
 SWEDEN BIOLOGY / GLUTINOSA       **ANIMAL** OF SCIENTIFIC INTEREST.] + *MYXINE GLUTINOS    5271F
A TRIDENTATUS U.S.A. BLOOD        **ANIMAL** PHYSIOLOGY / RINOGEN AND FIBRIN. + *LAMPETR    4594F
LLS OF FISHES.LL. + PISCES        **ANIMAL** PHYSIOLOGY BIOCHEMISTRY IONIC REGULATION /U    5307F
RTEBRATE OVIDUCT-UTERUS. +        **ANIMAL** PHYSIOLOGY GONADOGENESIS /AACTION ON THE VE    5390F
 ORGANS. /SUES AND ISOLATED       **ANIMAL** SUBORGANISMS,ORGANITES, TISSUES AND ISOLATE    5511
RT 1: FISHES. + ANTARCTICA        **ANIMAL** SYSTEMATICS /EE SOUTH AUSTRALIAN MUSEUM. PA    4718F
THENTERON JAPONICUM NORWAY        **ANIMAL** SYSTEMATICS /EVIATILIS *LAMPETRA PLANERI*LE    5395F
THENTERON JAPONICUM NORWAY        **ANIMAL** SYSTEMATICS /EVIATILIS *LAMPETRA PLANERI*LE    5469F
SZA BASIN.] + DISTRIBUTION        **ANIMAL** SYSTEMATICS /GE OF THE FISH FAUNA OF THE TI    5127F
METABOLISM ADULT AMMOCOETE        **ANIMAL** TECHNIQUES CYTOLOGY ENDOCRINOLOGY /HGAN. +    4916
 CIRCULATORY SYSTEM MUSCLE        **ANIMAL** TECHNIQUES PHYSIOLOGY HISTOLOGY //USCLES. +    5602
OSPHENUS TRIDENTATUS ADULT        **ANIMALS** ENDOCRINOLOGY / N THE LEOPARD FROG. + *ENT    5003F
RTEBRATES.] + *PETROMYZON M       **ANIMALS** NAMES WITH CORRESPONDING LATIN NAMES. A.VE    5469F
CTOTHERMIC AND ENDOTHERMIC        **ANIMALS.** + ADULT BIOCHEMISTRY PHYSIOLOGY /TES OF E    5612
CTOTHERMIC AND ENDOTHERMIC        **ANIMALS.** /IATE DEHYDROGENASE FROM THE MUSCLES OF E    5522
E MICROSCOPIC STRUCTURE OF        **ANIMALS.** /NRATIVE HISTOLOGY. AN INTRODUCTION TO TH    4829
         /REPORT OF THE           **ANNUAL** MEETING. /V                                    5499F
AND THE GREAT LAKES FISHER        **ANNUAL** REPORT TO THE DEPARTMENT OF THE ENVIRONMENT    4908F
AND THE GREAT LAKES FISHER        **ANNUAL** REPORT TO THE DEPARTMENT OF THE ENVIRONMENT    5236F
 AND THE GREAT LAKES FISHER       **ANNUAL** REPORT TO THE DEPARTMENT OF THE ENVIRONMENT    5237F
E SAULT STE. MARIE, ONTARIO       **ANNUAL** REPORT 1973 OF THE SEA LAMPREY CONROL CENTR    4783F
RE SAULT STE. MARIE, ONTARI       **ANNUAL** REPORT 1974 OF THE SEA LAMPREY CONTROL CENT    4784F
RE SAULT STE. MARIE, ONTARI       **ANNUAL** REPORT 1975 OF THE SEA LAMPREY CONTROL CENT    4844F
RE SAULT STE. MARIE, ONTARI       **ANNUAL** REPORT 1976 OF THE SEA LAMPREY CONTROL CENT    5194F
BAY BIOLOGICAL STATION. PP.       **ANNUAL** REPORT, CALENDAR YEAR 1976, OF THE HAMMOND     5366F
ANULARENDOPLASMIC RETICULUM       **ANNULATE** LAMELLAE AND CRYSTALLINE INCLUSIONS IN GR    4790F
ELS OF 3-TRIFLUOROMETHYL-4-       **ANODONTA** SP.) AS AN INDICATOR OF ENVIRONMENTAL LEV    5242F
HNIQUES /LOOD EVOLUTION TEC       **ANOMALIES.** + ADULT BIOCHEMISTRY BLOOD EVOLUTION TE    4715F
MUSEUM. PART 1: FISHES. +         **ANTARCTICA** ANIMAL SYSTEMATICS /EE SOUTH AUSTRALIAN    4718F
HE LAMPREY, LAMPETRA TRIDEN       **ANTERIOR** NEUROHYPOPHYSIS AND THE PARSDISTALIS OF T    4577F
         /NATURAL AND IMMUNE      **ANTIBODIES** TO RABBIT ERYTHROCYTEANTIGENS.ERI. /E     5338
  PETROMYZON MARINUS, EMBRYO      **ANTIMETABOLITES** ON DEVELOPMENT IN THE SEA LAMPREY,    5375F
ETTA KAWATAI. /STRACOD CYPR       **ANTIMYCIN** A, BAYER 73, AND TFM TO THE OSTRACOD CYP    5454F
ENTIC HABITAT. /         /        **ANTIMYCIN** AS A CONTROL FOR SEA LAMPREY LARVAE IN L    5425
OMETHYL-4-NITROPHENOL) AND        **ANTIMYCIN.** + CHEMISTRY MANAGEMENT ANIMAL ECOLOGY /    5403F
AMPETRA PLANERI UNITED KING       **ANTISERUM** TO CAERULEIN. + *LAMPETRA FLUVIATILIS *L    5562F
E? PP. 33-34. + ADULT HISTO       **ANY** STEROID HORMONE FORMATION IN THE OVARY OFMYXIN    4837F
         /FISHES OF THE           **APALACHICOLA** RIVER. /C                               5316
              /USE OF             **APESPORINE** OR AREMYXINE IN PEDIATRICS. /D              4976
NORWAY OLFACTION SENSE RECE       **APICAL** PART OF THE OLFACTORY EPITHELIUM. + SWEDEN     4808F
RNEI AND ASPATHRIA COMPLANA       **APPENDIX** STATISTICAL ANALYSES OF 2 SPECIES PILA WE    4544
G TO STUDIES OF GLOBULAR PR       **APPLICATION** OF LARGE ANGLE X-RAY DIFFUSE SCATTERIN    5349
 ACUTE TOXICITY STUDIES WIT       **APPLICATION** OF 24-HOUR POSTEXPOSURE OBSERVATION TO    5405F
 AND ITS EFFECTORS. + BIOCH       **APPLICATIONTO** THE OXYGEN EQUILIBRIUM OF HEMOGLOBIN    4552
IC DERIVATION OF ORGANS AN        **APPRAISAL** OF THE PROBLEMS. PP. 273-288. /S EMBRYON    5093
E OXYGEN EQUILIBRIUM OF HEM       **APPROACH** TO COOPERATIVITY AND ITS APPLICATIONTO TH    4552
TS-SAMHALLET. 104 PP. /ERHE       **APRIL,** 1972.GOTEBORG:KUNGL. VETENSKAPS-OCH VITTERH    4793F
THE SPAWNING OF RIVER AND B       **AQUARIUM** DES LAMPROIES FLUVIATILES ET DE PLANER. [    5357F

RI BELGIUM ADULT AMMOCOETE
INTEGUMENT /HISTOCHEMISTRY
HYL-4-NITROPHENOL (TFM) IN
/ FROG LARVE. + MANAGEMENT
DYNAMICS OF BAYER 2353 IN
HYL-4-NITROPHENOL (TFM) IN
NCHON AND MYRIOPHYLLUM SPIC
L METABOLISM /EMISTRY ANIMA
EMENT /S. + CHEMISTRY MANAG
ALLY ESPOSED LARVAE OF THE
ING ACUTELY SUBLETHAL TOXIC
E AND BAYER 73 TO SELECTED
CROFLORA AND MICROFAUNA OF
AN VERTEBRATES.] + *CYCLOST
THE SEA LAMPREY,PETROMYZON
IAN OF THE CANADIAN ARCTIC
OPROCTS. + MICROARCHITECTUR
E SILURIAN OF THE CANADIAN
REAT SLAVE LAKE, N.W.T. + C
/
/TRANSFER OF STORAGE
HE   RIVER LAMPREY LAMPETRA
RINUS CANADA AMMOCOETE BIOL
INE GLUTINOSA. + SWEDEN ADU
/USE OF APESPORINE OR
HAGFISH POLISTOTREMA STOUT
ICHTHYO FAUNA. + *ICHTHYOMY
AST MISSOURI AND NORTHEAST
ONIDAE). /                /1ST
/CHECKLIST OF
TEI /YPTERA *LAMPETRA LAMOT
TERA U.S.A. SYSTEMATICS ANI
/GAGEI DISTRIBUTION PISCES
F THE CURRENT RIVER WITHIN
HIKH PLZVONOCHNYKH. [ON THE
MIDDLE ELBE RIVER, WEST GE
USCLE /A.] + BIOCHEMISTRY M
PECULIARITIES OF PINOCYTOS
. ROBERTSON. + AMMOCOETE AD
IOLOGY /D BIOCHEMISTRY PHYS
CIES AND PARTIAL CHARACTERI
ARINUS, TROUT SALMO GAIRDNE
ORY STUDIES. + MANAGEMENT /
ON BAYER 73 AND ICI 24223
EN PHOSPHORYLASE EC-2.4.1.1
2 SPECIES PILA WERNEI AND
ON OFCHOLINORECEPTORS DURIN
NEUROSECRETIEI PREOPTICO-NE
ATILIS (L.): A CYTOARCHITEC
SIOLOGY ANIMAL NERVOUS SYST
S. + BLOOD ENDOCRINOLOGY NE
ATION IN THE LAMPREYS, LAMP
NK MUSCLE OF LOWER CHORDATE
SYNTHESIS. + *MYXINE GLUTIN
INDOCYANINE GREEN IN THE D
Y AND RAINBOW TROUT. + *PET
NULES OF THE INSULIN-PRODUC
ROHYPOSEAL NEUROSECRETION D
OF INSULIN. + *MYXINE GLUT
DSONI AMMOCOETE ADULT ENDOC
/COMPARATIVE
ONICA JAPAN AMMOCOETE ADULT
VELOPMENT OF A RADIOIMMUNE
N LOWER VERTEBRATES.] + *LA
MICRODISSECTED AREAS OF TH
PETROMYZON MARINUS CANADA G
5. + *PETROMYZON MARINUS C
. 5. + CANADA ADULT MANAGEM
*PETROMYZON MARINUS CANADA
ROMYZON MARINUS CANADA ADUL
OMYZON MARINUS MANAGEMENT M
. + *PETROMYZON MARINUS CAN
ADA MORPHOMETRY / 68. + CAN
CANADA ADULT MORPHOMETRY /
CHEMICALS. + PISCES CHEMIST
TROMYZON MARINUS CANADA GRE
DA ADULT MANAGEMENT /Y CANA
TILIS (CYCLOSTOMI). /FLUVIA
TILIS CYCLOSTOMI. + FINE ST
ON. / THYROID GLAND EVOLUTI
WTH MORTALITY BY /LAKES GRO

AQUARIUM.] + *LAMPETRA FLUVIATILIS *LAMPETRA PLANE          5357F
AQUATIC CHORDATES. + ANIMAL CYTOLOGY HISTOCHEMISTR          5148F
AQUATIC ENVIRONMENTS. /HSISTENCE OF 3-TRIFLUOROMET          5384
AQUATIC INVERTEBRATES AND FROG LARVE. + MANAGEMENT          4557F
AQUATIC INVERTEBRATES. + MANAGEMENT // AND RESIDUE          5444F
AQUATIC INVERTEBRATES. + MANAGEMENT /HTRIFLUOROMET          5144F
AQUATIC MACROPHYTES (ELODEA CANADENSIS (MICHX) PLA          5421F
AQUATIC MIDGE CHIRONOMUS TENTANS. + CHEMISTRY ANIM          4937F
AQUATIC MIDGE CHIRONOMUS TENTANS. + CHEMISTRY MANA          4647F
AQUATIC MIDGE CHIRONOMUS TENTANS. /NFM) IN SUBLETH          5457
AQUATIC ORGANISMS RESPONSE TO SEVERE STRESS FOLLOW          5410F
AQUATIC ORGANISMS. / THE ACUTE TOXICITY OF ROTENON          5460
AQUATIC SYSTEMS. PP. 109-115. /.SCICIDES ON THE MI          5441
ARCHI- AND NEOCORTEX IN PHYLOGENESIS OF SUBMAMMALI          5181F
ARCHINEPHRIC DUCT OF THE OPISTHONEPHRIC KIDNEY OF          4880F
ARCHIPELAGO. /.ORMES (HETEROSTRACI) FROM THE SILUR          5329
ARCHITECTURE OF BODY WALL OF EXTANT CYCLOSTOME ECT          5200
ARCTIC ARCHIPELAGO. /.ORMES (HETEROSTRACI) FROM TH          5329
ARCTIC LAMPREY LENTHENTERON JAPONICUM MARTENS OF G          4761F
ARE THERE LYMPHOCYTES IN HAGFISH?                          4957
AREA FROM ROOT RIVER TO ST. MARYS ISLAND. P. 174.          5497F
AREA IN THE GILLS OF THE MACROPHTHALMIA STAGE OF T          5101F
AREA, ST. MARY'S RIVER.PP. 88-91. + *PETROMYZON MA          4826F
AREAS OF THE ENDOCRINE PANCREAS IN THE HAGFISH MYX          5371F
AREMYXINE IN PEDIATRICS. /D                                4976
ARGININE VASOTOCIN IN THE PITUITARY OF THE PACIFIC          4558F
ARKANSAS FISHES WITH ADDITION OF 3 SPECIES TO THE          4944F
ARKANSAS U.S.A. /IE CANE CREEK WATERSHED IN SOUTHE          5287
ARKANSAS U.S.A. RECORDS OF LAMPETRA SPP. (PETROMYZ          5055
ARKANSAS USA FISHES. /F                                    4926
ARKANSAS USA. + *LAMPETRA AEPYPTERA *LAMPETRA LAMO          5032F
ARKANSAS USA. + *LAMPETRA LAMOTTEI *LAMPETRA AEPYP          5035F
ARKANSAS. + *ICHTHYOMYZON GAGEI DISTRIBUTION PISCE          5202F
ARKANSASUSA. /                /ICTHYO FAUNAL SURVEY O      5033
ARKHI- I NEOKORTEKSH V FILOGENEZE DOMLEKOPITAYUSHC          5181F
ARTEN, 1950-1975. [THE FISH FAUNA OF THE LOWER AND          5506F
ARTERIAL VESSELS OF CYCLOSTOMATA.] + BIOCHEMISTRY          5322F
ARTERIAL'NYKH SOSYDOV KRUGLOROTYKH. [MORPHOLOGICAL          5322F
ARTIFICIAL HATCHING OF LAMPREY EGGS; LETTER TO P.J          5143F
ART1FICIAL STRESS. + ANIMAL BLOOD BIOCHEMISTRY PHY          4545
ARYL TRANSFERASE DISTRIBUTION IN SEVERALMARINE SPE          4972
ASH CONTENT AND BODY WEIGHT IN LAMPREYPETROMYZON M          4727F
ASLAMPRICIDES IN LABORATORY STUDIES. + MANAGEMENT          5463F
ASMOLLUSCICIDES. / /SOME LABORATORY INVESTIGATIONS         5551
ASPARTATE AMINO TRANSFERASE EC-2.6.1.1. AND GLYCOG         5179
ASPATHRIA COMPLANATA. /NIX STATISTICAL ANALYSES OF         4544
ASPECT OF EVOLUTIONARY PHARMACOLOGY. OLIGOMERIZATI         4668
ASPECTELE CITOLOGICE ALE CELULELOR GONADOTROPE SI          5399F
ASPECTS ABOUT THE PROSENCEPHALON OF LAMPETRA FLUVI          4559F
ASPECTS OF BRAIN BARRIER SYSTEMS FOR IODIDE. + PHY          5513F
ASPECTS OF BRAIN BARRIER SYSTEMS FOR NONELECTROLTE          5515
ASPECTS OF FEEDING AND LIPID DEPOSITION AND UTILIZ          5142F
ASPECTS OF MUSCLE FIBER TYPES IN THE SEGMENTAL TRU          5265F
ASPECTS OF PROINSULIN AND INSULIN STRUCTURE ANDBIO          4918F
ASPECTS OF RENAL AND HEPATIC HANDLING OF PHENOLRED          4794F
ASPECTS OF TFM UPTAKE AND CONJUGATION IN SEALAMPRE          4751F
ASPECTS OF THE FINE STRUCTURE OF THE SECRETORY GRA          5605F
ASPECTS OF THE GONADOTROPIC CELLS AND PREOPTIC NEU          5399F
ASPECTS OF THE MOLECULAR STRUCTURE ANDBIOSYNTHESIS         4725F
ASPECTS OF THE NEUROHYPOPHYSIS. + *LAMPETRA RICHAR         5393
ASPECTS OF VITAMIN D TRANSPORT. PP. 405-407. /U            5099
ASREVEALED BY ELECTRON MICROSCOPY. + *LAMPETRA JAP         5356F
ASSAY FOR BAYER73. /                /STUDIES ON THE DE     5452
ASSAY OF ENZYMIC ACTIVITIES IN MUSCLE SARCOPLASM I         4803F
ASSAYS OF GLUTATHIONE ZINC COBALT AND MANGANESE IN        5371F
ASSESSMENT BARRIER OPERATIONS,LAKE HURON. P.5. + *        5246F
ASSESSMENT BARRIER OPERATIONS,LAKE HURON, 1974. P.        4907F
ASSESSMENT BARRIER OPERATIONS,LLAKE HURON: 1975. P        4969F
ASSESSMENT BARRIER OPERATIONS: 1972. PP. 95-96. +         5481
ASSESSMENT BARRIER OPERATIONS: 1973. PP. 5. + *PET        4866F
ASSESSMENT BARRIERS, LAKE HURON. PP.94-97. + *PETR        5260F
ASSESSMENT BARRIERS, LAKE HURON, 1972. PP. 160-170        5487
ASSESSMENT BARRIERS, LAKE HURON, 1974. P. 68. + CA        5088F
ASSESSMENT BARRIERS, LAKE HURON, 1975. PP.86-89. +        5226F
ASSESSMENT OF TOXICITY OF EFFICACY OF MIXTURES OF         5396F
ASSESSMENT WEIR AND TRAP OPERATIONS. PP.5-7. + *PE        5247F
ASSESSMENT WEIR, 1975. PP.96-98. + SEA LAMPREY CAN        5230F
ASSOCIATED NEURONS IN THE RETINA OF LAMPETRA FLUVI        5001F
ASSOCIATED NEURONS INTHE RETINA OF LAMPETRA FLUVIA        5001F
ASSOCIATED WITH THEPROBLEM OF THYROID GLAND EVOLUT        4997
ASSOCIATEDRATE OF HOST MORTALITY. + GREAT LAKES GR        5340F

AL-UND ADRENAL-SYSTEMS DES  
USE IN PREDICTING THE 3-DIM  
/RETUS STOUTI DISTRIBUTION  
TESTINAL EPITHELIUM OF THE  
.] + *LAMPETRA FLUVIATILIS  
N AND RELATED PROCESSES. +  
VOLUTION SYSTEMATICS /ION E  
VER LAMPREY IN THE EASTERN  
ECIES COMPOSITION OF FISH P  
. + U.S.A. NERVOUS SYSTEM H  
-83. + CANADA MANAGEMENT /3  
AGEMENT /OMYZON MARINUS MAN  
RVOUS SYSTEM CIRCULATORY SY  
MARINUS CANADA GREAT LAKES  
OMYZON MARINUS CANADA MANAG  
DA ADULT MANAGEMENT MORTALI  
 MARINUS CANADA ADULT MORTA  
INUS CANADA ADULT MORTALITY  
VOUS SYSTEM /OGY ANIMAL NER  
INOLOGY NERVOUS SYSTEM PHYS  
TROMYZON ANIMAL CIRCULATORY  
NUS MANAGEMENT MORPHOMETRY  
YZON MARINUS CANADA MORPHOM  
ETRY / 68. + CANADA MORPHOM  
T MORPHOMETRY / CANADA ADUL  
ON MARINUS CANADA ADULT BIO  
QUATIC CHORDATES. + ANIMAL  
LLOSTOMA MADE BY BASHFORD D  
.N.(S.) + SYSTEMATICS /8. Z  
TER DURING THE COURSE OF EV  
 + ANIMAL CYTOLOGY HISTOCHE  
ANS IN BDELLOSTOMA MADE BY  
HE FISH FAUNA OF THE TISZA  
 MORPHOMETRY SYSTEMATICS /Y  
HE POPRAD AND HORNAD RIVER  
 SOME COMPOUNDS TO STRIPED  
ON ADULT ANIMAL GROWTH /ATI  
TERPRETATION OF PALEONTOLOG  
STONE CANADA.] + *ENTOSPHEN  
 MARINUS U.S.A. GREAT LAKES  
 + *PETROMYZON MARINUS U.S.  
VERTEBRATES. + MANAGEMENT /  

BORATORY INVESTIGATIONS ON  
ATIC INVERTEBRATES. + MANAG  
ETHOD FOR DETERMINATION OF  
HEMISTRY MANAGEMENT TECHNIQ  
ERMINING CONCENTRATIONS OF  
TIC MIDGE CHIRONOMUS TENTAN  
3. PP.12. + *PETROMYZON MAR  
E TOXICITY OF ROTENONE AND  
-76. + CANADA AMMOCOETE CHE  
OMYZON MARINUS CANADA GREAT  
. + *PETROMYZON MARINUS *IC  
 + *PETROMYZON MARINUS *ICHT  
 *LAMPETRA LAMOTTEI *ICHTHY  
ETROMYZON MARINUS CANADA GR  
. + *PETROMYZON MARINUS CAN  
8 + *PETROMYZON MARINUS *IC  
. + SEA LAMPREY CANADA AMMO  
ORO-4-NITROSALICYLANILIDE (  
RATORY STUDIES. + MANAGEMEN  
RO-4'-NITROSALICYLANILIDE (  
RO-4'-NITROSALICYLANILIDE (  
NVIRONMENT. /ES IN A LAKE E  
 + MANAGEMENT /MUS TENTANS.  
GEMENT /S. + CHEMISTRY MANA  
RO-4'-NITROSALICYLANILIDE (  
VEN FISH SPECIES AND TO EGG  
/STRACOD CYPRETTA KAWATAI.  
OF A RADIOIMMUNE ASSAY FOR  
/PHOTOLYTIC SENSITIVITY TO  
T STAGES OF THE VARIEGATED  
E LITERATURE ON THE USE OF  
CTORS ON THE EFFICIENCY OF  
EMISTRY MANAGEMENT /S. + CH  
 BDELLOSTOMA CIRRHATUM. + *  
COETE DISTRIBUTION ECOLOGY  
OD EXCRETION /CIRRHATUM BLO  
RAWINGS OF THESE ORGANS IN  
OGY CHEMISTRY ECOLOGY FEEDI  

| Entry | No. |
|---|---|
| BACHNEUNAUGES (LAMPETRA PLANERI BLOCH). /PINTERREN | 5563 |
| BACKBONE CONFORMATION OF AMINO ACIDS IN PROTEINS. | 4626F |
| BAJA CALIFORNIA. + *EPTATRETUS STOUTI DISTRIBUTION | 5157F |
| BALTIC LAMPREY. /AND HISTOCHEMICAL STUDY OF THE IN | 5073 |
| BALTIC SEA ADULT ENDOCRINOLOGY TECHNIQUES /IBRATES | 4721F |
| BALTIC SEA ADULT METABOLISM HISTOLOGY / PHORYLATIO | 4613F |
| BALTIC SEA BLACK SEA ADULT AMMOCOETE DISTRIBUTION | 5358F |
| BALTIC. / /SPAWNING STOCK AND FISHERY OF RI | 4814 |
| BANK POLESYE OF THE UKRAINIAN-SSR USSR. PART 1. SP | 4873 |
| BARRIER AND VENTRICULAR SYSTEM OF MYXINE GLUTINOSA | 4805F |
| BARRIER DAMS AND STREAM IMPROVEMENT,1966-74. PP. 7 | 5134F |
| BARRIER DAMS. PP.100-101. + *PETROMYZON MARINUS MA | 5264F |
| BARRIER IN THE LAMPREY. + *LAMPETRA FLUVIATILIS NE | 4874 |
| BARRIER OPERATIONS,LAKE HURON. P.5. + *PETROMYZON | 5246F |
| BARRIER OPERATIONS,LAKE HURON, 1974. P. 5. + *PETR | 4907F |
| BARRIER OPERATIONS,LLAKE HURON: 1975. P. 5. + CANA | 4969F |
| BARRIER OPERATIONS: 1972. PP. 95-96. + *PETROMYZON | 5481 |
| BARRIER OPERATIONS: 1973. PP. 5. + *PETROMYZON MAR | 4866F |
| BARRIER SYSTEMS FOR IODIDE. + PHYSIOLOGY ANIMAL NE | 5513F |
| BARRIER SYSTEMS FOR NONELECTROLTES. + BLOOD ENDOCR | 5515 |
| BARRIER SYSTEMS IN CYCLOSTOMES. PP. 123-136. + *PE | 5117F |
| BARRIERS, LAKE HURON. PP.94-97. + *PETROMYZON MARI | 5260F |
| BARRIERS, LAKE HURON, 1972. PP. 160-170. + *PETROM | 5487 |
| BARRIERS, LAKE HURON, 1974. P. 68. + CANADA MORPHO | 5088F |
| BARRIERS, LAKE HURON, 1975. PP.86-89. + CANADA ADU | 5226F |
| BARRIERS: LAKE HURON, 1973. PP. 81-84. + *PETROMYZ | 4905F |
| BASAL LAMINA AND BASEMENT LAMELLA IN THE SKIN OF A | 5148F |
| BASED ON NOTES AND DRAWINGS OF THESE ORGANS IN BDE | 5584 |
| BASED ON THE TYPE GENUS PETROMYZON LINNAEUS,1758. | 4766F |
| BASED ON TRUE HOMOLOGY OR ON ADAPTATION BY THE LAT | 4983 |
| BASEMENT LAMELLA IN THE SKIN OF AQUATIC CHORDATES. | 5148F |
| BASHFORD DEAN. PP.67-102. /O DRAWINGS OF THESE ORG | 5584 |
| BASIN.] + DISTRIBUTION ANIMAL SYSTEMATICS /GE OF T | 5127F |
| BASIN.] + RUSSIA ADULT ANATOMY DISTRIBUTION ECOLOG | 4623F |
| BASINS. /QPLANERI WITH REGARD TO POPULATION FROM T | 5582 |
| BASS FINGERLINGS. + MANAGEMENT CHEMISTRY / CITY OF | 5426F |
| BASS MICROPTERUS SASMOIDES. + CANADA PISCES MIGRAT | 4727F |
| BATTERY POINT (DEVONIEN MOYEN), GRES DE GASPE. [IN | 5022F |
| BATTERY POINT FORMATION MIDDLE DEVONIAN GASPE SAND | 5022F |
| BAY BIOLOGICAL STATION. PP. 213-223. + *PETROMYZON | 5366F |
| BAY BIOLOGICAL STATION(HBBS), OCTOBER 29-30, 1975. | 4756F |
| BAYER 2353 IN AQUATIC INVERTEBRATES. + MANAGEMENT | 5444F |
| BAYER 73 - EXTRACTION AND CLEANUP. /D | 5427 |
| BAYER 73 AND ICI 24223 ASMOLLUSCICIDES. / /SOME LA | 5551 |
| BAYER 73 AND RESIDUE DYNAMICS OF BAYER 2353 IN AQU | 5444F |
| BAYER 73 IN NATURAL WATERS / /RAPID M | 5431 |
| BAYER 73 IN WATER DURINGLAMPRICIDE TREATMENTS. + C | 5462F |
| BAYER 73 INWATER. / /FIELD METHODS FOR DET | 5433 |
| BAYER 73 ON IN VIVO OXYGEN CONSUMPTION OF THE AQUA | 4937F |
| BAYER 73 SURVEYS OF LAKE HURON RIVER ESTUARIES,197 | 4863F |
| BAYER 73 TO SELECTED AQUATIC ORGANISMS. / THE ACUT | 5460 |
| BAYER 73 TREATMENT OF ST. MARYS RIVER, 1973. PP.75 | 4891F |
| BAYER 73 TREATMENTS, LAKE HURON. PP.86-92. + *PETR | 5257F |
| BAYER 73 TREATMENTS, LAKE HURON, 1972. PP. 154-161 | 5491F |
| BAYER 73 TREATMENTS, LAKE HURON, 1974. PP. 59-65. | 5104F |
| BAYER 73 TREATMENTS, LAKE HURON, 1975. PP.76-83. + | 5223F |
| BAYER 73 TREATMENTS, LAKE SUPERIOR. PP.78-85. + *P | 5256F |
| BAYER 73 TREATMENTS, LAKE SUPERIOR, 1973. PP 71-74 | 4892F |
| BAYER 73 TREATMENTS, LAKE SUPERIOR, 1974. PP. 51-5 | 5105F |
| BAYER 73 TREATMENTS, LAKE SUPERIOR, 1975. PP.68-75 | 5222F |
| BAYER 73) + CHEMISTRY MANAGEMENT /AIDE 2', 5-DICHL | 5442F |
| BAYER 73) AND A 98:2 MIXTURE ASLAMPRICIDES IN LABO | 5463F |
| BAYER 73) IN RAINBOW TROUT (SALMO GAIRDNERI). / LO | 5438 |
| BAYER 73) IN RAINBOW TROUT (SALMO GAIRDNERI). /HLO | 5440 |
| BAYER 73) ON BENTHIC MACROINVERTEBRATES IN A LAKE | 5459 |
| BAYER 73) TO LARVAEOF THE MIDGE CHIRONOMUS TENTANS | 5120F |
| BAYER 73) TO SEVERAL BIRD SPECIES. + CHEMISTRY MAN | 5430F |
| BAYER 73) TO 4 BIRD SPECIES. /OSALT OF 2',5-DICHLO | 5456 |
| BAYER 73), AND A 98:2 MIXTURE TO FINGERLINGS OF SE | 5095F |
| BAYER 73, AND TFM TO THE OSTRACOD CYPRETTA KAWATAI | 5454F |
| BAYER73. / /STUDIES ON THE DEVELOPMENT | 5452 |
| BAYLUSCIDE (R). / | 5434 |
| BAYLUSCIDE AND LABAYCID ON EGG-, JUVENILE AND ADUL | 5448 |
| BAYLUSCIDE IN FISHERIES. / /REVIEW OF TH | 5458 |
| BAYLUSCIDE. /LON THE INFLUENCE OF ENVIRONMENTAL FA | 5550 |
| BAYLUSCIDE, BAYER 73) TO SEVERAL BIRD SPECIES. + C | 5430F |
| BDELLESTOMA CIRRHATUM BLOOD EXCRETION /OHE HAGFISH | 4620 |
| BDELLIUM *ICHTHYOMYZON CASTANEUS U.S.A. ADULT AMMO | 4590F |
| BDELLOSTOMA CIRRHATUM. + *BDELLESTOMA CIRRHATUM BL | 4620 |
| BDELLOSTOMA MADE BY BASHFORD DEAN. PP.67-102. /O D | 5584 |
| BE PROTECTED. + U.S.A. AMMOCOETE ADULT ANIMAL BIOL | 4600F |

THYOMYZON UNICUSPIS *LAMPET
PETRA REISSNERI. + ADULT EN
EYS (LAMPETRA FLUVIATILIS)
+ BLOOD ENDOCRINOLOGY PHYS
IMMUNOLOG / HISTOCHEMISTRY
TEM RESPIRATION CIRCADIAN /
CANADA ADULT PARASITISM BY
ODUCTIONCOMOTION MIGRATION
UTION LOCOMOTION MIGRATION
PETRA FLUVIATILIS U.S.S.R.
EMENT MORTALITY BY SENSE RE
TEM RESPIRATION /ERVOUS SYS
PTATRETUS BURGERI. + JAPAN
YS (PETROMYZON MARINUS). +
. PP.90-93. + CANADA ADULT
N BEHAVIOUR FEEDING MIGRATI
NOSA CANADA ATLANTIC OCEAN
FEEDING GROWTH MORPHOMETRY
TINOSA. + SWEDEN OLFACTION
EVENTS IN THE EVOLUTION OF
            /PINEAL GLAND AND
(PETROMYZON MARINUS). + BE
GIE DES INTERRENAL-SYSTEMS
E AMPHIOXUS, BRANCHIOSTOMA
VIATILIS *LAMPETRA PLANERI
LIVER. + *LAMPETRA PLANERI
CHINOIDS OF CARRIE-BOW CAY
V SARKOPLAZME MYSHTS NIZSHI
ZOOGEOGRAPHIC ANALYSIS OF
ES IN A LAKE ENVIRONMENT. /
METERS. + *EPTATRETUS DEAN
MOGLOBINEN UND MYOGLOBINEN.
OMYZON MARINUS, TROUT SALMO
GROWTH.] + FRANCE AMMOCOET
THE EARLIEST LOWER VERTEBR
INS AND VERTEBRATE IMMUNOGL
NOSA L. CYCLOSTOMATA.] + NO
ING UPON LIGHT STIMULATION
EM IN EARLY SPERMATIDS OF M
LCELLS IN A TELEOST CYPRINU
+ *MYXINE GLUTINOSA L. ATLA
NS BASED ON TRUE HOMOLOGY O
C RETICULUM INTHE CARDIAC V
OF COAGULATION MECHANISMS I
AGLOUSTYCH I RYB. + POLAND
N THE LAMPREY. /HE KIDNEY I
E KIDNIN THE LAMPREY.] + RU
KI MINOGI LAMPETRAFLUVIATIL
ALITY /            /FISH
Y HISTOLOGY METABOLISM ENDO
/RECRUITMENT IN MAMMALIAN
. + SWEDEN ENDOCRINOLOGY HI
4-NITROPHENOL. /UOROMETHYL-
TION OF TFM GLUCURONIDE IN
NOSA *EPTATRETUS STOUTI *PE
DUS AND PROTOPTERUS AND THO
NE CHEMICALS. + HISTOCHEMIS
RI BELGIUM HISTOCHEMISTRY A
RBONATE (CARBARYL) AND 2',5
TLANTIC HAGFISH MYXINE GLUT
NG PROTEINS. / OF THE BINDI
HARIDESIN THE PRESENCE OF P
JAPAN BIOCHEMISTRY BLOOD E
ANIMAL SPECIES DATA FOR B-1
AND MOLECULAR SIZES OF THE
IOCHEMISTRY /NDOCRINOLOGY B
A VERY PRIMITIVE VERTEBRATE
TEMYXINE GLUTINOSA L. + BIO
LS SPECIFICITY OF RECEPTOR
/FORM CREATING FUNCTION OF
R NEURO TRANSMISSION. PP. 2
GERRAK. + BLOOD /OM THE SKA
SEASONAL VARIATIONS IN THE
IOCHEMISTRY /
TUDY ON MYXINE GLUTINOSA.]
ALYSES OFPRESUMED STEROID-P
OF THE VERTEBRATE BRAIN. +
NIN-C. PP. 65-67. + ANIMAL
A FLUVIATILIS ANIMAL BLOOD
S FRANCE BIOLOGY EVOLUTION
S FRANCE BIOLOGY EVOLUTION

BEFORE AND AFTER METAMORPHOSIS IN LAMPREYS. + *ICH      5466F
BEFORE AND AFTER METAMORPHOSIS IN THE LAMPREY, LAM      5559
BEFORE AND DURING PERIOD OF SEXUAL MATURATION. / R      5234
BEFORE AND DURING THE PERIOD OF SEXUAL MATURATION.      5062F
BEGIUM ADULT CYTOLOGY ENDOCRINOLOGY HISTOCHEMISTRY      5312
BEHAVIOR CIRCULATORY SYSTEM RESPIRATION CIRCADIAN       4719F
BEHAVIOUR /BNUS PREDATION ON FRESHWATERTELEOST. +       4754F
BEHAVIOUR /RFIC OCEAN ENDOCRINOLOGY HISTOLOGY REPR      5152F
BEHAVIOUR /UIATILIS SWEDEN ADULT CIRCADIAN DISTRIB      5141F
BEHAVIOUR BIOCHEMISTRY ENDOCRINOLOGY MUSCLE /E*LAM      4598F
BEHAVIOUR CHEMISTRY ENDOCRINOLOGY INNUNOLOGY MANAG      5366F
BEHAVIOUR CIRCULATORY SYSTEM LOCOMOTION NERVOUS SY      5464
BEHAVIOUR FEEDING MIGRATION CIRCADIAN /VE HAGFISHE      4699F
BEHAVIOUR LIGHT TEMPERATURE /E IN ADULT SEA LAMPRE      5298F
BEHAVIOUR MIGRATION /    STUDIES: LAKE ONTARIO, 1975   5228F
BEHAVIOUR OF THE HAGFISHEPTATRETUS BURGERI. + JAPA      4699F
BEHAVIOUR OLFACTIONPHYSIOLOGY SENSE RECEPTORS /ITI      5219F
BEHAVIOUR PARASITISM LIFE CYCLE /LOLOGY FECUNDITY       4974F
BEHAVIOUR SENSE RECEPTORS /TTHE HAGFISH MYXINE GLU      5100F
BEHAVIOUR. PP. 39-51. /              /CELLULAR          5532
BEHAVIOUR. PP. 697-721. /T                              5535
BEHAVIOURAL THERMOREGULATION IN ADULT SEA LAMPREYS      5298F
BEIM BACHNEUNAUGE (LAMPETRA PLANERI BLOCH). /FTOLO      5215
BELCHERII. + BIOCHEMISTRY ADULT /EIVE TISSUE OF TH      5041
BELGIUM ADULT AMMOCOETE CULTURE /E + *LAMPETRA FLU      5357F
BELGIUM HISTOCHEMISTRY AMMOCOETE ADULT /CNLAMPREY       4734F
BELIZE. + ANIMAL /                 /THE E               4819F
BELKOV I ISSLEDOVANIE FERMENTATISNYKH AKTIVNOSTEI       4803F
BENTHIC FISH FAUNA OF THE SCOTIA SEA. /ASITION AND      5597
BENTHIC MACROINVERTEBRATES IN A LAKE ENVIRONMENT.       5459
BENTHOPELAGIC FISHES: IN SITU MEASUREMENTS AT 1230      4575F
BESTIMMUNG PARTIELLER SPEZIFISCHER VOLUMINA VON HA      5128F
BETWEEN ASH CONTENT AND BODY WEIGHT IN LAMPREYPETR      4727F
BETWEEN BOTH THE EXPERIMENTAL AND NATURAL RATES OF      5352F
BETWEEN CARTILAGE AND BONE DURING THE PHYLOGENY OF      4879
BETWEEN NATURALLY OCCURING ECHINODERM HAEMOGGLUTIN      4602F
BETWEEN NEURO- AND ADENOHYPOPHYSIS IN MYXINE GLUTI      4809F
BETWEEN RETINA AND PINEAL ORGAN. PP. 137-144. /ISS      4950
BETWEEN THE CHROMATOID BODY AND THE ACROSOMAL SYST      5296F
BETWEEN THE PIGMENTARY RETINAL EPITHELIUM ANDVISUA      4665
BETWEEN THE PLASMA MEMBRANE AND THE LAMINA DENSA.       4563F
BETWEEN THE STRUCTURE OF ANIMAL ANDPLANT HEMOGLOBI      4983
BETWEENTHE MYOCARDIAL GRANULES AND THE SARCOPLASMI      4792F
BEZOBRATLYCH, OBRATLOVCU A CLOVEKA. [BRIEF REVIEW       5372F
BIBLIOGRAPHY /EICZEJ POLSKI. CZESC II. PASOZYTY KR      4688F
BICARBONATE IN THE PROXIMAL TUBULES OF THE KIDNEY       4569
BICARBONATE REABSORPTION IN PROXIMAL TUBULES OF TH      5507
BIKARBONATA NATRIYA V PROKSIMAL'NOM KANAL'TSE POCH      5507
BILE ANALYSIS. POSSIBLE AID IN MONITORING WATER QU      5402
BILE COMPOSITION IN MYXINE. PP. 89-92. + PHYSIOLOG      4582F
BILE DUCT FORMATION. + HISTOLOGY /A                     5392
BILE DUCT MUCOSA OF A CYCLOSTOME, MYXINE GLUTINOSA      4617F
BILE OF RAINBOW TROUT EXPOSED TO 3-TRIFLUOROMETHYL      5544
BILE OF TFM EXPOSED RAINBOW TROUT. /UND IDENTIFICA      5415
BILE SALTS IN FISHES. PP. 595-612. + *MYXINE GLUTI      4910F
BILE SALTS OF THE LUNGFISHES LEPIDOSIREN NEOCERATO      5156F
BILE: A PROPOSED MONITORING AID FOR SOME WATERBOUR      5382F
BILIARY ATRESIA INLAMPREY LIVER. + *LAMPETRA PLANE      4734F
BILIARY EXCRETION PRODUCTS OF 1-NAPHTHL-N-METHYLCA      5438
BINDING AFFINITY AND POTENCY OF INSULIN FROM THE A      5154F
BINDING CAPACITIES AND MOLECULAR SIZES OF THE BIND      5188
BINDING OF TOLUIDINE BLUE WITH ACID MUCO POLY SACC      5040
BINDING PROTEIN. + *ENTOSPHENUS JAPONICUS HOKKAIDO      5320
BINDING PROTEINS IN STOMACH AND SERUM FROM VARIOUS      5188
BINDING PROTEINS. /IA FOR B-12 BINDING CAPACITIES       5188
BINDING REGION OF INSULIN. + ANIMAL ENDOCRINOLOGY       5601
BINDING SPECIFICITY AND NEGATIVE COOPERATIVITY IN       5286
BINDING TO HAEMEOGLOBINS OF THE PRIMITIVE VERTEBRA      5244F
BINDING. / OF INSULIN WITH ISOLATED RAT LIVER CEL       5090
BIO MEMBRANES AND PHOTO RECEPTORDIFFERENTIATION. /      4914
BIOCHEMICAL AND HISTOCHEMICAL STUDIES ON VESTIBULA      5052
BIOCHEMICAL BLOOD PARAMETERS IN FISHES FROM THE SK      5113F
BIOCHEMICAL COMPOSITION OF CLARIAS-BATRACHUS. / /       5268
BIOCHEMICAL EVOLUTION OF THE VERTEBRATE BRAIN. + B      5512F
BIOCHEMICAL MECHANISMS OF THE HEART AUTOMATICITY S      4740
BIOCHEMICAL, HISTOCHEMICAL, AND ULTRASTRUCTURAL AN      5470F
BIOCHEMISTRY /              /BIOCHEMICAL EVOLUTION      5512F
BIOCHEMISTRY /    /IMMUNOCHEMICAL STUDIES OF TROPO      4671F
BIOCHEMISTRY / GLOBINS AND MYOGLOBINS.] + *LAMPETR      5128F
BIOCHEMISTRY / PTATRETUS STOUTI *PETROMYZON MARINU      5337F
BIOCHEMISTRY / ULINS CHAINS.] + *PETROMYZON MARINU      4602F

| | | |
|---|---|---|
| ECIES. + METABOLISM ANIMAL | BIOCHEMISTRY / YMES IN COASTAL MAINE USA MARINE SP | 4964F |
| HISTOCHEMISTRY MORPHOMETRY | BIOCHEMISTRY /A PP. 21-22. + ADULT ANATOMY MUSCLE | 4841F |
| IN. + ANIMAL ENDOCRINOLOGY | BIOCHEMISTRY /AHE RECEPTOR BINDING REGION OF INSUL | 5601 |
| LT EVOLUTION ENDOCRINOLOGY | BIOCHEMISTRY /CGLUTINOSA *LAMPETRA FLUVIATILIS ADU | 4725F |
| . 33-34. + ADULT HISTOLOGY | BIOCHEMISTRY /EFORMATION IN THE OVARY OFMYXINE? PP | 4837F |
| T ENDOCRINOLOGY TECHNIQUES | BIOCHEMISTRY /EICHAGFISH EPTATRETUS STOUTI. + ADUL | 5159F |
| HESIS. + *MYXINE GLUTINOSA | BIOCHEMISTRY /EIN AND INSULIN STRUCTURE ANDBIOSYNT | 4918F |
| S SYSTEM PHYSIOLOGY ANIMAL | BIOCHEMISTRY /ELIKE PEPTIDE "ENKEPHALIN". + NERVOU | 5158 |
| INOSA. + SWEDEN METABOLISM | BIOCHEMISTRY /ETE THE ATLANTIC HAGFISH MYXINE GLUT | 4654F |
| NOLOGY DIGESTION EVOLUTION | BIOCHEMISTRY /EUCTION. + *MYXINE GLUTINOSA ENDOCRI | 4744F |
| ES. + *LAMPETRA TRIDENTATA | BIOCHEMISTRY /HRILLARY ACIDIC PROTEIN IN VERTEBRAT | 4570F |
| ERTEBRATES. + ANIMAL BLOOD | BIOCHEMISTRY /I FIBRIN FROM THE MAJOR CLASSES OF V | 4549 |
| YSTEM SYNAPSE. + *LAMPETRA | BIOCHEMISTRY /IR AT A VERTEBRATE CENTRAL NERVOUS S | 4580F |
| CLOSTOMATA. + CANADA BLOOD | BIOCHEMISTRY /LACIFIC HAGFISH EPTATRETUS STOUTI CY | 4656F |
| T BLOOD CIRCULATORY SYSTEM | BIOCHEMISTRY /NRINUS *LAMPETRA FLUVIATILIS L. ADUL | 4760F |
| LOSUS. + *MYXINE GLUTINOSA | BIOCHEMISTRY /O FLUID IN THE MUDPUPPYNECTURUS MACU | 4627F |
| PARAMYXINE ATAMIPHYSIOLOGY | BIOCHEMISTRY /OA ENTOSPHENUS *LAMPETRA JAPONICUS * | 4919 |
| S AND MAN.] + ANIMAL BLOOD | BIOCHEMISTRY /OHANISMS ININVERTEBRATES, VERTEBRATE | 5372F |
| TECHNIQUES NERVOUS SYSTEM | BIOCHEMISTRY /PLUTINOSA CANADA ADULT ENDOCRINOLOGY | 4558F |
| LT ENDOCRINOLOGY HISTOLOGY | BIOCHEMISTRY /RG ELECTRON MICROSCOPY. + SWEDEN ADU | 4762F |
| OSA SWEDEN BLOOD HISTOLOGY | BIOCHEMISTRY /TM MYXINE GLUTINOSA + *MYXINE GLUTIN | 4945F |
| DUCTS. + *MYXINE GLUTINOSA | BIOCHEMISTRY /UTIVE BIOLOGY OF ISLET SECRETORY PRO | 5400F |
| *EPTATRETUS STOUTII BLOOD | BIOCHEMISTRY /V PP. 42-45. + *LAMPETRA FLUVIATILIS | 4834F |
| CLOSTOMATA ANIMAL CYTOLOGY | BIOCHEMISTRY /YACTORY ORGAN IN VERTEBRATES.] + *CY | 4634F |
| L ACTIVITY OF CERTAIN... + | BIOCHEMISTRY /YART I   NMR SPECTRA AND PHYSIOLOGICA | 4930 |
| OGY ENDOCRINOLOGY CYTOLOGY | BIOCHEMISTRY /YEY, PETROMYZON MARINUS L. + IMMUNOL | 5538F |
| VOLUTION ANATOMY HISTOLOGY | BIOCHEMISTRY /YINOSA. PP. 26-29. + ENDOCRINOLOGY E | 4838F |
| TRIDENTATUS U.S.A. ANIMAL | BIOCHEMISTRY ADULT /C ELASMOBRANCH. + *ENTOSPHENUS | 4729F |
| BRANCHIOSTOMA BELCHERII. + | BIOCHEMISTRY ADULT /EIVE TISSUE OF THE AMPHIOXUS, | 5041 |
| CTUREOF THE POLY PEPT... + | BIOCHEMISTRY ANIMAL /ICTING THE 3-DIMENSIONAL STRU | 4626F |
| L. PP. 30-32. + DIGESTION | BIOCHEMISTRY ANIMAL /YACTIVITY IN MYXINE GLUTINOSA | 4846F |
| STRIBUTION. + BLOOD ANIMAL | BIOCHEMISTRY BIOLOGY /AMA CARBOXY GLUTAMIC-ACID DI | 5028 |
| AMPETRA FLUVIATILIS RUSSIA | BIOCHEMISTRY BIOLOGY CYTOLOGY MUSCLE /HATES.] + *L | 5278 |
| .. + *LAMPETRA FLUVIATILIS | BIOCHEMISTRY BIOLOGY EVOLUTION /RIES. PHYLOGENIES. | 4982 |
| YXINE. PP. 86-88. + ANIMAL | BIOCHEMISTRY BLOOD /        /CARBONIC ANHYDRASE IN M | 4849F |
| EY. + CANADA NEW BRUNSWICK | BIOCHEMISTRY BLOOD  / IGRATING ANADROMOUS SEA LAMPR | 5096F |
| BRATEMYXINE GLUTINOSA L. + | BIOCHEMISTRY BLOOD /ALOBINS OF THE PRIMITIVE VERTE | 5244F |
| MMALIAN THROMBINS. + ADULT | BIOCHEMISTRY BLOOD /ITIONS SPLIT BY LAMPREY AND MA | 5146F |
| LOBIN AND ITS EFFECTORS. + | BIOCHEMISTRY BLOOD BIOLOGY /T EQUILIBRIUM OF HEMOG | 4552 |
| AMMALIAN INSULIN. + CANADA | BIOCHEMISTRY BLOOD ENDOCRINOLOGY /HUS STOUTI, TO M | 5326F |
| *EPTATRETUS STOUTII ANIMAL | BIOCHEMISTRY BLOOD ENDOCRINOLOGYIMMUNOLOGY /TW. + | 4708F |
| SS OF BLOOD COAGULATION. + | BIOCHEMISTRY BLOOD EVOLUTION /III DURING THE PROCE | 5302F |
| S JAPONICUS HOKKAIDO JAPAN | BIOCHEMISTRY BLOOD EVOLUTION PISCES ANIMAL /OPHENU | 5320 |
| UCTURAL ANOMALIES. + ADULT | BIOCHEMISTRY BLOOD EVOLUTION TECHNIQUES /OSOME STR | 4715F |
| ARITY TO SERUM PROTEINS. + | BIOCHEMISTRY BLOOD IMMUNOLOGY /E IMMUNOLOGIC SIMIL | 5018F |
| . + *ENTOSPHENUS JAPONICUS | BIOCHEMISTRY BLOOD TECHNIQUES /INE ELECTROPHORESIS | 5074 |
| OROMETHYL-4-NITROPHENOL. + | BIOCHEMISTRY CHEMISTRY MANAGEMENT / TE OR 3-TRIFLU | 5424 |
| BORATORY STUDIES. + ANIMAL | BIOCHEMISTRY CHEMISTRY MANAGEMENT /FSCLE TISSUE LA | 4985F |
| OROMETHYL-4-NITROPHENOL. + | BIOCHEMISTRY CHEMISTRY MANAGEMENT /XIONOF 3-TRIFLU | 4652F |
| RA FLUVIATILIS L.) + BLOOD | BIOCHEMISTRY CIRCULATORY SYSTEM. /ROGLOBINS(LAMPET | 4775F |
| RCULATORY SYSTEM HISTOLOGY | BIOCHEMISTRY CYTOLOGY /ILUTINOSA L.PP. 37-38. + CI | 4847F |
| STOUTI *PETROMYZON MARINUS | BIOCHEMISTRY DIGESTION /PNE GLUTINOSA *EPTATRETUS | 4910F |
| ROMYZON MARINUS L. + BLOOD | BIOCHEMISTRY ENDOCRINOLOGY /  THE SEA LAMPREY, PET | 5325F |
| TINOSA. PP. 82-83. + ADULT | BIOCHEMISTRY ENDOCRINOLOGY /.ALANINE IN MYXINE GLU | 4851F |
| NE GLUTINOSA. PP. 84-85. + | BIOCHEMISTRY ENDOCRINOLOGY /GA-D)IN ORGANS OF MYXI | 4850F |
| SOME LOWER VERTEBRATES. + | BIOCHEMISTRY ENDOCRINOLOGY /HLADDER CONTRACTION IN | 5048 |
| IATILIS.] + U.S.S.R. ADULT | BIOCHEMISTRY ENDOCRINOLOGY /RAMPREY, LAMPETRA FLUV | 4561F |
| ISH, EPTATRETUS BURGERI. + | BIOCHEMISTRY ENDOCRINOLOGY /SURE OVARY OF THE HAGF | 4609F |
| E GLUTINOSA UNITED KINGDOM | BIOCHEMISTRY ENDOCRINOLOGY /TISH INSULIN. + *MYXIN | 4629 |
| NESQUE. + EXCRETION ANIMAL | BIOCHEMISTRY ENDOCRINOLOGY CYTOLOGY HISTOLOGY /EFI | 4993 |
| + *LAMPETRA PLANERI ADULT | BIOCHEMISTRY ENDOCRINOLOGY MIGRATION PHYSIOLOGY /T | 4886F |
| IATILIS U.S.S.R. BEHAVIOUR | BIOCHEMISTRY ENDOCRINOLOGY MUSCLE /E*LAMPETRA FLUV | 4598F |
| Y /NERVOUS SYSTEM PHYSIOLOG | BIOCHEMISTRY ENDOCRINOLOGY NERVOUS SYSTEM PHYSIOLO | 4591F |
| Y CYTOLOGY /S HISTOCHEMISTR | BIOCHEMISTRY ENDOCRINOLOGY TECHNIQUES HISTOCHEMIST | 5196 |
| S. + *LAMPETRA FLUVIATILIS | BIOCHEMISTRY EVOLUTION MUSCLE /ISSUE OF VERTEBRATE | 5305 |
| TEBRATES. + ANIMAL BIOLOGY | BIOCHEMISTRY EVOLUTION NERVOUS SYSTEM /.NG THE VER | 4685 |
| .] + *LAMPETRA FLUVIATILIS | BIOCHEMISTRY EVOLUTION NERVOUS SYSTEM /MERTEBRATES | 4741F |
| BRATE FIBRINOGEN. + ANIMAL | BIOCHEMISTRY EVOLUTION PHYSIOLOGY /LUTION OF VERTE | 5097F |
| .] + *LAMPETRA FLUVIATILIS | BIOCHEMISTRY EXCRETION ANIMAL /ERON IN VERTEBRATES | 4742F |
| OF DNA OF CERTAIN FISH.] + | BIOCHEMISTRY GENETICS /DF THE NUCLEOTIDE SEQUENCE | 4548F |
| OCOETE ADULT ENDOCRINOLOGY | BIOCHEMISTRY HISTOCHEMISTRY /NETRA RICHARDSONI AMM | 5393 |
| LUVIATILIS U.S.S.R. ANIMAL | BIOCHEMISTRY HISTOCHEMISTRY PISCES /M+ *LAMPETRA F | 4803F |
| . + CANADA AMMOCOETE ADULT | BIOCHEMISTRY HISTOLOGY HISTOCHEMISTRY /YMARINUS L. | 4899F |
| MAINE USA. + ANIMAL ADULT | BIOCHEMISTRY HISTOLOGY METABOLISM EXCRETION /OSTAL | 4604F |
| TOSPHENUS JAPONICUS ANIMAL | BIOCHEMISTRY IMMUNOLOGY /EERTEBRATE SPECIES. + *EN | 5574F |
| PTATRETUS STOUTII. + BLOOD | BIOCHEMISTRY IMMUNOLOGY /OCUS OF PACIFIC HAGFISH E | 5465F |
| INS. + *PETROMYZON MARINUS | BIOCHEMISTRY IMMUNOLOGY /PHER VERTEBRATE TRANSFERR | 5004F |
| MA. + SWEDEN ENDOCRINOLOGY | BIOCHEMISTRY IMMUNOLOGY /UHE MYXINE ISLET PARENCHY | 5605F |
| NOSA *LAMPETRA FLUVIATILIS | BIOCHEMISTRY INTEGUMENT / IMAERA). + *MYXINE GLUTI | 5019F |
| + PISCES ANIMAL PHYSIOLOGY | BIOCHEMISTRY IONIC REGULATION /ULLS OF FISHES.LL. | 5307F |
| ETRA FLUVIATILIS. + RUSSIA | BIOCHEMISTRY METABOLISM /GCLES OF THE LAMPREY LAMP | 5313 |

SPIRATION /IS PHYSIOLOGY RE
Y LAMPETRA FLUVIATILIS.] +
TRA FLUVIATILIS.] + RUSSIA
FAT CELLS. + ADULT ANIMAL
ERI. + ADULT ENDOCRINOLOGY
EY LAMPETRA FLUVIATILIS. +
ESSELS OF CYCLOSTOMATA.] +
ERTEBRATE LIVERS. + ANIMAL
M. + *LAMPETRA FLUVIATILIS
X AND THEIR INTERACTION. +
WEDEN ADULT HISTOCHEMISTRY
S IN THE TESTES O... PP.99-
IAL STRESS. + ANIMAL BLOOD
SPECIES OF FISH. + ANIMAL
DOTHERMIC ANIMALS. + ADULT
GERI. + CYTOLOGY HISTOLOGY
9-41. + CIRCULATORY SYSTEM
S AND AMPHIBIANS. + ANIMAL
MARINE FISH + ANIMAL BLOOD
IMERIA CHALUMNAE. + ANIMAL
SHES.] PP. 59-66. + ANIMAL
ER IN GLOBULAR PROTEINS. +
NSITIVITY OF THE METHOD. +
T LIVER HISTONES. + ANIMAL
PETROMYZON MARINUS DENMARK
LDEHYDEREDUCTASE. + ANIMAL
LUVIATILIS. + SWEDEN ADULT
HA-1 3 COLLAGENS. + ANIMAL
AMPETRA FLUVIATILIS ANIMAL
COMPARATIVE PHYSIOLOGY AND
A SYMPOSIUM IN GOTEBORG 28
S *ENTOSPHENUS TRIDENTATUS
LA MYXINE. [ON THE BIOCHEMI
ROPOSED MONITORING AID FOR
THE HAGFISH, EPTATRETUS BUR
SSERVAZIONI SULLA PROLUNGAT
N. + *LAMPETRA FLUVIATILIS
ERCIALFISHERMEN, 1973. P. 8
ERCIAL FISHERMEN, 1975, P.8
ERCIAL FISHERMEN. P.98. + *
BERRIVER, 1973. PP. 85-87.
M ELECTRICAL ASSESSMENT BAR
M ELECTRICAL ASSESSMENT BAR
M ELECTRICAL ASSESSMENT BAR
THE HUMBER RIVER, 1972. P.
ER RIVER,1974. P. 70. + CAN
DIAN ELECTRICAL ASSESSMENT
TRICALASSESSMENT BARRIERS:
1975. P.90 + CANADA ADULT M
ER, LAKE ONTARIO. PP.98-99.
INUS U.S.A. GREAT LAKES ADU
PETROMYZON MARINUS U.S.A. M
+ *ICHTHYOMYZON SPP. ADULT
N MARINUS CANADA AMMOCOETE
MYZON MARINUS CANADA ADULT
+ *MYXINE GLUTINOSA SWEDEN
N MARINUS U.S.A. AMMOCOETE
BLOOD ANIMAL BIOCHEMISTRY
+ NERVOUS SYSTEM HISTOLOGY
THYOMYZON CANADA AMMOCOETE
. + NEW ZEALAND LIFE CYCLE
THYOMYZON CANADA AMMOCOETE
NTARIO TRIBUTARY. + CANADA
THYOMYZON CANADA AMMOCOETE
MYZON MARINUS CANADA ADULT
A DISTRIBUTION SYSTEMATICS
THYOMYZON CANADA AMMOCOETE
MYZON MARINUS CANADA ADULT
PARASITISM BYPARASITISM OF
TORS. + BIOCHEMISTRY BLOOD
N MARINUS CANADA AMMOCOETE
THE VERTEBRATES. + ANIMAL
IATILIS. + ADULT AMMOCOETE
IES ON PHOSPHORYLASE ISOZYM
ITY OF /G MANAGEMENT MORTAL
MINNOWS EXPOSED TO TFM. +
/ HISTOLOGY NERVOUS SYSTEM
ATILIS RUSSIA BIOCHEMISTRY
LAMPETRA FLUVIATILIS ADULT
RTEBRATES. + *CYCLOSTOMATA
A FLUVIATILIS BIOCHEMISTRY

| Entry | Ref |
|---|---|
| BIOCHEMISTRY METABOLISM METAMORPHOSIS PHYSIOLOGY R | 4678F |
| BIOCHEMISTRY METABOLISM MUSCLE / UES OF THE LAMPRE | 5191 |
| BIOCHEMISTRY METABOLISM MUSCLE /ATHE LAMPREY LAMPE | 5360F |
| BIOCHEMISTRY METABOLISM PHYSIOLOGY /F ISOLATED RAT | 5154F |
| BIOCHEMISTRY METABOLISM TECHNIQUES /EMPETRA REISSN | 5559 |
| BIOCHEMISTRY MUSCLE /RSE FROM MUSCLES OF THE LAMPR | 5276 |
| BIOCHEMISTRY MUSCLE /TH MUSCLE CELLS IN ARTERIAL V | 5322F |
| BIOCHEMISTRY MUSCLE BLOOD / /GLUCONEOGENESIS IN V | 5211 |
| BIOCHEMISTRY MUSCLE EVOLUTION /   THE ANIMAL KINGDO | 4668 |
| BIOCHEMISTRY MUSCLE PHYSIOLOGY /YE TROPONIN COMPLE | 4628F |
| BIOCHEMISTRY NERVOUS SYSTEM /DTRA FLUVIATILIS. + S | 4735F |
| BIOCHEMISTRY OF STEROID SYNTHESIZING CELLULAR SITE | 5190 |
| BIOCHEMISTRY PHYSIOLOGY /ATIONS INDUCED BY ART1FIC | 4545 |
| BIOCHEMISTRY PHYSIOLOGY /DTIES OF DNA FROM SEVERAL | 5147 |
| BIOCHEMISTRY PHYSIOLOGY /TES OF ECTOTHERMIC AND EN | 5612 |
| BIOCHEMISTRY PHYSIOLOGY /TF HAGFISH EPTATRETUS BUR | 5590 |
| BIOCHEMISTRY PHYSIOLOGY /TL VEINS OF MYXINE. PP. 3 | 4835F |
| BIOCHEMISTRY PHYSIOLOGY /YDEHYDROGENASE FROM FISHE | 5170F |
| BIOCHEMISTRY SPAWNING /TTY-ACIDS IN THE BLOOD OF | 5015 |
| BIOCHEMISTRY SYSTEMATICS /AE OF THE COELACANTH LAT | 5156F |
| BIOCHEMISTRY SYSTEMATICS /MS IN CYLLOSTOMES AND FI | 4896F |
| BIOCHEMISTRY TECHNIQUES /              /LONG RANGE ORD | 4813 |
| BIOCHEMISTRY TECHNIQUES /NY GLOBINS IN SOLUTION SE | 5070 |
| BIOCHEMISTRY TECHNIQUES /V STUDY OF HAGFISH AND RA | 4579 |
| BIOCHEMISTRY TECHNIQUES BLOOD / CORTICOTROPIN. + | 4574F |
| BIOCHEMISTRY TECHNIQUES EXCRETION /*PH-DEPENDANT A | 5570 |
| BIOCHEMISTRY TECHNIQUES GENETICS /EEY PETROMYZON F | 4562F |
| BIOCHEMISTRY TECHNIQUES VISION /I2,ALPHA-2 AND ALP | 4571 |
| BIOCHEMISTRY VISION /RN CERTAIN VERTEBRATES.] + *L | 4817F |
| BIOCHEMISTRY. VOL.4.NEW YORK: ACADEMIC PRESS. /BN | 4689 |
| BIOCHEMISTRY, PHYSIOLOGY AND STRUCTURE.REPORT FROM | 4793F |
| BIOCHEMISTRYPHYSIOLOGY / S. + *LAMPETRA TRIDENTATU | 4955F |
| BIOCHIMIQUES DE L'AUTOMATISME CARDIAQUE EXAMEN DE | 4740 |
| BIOCONCENTRATION OF XENOBIOTICS IN TROUT BILE: A P | 5382F |
| BIOCONVERSIONS OF STEROIDS IN THE MATURE OVARY OF | 4609F |
| BIOLOGIA DI LAMPETRA PLANERI (BLOCH) IN ABRUZZO. O | 4807F |
| BIOLOGICAL CONDITIONS IN THE BRISTOL CHANNEL REGIO | 5009 |
| BIOLOGICAL DATA FROM SEA LAMPREY COLLECTED BY COMM | 4902F |
| BIOLOGICAL DATA FROM SEA LAMPREY COLLECTED BY COMM | 5225F |
| BIOLOGICAL DATA FROM SEA LAMPREY COLLECTED BY COMM | 5261F |
| BIOLOGICAL DATA ON ADULT SEA  LAMPREY FROM THE HUM | 4900F |
| BIOLOGICAL DATA ON ADULT SEA LAMPREY COLLECTED FRO | 5088F |
| BIOLOGICAL DATA ON ADULT SEA LAMPREY COLLECTED FRO | 5226F |
| BIOLOGICAL DATA ON ADULT SEA LAMPREY COLLECTED FRO | 5260F |
| BIOLOGICAL DATA ON ADULT SEA LAMPREY COLLECTED IN | 5486F |
| BIOLOGICAL DATA ON ADULT SEA LAMPREY FROM THE HUMB | 5136F |
| BIOLOGICAL DATA ON SEA LAMPREY COLLECTED FROM CANA | 5487 |
| BIOLOGICAL DATA ON SEA LAMPREY COLLECTED FROM ELEC | 4905F |
| BIOLOGICAL DATA ON SEA LAMPREY FROM HUMBER RIVER, | 5227F |
| BIOLOGICAL DATA ON SEA LAMPREY FROM THE HUMBER RIV | 5262F |
| BIOLOGICAL STATION. PP. 213-223. + *PETROMYZON MAR | 5366F |
| BIOLOGICAL STATION(HBBS), OCTOBER 29-30, 1975. + * | 4756F |
| BIOLOGY / ED BY COMMERCIAL FISHERMEN, 1975, P.86. | 5225F |
| BIOLOGY / F LAKE ONTARIO 1973. PP.15. + *PETROMYZO | 4861F |
| BIOLOGY /+S: LAKE HURON, 1973. PP. 81-84. + *PETRO | 4905F |
| BIOLOGY /AISH: AN ANIMAL OF SCIENTIFIC INTEREST.] | 5271F |
| BIOLOGY /AKE ONTARIO, 1973. PP.15-18. + *PETROMYZO | 4867F |
| BIOLOGY /AMA CARBOXY GLUTAMIC-ACID DISTRIBUTION. + | 5028 |
| BIOLOGY /CE CELLS OF THE VERTEBRATE HYPOTHALAMUS. | 5000 |
| BIOLOGY /D3. PP. 12-15. + *PETROMYZON MARINUS *ICH | 4862F |
| BIOLOGY /EKOROKORO). GEOTRIA AUSTRALIS GRAY. 58-62 | 5232F |
| BIOLOGY /I,1973. PP.12. + *PETROMYZON MARINUS *ICH | 4863F |
| BIOLOGY /IAMPREYS PETROMYZON MARINUS FROM A LAKE O | 4759F |
| BIOLOGY /I1973.PP.7-12. + *PETROMYZON MARINUS *ICH | 4864F |
| BIOLOGY /MMERCIALFISHERMEN, 1973. P. 85. + *PETRO | 4902F |
| BIOLOGY /OCUM (LAMPETRA JAPONICA) *MYXINE GLUTINOS | 4770F |
| BIOLOGY /O1973. PP.6-7. + *PETROMYZON MARINUS *ICH | 4865F |
| BIOLOGY /PE HUMBERRIVER, 1973. PP. 85-87. + *PETRO | 4900F |
| BIOLOGY /T AMMOCOETE ADULT LIFECYCLE DISTRIBUTION | 4761F |
| BIOLOGY /T EQUILIBRIUM OF HEMOGLOBIN AND ITS EFFEC | 4552 |
| BIOLOGY /TST. MARY'S RIVER.PP. 88-91. + *PETROMYZO | 4826F |
| BIOLOGY BIOCHEMISTRY EVOLUTION NERVOUS SYSTEM /.NG | 4685 |
| BIOLOGY BLOOD HISTOLOGY LIFE CYCLE PHYSIOLOG /+LUV | 4716F |
| BIOLOGY BLOOD HISTOLOGY LIFE CYCLE PHYSIOLOG /STUD | 4717 |
| BIOLOGY CHEMISTRY ECOLOGY FEEDING MANAGEMENT MORTA | 4600F |
| BIOLOGY CHEMISTRY MANAGEMENT /+ODUCTION OF FATHEAD | 5379F |
| BIOLOGY CIRCULATORY SYSTEM HISTOLOGY NERVOUS SYSTE | 5504 |
| BIOLOGY CYTOLOGY MUSCLE /HATES.] + *LAMPETRA FLUVI | 5278 |
| BIOLOGY DISTRIBUTION MIGRATION /LROMYZON MARINUS * | 5506F |
| BIOLOGY ENDOCRINOLOGY /.AL PRINCIPLES AMONG THE VE | 5389F |
| BIOLOGY EVOLUTION /RIES. PHYLOGENIES... + *LAMPETR | 4982 |

```
*PETROMYZON MARINUS FRANCE
*PETROMYZON MARINUS FRANCE
*PETROMYZON ANATOMY ANIMAL
PHOSIS MIGRATION PARASITISM
IN A SEA LAMPREY + U.S.A.
NUS CANADA ADULT AMMOCOETE
H,1784). + AMMOCOETE ADULT
VIATILIS *MYXINE GLUTINOSA
AMPETRA FLUVIATILIS RUSSIA
TROMYZON MARINUS. + U.S.A.
          /VASCULAR SYSTEM
N MARINUS. + CANADA ADULT A
1784). + AMMOCOETE ADULT BI
INOSA BIOCHEMISTRY /NE GLUT
SERVATION ON THE PROLONGED
. /GENUS LAMPETRA IN OREGON
ARINUSIN NEW BRUNSWICK. + C
VISION TECHNIQUES CYTOLOGY
ADULTE, LAMPETRA PLANERI (
LARVES (AMMOCOETES) D'UNE
EY. /        /THYRO GLOBULIN
WATER LAMPREY LAMPETRA PLA
THE CYCLOSTOME MYXINE GLUT
AMORPHOSIS IN THE LAMPREY,
LAMPETRA PLANERI (BLOCH)]
OETES) OF A FRESH WATER LAM
4'-NITROSALICYLANILIDE BY C
HYL-4-NITROPHENOL (TFM) BY
INARY NOTESON ITS IN VITRO
UOROMETHYL-4-NITROPHENOL) I
CIDE, BAYER 73) TO SEVERAL
CYLANILIDE (BAYER 73) TO 4
RINE SPECIALIZATION OF THE
IS /) + *LAMPETRA FLUVIATIL
NVIRONMENTAL POLLUTANTS TO
OXICITIES OF PESTICIDES TO
ARVAE (AMMOCOETES) OF A FRE
STOMES HISTOLOGY VISION /LO
STEMATICS /ION EVOLUTION SY
ISION /LOSTOMES HISTOLOGY V
LAMPREY, LAMPETRA PLANERI
TIONS OF THE NUCLEAR DNA IN
RVIVAL OF SPECIMENS AT DIFF
OPRAVVIVENZA DI ESEMPLARI A
MPARISON AVEC LES CROISSANC
MMOCOTEN LAMPETRA PLANERI (
N ADULT LAMPREY, LAMPETRA P
RY OF MYOTUBE STRIATED MUSC
LAMPREYS LAMPETRA PLANERI (
ILIS AND LAMPETRA PLANERI (
METAMORPHOSIS RESPIRATION
NEUNAUGE (LAMPETRA PLANERI
EUNAUGES (LAMPETRA PLANERI
ND NATURAL RATES OF GROWTH.
M HISTOLOGY CYTOLOGYHISTOCH
LAMPREY, LAMPETRA PLANERI (
MYZON DANFORD *EUDONTOMYZON
ING /T BIOLOGY GROWTH SPAWN
IN ELASMOBRANCHS. + MYXINE
-88. + ANIMAL BIOCHEMISTRY
ANIMAL BIOCHEMISTRY MUSCLE
RK BIOCHEMISTRY TECHNIQUES
NEW BRUNSWICK BIOCHEMISTRY
AT AN INVARIANT RESIDUE. +
A PLANERI AMMOCOETE ANIMAL
SHES FROM THE SKAGERRAK. +
A) AND IN FISH(PISCES)]. +
LUTINOSA L. + BIOCHEMISTRY
T TECHNIQUES ENDOCRINOLOGY
N ENDOCRINOLOGY PHYSIOLOGY
BINS. + ADULT BIOCHEMISTRY
IS /ION FEEDING METAMORPHOS
EVOLUTION CIRCULATORY SYST
IONIC REGULATION /GULATION
TAMIC-ACID DISTRIBUTION. +
AMPETRA TRIDENTATUS U.S.A.
AMPETRA FLUVIATILIS ANIMAL
S OF VERTEBRATES. + ANIMAL
UTI CYCLOSTOMATA. + CANADA
EBRATES AND MAN.] + ANIMAL
ATILIS *EPTATRETUS STOUTII
```

```
BIOLOGY EVOLUTION BIOCHEMISTRY / PTATRETUS STOUTI        5337F
BIOLOGY EVOLUTION BIOCHEMISTRY / ULINS CHAINS.] +        4602F
BIOLOGY EVOLUTION OLFACTION NERVOUS SYSTEM /ONOSA        4771F
BIOLOGY FECUNDITY FEEDING GROWTH LIFECYCLE METAMOR       4710F
BIOLOGY FEEDING /        /ABNORMAL TOOTH DEVELOPMENT     5241F
BIOLOGY GROWTH HISTORY MANAGEMENT /GETROMYZON MARI       4908F
BIOLOGY GROWTH SPAWNING /A, LAMPETRA PLANERI (BLOC       4764F
BIOLOGY HISTOCHEMISTRY CYTOLOGY /I + *LAMPETRA FLU       5265F
BIOLOGY HISTOLOGY NERVOUS SYSTEM PHYSIOLOGY /D+ *L       5505
BIOLOGY INTEGUMENT HISTOLOGY /LTHE SEA LAMPREY, PE       5235F
BIOLOGY INTRODUCTION. PP.82-88. /L                       4932
BIOLOGY OF ADULT ANADROMOUS SEA LAMPREYS PETROMYZO       5140F
BIOLOGY OF BROOK LAMPREY, LAMPETRA PLANERI (BLOCH,       4764F
BIOLOGY OF ISLET SECRETORY PRODUCTS. + *MYXINE GLU       5400F
BIOLOGY OF LAMPETRA PLANERI (BLOCH) IN ABRUZZO. OB       4807F
BIOLOGY OF LAMPREYS OF THE GENUS LAMPETRA IN OREGO       5554
BIOLOGY OF THE ANADROMOUS SEA LAMPREY PETROMYZON M       4710F
BIOLOGY OLFACTION /.TRA PLANERI. + NERVOUS SYSTEM        5586
BIOSYNTHESE DE LA THYROGLOBULINE CHEZ UNE LAMPROIE       5025F
BIOSYNTHESE DE THYROGLOBULINE DANS L'ENDOSTYLE DES       5342F
BIOSYNTHESIS AND THYROID HORMONE FORMATION INLAMPR       5163
BIOSYNTHESIS IN A LARVAL AMMOCOETE AND ADULT FRESH       5603
BIOSYNTHESIS OF INSULIN IN A PRIMITIVE VERTEBRATE,       5330
BIOSYNTHESIS OF THYROGLOBULIN BEFORE AND AFTER MET       5559
BIOSYNTHESIS OF THYROGLOBULIN IN AN ADULT LAMPREY,       5025F
BIOSYNTHESIS OF THYROGLOBULIN IN THE LARVAE (AMMOC       5342F
BIOTRANSFORMATION OF THE LAMPRICIDE 2',5-DICHLORD-       5453F
BIOTRANSFORMATION OF THE LAMPRICIDE 3-TRIFLUOROMET       4647F
BIOTRANSFORMATION. + MANAGEMENT / FISH WITH PRELIM       5404F
BIOTRANSFORMATION, AND ELIMINATION OF TFM (3-TRIFL       5380
BIRD SPECIES. + CHEMISTRY MANAGEMENT /PLID (BAYLUS       5430F
BIRD SPECIES. /OSALT OF 2',5-DICHLORO-4'-NITROSALI       5456
BIRD. /         /SOME ENDOC                              5027
BIRDS IN A STREAM ECOSYSTEM.) + *LAMPETRA FLUVIATI       4672F
BIRDS. + CHEMISTRY MANAGEMENT /7RY TOXICITIES OF E       5498F
BIRDS. + MANAGEMENT /        /COMPARATIVE DIETARY T      5416F
BL. [ON THE BIOSYNTHESIS OF THYROGLOBULIN IN THE L       5342F
BLACK PIGMENTS IN BLACK SEA ELASMOBRANCHS.] + CYCL       5274
BLACK SEA ADULT AMMOCOETE DISTRIBUTION EVOLUTION S       5358F
BLACK SEA ELASMOBRANCHS.] + CYCLOSTOMES HISTOLOGY        5274
BLOCH + FRANCE ENDOCRINOLOGY METAMORPHOSIS /OWATER       5342F
BLOCH 1784) (CYCLOSTOMATA). [QUALITATIVE INVESTIGA       4700F
BLOCH) IN ABRUZZO. OBSERVATION ON THE PROLONGED SU       4807F
BLOCH) IN ABRUZZO. OSSERVAZIONI SULLA PROLUNGATA C       4807F
BLOCH) PAR UNE NOUVELLE METHODE D'ALIMENTATION. CO       5352F
BLOCH) UND EUDONTOMYZON DANFORDI (REGAN). /TCHEN A       5331
BLOCH). [ON THE BIOSYNTHESIS OF THYROGLOBULIN IN A       5025F
BLOCH). [RESEARCHON THE HISTOLOGY AND HISTOCHEMIST       4619F
BLOCH). + SWEDEN AMMOCOETE PARASITISM OF ADULT /U        5467F
BLOCH). + U.K. AMMOCOETE ADULT RESPIRATION /OUVIAT       4877F
BLOCH). + UNITED KINGDOM ADULT AMMOCOETE EVOLUTION       5300F
BLOCH). /FTOLOGIE DES INTERRENAL-SYSTEMS BEIM BACH       5215
BLOCH). /PINTERRENAL-UND ADRENAL-SYSTEMS DES BACHN       5563
BLOCH). COMPARISON BETWEEN BOTH THE EXPERIMENTAL A       5352F
BLOCH).] + ITALY AMMOCOETE ADULT CIRCULATORY SYSTE       4619F
BLOCH)] /OOSYNTHESIS OF THYROGLOBULIN IN AN ADULT        5025F
BLOCH, 1784) (CYCLOSTOMATA) IN ROMANIA. + *EUDONTO       4541F
BLOCH,1784). + AMMOCOETE ADULT BIOLOGY GROWTH SPAW       4764F
BLOOD /          /OSMOREGULATION                         5238F
BLOOD /        /CARBONIC ANHYDRASE IN MYXINE. PP. 86     4849F
BLOOD /        /GLUCONEOGENESIS IN VERTEBRATE LIVERS. +  5211
BLOOD / CORTICOTROPIN. + *PETROMYZON MARINUS DENMA       4574F
BLOOD / IGRATING ANADROMOUS SEA LAMPREY. + CANADA        5096F
BLOOD / NCTIONALLY SILENT AMINO ACID SUBSTITUTION        4971F
BLOOD /+IEREN MIT AUSNAHME VON SAUGERN. + *LAMPETR       5119F
BLOOD /*CAL AND BIOCHEMICAL BLOOD PARAMETERS IN FI       5113F
BLOOD /AEL OF GLYCEMIA IN CYCLOSTOMES (CYCLOSTOMAT       4681F
BLOOD /ALOBINS OF THE PRIMITIVE VERTEBRATEMYXINE G       5244F
BLOOD /IHE SEA LAMPREY, PETROMYZON MARINUS. + ADUL       5208
BLOOD /IN. + *PETROMYZON MARINUS SPAWNING MIGRATIO       5083F
BLOOD /ITIONS SPLIT BY LAMPREY AND MAMMALIAN THROM       5146F
BLOOD ADULT AMMOCOETE MIGRATION FEEDING METAMORPHO       5145
BLOOD AND LYMPH VESSELS. + HISTOLOGY ENDOCRINOLOGY       4640F
BLOOD AND MUSCLE CELLS. + PHYSIOLOGY OSMOREGULATIO       4767F
BLOOD ANIMAL BIOCHEMISTRY BIOLOGY /AMA CARBOXY GLU       5028
BLOOD ANIMAL PHYSIOLOGY / RINOGEN AND FIBRIN. + *L       4594F
BLOOD BIOCHEMISTRY / GLOBINS AND MYOGLOBINS.] + *L       5128F
BLOOD BIOCHEMISTRY /I FIBRIN FROM THE MAJOR CLASSE       4549
BLOOD BIOCHEMISTRY /LACIFIC HAGFISH EPTATRETUS STO       4656F
BLOOD BIOCHEMISTRY /OHANISMS ININVERTEBRATES, VERT       5372F
BLOOD BIOCHEMISTRY /V PP. 42-45. + *LAMPETRA FLUVI       4834F
```

LAMPETRA FLUVIATILIS L.) +
Y, PETROMYZON MARINUS L. +
FISH EPTATRETUS STOUTII. +
RT1FICIAL STRESS. + ANIMAL
OD OF MARINE FISH + ANIMAL
EFFECTORS. + BIOCHEMISTRY
NE GLUTINOSA. + U.S.A. NERV
UVIATILIS NERVOUS SYSTEM CI
AGFISH EPTATRETUS STOUTI:IM
UVIATILIS. + ADULT AMMOCOET
*LAMPETRA PLANERI SWEDEN A
OSA (L.). + SWEDEN BLOOD HI
APAN BLOOD HISTOLOGY CYTOLO
PETRA FLUVIATILIS L. ADULT
CHEMISTRY BLOOD EVOLUTION /

AT BRITIAN AMMOCOETE ADULT
LIN. + CANADA BIOCHEMISTRY
NTIC OCEAN ADULT AMMOCOETE
MPETRA FLUVIATILIS DENMARK
A (CYCLOSTOMATA). + CANADA
AND FISHES. PP. 251-269. +
A FLUVIATILIS RUSSIA ADULT
TEMS FOR NONELECTROLTES. +
OD OF SEXUAL MATURATION. +
EM OF MYXINE. PP. 57-64. +
TOUTII ANIMAL BIOCHEMISTRY
ING PATHWAY FOR GLOBINS. +
+ *MYXINE GLUTINOSA SWEDEN
OAGULATION. + BIOCHEMISTRY
IS ENGLAND AMMOCOETE ADULT
SPHENUS JAPONICUS. + JAPAN
OKKAIDO JAPAN BIOCHEMISTRY
LIES. + ADULT BIOCHEMISTRY
. + *BDELLESTOMA CIRRHATUM
EDEN BLOOD HISTOLOGY BIOCHE
ED BY ART1FICIAL STRESS. +
ED RIVER LAMPREYS LAMPETRA
NMARK ADULT ENDOCRINOLOGY M
LT ENDOCRINOLOGY EVOLUTION
MOCOETE ECOLOGY LIFE CYCLE
OLOGY AMMOCOETE LIFE CYCLE
RIN. + *PETROMYZON MARINUS
ETRA PLANERI SWEDEN ANIMAL
MYZON MARINUS U.S.A. ADULT
E GLUTINOSA (L.). + SWEDEN
T HISTOCHEMISTRY EVOLUTION
+ *MYXINE GLUTINOSA SWEDEN
*EPTATRETUS BURGERI JAPAN
*EPTATRETUS BURGERI JAPAN
TO AZUROPHIL LEUCOCYTES. +
+ ADULT AMMOCOETE BIOLOGY
HOSPHORYLASE ISOZYMES IN LO
M PROTEINS. + BIOCHEMISTRY
ID CELLS IN THE HAGFISH. +
/RESPIRATORY FUNCTION OF
S. + *LAMPETRA FLUVIATILIS
TROSA. + NORWAY PHYSIOLOGY
YXINE GLUTINOSA. + GERMANY
IS AND LAMPETRA PLANERI. +
STAGES OF SPAWNING MIGRATIO
ILIS. + *ICHTHYOMYZON UNICU
AWNING /OOD BIOCHEMISTRY SP
ER TREATMENT WITH MAMMALIAN
TREATMENT WITH MAMMALIANCOR
ADULT ENDOCRINOLOGY /UROPE
A ENDOCRINOLOGY /+ ADULT US
UTINOSA SWEDEN BLOOD EVOLUT
OOD /OM THE SKAGERRAK. + BL
PETRA FLUVIATILIS. + ADULT
BLOOD IONIC REGULATION /IS
RD METABOLIC RATE. + RESPIR
NTERRENAL, AND CHROMAFFIN T
NUS JAPONICUS BIOCHEMISTRY
. + *LAMPETRA JAPONICA CIRC
ATORY SYSTEM PHYSIOLOGY /UL
MYZON MARINUS U.S.A. ADULT
TRY /O USA. + U.S.A. CHEMIS
E OF PHENOLS STAINING TECHN
+ *LAMPETRA PLANERI AMMOCOE
APONICA JAPAN ADULT RESPIRA

| | |
|---|---|
| BLOOD BIOCHEMISTRY CIRCULATORY SYSTEM. /ROGLOBINS( | 4775F |
| BLOOD BIOCHEMISTRY ENDOCRINOLOGY /  THE SEA LAMPRE | 5325F |
| BLOOD BIOCHEMISTRY IMMUNOLOGY /OCUS OF PACIFIC HAG | 5465F |
| BLOOD BIOCHEMISTRY PHYSIOLOGY /ATIONS INDUCED BY A | 4545 |
| BLOOD BIOCHEMISTRY SPAWNING /TTTY-ACIDS IN THE BLO | 5015 |
| BLOOD BIOLOGY /T EQUILIBRIUM OF HEMOGLOBIN AND ITS | 4552 |
| BLOOD BRAIN BARRIER AND VENTRICULAR SYSTEM OF MYXI | 4805F |
| BLOOD BRAIN BARRIER IN THE LAMPREY. + *LAMPETRA FL | 4874 |
| BLOOD CAPILLARIES OF SOMEENDOCRINE GLANDS OF THE H | 4640F |
| BLOOD CELL FORMATION IN RIVER LAMPREY, LAMPETRA FL | 4716F |
| BLOOD CELLS OF FISHES AND CYCLOSTOMES. PP. 53-56 + | 4859F |
| BLOOD CELLS OF THE ATLANTIC HAGFISH, MYXINE GLUTIN | 5341F |
| BLOOD CELLS OF THE HAGFISH + *EPTATRETUS BURGERI J | 4796F |
| BLOOD CIRCULATORY SYSTEM BIOCHEMISTRY /NRINUS *LAM | 4760F |
| BLOOD COAGULATION. + BIOCHEMISTRY BLOOD EVOLUTION | 5302F |
| BLOOD COAGULATION. PP. 51-52. /E | 4860F |
| BLOOD CYTOLOGY /MPREYS LAMPETRA FLUVIATILIS. + GRE | 4649F |
| BLOOD ENDOCRINOLOGY /HUS STOUTI, TO MAMMALIAN INSU | 5326F |
| BLOOD ENDOCRINOLOGY FEEDING METAMORPHOSIS / A ATLA | 5303F |
| BLOOD ENDOCRINOLOGY GROWTHADULT GONADOGENESIS /MLA | 5126F |
| BLOOD ENDOCRINOLOGY HISTOLOGY /AH, MYXINE GLUTINOS | 4723F |
| BLOOD ENDOCRINOLOGY METABOLISM TECHNIQUES /ITOMES | 5124F |
| BLOOD ENDOCRINOLOGY MIGRATION /O27-133. + *LAMPETR | 5361F |
| BLOOD ENDOCRINOLOGY NERVOUS SYSTEM PHYSIOLOGY /,YS | 5515 |
| BLOOD ENDOCRINOLOGY PHYSIOLOGY /ED DURING THE PERI | 5062F |
| BLOOD ENDOCRINOLOGY PHYSIOLOGY /7HE LYMPHATIC SYST | 4857F |
| BLOOD ENDOCRINOLOGYIMMUNOLOGY /TW. + *EPTATRETUS S | 4708F |
| BLOOD ENERGETICS / /FOLD | 5314F |
| BLOOD EVOLUTION / OFFISHES FROM THE SCAGERAC SEA. | 5063F |
| BLOOD EVOLUTION /III DURING THE PROCESS OF BLOOD C | 5302F |
| BLOOD EVOLUTION LIFE CYCLE /UI *LAMPETRA FLUVIATIL | 4584F |
| BLOOD EVOLUTION PHYSIOLOGY /NPASEFROM LAMPREY ENTO | 5023F |
| BLOOD EVOLUTION PISCES ANIMAL /OPHENUS JAPONICUS H | 5320 |
| BLOOD EVOLUTION TECHNIQUES /OSOME STRUCTURAL ANOMA | 4715F |
| BLOOD EXCRETION /OHE HAGFISH BDELLOSTOMA CIRRHATUM | 4620 |
| BLOOD FROM MYXINE GLUTINOSA + *MYXINE GLUTINOSA SW | 4945F |
| BLOOD GLUCOSE IN CRAYFISH PART 2: VARIATIONS INDUC | 4545 |
| BLOOD GLUCOSE LEVELS IN INTACT AND HYPOPHYSECTOMIZ | 5062F |
| BLOOD GLUCOSE REGULATION, AND VITELLOGENESIS. + DE | 5509 |
| BLOOD GONADOGENESIS MIGRATION /R GREAT BRITAIN ADU | 5065F |
| BLOOD GROWTH METABOLISMMETAMORPHOSIS PHYSIOLOGY /A | 5616F |
| BLOOD GROWTH METABOLISMMETAMORPHOSIS PHYSIOLOGY /H | 4975F |
| BLOOD HISTOCHEMISTRY /  LAMPREY FIBRINOGEN AND FIB | 4960F |
| BLOOD HISTOCHEMISTRY /CLOSTOMES. PP. 53-56 + *LAMP | 4859F |
| BLOOD HISTOCHEMISTRY /SIS A GLYCOPEPTIDE. + *PETRO | 4954F |
| BLOOD HISTOLOGY /LS OF THE ATLANTIC HAGFISH, MYXIN | 5341F |
| BLOOD HISTOLOGY /TETROMYZON MARINUS AMMOCOETE ADUL | 4743F |
| BLOOD HISTOLOGY BIOCHEMISTRY /TM MYXINE GLUTINOSA | 4945F |
| BLOOD HISTOLOGY CYTOLOGY /FPLEEN OF THE HAGFISH. + | 4795F |
| BLOOD HISTOLOGY CYTOLOGY /I CELLS OF THE HAGFISH + | 4796F |
| BLOOD HISTOLOGY HISTOCHEMISTRY /SICULAR REFERENCE | 5131F |
| BLOOD HISTOLOGY LIFE CYCLE PHYSIOLOG /+LUVIATILIS. | 4716F |
| BLOOD HISTOLOGY LIFE CYCLE PHYSIOLOG /STUDIES ON P | 4717 |
| BLOOD IMMUNOLOGY /E IMMUNOLOGIC SIMILARITY TO SERU | 5018F |
| BLOOD IMMUNOLOGY HISTOLOGY /LYMPHO | 5114F |
| BLOOD IN FISHES. PP. 331-368. /T | 4909F |
| BLOOD IONIC REGULATION /EBLOOD SERUM OF VERTEBRATE | 5204 |
| BLOOD IONIC REGULATION OSMOREGULATION /NMAERA MONS | 4811F |
| BLOOD IONIC REGULATION OSMOREGULATION PHYSIOLOGY / | 5351F |
| BLOOD LIFE CYCLE PHYSIOLOGY /IS LAMPETRA FLUVIATIL | 5301F |
| BLOOD OF DIADROMOUS LAMPREYS ANDFISHES ON VARIOUS | 5361F |
| BLOOD OF LARVAL AND ADULTLAMPREYS LAMPETRA FLUVIAT | 4678F |
| BLOOD OF MARINE FISH + ANIMAL BLOOD BIOCHEMISTRY S | 5015 |
| BLOOD OF RIVER LAMPREYS (LAMPETRA FLUVIATILIS) AFT | 5527 |
| BLOOD OF RIVER LAMPREYSLAMPETRA FLUVIATILIS AFTER | 4574F |
| BLOOD OF THE LAMPREY LAMPETRA FLUVIATILIS. + EUROP | 4797F |
| BLOOD OF THE LAMPREY,PETROMYZON MARINUS. + ADULT U | 4780F |
| BLOOD OFFISHES FROM THE SCAGERAC SEA. + *MYXINE GL | 5063F |
| BLOOD PARAMETERS IN FISHES FROM THE SKAGERRAK. + B | 5113F |
| BLOOD PHYSIOLOGY RESPIRATION TECHNIQUES /IREY, LAM | 5002F |
| BLOOD SERUM OF VERTEBRATES. + *LAMPETRA FLUVIATILI | 5204 |
| BLOOD SUGAR CONCENTRATIONS IN VERTERATES TO STANDA | 5182F |
| BLOOD SUGAR LEVELS AND ON THEENDOCRINE PANCREAS, I | 5002F |
| BLOOD TECHNIQUES /INE ELECTROPHORESIS. + *ENTOSPHE | 5074 |
| BLOOD VASCULAR SYSTEM OF THE LAMPREY GILL FILAMENT | 5519 |
| BLOODCELLS OF MYXINE GLUTINOSA. PP. 49-50. + CIRCU | 4804F |
| BLOODCYTOLOGY EVOLUTION IMMUNOLOGY /ATOUTII *PETRO | 4550F |
| BLUE HOLE SANDUSKY COUNTY OHIO USA. + U.S.A. CHEMI | 4637F |
| BLUE WITH ACID MUCO POLY SACCHARIDESIN THE PRESENC | 5040 |
| BLUTES BEI WIRBELTIEREN MIT AUSNAHME VON SAUGERN. | 5119F |
| BODIES IN THELAMPREY GILL FILAMENTS. + *LAMPETRA J | 5540F |

| | | |
|---|---|---|
| 5. PP.31-44. + SEA LAMPREY | CANADA AMMOCOETE CHEMISTRY MANAGEMENT /N HURON, 197 | 5216F |
| 77. + *PETROMYZON MARINUS | CANADA AMMOCOETE /+REAM TREATMENTS, 1971-1973. PP. | 4901F |
| 78. + *PETROMYZON MARINUS | CANADA AMMOCOETE /EREAM TREATMENTS, 1971-1973. PP. | 4904F |
| MOTTEI *PETROMYZON MARINUS | CANADA AMMOCOETE /M1973. PP. 19-20. + *LAMPETRA LA | 4890F |
| 103. + *PETROMYZON MARINUS | CANADA AMMOCOETE /ORN RIVER SYSTEM, 1972. PP. 101- | 5477F |
| CHEMISTRY / HISTOLOGY HISTO | CANADA AMMOCOETE ADULT BIOCHEMISTRY HISTOLOGY HIST | 4899F |
| MISTRY / HISTOLOGY HISTOCHE | CANADA AMMOCOETE ADULT EXCRETION HISTOLOGY HISTOCH | 4898F |
| SITISM BYPARASITISM OF BIOL | CANADA AMMOCOETE ADULT LIFECYCLE DISTRIBUTION PARA | 4761F |
| .15. + *PETROMYZON MARINUS | CANADA AMMOCOETE BIOLOGY / F LAKE ONTARIO 1973. PP | 4861F |
| YZON MARINUS *ICHTHYOMYZON | CANADA AMMOCOETE BIOLOGY /D3. PP. 12-15. + *PETROM | 4862F |
| YZON MARINUS *ICHTHYOMYZON | CANADA AMMOCOETE BIOLOGY /I,1973. PP.12. + *PETROM | 4863F |
| YZON MARINUS *ICHTHYOMYZON | CANADA AMMOCOETE BIOLOGY /I1973.PP.7-12. + *PETROM | 4864F |
| -91. + *PETROMYZON MARINUS | CANADA AMMOCOETE BIOLOGY /O1973. PP.6-7. + *PETROM | 4865F |
| S RIVER, 1973. PP.75-76. + | CANADA AMMOCOETE BIOLOGY /TST. MARY'S RIVER.PP. 88 | 4826F |
| 5. PP.44-55. + SEA LAMPREY | CANADA AMMOCOETE CHEMISTRY /ITREATMENT OF ST. MARY | 4891F |
| UPERIOR, 1975. PP.16-30. + | CANADA AMMOCOETE CHEMISTRY MANAGEMENT / TARIO, 197 | 5220F |
| 5. PP.68-75. + SEA LAMPREY | CANADA AMMOCOETE CHEMISTRY MANAGEMENT /DTS, LAKE S | 4554F |
| THYOMYZON SPP. SEA LAMPREY | CANADA AMMOCOETE CHEMISTRY MANAGEMENT /IERIOR, 197 | 5222F |
| -31. + *PETROMYZON MARINUS | CANADA AMMOCOETE CHEMISTRY MANAGEMENT /NOTTEI *ICH | 5223F |
| ZON SP. *LAMPETRA LAMOTTEI | CANADA AMMOCOETE CHEMISTRY MORTALITY OF /M3.PP. 21 | 4889F |
| NTARIO, 1975. PP. 12-14. + | CANADA AMMOCOETE DISTRIBUTION GROWTH // *ICHTHYOMY | 4645F |
| E HURON, 1975. PP. 7-12. + | CANADA AMMOCOETE DISTRIBUTION MANAGEMENT // LAKE O | 4547F |
| SUPERIOR, 1975. PP. 6-7. + | CANADA AMMOCOETE DISTRIBUTION MANAGEMENT /B TO LAK | 4605F |
| ZON SP. *LAMPETRA LAMOTTII | CANADA AMMOCOETE DISTRIBUTION MANAGEMENT /NO LAKE | 5116F |
| ZON SP. *LAMPETRA LAMOTTII | CANADA AMMOCOETE GROWTH CHEMISTRY /INUS *ICHTHYOMY | 5104F |
| EMS. + *PETROMYZON MARINUS | CANADA AMMOCOETE GROWTH CHEMISTRY /TNUS *ICHTHYOMY | 5105F |
| ZON SP. *LAMPETRA LAMOTTEI | CANADA AMMOCOETE HISTORY. /CIN SEDIMENT-WATER SYST | 4736F |
| -52. + *PETROMYZON MARINUS | CANADA AMMOCOETE MANAGEMENT GROWTH / US *ICHTHYOMY | 5109F |
| -46. + *PETROMYZON MARINUS | CANADA AMMOCOETE MORTALITY OF CHEMISTRY /H. PP. 47 | 4893F |
| INOLOGY FEEDING METAMORPHOS | CANADA AMMOCOETE MORTALITY OF CHEMISTRY /S3. PP.32 | 4888F |
| Y / EXCRETION HISTOCHEMISTR | CANADA ATLANTIC OCEAN ADULT AMMOCOETE BLOOD ENDOCR | 5303F |
| IMMUNOLOGY / ENDOCRINOLOGY | CANADA ATLANTIC OCEAN ADULT EXCRETION HISTOCHEMIST | 5350F |
| SENSE RECEPTORS /HYSIOLOGY | CANADA ATLANTIC OCEAN ADULT HISTOLOGY ENDOCRINOLOG | 5539F |
| I, TO MAMMALIAN INSULIN. + | CANADA ATLANTIC OCEAN BEHAVIOUR OLFACTIONPHYSIOLOG | 5219F |
| LAKE ONTARIO TRIBUTARY. + | CANADA BIOCHEMISTRY BLOOD ENDOCRINOLOGY /HUS STOUT | 5326F |
| TUS STOUTI CYCLOSTOMATA. + | CANADA BIOLOGY /IAMPREYS PETROMYZON MARINUS FROM A | 4759F |
| LUTINOSA (CYCLOSTOMATA). + | CANADA BLOOD BIOCHEMISTRY /LACIFIC HAGFISH EPTATRE | 4656F |
| M BY PHYSIOLOGY / PARASITIS | CANADA BLOOD ENDOCRINOLOGY HISTOLOGY /AH, MYXINE G | 4723F |
| FISH, EPTATRETUS STOUTI. + | CANADA DIGESTION EXCRETION FEEDING GROWTH PARASITI | 4673F |
| Y, PETROMYZON MARINUS L. + | CANADA ENDOCRINOLOGY HISTOCHEMISTRY /I PACIFIC HAG | 4564F |
| PREY PETROMYZON MARINUS. + | CANADA ENDOCRINOLOGY HISTOLOGY /R ADULT SEA LAMPRE | 4781F |
| OGY /TION PHYSIOLOGY HISTOL | CANADA ENDOCRINOLOGY HISTOLOGY OSMOREGULATION /LAM | 4687F |
| .93. + *PETROMYZON MARINUS | CANADA ENDOCRINOLOGY REPRODUCTION PHYSIOLOGY HISTO | 4585F |
| EXCRETION HISTOCHEMISTRY / | CANADA GREAT LAKES ADULT DISTRIBUTION /AE HURON. P | 5258F |
| -99. + *PETROMYZON MARINUS | CANADA GREAT LAKES ADULT EXCRETION HISTOCHEMISTRY | 5045F |
| MANAGEMENT MORTALITY OF /N | CANADA GREAT LAKES ADULT MORPHOMETRY / ARIO. PP.98 | 5262F |
| MANAGEMENT MORTALITY OF /N | CANADA GREAT LAKES AMMOCOETE CHEMISTRY DISTRIBUTIO | 5253F |
| MANAGEMENT MORTALITY OF /N | CANADA GREAT LAKES AMMOCOETE CHEMISTRY DISTRIBUTIO | 5254F |
| MANAGEMENT MORTALITY OF /N | CANADA GREAT LAKES AMMOCOETE CHEMISTRY DISTRIBUTIO | 5256F |
| T /E DISTRIBUTION MANAGEMEN | CANADA GREAT LAKES AMMOCOETE CHEMISTRY DISTRIBUTIO | 5257F |
| T /E DISTRIBUTION MANAGEMEN | CANADA GREAT LAKES AMMOCOETE DISTRIBUTION MANAGEME | 5248F |
| /TE MANAGEMENTMORTALITY OF | CANADA GREAT LAKES AMMOCOETE DISTRIBUTION MANAGEME | 5250F |
| P.5. + *PETROMYZON MARINUS | CANADA GREAT LAKES AMMOCOETE MANAGEMENTMORTALITY O | 5252F |
| T MORTALITY OF /N MANAGEMEN | CANADA GREAT LAKES CANADA MANAGEMENT MIGRATION /C | 5246F |
| OCHEMISTRY /ADOGENESIS HIST | CANADA GREAT LAKES CHEMISTRY DISTRIBUTION MANAGEME | 5249F |
| ORTALITY OF /T MANAGEMENT M | CANADA GREAT LAKES ENDOCRINOLOGY GONADOGENESIS HIS | 5470F |
| MANAGEMENT MORTALITY OF /Y | CANADA GREAT LAKES LAKE SUPERIOR ADULT MANAGEMENT | 5247F |
| LAKE ONTARIO. PP. 84-90 + | CANADA GREAT LAKES U.S.A. ADULT AMMOCOETE CHEMISTR | 4844F |
| ARINUS ADULT ENDOCRINOLOGY | CANADA GROWTH MANAGEMENT /IEE STREAMS TRIBUTARY TO | 5133F |
| S STOUTI (CYCLOSTOMATA). + | CANADA HISTOLOGY /AEROL INJECTION. + *PETROMYZON M | 4881F |
| *EPTATRETUS BURGERI SWEDEN | CANADA HISTOLOGY ANATOMY NERVOUS SYSTEM /MPTATRETU | 4922F |
| ON /E ENDOCRINOLOGY EXCRETI | CANADA JAPAN ADULT HISTOLOGY ENDOCRINOLOGY /YOUTI | 4763F |
| WEIGHT AND TEMPERATURE. + | CANADA LAKE ONTARIO AMMOCOETE ENDOCRINOLOGY EXCRET | 4639F |
| ION. + *PETROMYZON MARINUS | CANADA LOCOMOTION RESPIRATION /MN RELATION TO BODY | 4755F |
| . 5. + *PETROMYZON MARINUS | CANADA MANAGEMENT /LHE GREAT LAKES FISHERY COMMISS | 5194F |
| MENT,1966-74. PP. 73-83. + | CANADA MANAGEMENT /NOPERATIONS,LAKE HURON, 1974. P | 4907F |
| UMBER RIVER,1974. P. 70. + | CANADA MANAGEMENT /OARRIER DAMS AND STREAM IMPROVE | 5134F |
| -26. + *PETROMYZON MARINUS | CANADA MANAGEMENT ADULT /KT SEA LAMPREY FROM THE H | 5136F |
| -39. + *PETROMYZON MARINUS | CANADA MANAGEMENT ADULT AMMOCOETE CHEMISTRY /A. 17 | 5108F |
| VER, 1971-74. PP. 91-93. + | CANADA MANAGEMENT ADULT AMMOCOETE CHEMISTRY /N. 26 | 5107F |
| ZON SP. *LAMPETRA LAMOTTEI | CANADA MANAGEMENT AMMOCOETE /R ON THE ST. MARYS RI | 5132F |
| -51. + *PETROMYZON MARINUS | CANADA MANAGEMENT AMMOCOETE GROWTH / US *ICHTHYOMY | 5111F |
| LAMOTTEI *ICHTHYOMYZON SP | CANADA MANAGEMENT CHEMISTRY AMMOCOETE ADULT /OP.39 | 5106F |
| MARINUS CANADA GREAT LAKES | CANADA MANAGEMENT GROWTH AMMOCOETE /NNUS *LAMPETRA | 5110F |
| LAKE ONTARIO. PP. 68-69. + | CANADA MANAGEMENT MIGRATION /C P.5. + *PETROMYZON | 5246F |
| LAKE HURON, 1974. P. 68. + | CANADA MORPHOMETRY /AD BY COMMERCIAL FISHERMEN IN | 4962F |
| 170. + *PETROMYZON MARINUS | CANADA MORPHOMETRY /MCTRICAL ASSESSMENT BARRIERS, | 5088F |
| UT SALVELINUS NAMAYCUSH. + | CANADA MORPHOMETRY /MS, LAKE HURON, 1972. PP. 160- | 5487 |
| -74. + *PETROMYZON MARINUS | CANADA MORTALITY BY PARASITISM BY /LUS ON LAKE TRO | 4769F |
| DIGESTION HISTOLOGY /OLOGY | CANADA MORTALITY OF CHEMISTRY AMMOCOETE /T3. PP 71 | 4892F |
| | CANADA NEW BRUNSWICK AMMOCOETE ANADROMOUS CYTOLOGY | 5518F |

NADA GREAT LAKES AMMOCOETE  
REAT LAKES ADULT AMMOCOETE  
NADA GREAT LAKES AMMOCOETE  
MARINUS U.S.A. GREAT LAKES  
REAT LAKES ADULT AMMOCOETE  
MARINUS CANADA GREAT LAKES  
G MANAGEMENT MORTALITY OF /  
OF SEA LAMPREYS. + ANIMAL  
LITY BY SENSE RECEPTORS /TA  
REAT LAKES ADULT AMMOCOETE  
RA LAMOTTICANADA AMMOCOETE  
TECHNIQUES /TRY MANAGEMENT  
GREEN EGGS OF SALMONIDS. +  
UPPER GREAT LAKES + ANIMAL  
S.A. GREAT LAKES AMMOCOETE  
A LAMPREY CANADA AMMOCOETE  
ITROPHENOL. + BIOCHEMISTRY  
OL) BY BOTTOM SEDIMENTS. +  
EXPOSED TO TFM. + BIOLOGY  
RES OF CHEMICALS. + PISCES  
ES IN 1975. PP. 189-195. +  
ALICYLANILIDE (BAYER 73) +  
ON TFM IN 1972. P. 310. +  
.16-30. + CANADA AMMOCOETE  
/TFM FIELD FORMULATION. +  
OROMETHYL-4-NITROPHENOL. +  
IES. + ANIMAL BIOCHEMISTRY  
/TFM. +  
A LAMPREY CANADA AMMOCOETE  
OXICANT EXPOSURE. + ANIMAL  
ND TFM IN RAINBOW TROUT. +  
IDGE CHIRONOMUS TENTANS. +  
TROUT (SALMO GAIRDNERI). +  
EA LAMPREY CANADA AMMCOETE  
IS AND MYXINE GLUTINOSA. +  
A LAMPREY CANADA AMMOCOETE  
TO SEVERAL BIRD SPECIES. +  
ES IN 1976. PP. 189-199. +  
R WITH ACTIVATED CARBON. +  
ITROPHENOL. + BIOCHEMISTRY  
HYL-4-NITROPHENOL (TFM). +  
TAL POLLUTANTS TO BIRDS. +  
ETED MACROINVERTEBRATES. +  
ODEL STREAM COMMUNITIES. +  
ODEL STREAM COMMUNITIES. +  
ROPHENOL) AND ANTIMYCIN. +  
RIOPHYLLUM SPICATUM L.). +  
ICAL TO THE ENVIRONMENT. +  
KES U.S.A. ADULT AMMOCOETE  
TROMYZON MARINUS AMMOCOETE  
MISTRY MANAGEMENT + PISCES  
NGLAMPRICIDE TREATMENTS. +  
HENOL IN NATURAL WATERS. +  
CHEMISTRY ADULTS AMMOCOETE  
N MARINUS CANADA AMMOCOETE  
MYZON MARINUS CANADA ADULT  
ITROPHENOL (TFM). + ANIMAL  
.A. AMMOCOETE ADULT ANIMAL  
AND VISUAL PIGMENTS IN BLA  
N WATERS OF THE MAGALLANES  
C REGULATION OSMOREGULATION  
IS BIOCHEMISTRY INTEGUMENT  
N / NERVOUS SYSTEM EVOLUTIO  
EASPIDES VERTEBRATA CYCLOST  
NEWBORN GUINEA PIGS, RATS,  
CHEMISTRY MANAGEMENT /IMAL  
ROMETHYL-4-NITROPHENOL) IN  
NOL) IN CHIRONOMID LARVAE (  
EMISTRY ANIMAL METABOLISM /  
ARVAE OF THE AQUATIC MIDGE  
-4'-NITROSALICYLANILIDE BY  
73) TO LARVAEOF THE MIDGE  
ARVAE OF THE AQUATIC MIDGE  
F THE KIDNEY IN THE LAMPREY  
OXIMAL TUBULES OF THE KIDNI  
Y CYTOLOGY / JAPAN HISTOLOG  
OF FISHES.] + METABOLISM /  
GLUTINOSA *LAMPETRA FLUVIAT  
LOWER VERTEBRATES. + BIOCH  
L. PP. 30-32. + DIGESTION B  
ARINE FISH + ANIMAL BLOOD B  

CHEMISTRY DISTRIBUTION MANAGEMENT MORTALITY OF /FA    5257F  
CHEMISTRY DISTRIBUTION MANAGEMENT MORTALITY OF /GG    5502F  
CHEMISTRY DISTRIBUTION MANAGEMENT MORTALITY OF /SA    5253F  
CHEMISTRY DISTRIBUTION MANAGEMENT MORTALITY OF /T     5255F  
CHEMISTRY DISTRIBUTION MANAGEMENT MORTALITY OF /TG    5368F  
CHEMISTRY DISTRIBUTION MANAGEMENT MORTALITY OF /Y     5249F  
CHEMISTRY ECOLOGY FEEDING MANAGEMENT MORTALITY OF     4600F  
CHEMISTRY ECOLOGY MANAGEMENT /REATMENT FOR CONTROL    5051F  
CHEMISTRY ENDOCRINOLOGY INNUNOLOGY MANAGEMENT MORT    5366F  
CHEMISTRY ENDOCRINOLOGY MANAGEMENT MIGRATION /A. G    5263F  
CHEMISTRY GROWTH MANAGEMENT /NHYOMYZON SP. *LAMPET    5491F  
CHEMISTRY MANAGEMENT + PISCES CHEMISTRY MANAGEMENT    5397F  
CHEMISTRY MANAGEMENT / ICITY OF FOUR TOXICANTS TO     5449  
CHEMISTRY MANAGEMENT / ND CHINOOK SALMON FROM THE     5189F  
CHEMISTRY MANAGEMENT / ON MARINUS *ICHTHYOMYZON U.    5251F  
CHEMISTRY MANAGEMENT / TARIO, 1975. PP.44-55. + SE    5220F  
CHEMISTRY MANAGEMENT / TE OR 3-TRIFLUOROMETHYL-4-N    5424  
CHEMISTRY MANAGEMENT / TRIFLUOROMETHYL-4-NITROPHEN    5411F  
CHEMISTRY MANAGEMENT /+ODUCTION OF FATHEAD MINNOWS    5379F  
CHEMISTRY MANAGEMENT /+XICITY OF EFFICACY OF MIXTU    5396F  
CHEMISTRY MANAGEMENT /ENTED RESEARCH ON LAMPRICID    5500F  
CHEMISTRY MANAGEMENT /AIDE 2', 5-DICHLORO-4-NITROS    5442F  
CHEMISTRY MANAGEMENT /DISTRATION-ORIENTED RESEARCH    5472F  
CHEMISTRY MANAGEMENT /DTS, LAKE SUPERIOR, 1975. PP    4554F  
CHEMISTRY MANAGEMENT /E    5377F  
CHEMISTRY MANAGEMENT /ERACUTE EXPOSURE TO 3-TRIFLU    5422F  
CHEMISTRY MANAGEMENT /FSCLE TISSUE LABORATORY STUD    4985F  
CHEMISTRY MANAGEMENT /I    5378F  
CHEMISTRY MANAGEMENT /IERIOR, 1975. PP.68-75. + SE    5222F  
CHEMISTRY MANAGEMENT /ILLOWING ACUTELY SUBLETHAL T    5410F  
CHEMISTRY MANAGEMENT /IUE DYNAMICS OF QUINALDINE A    4660F  
CHEMISTRY MANAGEMENT /K BY LARVAE OF THE AQUATIC M    4647F  
CHEMISTRY MANAGEMENT /LRLY LIFE STAGES OF RAINBOW    5080F  
CHEMISTRY MANAGEMENT /N HURON, 1975. PP.31-44. + S    5216F  
CHEMISTRY MANAGEMENT /NANS OF PETROMYZON FLUVIATIL    5447  
CHEMISTRY MANAGEMENT /NOTTEI *ICHTHYOMYZON SPP. SE    5223F  
CHEMISTRY MANAGEMENT /PLID (BAYLUSCIDE, BAYER 73)    5430F  
CHEMISTRY MANAGEMENT /TENTED RESEARCH ON LAMPRICID    5367F  
CHEMISTRY MANAGEMENT /UF TOXIC CHEMICALS FROM WATE    5407F  
CHEMISTRY MANAGEMENT /XIONOF 3-TRIFLUOROMETHYL-4-N    4652F  
CHEMISTRY MANAGEMENT /YNTINALIS) TO 3-TRIFLUOROMET    5577F  
CHEMISTRY MANAGEMENT /7RY TOXICITIES OF ENVIRONMEN    5498F  
CHEMISTRY MANAGEMENT ANIMAL /L-NITROPHENOL) TO SEL    5387F  
CHEMISTRY MANAGEMENT ANIMAL /T THE METABOLISM OF M    5386F  
CHEMISTRY MANAGEMENT ANIMAL ECOLOGY /AESIDUES IN M    5419F  
CHEMISTRY MANAGEMENT ANIMAL ECOLOGY /MMETHYL-4-NIT    5403F  
CHEMISTRY MANAGEMENT ECOLOGY /NHX) PLANCHON AND MY    5421F  
CHEMISTRY MANAGEMENT MORTALITY OF /MTION OF A CHEM    5183F  
CHEMISTRY MANAGEMENT MORTALITY OF /RANADA GREAT LA    4844F  
CHEMISTRY MANAGEMENT TECHNIQUE /OLAMPRICIDE. + *PE    5053F  
CHEMISTRY MANAGEMENT TECHNIQUES /CRINUS PISCES CHE    5397F  
CHEMISTRY MANAGEMENT TECHNIQUES /E73 IN WATER DURI    5462F  
CHEMISTRY MANAGEMENT TECHNIQUES /VROMETHL-4-NITROP    5408F  
CHEMISTRY MANAGEMENT. /Y-68. + SEA LAMPREY U.S.A.    5221F  
CHEMISTRY MORTALITY OF /M3.PP. 21-31. + *PETROMYZO    4889F  
CHEMISTRY MORTALITY OF AMMOCOETE /T. 1-V. + *PETRO    5237F  
CHEMISTRY TOXICITY MANAGEMENT /ORIFLUOROMETHYL-4-N    5077  
CHEMISTRYBIOCHEMISTRY MORTALITY OF /TN MARINUS U.S    4751F  
CHERNOMORSKIKH PLASTINOZHABERNYKH. [PHOTORECEPTORS    5274  
CHILE REGION. /INEW RECORD MYXINE PETROMYZONIDAE I    5571  
CHIMAERA MONSTROSA. + NORWAY PHYSIOLOGY BLOOD IONI    4811F  
CHIMAERA). + *MYXINE GLUTINOSA *LAMPETRA FLUVIATIL    5019F  
CHINA.] + ANATOMY EVOLUTION NERVOUS SYSTEM EVOLUTI    4709F  
CHINE). [ANATOMY AND TAXONOMIC POSITION OF THE GAL    4709F  
CHINESE HAMSTERS AND SPINY MICE. + MANAGEMENT / D    5365F  
CHINOOK SALMON FROM THE UPPER GREAT LAKES + ANIMAL    5189F  
CHIRONOMID LARVAE (CHIRONOMOUS TENTANS). /CTRIFLUO    5380  
CHIRONOMOUS TENTANS). /CTRIFLUOROMETHYL-4-NITROPHE    5380  
CHIRONOMUS TENTANS. + CHEMISTRY ANIMAL METABOLISM    4937F  
CHIRONOMUS TENTANS. + CHEMISTRY MANAGEMENT /K BY L    4647F  
CHIRONOMUS TENTANS. + MANAGEMENT /AE 2',5-DICHLORD    5453F  
CHIRONOMUS TENTANS. + MANAGEMENT /OLANILIDE (BAYER    5120F  
CHIRONOMUS TENTANS. /NFM) IN SUBLETHALLY ESPOSED L    5457  
CHLORIDE AND BICARBONATE IN THE PROXIMAL TUBULES O    4569  
CHLORIDE AND SODIUM BICARBONATE REABSORPTION IN PR    5507  
CHLORIDE CELLS. + *LAMPETRA JAPONICA JAPAN HISTOLO    4603F  
CHLORIDE SECRETING CELLS OF FISHES.] + METABOLISM    4616F  
CHLOROID PLEXUS: A PHYLOGENETIC SURVEY. + *MYXINE    5162  
CHOLECYSTOKININ ON GALLBLADDER CONTRACTION IN SOME    5048  
CHOLECYSTOKININ-LIKE ACTIVITY IN MYXINE GLUTINOSA    4846F  
CHOLESTEROL AND FREE FATTY-ACIDS IN THE BLOOD OF M    5015

GEL ELECTROPHORESIS. /MIDE

NITRALID (BAYLUSCIDE, BAYER

AL BIOCHEMISTRY METABOLISM

II MOLODI RYB. [METHODS OF

/FIELD METHODS FOR

KELETON. + *CYCLOSTOMATA AN

LUTINOSA L. (CYCLOSTOMATA.)

ING / /ABNORMAL TOOTH

RYOS. + EMBRYOLOGY /US, EMB

, EMBRYOS. + EMBRYOLOGY CHE

/STUDIES ON THE

E, BIOTRANSFORMATION, AND E

-HYPOPHYSIAL NEUROSECRETORY

IN RIVERS OF EASTERN KAZAKH

VIATILIS BIOCHEMISTRY MUSCL

IN RELATION TO ECOLOGY. + E

GLUTINOSA L. + AUSTRIA GON

PETROMYZON MARINUS. + U.S.A

S.RIVER LAMPREYS (LAMPETRA

ONTOGENESIS AND PHYLOGENES

FORMATION OF THE LARVA INTO

OHYPOPHYSIS. + *LAMPETRA RI

YZON MARINUS CHEMISTRY AMMO

. 174-175. /RRENT METER. PP

/LOWER

ON OF THE LUDLOW AND LOWER

PONICUS / + *ENTOSPHENUS JA

OUS SYSTEM EVOLUTION / NERV

VOLUTION MOUTH NERVOUS SYST

EOSTRACI), CEPHALASPIDE DU

TAXONOMIC POSITION OF THE

PALEONTOLOGY AND SEDIMENTO

SPAWNING MIGRATION.] PP.12

BIRDS. + CHEMISTRY MANAGEME

EMENT / /COMPARATIVE

ION BETWEEN RETINA AND PINE

H MYXINE GLUTINOSA. /HAGFIS

. ADULT AMMOCOETE ENDOCRINO

PETROMYZON MARINUS *LAMPETR

FERENCE TO THE TRANSFORMATI

LT ENDOCRINOLOGY /USSIA ADU

AGFISH (MYXINE GLUTINOSA L.

GFISH MYXINE GLUTINOSA + AD

PROXIMAL RE ABSORPTION IN

AND SPINAL CORD OF THE LAM

ITY OF THE METHOD. + BIOCHE

STRUCTURE IN SOLUTION. + *L

ROCYTE. /E VERTEBRATE ERYTH

PETRA FLUVIATILIS HISTOLOGY

DROMOUS HISTOLOGY CYTOLOGY

INOSA. + SWEDEN TECHNIQUES

GY HISTOCHEMISTRY CYTOLOGY

TION PHYSIOLOGY METABOLISM

TION PHYSIOLOGY METABOLISM

LOGY MUSCLE NERVOUS SYSTEM

ADULT LARVA ENDOCRINOLOGY

MYZON MARINUS BIOCHEMISTRY

GLUTINOSA L. PP. 30-32. +

CH). + GREAT BRITAIN ADULT

ANERI UNITED KINGDOM LARVA

NE GLUTINOSA ENDOCRINOLOGY

YSIOLOGY / PARASITISM BY PH

AMPETRA PLANERI. + ENGLAND

EN ENDOCRINOLOGY HISTOLOGY

OCOETE ANADROMOUS CYTOLOGY

PETRA FLUVIATILIS EVOLUTION

IS HISTOLOGY ENDOCRINOLOGY

TYPES IN THE ATLANTIC HAGFI

PREY PETROMYZON FLUVIATILIS

REY, LAMPETRA FLUVIATILIS L

1758. + GREAT BRITAIN ADULT

DIPLOSTOMUM IN THE FAUNA OF

S / THE USSR.] + SYSTEMATIC

ME. [STATUS OF LUNGFISHES (

FISHES (DIPNOI) AND THEIR S

ND THE SARCOPLASMIC RETICUL

76-387. / /

S EVOKED IN MOTONEURONS BY

GENETICS /ETRA FLUVIATILIS

ES (LAMPETRA PLANERI BLOCH)

AMMOCOETE /ARINUS. + ADULT

| | | |
|---|---|---|
| DETERMINATION | OFMOLECULAR WEIGHT BY POLY ACRYLAMID | 4572 |
| DETERMINE | THE TOXICITY OF A MIXTURE OF TFM AND CLO | 5430F |
| DETERMINED | IN ISOLATED RAT FAT CELLS. + ADULT ANIM | 5154F |
| DETERMINING | ACTIVE METABOLISM IN YOUNG FISH.] /MAN | 4924 |
| DETERMINING | CONCENTRATIONS OF BAYER 73 INWATER. | 5433 |
| DEVELOPMENT | AND HOMOLOGY OF THE CYCLOSTOME AXIAL S | 5161F |
| DEVELOPMENT | AND MATURATION OF THE EGGS IN MYXINE G | 4667F |
| DEVELOPMENT | IN A SEA LAMPREY + U.S.A. BIOLOGY FEED | 5241F |
| DEVELOPMENT | IN SEA LAMPREY, PETROMYZON MARINUS, EM | 5373 |
| DEVELOPMENT | IN THE SEA LAMPREY, PETROMYZON MARINUS | 5375F |
| DEVELOPMENT | OF A RADIOIMMUNE ASSAY FOR BAYER73. /H | 5452 |
| DEVELOPMENT | OF METHODS FOR INVESTIGATIONS OF UPTAK | 5380 |
| DEVELOPMENT | OF NEUROHEMAL PARTS OF THE HYPOTHALAMO | 4556F |
| DEVELOPMENT | OF SIBERIAN LAMPREY LAMPETRA KESSLERI | 5318 |
| DEVELOPMENT | OF THE ANIMAL KINGDOM. + *LAMPETRA FLU | 4668 |
| DEVELOPMENT | OF THE EYES OF CYCLOSTOMES AND FISHES | 5363 |
| DEVELOPMENT | OF THE FEMALE ATLANTIC HAGFISH, MYXINE | 5526F |
| DEVELOPMENT | OF THE INTEGUMENT OF THE SEA LAMPREY, | 5235F |
| DEVELOPMENT | OF THE MEMBRANOUS LABYRINTH IN LAMPREY | 5234 |
| DEVELOPMENT | OF VASOMOTOR ADRENERGIC INNERVATION IN | 5504 |
| DEVELOPMENT | WITH PARTICULAR REFERENCE TO THE TRANS | 4807F |
| DEVELOPMENTAL | AND EVOLUTIONARY ASPECTS OF THE NEUR | 5393 |
| DEVELOPMENTAL | STAGES OF THE SEA LAMPREY. + *PETROM | 5050F |
| DEVICE | OF THE SMALL, TYPE-A,PRICE CURRENT METER. P | 5494F |
| DEVONIAN | AGNATHANS OF YUNNAN AND SICHUAN. /I | 5417 |
| DEVONIAN | DEPOSITS IN EASTERN EUROPE.] PP. 63-70. / | 5087 |
| DEVONIAN | GASPE SANDSTONE CANADA.] + *ENTOSPHENUS J | 5022F |
| DEVONIAN | OF YUNNAN CHINA.] + ANATOMY EVOLUTION NER | 4709F |
| DEVONIEN | INFERIEUR DU SPITSBERG. + ADULT ANATOMY E | 5346F |
| DEVONIEN | INFERIEUR DU SPITSBERG. /ICYCLOSTOMI, OST | 4578 |
| DEVONIEN | INTERIEUR DU YUNNAN (CHINE). [ANATOMY AND | 4709F |
| DEVONIEN | MOYEN), GRES DE GASPE. [INTERPRETATION OF | 5022F |
| DIADROMOUS | LAMPREYS ANDFISHES ON VARIOUS STAGES OF | 5361F |
| DIETARY | TOXICITIES OF ENVIRONMENTAL POLLUTANTS TO | 5498F |
| DIETARY | TOXICITIES OF PESTICIDES TO BIRDS. + MANAG | 5416F |
| DIFFERENCES | IN DATA PROCESSING UPON LIGHT STIMULAT | 4950 |
| DIFFERENT | MUSCLE FIBER TYPES IN THE ATLANTIC HAGFI | 5043 |
| DIFFERENT | STAGES IN ITS LIFE CYCLE. + CANADA U.S.A | 4758F |
| DIFFERENT | STAGES IN THE LIFE CYCLE OFLAMPREYS. + * | 4651F |
| DIFFERENT | STAGES OF DEVELOPMENT WITH PARTICULAR RE | 4807F |
| DIFFERENT | STAGES OF THEIR LIFE CYCLE.] + RUSSIA AD | 5323F |
| DIFFERENT | TYPES OF MUSCLE FIBERS IN THE ATLANTIC H | 5177F |
| DIFFERENT | TYPESOF MUSCLE FIBERS IN THE ATLANTIC HA | 4610F |
| DIFFERENT | VERTEBRATE CLASSES. / HRON STRUCTURE AND | 5165 |
| DIFFERENTIATION | OF MONO AMINES IN THE HYPOTHALAMUS | 4940F |
| DIFFUSE | SCATTERING BY GLOBINS IN SOLUTION SENSITIV | 5070 |
| DIFFUSE | SCATTERING TO STUDIES OF GLOBULAR PROTEIN | 5349 |
| DIFFUSION | PATHWAY FOR AMIDESIN THE VERTEBRATE ERYT | 4968 |
| DIGESTIF | DES CYCLOSTOMES. + *MYXINE GLUTINOSA *LAM | 5558 |
| DIGESTION | / ON MARINUS. + CANADA NEW BRUNSWICK ANA | 5516F |
| DIGESTION | /AAL MUCOSA OF A CYCLOSTOME, MYXINE GLUT | 4870F |
| DIGESTION | /CMYXINE GLUTINOSA. + SWEDEN ENDOCRINOLO | 4786F |
| DIGESTION | /IOTHERMS. + *LAMPETRA FLUVIATILIS EVOLU | 4947F |
| DIGESTION | /LE HAGFISHPOLISTOTREME STOUTII. + EVOLU | 4724 |
| DIGESTION | /M GLUTINOSA L. PP. 24-25. + ADULT HISTO | 4839F |
| DIGESTION | /NA PLANERI *LAMPETRA FLUVIATILIS SWEDEN | 5561F |
| DIGESTION | /PNE GLUTINOSA *EPTATRETUS STOUTI *PETRO | 4910F |
| DIGESTION | BIOCHEMISTRY ANIMAL /YACTIVITY IN MYXINE | 4846F |
| DIGESTION | ECOLOGY FEEDING GROWTH /RTRA PLANERI(BLO | 5142F |
| DIGESTION | ENDOCRINOLOGY /UFLUVIATILIS *LAMPETRA PL | 5562F |
| DIGESTION | EVOLUTION BIOCHEMISTRY /EUCTION. + *MYXI | 4744F |
| DIGESTION | EXCRETION FEEDING GROWTH PARASITISM BY P | 4673F |
| DIGESTION | FEEDING GROWTH /. OF THE BROOK LAMPREY L | 4776F |
| DIGESTION | HISTOCHEMISTRY CYTOLOGY /BTINOSA. + SWED | 4617F |
| DIGESTION | HISTOLOGY /O. + CANADA NEW BRUNSWICK AMM | 5518F |
| DIGESTION | IN POIKLOTHERMS AND HOMOIOTHERMS. + *LAM | 4947F |
| DIGESTION | PHYSIOLOGY /SUTINOSA *LAMPETRA FLUVIATIL | 5558 |
| DIMENSION | OF THE SYSTEM IN DIFFERENT MUSCLE FIBER | 5043 |
| DIMERIC | STRUCTURE OF AN ACID PHOSPHATASEIN THE LAM | 4562F |
| DIMORPHISM | IN THE GILLS OF THE SPAWNING RIVER LAMP | 5169F |
| DIPLOSTOMULUM | FROM LAMPETRA FLUVIATILIS LINNAEUS, | 5061F |
| DIPLOSTOMUM | FAUNY SSSR. [METACERCARIA OFTHE GENUS | 5153 |
| DIPLOSTOMUM | IN THE FAUNA OF THE USSR.] + SYSTEMATI | 5153 |
| DIPNOI) | AND THEIR SYSTEMATIC POSITION.] / EV SISTE | 4567F |
| DIPNOI) | I IKH POLOZHENIEV SISTEME. [STATUS OF LUNG | 4567F |
| DIRECT | CONNECTION BETWEENTHE MYOCARDIAL GRANULES A | 4792F |
| DIRECT | PHOTO-SENSORY ACTIVITY OF THE PINEAL. PP. 3 | 4986 |
| DIRECT | STIMULATION OF SINGLE PRESYNAPTIC FIBERS. / | 5353 |
| DIRECTION, | MECHANISM, RATE. + *LAMPETRA FLUVIATILI | 5401F |
| DIS | INTERRENAL-UND ADRENAL-SYSTEMS DES BACHNEUNAUG | 5563 |
| DISC | OF THE SEA LAMPREY PETROMYZON MARINUS. + ADUL | 4702F |

IN SEA LAMPREY, PETROMYZON
LS INTHE SYSTEMATIC AND POR
AND VITELLOGENESIS. + DENM
MPETRA REISSNERI. + JAPAN A
T LA GLYCOGENE-SYNTHETASE H
) AS A LAMPRICIDE. + *PETRO
RY MANAGEMENT /SCES CHEMIST
2', 5-DICHLORO-4'-NITROSAL
VIRONMENTAL FACTORS ON THE
TIGATIONS.] + REPRODUCTION
PERCHES TILAPIA LEUCOSTICTA
SEVEN FISH SPECIES AND TO
ND ELECTRON MICROSCOPICAL I
OF FOUR TOXICANTS TO GREEN
ULT EMBRYOLOGY ENDOCRINOLOG
CULTURE EMBRYOLOGY FECUNDIT
- UND ELEKTRONENMIKROSKOPIS
YPOPHYSE BEI MYXINE GLUTINO
HWARZEN MEERES IN DER NORDO
RY CELLS ON THE TENTACLES O
N-SEE BEI POTSDAM. /M GOTTI
CHEN TURKEI. /DER NORDOSTLI
             /MATERIALY PO
[MATERIAL CONCERNING THE EC
ON NUCLEOTIDE COMPLEXES IN
MAL BIOCHEMISTRY ADULT /ANI
     /OSMOREGULATION IN
LACK PIGMENTS IN BLACK SEA
75.] + *PETROMYZON MARINUS
UNA OF THE LOWER AND MIDDLE
ROBLEM OF THE EVOLUTION OF
SYSTEMICHEART AND THE PORT
TENTIALS EVOKED IN LAMPREY
EVOLUTION OF THE CENTRAL NE
PSESIN LAMPREY SPINAL CORD.
N, 1974. P. 5. + *PETROMYZO
N. P.5. + *PETROMYZON MARIN
ON: 1975. P. 5. + CANADA AD
97. + *PETROMYZON MARINUS M
PP. 160-170. + *PETROMYZON
P. 68. + CANADA MORPHOMETRY
PP.86-89. + CANADA ADULT MO
HE SPINAL CORD OF THE LAMPR
P. 81-84. + *PETROMYZON MAR
             /

HEARTS OF MYXINE. PP. 35-3
RESHWATER FISH AND AMPHIBIA
CATION OF THE RECEPTORS BY
OFTHE LAMPREY ENTOSPHENUS
HELIUM OF THE HAGFISH GILLS
AL TISSUES IN THE SCANNING
ON OF THYROGLOBULIN IN THE
N 2 TYPESOF GILL EPITHELIAL
S OF THEHAGFISH EPTATRETUS
RVAL LAMPREY LAMPETRA JAPON
OF THEHAGFISH EPTATRETUS BU
IN THELAMPREY GILL FILAMEN
REY, LAMPETRA FLUVIATILIS (
UNCTION IN THE MYOTOMES OF
IN VERTEBRATES.] + *CYCLOST
HILIC GRANULOCYTES IN THE I
ION EGG CYTOLOGY /REPRODUCT
TEETH. + HISTOLOGY /INFISH
THE LAMPREY LAMPETRA JAPONI
OCOETE ADULT HISTOLOGY /AMM
ISTOLOGY BIOCHEMISTRY /GY H
OLYMERIZATION PRODUCTS VIA
RT 2: LIGHT MICROSCOPY AND
N NEURO- AND ADENOHYPOPHYSI
P. 46. + *MYXINE GLUTINOSA
IS HISTOLOGY /+ GONADOGENES
S OF THE ATLANTIC HAGFISH,
HE LAMPREY LAMPETRA FLUVIAT
LY RELATED LAMPREYS. + *LAM
TRY BLOOD TECHNIQUES /HEMIS
GHT BY POLY ACRYLAMIDE GEL
ND THEIR ENZYMIC PROPERTIES
ON OF THEINTRA MEDULLARY PR
ROPHORETIC STUDIES ON SARCO
Y STAGES OF THE OOGENESIS I
RESSIVE DEVELOPMENT AND MAT

EFFECTS OF 6-METHYL MERCAPTOPURINE ON DEVELOPMENT          5373
EFFECTS OF 6-OH-DOPAMINE ON THE CATECHOLAMINE LEVE         4835F
EFFECTS ON REGENERATION, BLOOD GLUCOSE REGULATION,         5509
EFFERENT GILL DUCT EPITHELIAL CELLS OF LAMPREYS LA         5180
EFFETS DU GLUCAGON SUR LA GLYCEMIE, LE GLYCOGENE E         5468F
EFFICACITY OF 3-TRIFLUOROMETHYL-4-NITROPHENOL (TFM         5053F
EFFICACY OF MIXTURES OF CHEMICALS. + PISCES CHEMIS         5396F
EFFICACY OF 3-TRIFLUOROMETHYL-4-NITROPHENOL (TFM),         5463F
EFFICIENCY OF BAYLUSCIDE. /LON THE INFLUENCE OF EN         5550
EGG CYTOLOGY /CHT AND ELECTRON MICROSCOPICAL INVES         4667F
EGG-, JUVENILE AND ADULT STAGES OF THE VARIEGATED          5448
EGGS AND FRY OF COHO SALMON. + MANAGEMENT /FNGS OF         5095F
EGGS IN MYXINE GLUTINOSA L. (CYCLOSTOMATA.)LIGHT A         4667F
EGGS OF SALMONIDS. + CHEMISTRY MANAGEMENT / ICITY          5449
EGGS OF THE HAGFISH EPTATRETUS BURGERI. + JAPAN AD         4703F
EGGS; LETTER TO P.J. ROBERTSON. + AMMOCOETE ADULT          5143F
EIER VON MYXINE GLUTINOSA L. (CYCLOSTOMATA). LICHT         4667F
EINE NERVOSE VERBINDUNG ZWISCHEN NEURO- UND ADENOH         4809F
EINE NEUE NEUNAGENART AUS DEM EINZUGSGEBIET DES SC         5581F
EINE RASTERELEKTRONENOPTISCHE UNTERSUCHUNG. [SENSO         5543F
EINES MEERNEUNAGES (PETROMYZON MARINUS L.) IM GOTT         5580F
EINZUGSGEBIET DES SCHWARZEN MEERES IN DER NORDOSTL         5581F
EKOLOGII KHARIUSA V REKAKH ZAKARPATSKOI OBLASTI /O         5573F
EKOLOGII SHCHUKI ESOX LUCIUS L. NIZOV'EV R. UMBY.          4622F
ELASMOBRANCH AND MARINE TELEOST RED CELLS. /TND IR         5168
ELASMOBRANCH. + *ENTOSPHENUS TRIDENTATUS U.S.A. AN         4729F
ELASMOBRANCHS. + MYXINE BLOOD /T                          5238F
ELASMOBRANCHS.] + CYCLOSTOMES HISTOLOGY VISION /MB         5274
ELBE RIVER, WEST GERMANY. THE SPECIES USED 1950-19         5506F
ELBE: DIE GENUTZTEN ARTEN, 1950-1975. [THE FISH FA         5506F
ELECTRIC ORGANS OF FISHES. /              /THE P          4987
ELECTRICAL ACTIVITY AND MECHANICAL RESPONSE IN THE         4573F
ELECTRICAL AND CHEMICAL EXCITATORY POSTSYNAPTIC PO         5521
ELECTRICAL AND CHEMICAL MODES OF TRANSMISSION AND          5505
ELECTRICAL AND CHEMICAL TRANSMISSION AT LARGE SYNA         5333F
ELECTRICAL ASSESSMENT BARRIER OPERATIONS,LAKE HURO         4907F
ELECTRICAL ASSESSMENT BARRIER OPERATIONS,LAKE HURO         5246F
ELECTRICAL ASSESSMENT BARRIER OPERATIONS,LLAKE HUR         4969F
ELECTRICAL ASSESSMENT BARRIERS, LAKE HURON. PP.94-         5260F
ELECTRICAL ASSESSMENT BARRIERS, LAKE HURON, 1972.          5487
ELECTRICAL ASSESSMENT BARRIERS, LAKE HURON, 1974.          5088F
ELECTRICAL ASSESSMENT BARRIERS, LAKE HURON, 1975.          5226F
ELECTRICAL CHARACTERISTICS OF A GIANT SYNAPSE IN T         4643F
ELECTRICALASSESSMENT BARRIERS: LAKE HURON, 1973. P         4905F
ELECTRO RETINOGRAM OF LAMPREYS. /I                        4956
ELECTROCARDIOGRAM (ECG) OF THE SYSTEMIC AND PORTAL         4836F
ELECTROLYTE TRANSPORT ACROSS THE BODY SURFACE OF F         4934
ELECTRON MICROGRAFIC AND ABLATION STUDIES. /±NTIFI         4536
ELECTRON MICROSCOPE STUDY OF THE OTOLITHIC MACULAE         4663F
ELECTRON MICROSCOPE STUDY ON THE HIGH WALLED ENDOT         5293F
ELECTRON MICROSCOPE.PP. 287-294. / NTERNALBIOLOGIC         4977
ELECTRON MICROSCOPIC IMMUNOCYTOCHEMICAL LOCALIZATI         5539F
ELECTRON MICROSCOPIC STUDIES OF COATED MEMBRANES I         5355
ELECTRON MICROSCOPIC STUDIES ON THE ADENOHYPOPHYSI         4608
ELECTRON MICROSCOPIC STUDIES ON THE MYOTOMES OF LA         5185F
ELECTRON MICROSCOPIC STUDIES ON THE THYROID GLAND          4690F
ELECTRON MICROSCOPIC STUDY OF THE CAVERNOUS BODIES         5540F
ELECTRON MICROSCOPIC STUDY OF THE GILLS OF THELAMP         5587
ELECTRON MICROSCOPIC STUDY OF THE NEURO MUSCULAR J         4884
ELECTRON MICROSCOPIC STUDY OF THE OLFACTORY ORGAN          4634F
ELECTRON MICROSCOPICAL CHARACTERIZATION OF HETEROP         4800F
ELECTRON MICROSCOPICAL INVESTIGATIONS.] + REPRODUC         4667F
ELECTRON MICROSCOPY OF ENAMELOID AND DENTIN INFISH         4875F
ELECTRON MICROSCOPY OF THE 3RD VENTRICULAR WALL IN         5187F
ELECTRON MICROSCOPY. + *LAMPETRA JAPONICA JAPAN AM         5356F
ELECTRON MICROSCOPY. + SWEDEN ADULT ENDOCRINOLOGY          4762F
ELECTRON MICROSCOPY. /AOGEN INVESTIGATION OF THE P         5167
ELECTRON MICROSCOPY. /TD IN EPTATRETUS BURGERI. PA         4581
ELECTRON OPTICAL INDICATIONS OF CONNECTIONS BETWEE         4809F
ELECTRON PARAMAGNETICRESONANCE (EPR) SPECTROSCOPY.         4832F
ELECTRONMICROSCOPIC INVESTIGATIONS.]. + GONADOGENE         4589F
ELECTRONMICROSCOPIC OBSERVATIONS ON THE BLOOD CELL         5341F
ELECTRONMICROSCOPIC STUDIES OF PHOTORECEPTORS IN T         5173
ELECTROPHEROGRAMS DURING THE LIFE CYCLE OF 2 CLOSE         4584F
ELECTROPHORESIS. + *ENTOSPHENUS JAPONICUS BIOCHEMI         5074
ELECTROPHORESIS. /ND DETERMINATION OFMOLECULAR WEI         4572
ELECTROPHORETIC STUDIES ON SARCOPLASMIC PROTEINS A         4896F
ELECTROPHYSIOLOGICAL INVESTIGATION OF THE PROJECTI         5295F
ELEKTROFOREZA. [COMBINED CHROMATOGRAPHIC AND ELECT         4896F
ELEKTRONENMIKROSKOPISCHE UNTERSUCHUNGEN. [THE EARL         4589F
ELEKTRONENMIKROSKOPISCHE UNTERSUCHUNGEN. [THE PROG         4667F

UVIATILIS UNDER THE INFLUEN
FLUVIATILIS RUSSIA ANATOMY
 PARASITISM BY PHYSIOLOGY /
ANADA ATLANTIC OCEAN ADULT
S CANADA GREAT LAKES ADULT
EY,PETROMYZON MARINUS L. +
 . + CANADA AMMOCOETE ADULT
 MYXINE GLUTINOSA. + ADULT
TERACUTE EXPOSURE TO 3-TRIF
XINE ATAMI ADULT AMMOCOETE
H, FLOUNDER AND HAGFISH. +
                          /
THEHAGFISH MYXINE GLUTINOSA
NBOW TROUT. /OPHENOL IN RAI
NIN THE LAMPREY.] + RUSSIA
(CARBARYL) AND 2',5-DICHLOR
AL URINARY SPACE. + ANIMAL
CHEMISTRY CHEMISTRY MANAGEM
OF QUINALDINE SULFATE OR 3-
-CONTAINING NEURONS IN THE
ELEMENTS COMPARABLES AUX CE
N FOSSOR *PETROMYZON MARINU
HRIC KIDNEY OF THE SEA LAMP
OPISTHONEPHRIC KIDNEY OF TH
DAE IN WATERS OF THE MAGALL
E AMMOCOETE CULTURE FEEDING
FUSE SCATTERING BY GLOBINS
IPLOSTOMULUM FROM LAMPETRA
96-98. + SEA LAMPREY CANADA
            /PROGRESS IN
            /RESPONSE OF THE
GLUCURONIDE IN BILE OF TFM
DUCTION OF FATHEAD MINNOWS
FROM BILE OF RAINBOW TROUT
ISTRY MANAGEMENT /L. + CHEM
ACUTELY SUBLETHAL TOXICANT
O 3-TRIFLUOROMETHYL-4-NITRO
OF BODY WALL OF CYCLOSTOME
+ *MYXINE ANATOMY EVOLUTION
+ *PETROMYZON MARINUS *LAMP
RATIONS IN OSMO CONFORMERS,
            /BAYER 73 -
UBMAMMALIAN CHORDATES. + *P
E MIDBRAIN OF LAMPREYS. + N
LAMPETRA PLANERII ADULT CYT
XTRAHYPOTHALAMIC PEPTIDERGI
CLE. /RECEPTOR FROM RAT MUS
AND NEGATIVE COOPERATIVITY
 RESPONSES FROM THE PINEAL
OTE: THE VERTEBRATE MEDIAN
OBRANCH AND TO THE SWIMBLAD
YXINEAND MYXINE) - A CASE O
GY. + EVOLUTION PISCES /OLO
THE ORIGIN OF MYXINOIDS.] +
SH BROOK LAMPREYS LAMPETRA
REGION NORWAY. /F THE OSLO
EBRATE ERYTHROCYTE. /E VERT
NAL NEURONS OF LAMPREY. + *
SS OF BLOOD COAGULATION. +
AMPREY FIBRIN WITH LAMPREY
N. + BIOCHEMISTRY BLOOD EVO
PREY FACTOR XIII AND HUMAN
Y PLASMA TRANSGLUTAMINASE,
FIC HAGFISH EPTATRETUS STOU
NIN RIVER LAMPREYS (LAMPETR
INFLUENCE OF ENVIRONMENTAL
 A STREAM ECOSYSTEM.) + *LA
            /THE LAMPREY AT KETTLE
ES AND THE SPECIES OF FRESH
MISPHERE LAMPREYS OF THE FA
AMILLE PETROMYZONIDE. [SUBD
T AT CLASSIFICATION OF THE
Y TO THE 5-7 GILLED EPTATRE
TED MAXIMUM PARSIMONY GENET
ZON LINNAEUS,1758. Z.N.(S.)
AMMOCOETE ADULT ANIMAL BIOL
MYZONINAE *ENTOSPHENINAE *L
GOTTIN-SEE BEI POTSDAM. /M
ETROMYZON MARINUS *LAMPETRA
ISH. + *MYXINE NORWAY MUSCL
PREY. + *LAMPETRA FLUVIATIL

| | | |
|---|---|---|
| EXCRETION BY THE KIDNEY OF THE LAMPREY LAMPETRA FL | 5210F |
| EXCRETION EVOLUTION / TURE STUDIES.)] + *LAMPETRA | 5064F |
| EXCRETION FEEDING GROWTH PARASITISM BY PHYSIOLOGY | 4673F |
| EXCRETION HISTOCHEMISTRY /E PETROMYZON MARINUS + C | 5350F |
| EXCRETION HISTOCHEMISTRY /EN. + *PETROMYZON MARINU | 5045F |
| EXCRETION HISTOLOGY /NHRIC KIDNEY OF THE SEA LAMPR | 4880F |
| EXCRETION HISTOLOGY HISTOCHEMISTRY / ON MARINUS L. | 4898F |
| EXCRETION HISTOLOGY IONIC REGULATION /THE HAGFISH | 4773F |
| EXCRETION IN COHO SALMON (ONCORHYNCHUS KISUTCH) AF | 5422F |
| EXCRETION INTEGUMENT ENDOCRINOLOGY PHYSIOLOGY / MY | 4705F |
| EXCRETION METABOLISM /ICYANINE GREEN IN THE DOGFIS | 4794F |
| EXCRETION OF LAMPRICIDES BY FISH /A | 5455 |
| EXCRETION OF PROTEIN AND NITROGEN END PRODUCTS IN | 5351F |
| EXCRETION OF 3-TRIFLUOROMETHYL-4-NITROPHENOL IN RA | 5412F |
| EXCRETION PHYSIOLOGY /BPROXIMAL TUBULES OF THE KID | 5507 |
| EXCRETION PRODUCTS OF 1-NAPHTHL-N-METHYLCARBONATE | 5438 |
| EXCRETION TECHNIQUES CYTOLOGY /EOPHAGES IN THE REN | 5593 |
| EXCRETIONOF 3-TRIFLUOROMETHYL-4-NITROPHENOL. + BIO | 4652F |
| EXCRETIONS IN CHANNEL CATFISH FOLLOWING INJECTION | 5424 |
| EXISTENCE OF SEROTONIN-,DOPAMIN-, ANDNORADRENALINE | 4735F |
| EXISTENCE, DANS L'INTESTINE MOYEN DES AGNATHES, D' | 5437 |
| EXITABILITY.J. PHYSIOL., 266:69-89. + *ICHTHYOMYZO | 5206F |
| EXOGENOUS HORSERADISH PEROXIDASE IN THE OPISTHONEP | 5045F |
| EXOGENOUS PROTEIN IN THE ARCHINEPHRIC DUCT OF THE | 4880F |
| EXOMEGAS MACROSTOMUS NEW RECORD MYXINE PETROMYZONI | 5571 |
| EXPERIMENTAL AND NATURAL RATES OF GROWTH.] + FRANC | 5352F |
| EXPERIMENTAL AND THEORETICAL LARGE ANGLE X-RAY DIF | 5070 |
| EXPERIMENTAL DEMONSTRTION OF THE LIFE CYCLE OF A D | 5061F |
| EXPERIMENTAL MECHANICAL ASSESSMENT WEIR, 1975. PP. | 5230F |
| EXPERIMENTAL TUMOUR RESEARCH. V. 20. /O | 5282 |
| EXPOSED DENTAL PULP TO A CORTICO STEROID PASTE. /9 | 4827 |
| EXPOSED RAINBOW TROUT. /UND IDENTIFICATION OF TFM | 5415 |
| EXPOSED TO TFM. + BIOLOGY CHEMISTRY MANAGEMENT /+O | 5379F |
| EXPOSED TO 3-TRIFLUOROMETHYL-4-NITROPHENOL. /EIDE | 5544 |
| EXPOSURE TO 3-TRIFLUOROMETHYL-4-NITROPHENOL. + CHE | 5422F |
| EXPOSURE. + ANIMAL CHEMISTRY MANAGEMENT /ILLOWING | 5410F |
| EXPOSURES OF BROOK TROUT (SALVELINUS FONTINALIS) T | 5577F |
| EXTANT CYCLOSTOME ECTOPROCTS. + MICROARCHITECTURE | 5200 |
| EXTANT CYCLOSTOMES WITH REMARKS ONAMPHIASPIFORMS. | 4675F |
| EXTRACARDIAC CHROMAFFIN CELLS OF LARVAL LAMPREYS. | 4998F |
| EXTRACELLULAR SPACE AND INTRA CELLULAR ION CONCENT | 4931 |
| EXTRACTION AND CLEANUP. /D | 5427 |
| EXTRAHYPOTHALAMIC BRAIN TISSUES OF MAMMALIAN AND S | 5056F |
| EXTRAHYPOTHALAMIC PEPTIDERGIC NEUROSECRETION IN TH | 4927 |
| EXTRAHYPOTHALAMIC PEPTIDERGIC NEUROSECRETION.] + * | 4747F |
| EXTRAHYPOTHALAMISCHE PEPTIDERGE NEUROSEKRETION. [E | 4747F |
| EXTRAJUNCTIONAL ACETYLCHOLINE RECEPTOR FROM RAT MU | 5014 |
| EXTRAORDINARY CONSERVATION OF BINDING SPECIFICITY | 5286 |
| EYE OF THE LAMPREY(PETROMYZON FLUVIATILIS). /RSORY | 5585 |
| EYE. + ANIMAL EVOLUTION CYTOLOGY SENSE RECEPTORS / | 4555F |
| EYE. PART 2:DISTRIBUTION AND RELATION TO THE PSEUD | 4961F |
| EYES IN THREE GENERA OF HAGFISH (EPTATRETUS, PARAM | 4631F |
| EYES OF CYCLOSTOMES AND FISHES IN RELATION TO ECOL | 5363 |
| EYES OF THE FOSSIL CYCLOSTOMES AND THE PROBLEM OF | 4648F |
| FABRICUS, 1794) (NEMATODA CUCULLANICAE) FROM SWEDI | 5467F |
| FACIES ANALYSIS OF THE RINGERIKE GROUP OF THE OSLO | 5031 |
| FACILITATED DIFFUSION PATHWAY FOR AMIDESIN THE VER | 4968 |
| FACILITATION OF SYNAPTIC POTENTIALS IN RETICULOSPI | 5283F |
| FACTOR XIII AND HUMAN FACTOR XIII DURING THE PROCE | 5302F |
| FACTOR XIII AND HUMAN FACTOR XIII. / PATTERN OF L | 4682 |
| FACTOR XIII DURING THE PROCESS OF BLOOD COAGULATIO | 5302F |
| FACTOR XIII. / PATTERN OF LAMPREY FIBRIN WITH LAM | 4682 |
| FACTOR XIII. /DBRIN: CROSS-LINKING OF THE FIBRIN B | 4738 |
| FACTORS AFFECTING GLOMERULAR FUNCTIONS IN THE PACI | 5536F |
| FACTORS INVOLVED IN INITIATION OF SEXUAL MATURATIO | 4868F |
| FACTORS ON THE EFFICIENCY OF BAYLUSCIDE. /LON THE | 5550 |
| FAGLAR I ETT STROMEKOSYSTEM. [FISH-EATING BIRDS IN | 4672F |
| FALLS. + ADULT MIGRATION MANAGEMENT /N | 5129F |
| FAMILIES AND GENERA OF PENNSYLVANIA FRESHWATERFISH | 5068F |
| FAMILIE PETROMYZONIDE. [SUBDIVISION OF NORTHERN HE | 4753F |
| FAMILLES DES LAMPROIES DE L'HEMISPHERENORD DE LA F | 4753F |
| FAMILY CUCULLANIDAE. / /ATTEMP | 4546 |
| FAMILY EPTATRETIDAE FROM SOUTH AUSTRALIA WITH A KE | 4992 |
| FAMILY GENES: CONCORDANCE OF STOCHASTIC AND AUGMEN | 4982 |
| FAMILY GROUP NAMES BASED ON THE TYPE GENUS PETROMY | 4766F |
| FAMILY PETROMYZONIDAE MUST BE PROTECTED. + U.S.A. | 4600F |
| FAMILY PETROMYZONIDEA INTO3 SUBFAMILIES.] + *PETRO | 4753F |
| FANG EINES MEERNEUNAGES (PETROMYZON MARINUS L.) IM | 5580F |
| FAR OF THEIR INVESTIGATION. + *LAMPETRA PLANERI *P | 5186F |
| FAST AND SLOW CRANIAL MUSCLES OF THE ATLANTIC HAGF | 5336F |
| FAST FIBERS OF THE TRUNK MUSCULATURE OF LARVAL LAM | 5317 |

ZON MARINUS *LAMPETRA TRIDE  
LE.] + RUSSIA ADULT ENDOCRI  
S OF THE LAMPREY (LAMPETRA  
HE RIVER LAMPREY (LAMPETRA  
GY VISION /HEMISTRY HISTOLO  
FEEDING METABOLISM /NOLOGY  
OF THE LAMPREY(PETROMYZON  
E RIVER LAMPREYS (LAMPETRA  
N RIVER LAMPREYS (LAMPETRA  
LT GONADOGENESIS HISTOLOGY  
NICKS.] + ENDOCRINOLOGY PHY  
[THE EFFECT OF VARIOUS INS  
TROMYZON MARINUS *LAMPETRA  
TROMYZON MARINUS *LAMPETRA  
THE RIVER LAMPREY LAMPETRA  
PLANERI U.S.S.R. FISHERIES  
OCERCAL TAIL ANDLATERALFIN  
                          /  
ENTATUS *ENTOSPHENUS MINIMU  
PREY LAMPETRA JAPONICA. + J  
INOSA STUDIED BY SCANNING E  
ID SPECIES + *MYXINE GLUTIN  
ISH AND LAMPREY. + ADULT EN  
NIMAL CHEMISTRY MANAGEMENT  
HOOCHEE RIVER, ALABAMA GEOR  
FLUOROMETHYL-4-NITROPHENOL.  
M). + CHEMISTRY MANAGEMENT  
TO BROOK TROUT (LALVELINU  
EYPETROMYZON MARINUS. + CAN  
HE SEA LAMPREY, PETROMYZON  
AMPREY ICHYHYOMYZON GAGEI I  
ECEPTORDIFFERENTIATION. / R  
C. /              /CAPE STORM  
DE GASPE. [INTERPRETATION O  
ON THE ACUTE TOXICITY CONJ  
+ ADULT AMMOCOETE BIOLOGY B  
+ ADULT AMMOCOETE BIOLOGY B  
MPREY, PETROMYZON MARINUS L  
HISTOLOGY BIOCHEMISTRY /LT  
ETHYL-4-NITROPHENOL (TFM) F  
THESIS AND THYROID HORMONE  
+ *ENTOSPHENUS JAPONICUS /  
ONE BY THE SEA LAMPREY, PET  
N OF PALEO- ARCHI- AND NEOC  
ORTHWEST TERRITORIES CANADA  
ENT IN MAMMALIAN BILE DUCT  
YZON SPP. ANATOMY OLFACTION  
ATSII PALEO-, ARKHI- I NEOK  
AND RECENT PRIMITIVE BRAIN  
                  /TFM FIELD  
ON MYXINE GLUTINOSA L. (CYC  
TYKH RY'. [PHOSPHORYLASE IS  
ASPIFORMS. + *MYXINE ANATOM  
F MYXINOIDS.] + *EPTATRETUS  
YLOGENY AND SYSTEMATICS OF  
THE VERTEBRATES. /ESTRY OF  
UTION OF MYOGLOBIN AND THE  
. [THE EYES OF THE FOSSIL C  
266:69-89. + *ICHTHYOMYZON  
) LAMOTTENII *ICHTHYOMYZON  
T MICROSCOPIC AND ELECTRONM  
H PLASTINOZHABERNYKH. [PHOT  
Y INTERNEURONES. + *ICHTHYO  
5-406. + *ENTOSPHENUS JAPAN  
         /NEW MORPHOLOGICAL  
IN VITRO BIOTRANSFORMATION.  
TRY MANAGEMENT /S. + CHEMIS  
MYZON MARINUS. + ADULT USA  
ATION OFMOLECULAR WEIGHT BY  
ACTIVITIES IN MUSCLE SARCOP  
RIZATION IN HEPATICSOLUBLE  
A THIN SECTION AND FREEZE  
ALBIOLOGICAL TISSUES IN THE  
SNYKH AKTIVNOSTEI V SARKOPL  
ATURAL RATES OF GROWTH.] +  
CAL RESPONSE. PP. 69-74. +  
STOUTI *PETROMYZON MARINUS  
NS.] + *PETROMYZON MARINUS  
, LAMPETRA PLANERI BLOCH +  
SERVATIONS ON RARE FISH IN  
F SUBLAINES INDRE-ET-LOIRE

UMOF THE BLOOD AND MUSCLE C
SEA LAMPREY. + CANADA NEW B
MAL BLOOD BIOCHEMISTRY SPAW
PETROMYZON MARINUS U.S.A. A
TINOSA, A THIN SECTION AND
IATILIS) DURING THEIR SPAWN
LAMPETRA FLUVIATILIS. + AD
/NATIVE
/THE
MUSCLES OF SOME MARINE AND
CTS OF 2 LAMPRICIDES TO 21
LARVAL AMMOCOETE AND ADULT
CE ENDOCRINOLOGY METAMORPHO
ERGIA AEPYPTERA, FROM THE D
S PETROMYZON MARINUS. + CAN
ACROSS THE BODY SURFACE OF
E ABOVE CONOWINGO DAM. + *I
ARASITES ON NORTH AMERICAN
PETRA PLANERI *LAMPETRA MAR
PETRA LAMOTTEI (PISCES: PE
F ENVIRONMENTAL LEVELS OF 3
HES OF THE SUSQUEHANNA RIVE
AVIOUR /T PARASITISM BY BEH
NT /ENTATUS ANIMAL INTEGUME
AQUATIC INVERTEBRATES AND
CRINOLOGY /ULT ANIMALS ENDO
, PETROMYZON MARINUS L. + B
AMPREYS PETROMYZON MARINUS
YXINE GLUTINOSA. + SWEDEN M
EBORG:KUNGL. VETENSKAPS-OCH
THYL-4-NITROPHENOL. /UOROME
STATISTICAL ANALYSES OF 2
HURON, 1972. PP. 160-170.
Y HISTOLOGY METABOLISM EXCR
.A. /HE GULF OF MEXICO, U.S
P.94-97. + *PETROMYZON MARI
974. P. 68. + CANADA MORPHO
975. PP.86-89. + CANADA ADU
73. PP. 81-84. + *PETROMYZO
HYSIOLOGY /L BIOCHEMISTRY P
FLUVIATILIS DURING THE PRE
PETRA FLUVIATILIS. + RUSSIA
COLOURLESS EUGLENOPHYCEAE
ION / + CANADA ADULT MIGRAT
, (PISCES: PETROMYZONIDAE)
ELAND ADULT /ZON MARINUS IR
NUS FOLLETTI *ENTOSPHENUS L
BRITAIN ADULT PARASITISM OF
) (NEMATODA: CUCULLANIDAE)
/INTRA CELLULAR POTENTIALS
N HEPATICSOLUBLE FRACTIONS
S. + ADULT BIOCHEMISTRY PHY
BIOCHEMISTRY MUSCLE /IS. +
+ RUSSIA BIOCHEMISTRY METAB
LOOD HISTOLOGY BIOCHEMISTRY
R SPECIES OF THE SAME GENUS
/ORDOVICIAN VERTEBRATES
/RON JAPONICUM MORPHOMETRY
OPHAGA U.S.A. SYSTEMATICS M
NAL ACETYLCHOLINE RECEPTOR
/TRANSFER OF STORAGE AREA
, 1975, P.86. + *ICHTHYOMYZ
. P.98. + *PETROMYZON MARIN
1973. P. 85. + *PETROMYZON
PHYSIOLOGY /L BIOCHEMISTRY
H.] + *LAMPETRA FLUVIATILIS
/CYCLOSTOMES AND TELEOSTS.
PTATRETIDAE. / 5-7 GILLED E
ENTOSPENUS PETROMYZONIDAE
EM PHYSIOLOGY /NERVOUS SYST
H). + SWEDEN AMMOCOETE PARA
NED IN ISOLATED RAT FAT CEL
HIOSTOMA BELCHERII. + BIOCH
/SEA LAMPREY
ATES). + ADULT DISTRIBUTION
U.S.A. AMMOCOETEADULT META
ABUNDANCE OF FISH SPECIES
PETROMYZON MARINUS CANADA G
ENT ADULT /+ CANADA MANAGEM
ON MARINUS CANADA ADULT BIO
SPHORYLATION AND RELATED PR

FREE AMINO ACIDS, TRIMETHYLAMINE OXIDE AND POTASSI   4767F
FREE FATTY ACID LEVEL IN THE MIGRATING ANADROMOUS   5096F
FREE FATTY-ACIDS IN THE BLOOD OF MARINE FISH + ANI   5015
FREEZE CLEAVE STUDYON THE LAMPREY SPINAL CORD. + *   4595F
FREEZE FRACTURE STUDY. /SOF THE HAGFISH MYXINE GLU   5086
FREQUENCY AND HEART RATE OFLAMPREYS (LAMPETRA FLUV   5138F
FREQUENCY ANDHEART RATE IN THE ADULT RIVER LAMPREY   4719F
FRESH WATER FISH. /H   4680
FRESH WATER FISHES OF THE NETHERLANDS. /O   4825
FRESH WATER FISHES. /ERETICULUM AND THE SYSTEM IN   4583
FRESH WATER INVERTEBRATES. / /ACUTE TOXIC EFFE   5076F
FRESH WATER LAMPREY LAMPETRA PLANERI. /AESIS IN A   5603
FRESH WATER LAMPREY, LAMPETRA PLANERI BLOCH + FRAN   5342F
FRESH WATER LAMPREYS LAMPETRA LAMOTTENII ANDOKKELB   4720F
FRESHWATER BIOLOGY OF ADULT ANADROMOUS SEA LAMPREY   5140F
FRESHWATER FISH AND AMPHIBIA. PP. 447-463. /YPORT   4934
FRESHWATER FISHES OF THE SUSQUEHANNA RIVER DRAINAG   5068F
FRESHWATER FISHES. PP. 1622-1627. / T OF CERTAIN P   4662
FRESHWATER LAMPREYS IN WATERS OF THE BSSR.] + *LAM   5358F
FRESHWATER LAMPREYS, OKKELBERGIA AEPYPTERA AND LAM   5139F
FRESHWATER MUSSEL (ANODONTA SP.) AS AN INDICATOR O   5242F
FRESHWATERFISHES AND THE SPECIES OF FRESHWATER FIS   5068F
FRESHWATERTELEOST. + CANADA ADULT PARASITISM BY BE   4754F
FRICTION. + *ENTOSPENUS TRIDENTATUS ANIMAL INTEGUM   4693F
FROG LARVE. + MANAGEMENT /NPHENOL(TFM) TO SELECTED   4557F
FROG. + *ENTOSPHENUS TRIDENTATUS ADULT ANIMALS END   5003F
FROM [1,2,6,7,-3H]-PROGESTERONE BY THE SEA LAMPREY   5325F
FROM A LAKE ONTARIO TRIBUTARY. + CANADA BIOLOGY /I   4759F
FROM A PRIMITIVE VERTEBRATE THE ATLANTIC HAGFISH M   4654F
FROM A SYMPOSIUM IN GOTEBORG 28-29 APRIL, 1972.GOT   4793F
FROM BILE OF RAINBOW TROUT EXPOSED TO 3-TRIFLUOROM   5544
FROM BORNU PROVINCE NORTHERN NIGERIA WITH APPENDIX   4544
FROM CANADIAN ELECTRICAL ASSESSMENT BARRIERS, LAKE   5487
FROM COASTAL MAINE USA. + ANIMAL ADULT BIOCHEMISTR   4604F
FROM EASTERN TRIBUTARIES OF THE GULF OF MEXICO, U.   5150
FROM ELECTRICAL ASSESSMENT BARRIERS, LAKE HURON. P   5260F
FROM ELECTRICAL ASSESSMENT BARRIERS, LAKE HURON, 1   5088F
FROM ELECTRICAL ASSESSMENT BARRIERS, LAKE HURON, 1   5226F
FROM ELECTRICALASSESSMENT BARRIERS: LAKE HURON, 19   4905F
FROM FISHES AND AMPHIBIANS. + ANIMAL BIOCHEMISTRY   5170F
FROM GLYCEROL IN THE ORGANS OF THE LAMPREY LMPETRA   5060
FROM GLYCEROL IN VARIOUS ORGANS OF THE LAMPREY LAM   5218F
FROM HOKKAIDO JAPAN. / /NOTES ON SOME SPECIES OF   4938
FROM HUMBER RIVER, 1975. P.90 + CANADA ADULT MIGRA   5227F
FROM ILLINOIS. + U.S.A. ADULT DISTRIBUTION /DPTERA   5523F
FROM IRISH WATERS IN 1973. + PETROMYZON MARINUS I   4686F
FROM KLAMATH RIVER SYSTEM, CALIFORNIA. + *ENTOSPHE   4750F
FROM LAMPETRA FLUVIATILIS LINNAEUS, 1758. + GREAT   5061F
FROM LAMPETRA LAMOTTENII (LESUEUR, 1827). /O, ·1910   5327F
FROM LAMPREY OLFACTORY RECEPTORS. /   4728
FROM LITTLE SKATE RAJAERINACEA LIVER. / RIZATION I   4972
FROM MUSCLES OF ECTOTHERMIC AND ENDOTHERMIC ANIMAL   5612
FROM MUSCLES OF THE LAMPREY LAMPETRA FLUVIATILIS.   5276
FROM MUSCLES OF THE LAMPREY LAMPETRA FLUVIATILIS.   5313
FROM MYXINE GLUTINOSA + *MYXINE GLUTINOSA SWEDEN B   4945F
FROM NORTHWESTERN NORTH AMERICA WITH NOTES ON OTHE   5542F
FROM ONTARIO CANADA. /L   4615
FROM ONTARIO. + *LETHENTERON JAPONICUM MORPHOMETRY   5553
FROM OREGON. + *LAMPETRA TRIDENTATA *LAMPETRA LETH   5555
FROM RAT MUSCLE. /I AND IODINATION OF EXTRAJUNCTIO   5014
FROM ROOT RIVER TO ST. MARYS ISLAND. P. 174. /   5497F
FROM SEA LAMPREY COLLECTED BY COMMERCIAL FISHERMEN   5225F
FROM SEA LAMPREY COLLECTED BY COMMERCIAL FISHERMEN   5261F
FROM SEA LAMPREY COLLECTED BY COMMERCIALFISHERMEN,   4902F
FROM SEVERAL SPECIES OF FISH. + ANIMAL BIOCHEMISTR   5147
FROM SKELETAL MUSCLES OF CYCLOSTOMATA AND BONY FIS   4887F
FROM SKELETAL MUSCLES OF CYCLOSTOMES AND TELEOSTS.   4989
FROM SOUTH AUSTRALIA WITH A KEY TO THE 5-7 GILLED   4992
FROM SOUTH CENTRAL CALIFORNIA U.S.A. /NF THE GENUS   4821
FROM STIMULATION OF OLFACTORY NERVE. + NERVOUS SYS   4739
FROM SWEDISH BROOK LAMPREYS LAMPETRA PLANERI (BLOC   5467F
FROM THE ATLANTIC HAGFISH MYXINE GLUTINOSA DETERMI   5154F
FROM THE CONNECTIVE TISSUE OF THE AMPHIOXUS, BRANC   5041
FROM THE DEEP OCEAN. /I   5125F
FROM THE DELMARVA PENINSULA (EAST COAST, UNITED ST   4720F
FROM THE DELMARVA PENINSULA. + *PETROMYZON MARINUS   4676F
FROM THE HANFORD REACH OF THE COLUMBIA RIVER. / VE   5569
FROM THE HUMBER RIVER, LAKE ONTARIO. PP.98-99. + *   5262F
FROM THE HUMBER RIVER,1974. P. 70. + CANADA MANAGE   5136F
FROM THE HUMBERRIVER, 1973. PP. 85-87. + *PETROMYZ   4900F
FROM THE LAMPREY LAMPETRA FLUVIATILIS OXIDATIVEPHO   4613F

SMOBRANCHS.] + CYCLOSTOMES
TURE OF ANIMALS. /PIC STRUC
REAT BRITIAN ENDOCRINOLOGY
Y OF HAGFISH AND RAT LIVER
AGFISH MYXINE GLUTINOSA. A
ATRETUS BULGERI *PETROMYZON
AL EVOLUTION CYTOLOGY SENSE
AND CERTAIN OTHER RIVERS IN
THENTERON JAPONICUM MARTENS
T AMMOCOETE BIOLOGY GROWTH
EMBER 1976. + DISTRIBUTION
AMMOCOETE ADULT MANAGEMENT
D AT THE MUSEUM OF NATURAL
          /                    /
PYPTERA AND LAMPETRA LAMOTT
ARINUS. + U.S.A. GREAT LAKE
N MARINUS CANADA AMMOCOETE
OF SEA LAMPREY EARLY LIFE
TAXONOMIC CONSIDERATIONS. +
ANIMAL /D EVOLUTION PISCES
URLESS EUGLENOPHYCEAE FROM
BBS, 1922, (PETROMYZONIDAE)
*PETROMYZON ICHTHYOMYZON *T
O USA. + U.S.A. CHEMISTRY /
ARS DISTALIS INCYCLOSTOME,
RUS) AND IN THE SEA LAMPREY
YSIOLOGY METABOLISM DIGESTI
  RAFINESQUE. + EXCRETION AN
MAL TECHNIQUES CYTOLOGY END
OSTOMATA ANATOMY EVOLUTION
COURSE OF EVOLUTION? / THE
YMES AS AN EVIDENCE OF ITS
WER VERTEBRATES.] /ISMIN LO
*MYXINE GLUTINOSA *LAMPETRA
TRAHYPOTHALAMIC BRAIN TISSU
 ANIMAL PHYSIOLOGY GONADOGE
OF A CYCLOSTOME, MYXINE GLU
. + ADULT HISTOLOGY BIOCHEM
N BIOSYNTHESIS AND THYROID
ILE DUCT MUCOSA OF A CYCLOS
TARY THYROID TISSUES IN VIT
OLICHOGNATHUS, AND A HAGFIS
SH MYXINE GLUTINOSA. + ANIM
/R VERTEBRATES. PP. 51-62.
 /UTII U.S.A. ENDOCRINOLOGY
SONS. [CIRCULATING THYROID
ETRA PLANERI *LAMPETRA FLUV
VE GNATHOSTOMES, PP. 141-15
WITH ANTISERUM TO CAERULEIN
*LAMPETRA FLUVIATILIS GREAT
VER AND MUSCLES OF THE LAMP
-418. /          /ACTIONS OF
OME ET DES POISSONS. [CIRCU
DENMARK BLOOD ENDOCRINOLOGY
LATION FROM THE POPRAD AND
ION HISTOLOGY /NERVE. + VIS
S OF HAGFISH EPTATRETUS BUR
Y OF THE SEA LAMPREY. PART
VITY AND LENGTH/WEIGHT OF T
LAKES GROWTH MORTALITY BY
INUS. /PREY, PETROMYZON MAR
LAMPETRA AEPYPTERA U.S.A. S
RTH AMERICA WITH NOTES ON O
 PLANERI UNITED KINGDOM ADU
ICHTHYOMYZON CASTANEUS U.S.
ULATION. + BIOCHEMISTRY BLO
TH LAMPREY FACTOR XIII AND
N MARINUS CANADA GREAT LAKE
ANADA ADULT MANAGEMENT /S C
 + CANADA ADULT MIGRATION /
 /+ CANADA MANAGEMENT ADULT
CANADA ADULT BIOLOGY /INUS
 DISTRIBUTION PARASITISM BY
TECHNIQUES /9-80. + CANADA
ARINUS *ICHTHYOMYZON CANADA
ROMYZON MARINUS CANADA AMMO
ES CANADA MANAGEMENT MIGRAT
ES ADULT DISTRIBUTION / LAK
T LAKES AMMOCOETE CHEMISTRY
T LAKES AMMOCOETE CHEMISTRY

| | | |
|---|---|---|
| LAKES CHEMISTRY DISTRIBUTI | HURON. PP.9-11. + *PETROMYZON MARINUS CANADA GREAT | 5249F |
| ORPHOMETRY /US MANAGEMENT M | HURON. PP.94-97. + *PETROMYZON MARINUS MANAGEMENT | 5260F |
| CHTHYOMYZON SP. *LAMPETRA L | HURON, 1972. PP. 154-161. + *PETROMYZON MARINUS *I | 5491F |
| ADA MORPHOMETRY /ARINUS CAN | HURON, 1972. PP. 160-170. + *PETROMYZON MARINUS CA | 5487 |
| A ADULT BIOLOGY /INUS CANAD | HURON, 1973. PP. 81-84. + *PETROMYZON MARINUS CANA | 4905F |
| A AMMOCOETE MORTALITY OF CH | HURON, 1973. PP.32-46. + *PETROMYZON MARINUS CANAD | 4888F |
| AGEMENT /MARINUS CANADA MAN | HURON, 1974. P. 5. + *PETROMYZON MARINUS CANADA MA | 4907F |
| ASSESSMENT BARRIERS, LAKE | HURON, 1974. P. 68. + CANADA MORPHOMETRY /MCTRICAL | 5088F |
| DA MANAGEMENT ADULT AMMOCOE | HURON, 1974. PP. 26-39. + *PETROMYZON MARINUS CANA | 5107F |
| THYOMYZON SP. *LAMPETRA LAM | HURON, 1974. PP. 59-65. + *PETROMYZON MARINUS *ICH | 5104F |
| ION MANAGEMENT /E DISTRIBUT | HURON, 1975. PP. 7-12. + CANADA AMMOCOETE DISTRIBU | 4605F |
| E CHEMISTRY MANAGEMENT /OET | HURON, 1975. PP.31-44. + SEA LAMPREY CANADA AMMCOE | 5216F |
| YOMYZON SPP. SEA LAMPREY CA | HURON, 1975. PP.76-83. + *LAMPETRA LAMOTTEI *ICHTH | 5223F |
| CANADA ADULT MORPHOMETRY / | HURON, 1975. PP.86-89. + CANADA ADULT MORPHOMETRY | 5226F |
| MYZON CANADA AMMOCOETE BIOL | HURON,1973.PP.7-12. + *PETROMYZON MARINUS *ICHTHYO | 4864F |
| YOMYZON SP. *LAMPETRA LAMOT | HURON,1974.PP. 10-14. + *PETROMYZON MARINUS *ICHTH | 4645F |
| ITY OF /T MANAGEMENT MORTAL | HURON: 1975. P. 5. + CANADA ADULT MANAGEMENT MORTA | 4969F |
| ES OF FISH. + ANIMAL BIOCHE | HYBRIDIZATION PROPERTIES OF DNA FROM SEVERAL SPECI | 5147 |
| CHLORO-4-NITROSALICYLANILID | HYDROLYSIS AND PHOTOSIS OF THE LAMPRICIDE 2', 5-DI | 5442F |
| TOLOGY / HISTOCHEMISTRY HIS | HYDROPHOBIC CORES OF GLOBINS.] + HISTOCHEMISTRY HI | 4632F |
| , PETROMYZON, DURING SUCTIO | HYDROSTATIC PRESSURES AND MOVEMENTS OF THE LAMPREY | 5461 |
| AND THE UPTAKE OF TRITATED | HYDROXYINDOLE-O-METHYLTRANSFERASE (HIOMT) ACTIVITY | 5209F |
| PTAKE INTO FIBROBLAST CELL | HYDROXYTRYPTAMINE: AUTORADIOGRAPHIC EVIDENCE FOR U | 5171 |
| , HYPO OSMO REGULATORS AND | HYPER OSMO REGULATORS. / ATIONS IN OSMO CONFORMERS | 4931 |
| A FLUVIATILIS DENMARK BLOOD | HYPERGLYCEMIC HORMONES. PP. 285-290.PP. + *LAMPETR | 5126F |
| OF AXONS OF LAMPREY. + *PE | HYPERPOLARIZING POTENTIALS IN RETICULOSPINAL AXONS | 5608F |
| ATIONS IN OSMO CONFORMERS, | HYPO OSMO REGULATORS AND HYPER OSMO REGULATORS. / | 4931 |
| CODERMS. / OF ANASPID OSTRA | HYPOCERCAL TAIL ANDLATERALFIN FOLD OF ANASPID OSTR | 4537 |
| ND ACTINOPTERYGIANS. + *PET | HYPOPHYSEAL ROSTRAL PARS DISTALIS OF CYCLOSTOMES A | 4779F |
| .S.A. PACIFIC OCEAN ENDOCRI | HYPOPHYSECTOMIZED HAGFISH, EPTATRETUS STOUTII. + U | 5184F |
| LIS TREATED WITH INSULIN ST | HYPOPHYSECTOMIZED RIVER LAMPREYS LAMPETRA FLUVIATI | 5062F |
| TIONS. + *EPTATRETUS STOUTI | HYPOPHYSECTOMY OF THE PACIFIC HAGFISH: 1ST OBSERVA | 4585F |
| RETUS BURGERI GIRARD (CYCLO | HYPOPHYSECTOMY ON THE TESTIS OF THE HAGFISH, EPTAT | 5525F |
| /ONTOMYZON DANFORDI REGAN. | HYPOPHYSIAL COMPLEX OF EUDONTOMYZON DANFORDI REGAN | 4878F |
| DANFORDI REGAN.] + ROMANIA | HYPOPHYSIAL NEUROSECRETORY SYSTEM OF EUDONTOMYZON | 5072F |
| S AND FISHES). /(CYCLOSTOME | HYPOPHYSIAL REGION IN LOWER VERTEBRATES (CYCLOSTOM | 4979 |
| S L.): EFFECTS ON REGENERAT | HYPOSECTOMIZED RIVER LAMPREYS (LAMPETRA FLUVIATILI | 5509 |
| F MAMMALIAN AND SUBMAMMALIA | HYPOTHALAMIC AND EXTRAHYPOTHALAMIC BRAIN TISSUES O | 5056F |
| MONE SECRETION IN LOWER VER | HYPOTHALAMIC CONTROL OF MELANOCYTE STIMULATING HOR | 5534 |
| NFORDI REGAN. /ONTOMYZON DA | HYPOTHALAMIC HYPOPHYSIAL COMPLEX OF EUDONTOMYZON D | 4878F |
| LUVIATILIS. + RUSSIA ADULT | HYPOTHALAMIC NERVE CELLS OF THE LAMPREY LAMPETRA F | 4973 |
| OF EUDONTOMYZON DANFORDI RE | HYPOTHALAMO AND HYPOPHYSIAL NEUROSECRETORY SYSTEM | 5072F |
| NTO- AND PHYLOGENESIS OF VE | HYPOTHALAMO-HYPOPHYSIAL NEUROSECRETORY SYSTEM IN O | 4556F |
| PETRA JAPONICA. + *LAMPETRA | HYPOTHALAMO-HYPOPHYSIAL REGION OF THE LAMPREY, LAM | 5029F |
| ETRA FLUVIATILIS. / /THE | HYPOTHALAMO-HYPOPHYSIAL SYSTEM OF THE LAMPREY LAMP | 4586 |
| PETRA JAPONICA. + JAPAN ADU | HYPOTHALAMO-HYPOPHYSIAL SYSTEM OF THE LAMPREY, LAM | 4882F |
| AMPETRA FLUVIATILIS. + BEGI | HYPOTHALAMO-HYPOPHYSIAL VASOTOCINERGIC SYSTEM OF L | 5312 |
| A NEW FILTER SYSTEM. + JAPA | HYPOTHALAMUS AND SPINAL CORD OF THE LAMPREY USING | 4940F |
| NTACTING NERVECELLS OF THE | HYPOTHALAMUS IN VERTEBRATES.] /DS IN THE LIQUOR CO | 4606F |
| . + ADULT ENDOCRINOLOGY HIS | HYPOTHALAMUS OF THE LAMPREY LAMPETRA FLUVIATILIS L | 4941F |
| ADULT CYTOLOGY NERVOUS SYST | HYPOTHALAMUS OF THE LAMPREY, LAMPETRA JAPONICA. + | 5596 |
| VE CELLS OF THE VERTEBRATE | HYPOTHALAMUS. + NERVOUS SYSTEM HISTOLOGY BIOLOGY / | 5000 |
| MMALIANVERTEBRATES. PP. 113 | HYPOTHALMIC CONTROL OF PITUITARY FUNCTION IN SUBMA | 4657 |
| PANCREAS, INTERRENAL, AND C | HYPOXIA ON BLOOD SUGAR LEVELS AND ON THEENDOCRINE | 5002F |
| RY EPITHELIUM. + SWEDEN NOR | I.FINE STRUCTURE OF THE APICAL PART OF THE OLFACTO | 4808F |
| BIOLOGY GROWTH HISTORY MAN | I-VI. + *PETROMYZON MARINUS CANADA ADULT AMMOCOETE | 4908F |
| THE CYCLOSTOMEBRAIN. PART | I: LAMPETRA PLANERI. /S OF ACETYLCHOLINESTERASE IN | 4641 |
| TICS / GAGEI U.S.A. SYSTEMA | ICHTHYO FAUNA. + *ICHTHYOMYZON GAGEI U.S.A. SYSTEM | 4944F |
| IBUTIONSYSTEMATICS /. DISTR | ICHTHYOMYZON *PETROMYZONIDAE /*LAMPETRA U.S.A. DIST | 5068F |
| *EUDONTOMYZON *LETHENTERON | ICHTHYOMYZON *TETRAPLEURODON *ENTOSPENUS *LAMPETRA | 5391F |
| A. ADULT AMMOCOETE DISTRIBU | ICHTHYOMYZON BDELLIUM *ICHTHYOMYZON CASTANEUS U.S. | 4590F |
| 15. + *PETROMYZON MARINUS * | ICHTHYOMYZON CANADA AMMOCOETE BIOLOGY /D3. PP. 12- | 4862F |
| 12. + *PETROMYZON MARINUS * | ICHTHYOMYZON CANADA AMMOCOETE BIOLOGY /I,1973. PP. | 4863F |
| 12. + *PETROMYZON MARINUS * | ICHTHYOMYZON CANADA AMMOCOETE BIOLOGY /I1973.PP.7- | 4864F |
| -7. + *PETROMYZON MARINUS * | ICHTHYOMYZON CANADA AMMOCOETE BIOLOGY /O1973. PP.6 | 4865F |
| A. /HYOMYZON UNICUSPIS U.S. | ICHTHYOMYZON CASTANEUS *ICHTHYOMYZON UNICUSPIS U.S | 5197F |
| US SYSTEM PHYSIOLOGY ADULT | ICHTHYOMYZON CASTANEUS U.S.A AMMOCOETEMUSCLE NERVO | 4674F |
| IBUTION ECOLOGY /OETE DISTR | ICHTHYOMYZON CASTANEUS U.S.A. ADULT AMMOCOETE DIST | 4590F |
| TION MORTALITY BY /DISTRIBU | ICHTHYOMYZON CASTANEUS U.S.A. ADULT ANIMAL DISTRIB | 4812F |
| CHEMISTR /RINUS ADULT HISTO | ICHTHYOMYZON FOSSOR *PETROMYZON MARINUS ADULT HIST | 5206F |
| (LETHENTERON) LAMOTTENII * | ICHTHYOMYZON FOSSOR AMMOCOETE PHYSIOLOGY /YAMPETRA | 4565F |
| VER SYSTEM IN ARKANSAS. + * | ICHTHYOMYZON GAGEI DISTRIBUTION PISCES /IACHITA RI | 5202F |
| S TO THE ICHTHYO FAUNA. + * | ICHTHYOMYZON GAGEI U.S.A. SYSTEMATICS /YF 3 SPECIE | 4944F |
| ATA *LAMPETRA PLANERI UNITE | ICHTHYOMYZON HUBBSI *LAMPETRA (ENTOSPENUS) TRIDENT | 4678F |
| ON BDELLIUM *ICHTHYOMYZON C | ICHTHYOMYZON HUBBSI *LAMPETRA LAMOTTEI *ICHTHYOMYZ | 4590F |
| /NAGEMENT GROWTH AMMOCOETE | ICHTHYOMYZON SP CANADA MANAGEMENT GROWTH AMMOCOETE | 5110F |
| E DISTRIBUTION GROWTH /COET | ICHTHYOMYZON SP. *LAMPETRA LAMOTTEI CANADA AMMOCOE | 4645F |
| E MANAGEMENT GROWTH /MOCOET | ICHTHYOMYZON SP. *LAMPETRA LAMOTTEI CANADA AMMOCOE | 5109F |
| NT AMMOCOETE GROWTH /NAGEME | ICHTHYOMYZON SP. *LAMPETRA LAMOTTEI CANADA MANAGEM | 5111F |
| CHEMISTRY GROWTH MANAGEMEN | ICHTHYOMYZON SP. *LAMPETRA LAMOTTICANADA AMMOCOETE | 5491F |
| E GROWTH CHEMISTRY /MMOCOET | ICHTHYOMYZON SP. *LAMPETRA LAMOTTII CANADA AMMOCOE | 5104F |

| | | |
|---|---|---|
| *TETRAPLEURODON SPADICEUS * | LAMPETRA *LETHENTERON ANATOMY SYSTEMATICS /MTATUS | 4752F |
| FISHES OF ARKANSAS USA. + * | LAMPETRA AEPYPTERA *LAMPETRA LAMOTTEI /IHREATENED | 5032F |
| LT FEEDING PHYSIOLOGY / ADU | LAMPETRA AEPYPTERA *LAMPETRA LAMOTTEI AMMOCOETE AD | 5466F |
| LOGY NERVOUS SYSTEM /PHYSIO | LAMPETRA AEPYPTERA *LAMPETRA LAMOTTEI U.S.A. PHYSI | 5606F |
| VIRGINIA U.S.A. /UNTY WEST | LAMPETRA AEPYPTERA IN LYNN CREEK WAYNE COUNTY WEST | 5284 |
| ER WEST VIRGINIA U.S.A. + * | LAMPETRA AEPYPTERA PISCES /OE FISHES OF GAULEY RIV | 4994F |
| ITATS AND DISTRIBUTION. + * | LAMPETRA AEPYPTERA U.S.A. AMMOCOETE DISTRIBUTION / | 5089F |
| /RESPIRATIONNERVOUS SYSTEM | LAMPETRA AEPYPTERA U.S.A. RESPIRATIONNERVOUS SYSTE | 4788F |
| IDAE) A DISTINCT TAXON? + * | LAMPETRA AEPYPTERA U.S.A. SYSTEMATICS /OPETROMYZON | 4768F |
| USA. + *LAMPETRA LAMOTTEI * | LAMPETRA AEPYPTERA U.S.A. SYSTEMATICS ANIMAL /LAS | 5035F |
| NSTRUCTION OF THE LAMPREY, | LAMPETRA AEPYPTERA. + U.S.A. SPAWNING ADULT /IT CO | 4653F |
| TISM BY /ATUS PISCES PARASI | LAMPETRA AYRESI *LAMPETRA TRIDENTATUS PISCES PARAS | 5364F |
| RGIA BY THE RIVER LAMPREY ( | LAMPETRA AYRESI). + AMMOCOETE ENDOCRINOLOGY / GEO | 5343F |
| STOCKS. + ADULT DISTRIBUTIO | LAMPETRA AYRESII) PREDATION ON SALMON AND HERRING | 4633F |
| NERVOUS SYSTEM SYNAPSE. + * | LAMPETRA BIOCHEMISTRY /IR AT A VERTEBRATE CENTRAL | 4580F |
| NE ATAMIPHYSIOLOGY BIOCHEMI | LAMPETRA ENTOSPHENUS *LAMPETRA JAPONICUS *PARAMYXI | 4919 |
| D NEURONS IN THE RETINA OF | LAMPETRA FLUVIATILIS (CYCLOSTOMI). / AND ASSOCIATE | 5001F |
| F THE GILLS OF THELAMPREY, | LAMPETRA FLUVIATILIS (L.) /SON MICROSCOPIC STUDY O | 5587 |
| RIVATIVE, LAMPETRA PLANERI | LAMPETRA FLUVIATILIS (L.) AND ITS NON-PARASITIC LA | 5300F |
| CH). + GREAT BRITAIN ADULT | LAMPETRA FLUVIATILIS (L.) AND LAMPETRA PLANERI(BLO | 5142F |
| ON. + SWEDEN ADULT ECOLOGY | LAMPETRA FLUVIATILIS (L.) DURING THE SPAWNING SEAS | 5304F |
| FERENCE TO AZUROPHIL LEUCOC | LAMPETRA FLUVIATILIS (L.) GRAY) WITH PARTICULAR RE | 5131F |
| ROVINCE OF VASTERBOTTEN, SW | LAMPETRA FLUVIATILIS (L.)) IN THE RICKLED RIVER, P | 5141F |
| PAWNING MIGRATION. + MIGRAT | LAMPETRA FLUVIATILIS (L.), DURING THE ANADROMOUS S | 5115F |
| COMPARATIVE STUDY. + AMMOC | LAMPETRA FLUVIATILIS (L.): A CYTOARCHITECTURAL AND | 4559F |
| .S.R. FISHERIES SYSTEMATICS | LAMPETRA FLUVIATILIS (LINNE) *LAMPETRA PLANERI U.S | 5321F |
| ETRA JAPONICA) *MYXINE GLUT | LAMPETRA FLUVIATILIS * LETHENTERON JAPONICUM (LAMP | 4770F |
| N MARINUS *MORDACIA MORDAX | LAMPETRA FLUVIATILIS *EPTATRETUS BURGERI*PETROMYZO | 4706F |
| HEMISTRY /TOUTII BLOOD BIOC | LAMPETRA FLUVIATILIS *EPTATRETUS STOUTII BLOOD BIO | 4834F |
| STOLOGY MUSCLE CYTOLOGY /HI | LAMPETRA FLUVIATILIS *LAMPETRA PLANERI AMMOCOETE A | 4884 |
| T AMMOCOETE CULTURE /M ADUL | LAMPETRA FLUVIATILIS *LAMPETRA PLANERI BELGIUM ADU | 5357F |
| DOM LARVA DIGESTION ENDOCRI | LAMPETRA FLUVIATILIS *LAMPETRA PLANERI UNITED KING | 5562F |
| N *LAMPETRA LAMOTTENI *LAMP | LAMPETRA FLUVIATILIS *LAMPETRA PLANERI*ICHTHYOMYZO | 4713 |
| JAPONICUM NORWAY ANIMAL SY | LAMPETRA FLUVIATILIS *LAMPETRA PLANERI*LETHENTERON | 5395F |
| JAPONICUM NORWAY ANIMAL SY | LAMPETRA FLUVIATILIS *LAMPETRA PLANERI*LETHENTERON | 5469F |
| N CANADA ADULT AMMOCOETE CI | LAMPETRA FLUVIATILIS *LAMPETRA PLANERIGREAT BRITIA | 4651F |
| ATAMI JA /NOSA *PARAMYXINE | LAMPETRA FLUVIATILIS *MYXINE GLUTINOSA *PARAMYXINE | 5349 |
| OCHEMISTRY CYTOLOGY /Y HIST | LAMPETRA FLUVIATILIS *MYXINE GLUTINOSA BIOLOGY HIS | 5265F |
| ATAMI JAPAN GONADOGENESIS | LAMPETRA FLUVIATILIS *MYZINE GLUTINOSA *PARAMYXINE | 5347F |
| PLANERIANATOMY OLFACTION S | LAMPETRA FLUVIATILIS *PETROMYZON MARINUS *LAMPETRA | 5047F |
| ADULTMUSCLE ANATOMY HISTOL | LAMPETRA FLUVIATILIS *PETROMYZON MARINUS AMMOCOETE | 4840F |
| IN A STREAM ECOSYSTEM.) + * | LAMPETRA FLUVIATILIS /NSYSTEM. [FISH-EATING BIRDS | 4672F |
| RATION /GY DISTRIBUTION MIG | LAMPETRA FLUVIATILIS ADULT BIOLOGY DISTRIBUTION MI | 5506F |
| BIOCHEMISTRY /NDOCRINOLOGY | LAMPETRA FLUVIATILIS ADULT EVOLUTION ENDOCRINOLOGY | 4725F |
| S IN THE RIVER LAMPREY. + * | LAMPETRA FLUVIATILIS ADULT GONADOGENESIS /GOGENESI | 4915F |
| / NERVOUS SYSTEM EVOLUTION | LAMPETRA FLUVIATILIS ADULT NERVOUS SYSTEM EVOLUTIO | 4540F |
| /OMY ANIMAL NERVOUS SYSTEM | LAMPETRA FLUVIATILIS ANATOMY ANIMAL NERVOUS SYSTEM | 4697F |
| + U.K. AMMOCOETE ADULT RES | LAMPETRA FLUVIATILIS AND LAMPETRA PLANERI (BLOCH). | 4877F |
| LIFE CYCLE PHYSIOLOGY /OOD | LAMPETRA FLUVIATILIS AND LAMPETRA PLANERI. + BLOOD | 5301F |
| ILLS OF AMMOCOETE LAMPREYS | LAMPETRA FLUVIATILIS AND LAMPETRA PLANERI. /R THEG | 4991 |
| PILLARIES. + *LAMPETRA FLUV | LAMPETRA FLUVIATILIS AND THEIR RELATION TOBLOOD CA | 4876F |
| N CERTAIN VERTEBRATES.] + * | LAMPETRA FLUVIATILIS ANIMAL BIOCHEMISTRY VISION /R | 4817F |
| LOBINS AND MYOGLOBINS.] + * | LAMPETRA FLUVIATILIS ANIMAL BLOOD BIOCHEMISTRY / G | 5128F |
| OUS SYSTEM PHYSIOLOGY /NERV | LAMPETRA FLUVIATILIS ANIMAL CIRCULATORY SYSTEM NER | 5162 |
| THE VERTEBRATE KIDNEY.] + * | LAMPETRA FLUVIATILIS ANIMAL IONIC REGULATION /AIN | 4670F |
| HISTOLOGY LOCOMOTION ANATOM | LAMPETRA FLUVIATILIS ANIMAL MUSCLE NERVOUS SYSTEM | 5079F |
| ROXIDASE ALONG THE OPTIC NE | LAMPETRA FLUVIATILIS AS REVEALED BY HORSERADISH PE | 5279 |
| TECHNIQUES / ENDOCRINOLOGY | LAMPETRA FLUVIATILIS BALTIC SEA ADULT ENDOCRINOLOG | 4721F |
| /EMISTRY BIOLOGY EVOLUTION | LAMPETRA FLUVIATILIS BIOCHEMISTRY BIOLOGY EVOLUTIO | 4982 |
| /HEMISTRY EVOLUTION MUSCLE | LAMPETRA FLUVIATILIS BIOCHEMISTRY EVOLUTION MUSCLE | 5305 |
| SYSTEM / EVOLUTION NERVOUS | LAMPETRA FLUVIATILIS BIOCHEMISTRY EVOLUTION NERVOU | 4741F |
| /HEMISTRY EXCRETION ANIMAL | LAMPETRA FLUVIATILIS BIOCHEMISTRY EXCRETION ANIMAL | 4742F |
| ERA). + *MYXINE GLUTINOSA * | LAMPETRA FLUVIATILIS BIOCHEMISTRY INTEGUMENT / IMA | 5019F |
| /HEMISTRY MUSCLE EVOLUTION | LAMPETRA FLUVIATILIS BIOCHEMISTRY MUSCLE EVOLUTION | 4668 |
| D SERUM OF VERTEBRATES. + * | LAMPETRA FLUVIATILIS BLOOD IONIC REGULATION /EBLOO | 5204 |
| TE ADULT SPAWNING LIFE CYCL | LAMPETRA FLUVIATILIS CANADA OSMOREGULATION AMMOCOE | 4599F |
| OF PHOTORECEPTORS AND ASSOC | LAMPETRA FLUVIATILIS CYCLOSTOMI. + FINE STRUCTURE | 5001F |
| STOLOGY ADULT /EN VISION HI | LAMPETRA FLUVIATILIS CYCLOSTOMI. + SWEDEN VISION H | 4655F |
| OWTHADULT GONADOGENESIS /GR | LAMPETRA FLUVIATILIS DENMARK BLOOD ENDOCRINOLOGY G | 5126F |
| EVOLUTION LIFE CYCLE /LOOD | LAMPETRA FLUVIATILIS ENGLAND AMMOCOETE ADULT BLOOD | 4584F |
| M DIGESTION /LOGY METABOLIS | LAMPETRA FLUVIATILIS EVOLUTION PHYSIOLOGY METABOLI | 4947F |
| /ND ADULT IONIC REGULATION | LAMPETRA FLUVIATILIS FINLAND ADULT IONIC REGULATIO | 4952F |
| ISTRY MORPHOLOGY /HISTOCHEM | LAMPETRA FLUVIATILIS FINLAND ADULT MUSCLE HISTOCHE | 4876F |
| CTION, MECHANISM, RATE. + * | LAMPETRA FLUVIATILIS GENETICS /INA STRUCTURE: DIRE | 5401F |
| STOLOGYHISTOCHEMISTRY /Y HI | LAMPETRA FLUVIATILIS GREAT BRITIAN ENDOCRINOLOGY H | 4748F |
| ION PHYSIOLOGY /LOGY DIGEST | LAMPETRA FLUVIATILIS HISTOLOGY ENDOCRINOLOGY DIGES | 5558 |
| CHEMISTRY /OCRINOLOGY HISTO | LAMPETRA FLUVIATILIS L. + ADULT ENDOCRINOLOGY HIST | 4941F |
| OLOGY IONIC REGULATION REPR | LAMPETRA FLUVIATILIS L. + GREAT BRITAIN ADULT HIST | 5169F |
| RY /LAND ADULT HISTOCHEMIST | LAMPETRA FLUVIATILIS L. + POLAND ADULT HISTOCHEMIS | 4778F |
| CRINOLOGY MIGRATION /T ENDO | LAMPETRA FLUVIATILIS L. + UNITED KINGDOM ADULT END | 5176F |
| TEM BIOCHEMISTRY /ATORY SYS | LAMPETRA FLUVIATILIS L. ADULT BLOOD CIRCULATORY SY | 4760F |

BLOOD GLUCOSE REGULATION,
N ADULT PARASITISM OF /ITAI
STOMATA AND BONY FISH.] + *
STEM PHYSIOLOGY CYTOLOGY AD
RELATED PROCESSES. + BALTIC
GY MIGRATION /D ENDOCRINOLO
A CARPE ET LA LAMPROIE. + *
CLE NERVOUS SYSTEM /OGY MUS
TION /ATOMY EXCRETION EVOLU
TOLOGY MUSCLE /Y BIOLOGY CY
STEM HISTOLOGY NERVOUS SYST
US SYSTEM PHYSIOLOGY /NERVO
M MUSCLE /INOLOGY METABOLIS
RIVER LAMPREY KIDNEY.] + *
EM PHYSIOLOGY /NERVOUS SYST
BUTION LOCOMOTION MIGRATION
 SYSTEMVISION /LOGY NERVOUS
GY DIGESTION /A ENDOCRINOLO
SE RECEPTORS /US SYSTEM SEN
-152. + *MYXINE GLUTINOSA *
YTOLOGYSWEDEN ADULT /STEM C
RGLUCOSE BEFORE AND DURING
ISTOCHEMISTRY PISCES /TRY H
Y ENDOCRINOLOGY MUSCLE /STR
LOGY /NERVOUS SYSTEM PHYSIO
ICS.] + RUSSIA IONIC REGULA
GROWTH LIFE CYCLE MIGRATION
INFAST MUSCLE FIBERS OF THE
RONMICROSCOPIC STUDIES OF P
HE NEPHRON IN THE LAMPREY L
LYCEROL IN ISOLATED TISSUES
NAMYLASE ACTIVITY IN THE LI
CHTHYOMYZON HUBBSI *LAMPETR
OCHEMISTRY ENDOCRINOLOGY MI
OOD HISTOLOGY LIFE CYCLE PH
OOD HISTOLOGY LIFE CYCLE PH
 SYSTEM RESPIRATION CIRCADI
IRATION TECHNIQUES /GY RESP
ION / HISTOLOGY OSMOREGULAT
CRINOLOGY HISTOCHEMISTRY IM
ROM MUSCLES OF THE LAMPREY
SIOLOGY /LOGY HISTOLOGY PHY
GRATION TECHNIQUES ADULT /I
/UROPE ADULT ENDOCRINOLOGY
INOLOGY EVOLUTION BLOOD GON
LT BLOOD CYTOLOGY /OETE ADU
IOLOGY /DULT MIGRATION PHYS
 HISTOLOGY / NERVOUS SYSTEM
ISM /A BIOCHEMISTRY METABOL
SIAL SYSTEM OF THE LAMPREY
ORPHIC LAMPREYS OF THE
... PP. 217-250. + SWEDEN H
USCLE HISTOCHEMISTRY MORPHO
KE VENOM POLYPEPTIDES ON CH
SCLE /EMISTRY METABOLISM MU
Y HISTOLOGY NERVOUS SYSTEM
HEMISTRY /ULT VISION HISTOC
LISM MUSCLE /EMISTRY METABO
Y ENDOCRINOLOGY /IOCHEMISTR
/.S.S.R. ANATOMY EXCRETION
LOGY ANIMAL / SYSTEM PHYSIO
LUTINOSA *EPTATRETUS STOUTI
N CORTICOTROPIN. / MAMMALIA
EXUAL MATURATION. /IOD OF S
 *PETROMYZON MARINUS *LAMPE
 LIFE CYCLE.] + RUSSIA ADUL
GY /TOMY HISTOLOGY PHYSIOLO
VISION / SYSTEM PHYSIOLOGY
Y HISTOLOGY VISION /HEMISTR
CRINOLOGY FEEDING METABOLIS
AND FEMALE RIVER LAMPREYS (
TURITY /57. + ADULT LIFE MA
WEDEN ADULT GONADOGENESIS H
US, AND CNICKS.] + ENDOCRIN
TSYPLYAT. [THE EFFECT OF VA
INOSA *PETROMYZON MARINUS *
ANERI *PETROMYZON MARINUS *
FAUNA OF THE RIVER LAMPREY
LAMPETRA PLANERI U.S.S.R. F
Y OF LAMPREYS OF THE GENUS
AMPETRA FLUVIATILIS (LINNE)

|  | |
|---|---|
| LAMPETRA FLUVIATILIS L.): EFFECTS ON REGENERATION, | 5509 |
| LAMPETRA FLUVIATILIS LINNAEUS, 1758. + GREAT BRITA | 5061F |
| LAMPETRA FLUVIATILIS MUSCLE METABOLISM /R OF CYCLO | 4887F |
| LAMPETRA FLUVIATILIS NERVOUS SYSTEM CIRCULATORY SY | 4874 |
| LAMPETRA FLUVIATILIS OXIDATIVEPHOSPHORYLATION AND | 4613F |
| LAMPETRA FLUVIATILIS RUSSIA ADULT BLOOD ENDOCRINOL | 5361F |
| LAMPETRA FLUVIATILIS RUSSIA ADULT ENDOCRINOLOGY / | 5468F |
| LAMPETRA FLUVIATILIS RUSSIA AMMOCOETE HISTOLOGY MU | 5317 |
| LAMPETRA FLUVIATILIS RUSSIA ANATOMY EXCRETION EVOL | 5064F |
| LAMPETRA FLUVIATILIS RUSSIA BIOCHEMISTRY BIOLOGY C | 5278 |
| LAMPETRA FLUVIATILIS RUSSIA BIOLOGY CIRCULATORY SY | 5504 |
| LAMPETRA FLUVIATILIS RUSSIA BIOLOGY HISTOLOGY NERV | 5505 |
| LAMPETRA FLUVIATILIS RUSSIA ENDOCRINOLOGY METABOLI | 5123F |
| LAMPETRA FLUVIATILIS RUSSIA EXCRETION /PULE OF THE | 5354F |
| LAMPETRA FLUVIATILIS RUSSIA METABOLISM NERVOUS SYS | 5281F |
| LAMPETRA FLUVIATILIS SWEDEN ADULT CIRCADIAN DISTRI | 5141F |
| LAMPETRA FLUVIATILIS SWEDEN ADULT HISTOLOGY NERVOU | 5214F |
| LAMPETRA FLUVIATILIS SWEDEN ADULT LARVA ENDOCRINOL | 5561F |
| LAMPETRA FLUVIATILIS SWEDEN ADULTNERVOUS SYSTEM SE | 4949F |
| LAMPETRA FLUVIATILIS SWEDEN ENDOCRINOLOGY /NP. 141 | 5288F |
| LAMPETRA FLUVIATILIS SWEDEN VISION NERVOUS SYSTEM | 5039 |
| LAMPETRA FLUVIATILIS TREATED WITH INSULIN STRESS O | 5062F |
| LAMPETRA FLUVIATILIS U.S.S.R. ANIMAL BIOCHEMISTRY | 4803F |
| LAMPETRA FLUVIATILIS U.S.S.R. BEHAVIOUR BIOCHEMIST | 4598F |
| LAMPETRA FLUVIATILIS U.S.S.R. NERVOUS SYSTEM PHYSI | 4601F |
| LAMPETRA FLUVIATILIS UNDER THE INFLUENCE OF DIURET | 5210F |
| LAMPETRA FLUVIATILIS UNITED KINGDOM ADULT ECOLOGY | 5009 |
| LAMPETRA FLUVIATILIS. [ACETYLCHOLINE CONTRACTURES | 5094F |
| LAMPETRA FLUVIATILIS. [LIGHT MICROSCOPIC AND ELECT | 5173 |
| LAMPETRA FLUVIATILIS. [MICRO DISSECTION STUDY ON T | 4694F |
| LAMPETRA FLUVIATILIS. [SYNTHESIS OF GLYCOGEN FROMG | 5360F |
| LAMPETRA FLUVIATILIS. [THE EFFECT OF EPINEPHRINE O | 4561F |
| LAMPETRA FLUVIATILIS. + *ICHTHYOMYZON UNICUSPIS *I | 4678F |
| LAMPETRA FLUVIATILIS. + *LAMPETRA PLANERI ADULT BI | 4886F |
| LAMPETRA FLUVIATILIS. + ADULT AMMOCOETE BIOLOGY BL | 4716F |
| LAMPETRA FLUVIATILIS. + ADULT AMMOCOETE BIOLOGY BL | 4717 |
| LAMPETRA FLUVIATILIS. + ADULT BEHAVIOR CIRCULATORY | 4719F |
| LAMPETRA FLUVIATILIS. + ADULT BLOOD PHYSIOLOGY RES | 5002F |
| LAMPETRA FLUVIATILIS. + ADULT HISTOLOGY OSMOREGULA | 5101F |
| LAMPETRA FLUVIATILIS. + BEGIUM ADULT CYTOLOGY ENDO | 5312 |
| LAMPETRA FLUVIATILIS. + BIOCHEMISTRY MUSCLE /RSE F | 5276 |
| LAMPETRA FLUVIATILIS. + ENDOCRINOLOGY HISTOLOGY PH | 5517 |
| LAMPETRA FLUVIATILIS. + ENDOCRINOLOGY PHYSIOLOGY M | 5005 |
| LAMPETRA FLUVIATILIS. + EUROPE ADULT ENDOCRINOLOGY | 4797F |
| LAMPETRA FLUVIATILIS. + GREAT BRITAIN ADULT ENDOCR | 5065F |
| LAMPETRA FLUVIATILIS. + GREAT BRITIAN AMMOCOETE AD | 4649F |
| LAMPETRA FLUVIATILIS. + RUSSIA ADULT MIGRATION PHY | 5218F |
| LAMPETRA FLUVIATILIS. + RUSSIA ADULT NERVOUS SYSTE | 4973 |
| LAMPETRA FLUVIATILIS. + RUSSIA BIOCHEMISTRY METABO | 5313 |
| LAMPETRA FLUVIATILIS. / /THE HYPOTHALAMO-HYPOPHY | 4586 |
| LAMPETRA FLUVIATILIS. /Y IN THYROIDS OF POST METAM | 5598 |
| LAMPETRA FLUVIATILIS. A LIGHT MICROSCOPICAL STUDYW | 5285F |
| LAMPETRA FLUVIATILIS. PP. 21-22. + ADULT ANATOMY M | 4841F |
| LAMPETRA FLUVIATILIS.[THE INHIBITORY EFFECT OF SNA | 4636F |
| LAMPETRA FLUVIATILIS.] + BIOCHEMISTRY METABOLISM M | 5191 |
| LAMPETRA FLUVIATILIS.] + RUSSIA ADULT ENDOCRINOLOG | 5071F |
| LAMPETRA FLUVIATILIS.] + RUSSIA ADULT VISION HISTO | 5173 |
| LAMPETRA FLUVIATILIS.] + RUSSIA BIOCHEMISTRY METAB | 5360F |
| LAMPETRA FLUVIATILIS.] + U.S.S.R. ADULT BIOCHEMIST | 4561F |
| LAMPETRA FLUVIATILIS.] + U.S.S.R. ANATOMY EXCRETIO | 4694F |
| LAMPETRA FLUVIATILIS.] + USSR NERVOUS SYSTEM PHYSI | 4636F |
| LAMPETRA FLUVIATILIS*LAMPETRA TRIDENTATA *MYXINE G | 4705F |
| LAMPETRA FLUVIATILIS) AFTER TREATMENT WITH MAMMALI | 5527 |
| LAMPETRA FLUVIATILIS) BEFORE AND DURING PERIOD OF | 5234 |
| LAMPETRA FLUVIATILIS) DURING THEIR SPAWNING RUN. + | 5138F |
| LAMPETRA FLUVIATILIS) ON DIFFERENT STAGES OF THEIR | 5323F |
| LAMPETRA FLUVIATILIS). + ANATOMY HISTOLOGY PHYSIOL | 5233F |
| LAMPETRA FLUVIATILIS). + NERVOUS SYSTEM PHYSIOLOGY | 5299F |
| LAMPETRA FLUVIATILIS). + SWEDEN ADULT HISTOCHEMIST | 5310F |
| LAMPETRA FLUVIATILIS). + UNITED KINGDOM ADULT ENDO | 5306F |
| LAMPETRA FLUVIATILIS). /UTOSTERONE ON INTACT MALE | 5607F |
| LAMPETRA FLUVIATILIS). PP. 156-157. + ADULT LIFE M | 4868F |
| LAMPETRA FLUVIATILIS, DURING GONAD MATURATION. + S | 4646F |
| LAMPETRA FLUVIATILIS, SCORPION-FISH SCORPAENA PORC | 5267F |
| LAMPETRA FLUVIATILIS, SKORPENY SCORPAENA PORCUS I | 5267F |
| LAMPETRA FLUVIATILISADULT VISION EVOLUTION /L GLUT | 4961F |
| LAMPETRA FLUVIATILISEAST GERMANY DISTRIBUTION /RPL | 5186F |
| LAMPETRA FLUVIATILISFROM THE KURSKI. / /PARASITE | 4664 |
| LAMPETRA FLUVIATIS *LAMPETRA FLUVIATILIS (LINNE) * | 5321F |
| LAMPETRA IN OREGON. /MION, DISTRIBUTION AND BIOLOG | 5554 |
| LAMPETRA JAPONICA (KESSLERI) LAMPETRA FLUVIATIS *L | 5321F |

SSLERI) LAMPETRA FLUVIATIS | **LAMPETRA** JAPONICA (MARTENS) *LAMPETRA JAPONICA (KE | 5321F
ART OF THE RIVER UMBRA. + * | **LAMPETRA** JAPONICA (MARTENS) CYTOLOGY METABOLISM /. | 4622F
LAMPREY GILL FILAMENT. + * | **LAMPETRA** JAPONICA CIRCULATORY SYSTEM ADULT /CF THE | 5519
GY /T HISTOLOGY ENDOCRINOLO | **LAMPETRA** JAPONICA JAPAN ADULT HISTOLOGY ENDOCRINOL | 5029F
/ULT RESPIRATION HISTOLOGY | **LAMPETRA** JAPONICA JAPAN ADULT RESPIRATION HISTOLOG | 5540F
AMMOCOETE ADULT HISTOLOGY / | **LAMPETRA** JAPONICA JAPAN AMMOCOETE ADULT HISTOLOGY | 5356F
LAMPREY CHLORIDE CELLS. + * | **LAMPETRA** JAPONICA JAPAN HISTOLOGY CYTOLOGY /DS OF | 4603F
H BASIN.] + RUSSIA ADULT AN | **LAMPETRA** JAPONICA KESSLERI (ANIKIN) OF UPPER IRTYS | 4623F
NEGO IRTYSHA. [MORPHOLOGICA | **LAMPETRA** JAPONICA KESSLERI (ANIKIN) VODOEMOV VERKH | 4623F
RA PLANERI AMMOCOETE HISTOL | **LAMPETRA** JAPONICA. + *LAMPETRA FLUVIATILIS *LAMPET | 4884
HISTOLOGY ENDOCRINOLOGY /T | **LAMPETRA** JAPONICA. + *LAMPETRA JAPONICA JAPAN ADUL | 5029F
HISTOLOGY / NERVOUS SYSTEM | **LAMPETRA** JAPONICA. + ADULT CYTOLOGY NERVOUS SYSTEM | 5596
RICULAR WALL INTHE LAMPREY | **LAMPETRA** JAPONICA. + ADULT HISTOLOGY /THE 3RD VENT | 5187F
S /NERVOUS SYSTEM TECHNIQUE | **LAMPETRA** JAPONICA. + ADULT NERVOUS SYSTEM TECHNIQU | 5572F
OLOGY /T ENDOCRINOLOGY HIST | **LAMPETRA** JAPONICA. + JAPAN ADULT ENDOCRINOLOGY HIS | 4587F
OCRINOLOGY /VOUS SYSTEM END | **LAMPETRA** JAPONICA. + JAPAN ADULT NERVOUS SYSTEM EN | 4882F
ISTOLOGY /TE ADULT MUSCLE H | **LAMPETRA** JAPONICA. + JAPAN AMMOCOETE ADULT MUSCLE | 5185F
TAL MUSCLE OF THE LAMPREY, | **LAMPETRA** JAPONICA. + JAPAN MUSCLE HISTOLOGY /TKELE | 4644F
YSTEMATICS BIOLOGY /UTION S | **LAMPETRA** JAPONICA) *MYXINE GLUTINOSA DISTRIBUTION | 4770F
HEMISTRY /MIPHYSIOLOGY BIOC | **LAMPETRA** JAPONICUS *PARAMYXINE ATAMIPHYSIOLOGY BIO | 4919
.S.S.R. /STERN KAZAKH-SSR U | **LAMPETRA** KESSLERI IN RIVERS OF EASTERN KAZAKH-SSR | 5318
E DELMARVA PENINSULA (EAST | **LAMPETRA** LAMOTTEI (PISCES: PETROMYZONIDAE), ON TH | 5139F
YZON CASTANEUS U.S.A. ADULT | **LAMPETRA** LAMOTTEI *ICHTHYOMYZON BDELLIUM *ICHTHYOM | 4590F
T GROWTH AMMOCOETE /NAGEMEN | **LAMPETRA** LAMOTTEI *ICHTHYOMYZON SP CANADA MANAGEME | 5110F
ANADA AMMOCOETE CHEMISTRY M | **LAMPETRA** LAMOTTEI *ICHTHYOMYZON SPP. SEA LAMPREY C | 5223F
ATICS ANIMAL /U.S.A. SYSTEM | **LAMPETRA** LAMOTTEI *LAMPETRA AEPYPTERA U.S.A. SYSTE | 5035F
ETE / MARINUS CANADA AMMOCO | **LAMPETRA** LAMOTTEI *PETROMYZON MARINUS CANADA AMMOC | 4890F
SA. + *LAMPETRA AEPYPTERA * | **LAMPETRA** LAMOTTEI /IHREATENED FISHES OF ARKANSAS U | 5032F
Y / ADULT FEEDING PHYSIOLOG | **LAMPETRA** LAMOTTEI AMMOCOETE ADULT FEEDING PHYSIOLO | 5466F
WTH /COETE DISTRIBUTION GRO | **LAMPETRA** LAMOTTEI CANADA AMMOCOETE DISTRIBUTION GR | 4645F
H /MOCOETE MANAGEMENT GROWT | **LAMPETRA** LAMOTTEI CANADA AMMOCOETE MANAGEMENT GROW | 5109F
H /NAGEMENT AMMOCOETE GROWT | **LAMPETRA** LAMOTTEI CANADA MANAGEMENT AMMOCOETE GROW | 5111F
/PHYSIOLOGY NERVOUS SYSTEM | **LAMPETRA** LAMOTTEI U.S.A. PHYSIOLOGY NERVOUS SYSTEM | 5606F
MMOCOETE SYSTEMATICS MORPHO | **LAMPETRA** LAMOTTENI *LAMPETRA LETHENTERON BRITAIN A | 4713
MATODA: CUCULLANIDAE) FROM | **LAMPETRA** LAMOTTENII (LESUEUR, 1827). /O, 1910) (NE | 5327F
THE DELMARVA PENINSULA (EA | **LAMPETRA** LAMOTTENII ANDOKKELBERGIA AEPYPTERA, FROM | 4720F
SITISM OF /T AMMOCOETE PARA | **LAMPETRA** LAMOTTENII). + CANADA ADULT AMMOCOETE PAR | 5541F
ANAGEMENT /EMISTRY GROWTH M | **LAMPETRA** LAMOTTICANADA AMMOCOETE CHEMISTRY GROWTH | 5491F
/MMOCOETE GROWTH CHEMISTRY | **LAMPETRA** LAMOTTII CANADA AMMOCOETE GROWTH CHEMISTR | 5104F
/MMOCOETE GROWTH CHEMISTRY | **LAMPETRA** LAMOTTII CANADA AMMOCOETE GROWTH CHEMISTR | 5105F
EINZUGSGEBIET DES SCHWARZE | **LAMPETRA** LANCEOLATA, EINE NEUE NEUNAGENART AUS DEM | 5581F
MORPHOMETRY PIGMENTATION M | **LAMPETRA** LETHENTERON BRITAIN AMMOCOETE SYSTEMATICS | 4713
MUSCLE /MATICS MORPHOMETRY | **LAMPETRA** LETHOPHAGA U.S.A. SYSTEMATICS MORPHOMETRY | 5555
THE UPPER PRIPYAT RIVER. + | **LAMPETRA** MARIAE ECOLOGY /SON OF FISH POPULATION IN | 4873
SEA ADULT AMMOCOETE DISTRIB | **LAMPETRA** MARIAE U.S.S.R. UKRAINE BALTIC SEA BLACK | 5358F
LITATIVE INVESTIGATIONS OF | **LAMPETRA** PLANERI (BLOCH 1784) (CYCLOSTOMATA). [QUA | 4700F
N THE PROLONGED SURVIVAL OF | **LAMPETRA** PLANERI (BLOCH) IN ABRUZZO. OBSERVATION O | 4807F
SULLA PROLUNGATA SOPRAVVIVE | **LAMPETRA** PLANERI (BLOCH) IN ABRUZZO. OSSERVAZIONI | 4807F
D'ALIMENTATION. COMPARISON | **LAMPETRA** PLANERI (BLOCH) PAR UNE NOUVELLE METHODE | 5352F
(REGAN). /TOMYZON DANFORDI | **LAMPETRA** PLANERI (BLOCH) UND EUDONTOMYZON DANFORDI | 5331
THYROGLOBULIN IN AN ADULT L | **LAMPETRA** PLANERI (BLOCH). [ON THE BIOSYNTHESIS OF | 5025F
Y AND HISTOCHEMISTRY OF MYO | **LAMPETRA** PLANERI (BLOCH). [RESEARCHON THE HISTOLOG | 4619F
TISM OF ADULT /COETE PARASI | **LAMPETRA** PLANERI (BLOCH). + SWEDEN AMMOCOETE PARAS | 5467F
SPIRATION /MOCOETE ADULT RE | **LAMPETRA** PLANERI (BLOCH). + U.K. AMMOCOETE ADULT R | 4877F
MMOCOETE EVOLUTION METAMORP | **LAMPETRA** PLANERI (BLOCH). + UNITED KINGDOM ADULT A | 5300F
THE EXPERIMENTAL AND NATURA | **LAMPETRA** PLANERI (BLOCH). COMPARISON BETWEEN BOTH | 5352F
CIRCULATORY SYSTEM HISTOLO | **LAMPETRA** PLANERI (BLOCH).] + ITALY AMMOCOETE ADULT | 4619F
BULIN IN AN ADULT LAMPREY, | **LAMPETRA** PLANERI (BLOCH)] /OOSYNTHESIS OF THYROGLO | 5025F
OMANIA. + *EUDONTOMYZON DAN | **LAMPETRA** PLANERI (BLOCH, 1784) (CYCLOSTOMATA) IN R | 4541F
OLOGY GROWTH SPAWNING /T BI | **LAMPETRA** PLANERI (BLOCH,1784). + AMMOCOETE ADULT B | 4764F
DULT AMMOCOETE ECOLOGY FECU | **LAMPETRA** PLANERI *ENTOSPHENUS TRIDENTATUS CANADA A | 4974F
ZONVLADYKOV BRATISLAVA DIST | **LAMPETRA** PLANERI *EUDONTOMYZON DANFORDI *EUDONTOMY | 5175F
REGULATION AMMOCOETE ADULT | **LAMPETRA** PLANERI *LAMPETRA FLUVIATILIS CANADA OSMO | 4599F
OCOETE ADULT BLOOD EVOLUTIO | **LAMPETRA** PLANERI *LAMPETRA FLUVIATILIS ENGLAND AMM | 4584F
T LARVA ENDOCRINOLOGY DIGES | **LAMPETRA** PLANERI *LAMPETRA FLUVIATILIS SWEDEN ADUL | 5561F
IDENTATA *MYXINE GLUTINOSA | **LAMPETRA** PLANERI *LAMPETRA FLUVIATILIS*LAMPETRA TR | 4705F
BALTIC SEA BLACK SEA ADULT | **LAMPETRA** PLANERI *LAMPETRA MARIAE U.S.S.R. UKRAINE | 5358F
THENTERON) LAMOTTENII *ICHT | **LAMPETRA** PLANERI *PETROMYZON MARINUS *LAMPETRA (LE | 4565F
VIATILISEAST GERMANY DISTRI | **LAMPETRA** PLANERI *PETROMYZON MARINUS *LAMPETRA FLU | 5186F
IGRATION PHYSIOLOGY /LOGY M | **LAMPETRA** PLANERI ADULT BIOCHEMISTRY ENDOCRINOLOGY | 4886F
T AUSNAHME VON SAUGERN. + * | **LAMPETRA** PLANERI AMMOCOETE ANIMAL BLOOD /+IEREN MI | 5119F
Y /HISTOLOGY MUSCLE CYTOLOG | **LAMPETRA** PLANERI AMMOCOETE HISTOLOGY MUSCLE CYTOLO | 4884
] + *LAMPETRA FLUVIATILIS * | **LAMPETRA** PLANERI BELGIUM ADULT AMMOCOETE CULTURE / | 5357F
DULT /CHEMISTRY AMMOCOETE A | **LAMPETRA** PLANERI BELGIUM HISTOCHEMISTRY AMMOCOETE | 4734F
GLOBULIN IN THE LARVAE (AMM | **LAMPETRA** PLANERI BL. [ON THE BIOSYNTHESIS OF THYRO | 5342F
ORPHOSIS /DOCRINOLOGY METAM | **LAMPETRA** PLANERI BLOCH + FRANCE ENDOCRINOLOGY META | 5342F
SYSTEMS BEIM BACHNEUNAUGE ( | **LAMPETRA** PLANERI BLOCH). /FTOLOGIE DES INTERRENAL- | 5215
SYSTEMS DES BACHNEUNAUGES ( | **LAMPETRA** PLANERI BLOCH). /PINTERRENAL-UND ADRENAL- | 5563
ERVOUS SYSTEM PHYSIOLOGY CY | **LAMPETRA** PLANERI CYCLOSTOMATA.] + HISTOCHEMISTRY N | 4700F
TION / HISTOCHEMISTRY EVOLU | **LAMPETRA** PLANERI ENDOCRINOLOGY HISTOCHEMISTRY EVOL | 4706F
OMES. + *MYXINE GLUTINOSA * | **LAMPETRA** PLANERI ENDOCRINOLOGY HISTOLOGY /DCYCLOST | 4765F

RVAE OF TRICHOPTERA. /US LA
/IMAL BLOOD HISTOCHEMISTRY
PETRA FLUVIATILIS (LINNE) *
OCHEMISTRY METABOLISM METAM
OCRINOLOGY /A DIGESTION END
E POPRAD AND HORNAD RIVER B
S LAMPETRA FLUVIATILIS AND
H / DIGESTION FEEDING GROWT
ADULT FRESH WATER LAMPREY
TTAE IN THE BROOK LAMPREY,
S LAMPETRA FLUVIATILIS AND
E CYCLOSTOMEBRAIN. PART I:
OMYZONTIDAE). + POLAND PIGM
ESTION ECOLOGY FEEDING GROW
*LAMPETRA LETHENTERON BRITA
L SYSTEMATICS /NORWAY ANIMA
L SYSTEMATICS /NORWAY ANIMA
OLFACTION SENSE RECEPTORS /
E CIRCULATORY SYSTEM /OCOET
TOLOGY METAMORPHOSIS NERVOU
FISH FAUNA OF MONGOLIA. + *
STRY METABOLISM TECHNIQUES
GY /AMMOCOETE ADULT HISTOLO
BIOCHEMISTRY HISTOCHEMISTR
THE WESTERN BROOK LAMPREY,
A. /S IN THE AMMOCOETE LARV
ARKANSAS U.S.A. RECORDS OF
STEMATICS MORPHOMETRY MUSCL
PROTEIN IN VERTEBRATES. + *
GRATION /DULT METABOLISM MI
M LOCOMOTION NERVOUS SYSTEM
STOLOGY NERVOUS SYSTEM / HI
US SYSTEM ENDOCRINOLOGY /VO
EMISTRYPHYSIOLOGY /US BIOCH
1974. + *LAMPETRA AYRESI *
/. BLOOD ANIMAL PHYSIOLOGY
THYOMYZON *PETROMYZONIDAE *
LOR CHANGE IN THE LAMPREY,
ATILIS SWEDEN ADULT HISTOLO
OCRINOLOGY HISTOLOGY GONADO
M CHLORIDE AND SODIUM BICAR
OMYZONINAE *ENTOSPHENINAE *
ROMYZON MARINUS AMMOCOETE A
+ NEW ZEALAND LIFE CYCLE B
Y /+ AMMOCOETE ENDOCRINOLOG
HERRING STOCKS. + ADULT DI
RIVER, PROVINCE OF VASTERB
PHYSIOLOGY /TOMY HISTOLOGY
YSIOLOGY VISION / SYSTEM PH
TOCHEMISTRY HISTOLOGY VISIO
DULT ENDOCRINOLOGY FEEDING
OETE PARASITISM OF /T AMMOC
E OF HOST MORTALITY. + GREA
SALINITY. /TION TO AMBIENT
S SYSTEM VISION /ULT NERVOU
TOOTH DEVELOPMENT IN A SEA
PETROMYZON MARINUS ADULT EN
/AL ANATOMY HISTOCHEMISTRY
ATMENTS, 1971-1973. PP. 78.
TREATMENTS, 1971-1973. PP.
RA FLUVIATILIS BIOCHEMISTRY
UVIATILIS SWEDEN VISION NER
IDENTATUS ADULT ANIMALS END
                      /THE SEA
RY BLOOD / ADULT BIOCHEMIST
U.S.S.R. BEHAVIOUR BIOCHEM
ION TO THE FISHERY THE SEA
5-96. + *PETROMYZON MARINUS
. + *PETROMYZON MARINUS CAN
ENT /                  /THE
CULATORY SYSTEM CYTOLOGY HI
/SPECIFIC GRANULES IN THE
4. PP. 73-83. + CANADA MANA
INUS MANAGEMENT /OMYZON MAR
OF THE SPINAL CORD. /PARTS
EIR, 1975. PP.96-98. + SEA
RON, 1975. PP.31-44. + SEA
RIO, 1975. PP.44-55. + SEA
IOR, 1975. PP.68-75. + SEA
TEI *ICHTHYOMYZON SPP. SEA
CHEMISTRY MANAGEMENT MORTAL

| | | |
|---|---|---|
| LAR CYTOPLASMIC TUBULES OF | **LAMPREY** CELLS.HLORIDE CELLS. /OSTRUCTURE OF AGRANU | 4722 |
| HISTOLOGY CYTOLOGY / JAPAN | **LAMPREY** CHLORIDE CELLS. + *LAMPETRA JAPONICA JAPAN | 4603F |
| ONTARIO. PP. 68-69. + CANAD | **LAMPREY** COLLECTED BY COMMERCIAL FISHERMEN IN LAKE | 4962F |
| *PETROMYZON MARINUS GREAT | **LAMPREY** COLLECTED BY COMMERCIAL FISHERMEN. P.98. + | 5261F |
| .86. + *ICHTHYOMYZON SPP. A | **LAMPREY** COLLECTED BY COMMERCIAL FISHERMEN, 1975, P | 5225F |
| 85. + *PETROMYZON MARINUS | **LAMPREY** COLLECTED BY COMMERCIALFISHERMEN, 1973. P. | 4902F |
| ENT BARRIERS, LAKE HURON, 1 | **LAMPREY** COLLECTED FROM CANADIAN ELECTRICAL ASSESSM | 5487 |
| ERS, LAKE HURON, 1974. P. 6 | **LAMPREY** COLLECTED FROM ELECTRICAL ASSESSMENT BARRI | 5088F |
| ERS, LAKE HURON, 1975. PP.8 | **LAMPREY** COLLECTED FROM ELECTRICAL ASSESSMENT BARRI | 5226F |
| ERS, LAKE HURON. PP.94-97. | **LAMPREY** COLLECTED FROM ELECTRICAL ASSESSMENT BARRI | 5260F |
| RS: LAKE HURON, 1973. PP. 8 | **LAMPREY** COLLECTED FROM ELECTRICALASSESSMENT BARRIE | 4905F |
| 1. + *PETROMYZON MARINUS CA | **LAMPREY** COLLECTED IN THE HUMBER RIVER, 1972. P. 17 | 5486F |
| THE GREAT LAKES FISHERY CO | **LAMPREY** CONROL CENTRE SAULT STE. MARIE, ONTARIO TO | 4783F |
| TO THE GREAT LAKES FISHERY | **LAMPREY** CONTROL CENTRE SAULT STE. MARIE, ONTARIO, | 4784F |
| TO THE GREAT LAKES FISHERY | **LAMPREY** CONTROL CENTRE SAULT STE. MARIE, ONTARIO, | 4844F |
| TO THE GREAT LAKES FISHERY | **LAMPREY** CONTROL CENTRE SAULT STE. MARIE, ONTARIO, | 5194F |
| + *PETROMYZON MARINUS U.S. | **LAMPREY** CONTROL IN THE UNITED STATES. PP. 145-188. | 5368F |
| + *PETROMYZON MARINUS U.S. | **LAMPREY** CONTROL IN THE UNITED STATES. PP. 153-181. | 5502F |
| TORY MANAGEMENT MORTALITY O | **LAMPREY** CONTROL, DECEMBER 1976. + DISTRIBUTION HIS | 5231F |
| RY, 1960-72. PP. 223-271. / | **LAMPREY** EARLY LIFE HISTORY, 1960-72. PP. 223-271. | 5473F |
| E ADULT CULTURE EMBRYOLOGY | **LAMPREY** EGGS; LETTER TO P.J. ROBERTSON. + AMMOCOET | 5143F |
| AKE HURON, 1974. P. 5. + *P | **LAMPREY** ELECTRICAL ASSESSMENT BARRIER OPERATIONS,L | 4907F |
| LAKE HURON: 1975. P. 5. + C | **LAMPREY** ELECTRICAL ASSESSMENT BARRIER OPERATIONS,L | 4969F |
| AKE HURON. P.5. + *PETROMYZ | **LAMPREY** ELECTRICAL ASSESSMENT BARRIER OPERATIONS,L | 5246F |
| OMY HISTOLOGY CYTOLOGY /NAT | **LAMPREY** ENTOSPHENUS JAPONICUS. + ADULT HEARING ANA | 4663F |
| ION PHYSIOLOGY /LOOD EVOLUT | **LAMPREY** ENTOSPHENUS JAPONICUS. + JAPAN BLOOD EVOLU | 5023F |
| HE PROCESS OF BLOOD COAGULA | **LAMPREY** FACTOR XIII AND HUMAN FACTOR XIII DURING T | 5302F |
| ERN OF LAMPREY FIBRIN WITH | **LAMPREY** FACTOR XIII AND HUMAN FACTOR XIII. / PATT | 4682 |
| ACTOR XIII. /II AND HUMAN F | **LAMPREY** FIBRIN WITH LAMPREY FACTOR XIII AND HUMAN | 4682 |
| ION OF THE JUNCTIONS SPLIT | **LAMPREY** FIBRIN PEPTIDES A AND B AND CHARACTERIZAT | 5146F |
| ATUS U.S.A. BLOOD ANIMAL PH | **LAMPREY** FIBRINOGEN AND FIBRIN. + *LAMPETA TRIDENT | 4594F |
| S BLOOD HISTOCHEMISTRY /INU | **LAMPREY** FIBRINOGEN AND FIBRIN. + *PETROMYZON MARIN | 4960F |
| E FIBRIN BY PLASMA TRANSGLU | **LAMPREY** FIBRINOGEN AND FIBRIN: CROSS-LINKING OF TH | 4738 |
| O ACID SEQUENCE STUDIES ON | **LAMPREY** FIBRINOGEN FIBRINAND FIBRINO PEPTIDES. /IN | 4683 |
| TION PRODUCTS VIA ELECTRON | **LAMPREY** FIBRINOGEN INVESTIGATION OF THE POLYMERIZA | 5167 |
| /CHARACTERIZATION OF | **LAMPREY** FIBRINOGEN. /I | 5011 |
| OME STRUCTURAL ANOMALIES. + | **LAMPREY** FIBRINOGEN. EVOLUTIONARY SIGNIFICANCE OF S | 4715F |
| E INTRINSIC LAMPREY FACTOR | **LAMPREY** FIBRINOGREN AND FIBRIN BY THE ACTION OF TH | 5302F |
| ROMYZON MARINUS U.S.A. ADUL | **LAMPREY** FIBRINOPEPTIDE B IS A GLYCOPEPTIDE. + *PET | 4954F |
| T MIGRATION / + CANADA ADUL | **LAMPREY** FROM HUMBER RIVER, 1975. P.90 + CANADA ADU | 5227F |
| TRA LETHOPHAGA U.S.A. SYSTE | **LAMPREY** FROM OREGON. + *LAMPETRA TRIDENTATA *LAMPE | 5555 |
| /SEA | **LAMPREY** FROM THE DEEP OCEAN. /I | 5125F |
| -99. + *PETROMYZON MARINUS | **LAMPREY** FROM THE HUMBER RIVER, LAKE ONTARIO. PP.98 | 5262F |
| MANAGEMENT ADULT /+ CANADA | **LAMPREY** FROM THE HUMBER RIVER,1974. P. 70. + CANAD | 5136F |
| PETROMYZON MARINUS CANADA A | **LAMPREY** FROM THE HUMBERRIVER, 1973. PP. 85-87. + * | 4900F |
| (PETROMYZONIDAE) FROM NORT | **LAMPREY** GENUS LETHENTERON CREASER AND HUBBS, 1922, | 5542F |
| ERN TRIBUTARIES OF THE GULF | **LAMPREY** GENUS LETHENTERON PETROMYZONIDAE FROM EAST | 5150 |
| TE ENDOCRINOLOGY /A AMMOCOE | **LAMPREY** GEOTRIA-AUSTRALIS GRAY. + AUSTRALIA AMMOCO | 5209F |
| TORY SYSTEM ADULT / CIRCULA | **LAMPREY** GILL FILAMENT. + *LAMPETRA JAPONICA CIRCUL | 5519 |
| ISTOL CHANNEL REGION. + *LA | **LAMPREY** GROWTH AND BIOLOGICAL CONDITIONS IN THE BR | 5009 |
| TRUCTURE OF THE CYCLOSTOME | **LAMPREY** HEART. + HISTOLOGY CIRCULATORY SYSTEM /G S | 5294F |
| RA FLUVIATILIS L. ADULT BLO | **LAMPREY** HEMOGLOBIN.] + *PETROMYZON MARINUS *LAMPET | 4760F |
| ATERSHED. /STTEXAS U.S.A. W | **LAMPREY** ICHYHYOMYZON GAGEI IN AN EASTTEXAS U.S.A. | 5207 |
| /THE BRAIN OF THE | **LAMPREY** IN A COMPARATIVE PERSPECTIVE. PP.97-145. / | 5592 |
| NUS GERMANY DISTRIBUTION AD | **LAMPREY** IN MIDWATER OFF EUROPE. + *PETROMYZON MARI | 4733F |
| TROMYZON MARINUS CANADA U.S | **LAMPREY** IN SABLE RIVER LAKE SUPERIOR. P.100. + *PE | 5263F |
| ROMYZON MARINUS CANADA ADUL | **LAMPREY** IN ST. MARY'S RIVER1973. PP. 80-81. + *PET | 4906F |
| A ADULT DISTRIBUTION /CANAD | **LAMPREY** IN ST. MARYS RIVER, 1975. PP.83-85. + CANA | 5224F |
| ZON MARINUS CANADA ADULT /Y | **LAMPREY** IN ST. MARYS RIVER,1972. P. 166. + *PETROM | 5488F |
| OMYZON MARINUS CANADA /PETR | **LAMPREY** IN ST. MARYS RIVER,1974. PP. 66-67. + *PET | 5103F |
| STOCK AND FISHERY OF RIVER | **LAMPREY** IN THE EASTERN BALTIC. / /SPAWNING | 4814 |
| TOCK AND FISHERY FOR RIVER | **LAMPREY** IN THE EASTERNBALTIC 1973. / /SPAWNING S | 5012 |
| *PETROMYZON MARINUS CANADA | **LAMPREY** IN THE ST MARYS RIVER, LAKE HURON. P.93. + | 5258F |
| LAMPETRA AEPYPTERA *LAMPETR | **LAMPREY** INTERNEURONES. + *ICHTHYOMYZON UNICUSPIS * | 5606F |
| NERVOUS SYSTEM TECHNIQUES / | **LAMPREY** INTERNEURONS. + NERVOUS SYSTEM TECHNIQUES | 5112F |
| RETION /VIATILIS RUSSIA EXC | **LAMPREY** KIDNEY.] + *LAMPETRA FLUVIATILIS RUSSIA EX | 5354F |
| TY WEST VIRGINIA U.S.A. /UN | **LAMPREY** LAMPETRA AEPYPTERA IN LYNN CREEK WAYNE COU | 5284 |
| ING SEASON. + SWEDEN ADULT | **LAMPREY** LAMPETRA FLUVIATILIS (L.) DURING THE SPAWN | 5304F |
| ADISH PEROXIDASE ALONG THE | **LAMPREY** LAMPETRA FLUVIATILIS AS REVEALED BY HORSER | 5279 |
| ISION HISTOLOGY ADULT /EN V | **LAMPREY** LAMPETRA FLUVIATILIS CYCLOSTOMI. + SWEDEN | 4655F |
| GY HISTOCHEMISTRY /OCRINOLO | **LAMPREY** LAMPETRA FLUVIATILIS L. + ADULT ENDOCRINOL | 4941F |
| ION AND RELATED PROCESSES. | **LAMPREY** LAMPETRA FLUVIATILIS OXIDATIVEPHOSPHORYLAT | 4613F |
| F DIURETICS.] + RUSSIA IONI | **LAMPREY** LAMPETRA FLUVIATILIS UNDER THE INFLUENCE O | 5210F |
| CULATORY SYSTEM RESPIRATION | **LAMPREY** LAMPETRA FLUVIATILIS. + ADULT BEHAVIOR CIR | 4719F |
| OREGULATION / HISTOLOGY OSM | **LAMPREY** LAMPETRA FLUVIATILIS. + ADULT HISTOLOGY OS | 5101F |
| /IS. + BIOCHEMISTRY MUSCLE | **LAMPREY** LAMPETRA FLUVIATILIS. + BIOCHEMISTRY MUSCL | 5276 |
| LOGY PHYSIOLOGY /LOGY HISTO | **LAMPREY** LAMPETRA FLUVIATILIS. + ENDOCRINOLOGY HIST | 5517 |
| IOLOGY MIGRATION TECHNIQUES | **LAMPREY** LAMPETRA FLUVIATILIS. + ENDOCRINOLOGY PHYS | 5005 |
| INOLOGY /UROPE ADULT ENDOCR | **LAMPREY** LAMPETRA FLUVIATILIS. + EUROPE ADULT ENDOC | 4797F |
| ION PHYSIOLOGY /DULT MIGRAT | **LAMPREY** LAMPETRA FLUVIATILIS. + RUSSIA ADULT MIGRA | 5218F |
| S SYSTEM HISTOLOGY / NERVOU | **LAMPREY** LAMPETRA FLUVIATILIS. + RUSSIA ADULT NERVO | 4973 |

```
METABOLISM /A BIOCHEMISTRY    LAMPREY LAMPETRA FLUVIATILIS. + RUSSIA BIOCHEMISTR      5313
-HYPOPHYSIAL SYSTEM OF THE     LAMPREY LAMPETRA FLUVIATILIS. /      /THE HYPOTHALAMO      4586
L STUDYW... PP. 217-250. +     LAMPREY LAMPETRA FLUVIATILIS. A LIGHT MICROSCOPICA      5285F
OLISM MUSCLE /EMISTRY METAB    LAMPREY LAMPETRA FLUVIATILIS.] + BIOCHEMISTRY META      5191
N HISTOCHEMISTRY /ULT VISIO    LAMPREY LAMPETRA FLUVIATILIS.] + RUSSIA ADULT VISI      5173
Y METABOLISM MUSCLE /EMISTR    LAMPREY LAMPETRA FLUVIATILIS.] + RUSSIA BIOCHEMIST      5360F
XCRETION /.S.S.R. ANATOMY E    LAMPREY LAMPETRA FLUVIATILIS.] + U.S.S.R. ANATOMY      4694F
M PHYSIOLOGY ANIMAL / SYSTE    LAMPREY LAMPETRA FLUVIATILIS.] + USSR NERVOUS SYST      4636F
ENA PORCUS, AND CNICKS.] +     LAMPREY LAMPETRA FLUVIATILIS, SCORPION-FISH SCORPA      5267F
ARASITE FAUNA OF THE RIVER     LAMPREY LAMPETRA FLUVIATILISFROM THE KURSKI. /  /P      4664
ER IRTYSH BASIN.] + RUSSIA     LAMPREY LAMPETRA JAPONICA KESSLERI (ANIKIN) OF UPP      4623F
 *LAMPETRA PLANERI AMMOCOET    LAMPREY LAMPETRA JAPONICA. + *LAMPETRA FLUVIATILIS      4884
3RD VENTRICULAR WALL INTHE     LAMPREY LAMPETRA JAPONICA. + ADULT HISTOLOGY /THE       5187F
OGY HISTOLOGY /T ENDOCRINOL    LAMPREY LAMPETRA JAPONICA. + JAPAN ADULT ENDOCRINO      4587F
MUSCLE HISTOLOGY /TE ADULT     LAMPREY LAMPETRA JAPONICA. + JAPAN AMMOCOETE ADULT      5185F
KH-SSR U.S.S.R. /STERN KAZA    LAMPREY LAMPETRA KESSLERI IN RIVERS OF EASTERN KAZ      5318
NG GROWTH / DIGESTION FEEDI    LAMPREY LAMPETRA PLANERI. + ENGLAND DIGESTION FEED       4776F
OETE AND ADULT FRESH WATER     LAMPREY LAMPETRA PLANERI. /AESIS IN A LARVAL AMMOC       5603
NOLOGY HISTOLOGY NERVOUS SY    LAMPREY LAMPETRA TRIDENTATA. + JAPAN ADULT ENDOCRI      5042F
D GLAND EVOLUTION. / THYROI    LAMPREY LARVAE ASSOCIATED WITH THEPROBLEM OF THYRO      4997
MYCIN AS A CONTROL FOR SEA     LAMPREY LARVAE IN LENTIC HABITAT. /       /ANTI       5425
PYPTERA U.S.A. RESPIRATIONN    LAMPREY LARVAE. + *PETROMYZON MARINUS *LAMPETRA AE      4788F
GROWTH /TE CULTURE FEEDING     LAMPREY LARVAE. + U.S.A. AMMOCOETE CULTURE FEEDING      4970F
IN SEDIMENT-WATER SYSTEMS.     LAMPREY LARVICIDE 3-TRIFLUOROMETHYL-4-NITROPHENOL       4736F
(TFM). + ANIMAL CHEMISTRY T    LAMPREY LARVICIDE 3-TRIFLUOROMETHYL-4-NITROPHENOL       5077
NITROPHENOL AS A SELECTIVE     LAMPREY LARVICIDE. /LILIDE AND 3-TRIFLUOROMETHL-4-      5552
        /TOXICITY OF TWO       LAMPREY LARVICIDES TO SELECTED INVERTEBRATES /G         5428
AVE LAKE, N.W.T. + CANADA A    LAMPREY LENTHENTERON JAPONICUM MARTENS OF GREAT SL      4761F
 *LAMPETRA JAPONICA JAPAN A    LAMPREY LIVER ASREVEALED BY ELECTRON MICROSCOPY. +      5356F
PERIOD. /G THE PRESPAWNING     LAMPREY LMPETRA FLUVIATILIS DURING THE PRESPAWNING      5060
RACT AND DORSAL ROOT AFFERE    LAMPREY MOTONEURONS BY STIMULATION OF DESCENDING T      5521
STIMULATION.] + *LAMPETRA      LAMPREY MOTONEURONS BY SUPRASPINAL ANDINTRA SPINAL      4601F
ICHTHYOMYZON UNICUSPIS *ENT    LAMPREY OF NORTH AMERICA.] + *PETROMYZON MARINUS *      4752F
KES AND THE ANADROMOUS SEA     LAMPREY OF THE ATLANTIC, PETROMYZON MARINUS L. /SA      4726
M SOUTH CENTRAL CALIFORNIA     LAMPREY OF THE GENUS ENTOSPENUS PETROMYZONIDAE FRO      4821
LAMPREY OF THE ATLANTIC, PE    LAMPREY OF THE GREAT LAKES AND THE ANADROMOUS SEA       4726
ELMARVA PENINSULA. + *PETRO    LAMPREY OKKELBERGIA AEPYPTERA (ABBOTT), FROM THE D      4676F
A CELLULAR POTENTIALS FROM     LAMPREY OLFACTORY RECEPTORS. /      /INTR       4728
CHEMISTRY TECHNIQUES GENETI    LAMPREY PETROMYZON FLUVIATILIS. + SWEDEN ADULT BIO      4562F
ADULT EXCRETION HISTOCHEMI     LAMPREY PETROMYZON MARINUS + CANADA ATLANTIC OCEAN      5350F
ITS LIFE CYCLE. + CANADA U.    LAMPREY PETROMYZON MARINUS AT DIFFERENT STAGES IN       4758F
ART 1: LIGHT MICROSCOPY AND    LAMPREY PETROMYZON MARINUS DURING METAMORPHOSIS. P      5174F
IFE STAGE. + HISTOCHEMISTRY    LAMPREY PETROMYZON MARINUS DURING THEIR PARASTIC L      4659F
970. + NORTH AMERICA AMMOCO    LAMPREY PETROMYZON MARINUS IN LAKE SUPERIOR 1953-1      4630F
ND TEMPERATURE. + CANADA AD    LAMPREY PETROMYZON MARINUS IN RELATION TO WEIGHT A      4593F
NERVOUS SYSTEM VISION /OGY     LAMPREY PETROMYZON MARINUS L. + CYTOLOGY HISTOLOGY      5245F
TELEOST. + CANADA ADULT PAR    LAMPREY PETROMYZON MARINUS PREDATION ON FRESHWATER      4754F
N THE ORAL DISC OF THE SEA     LAMPREY PETROMYZON MARINUS. + ADULT AMMOCOETE /Y O      4702F
HISTOLOGY OSMOREGULATION /     LAMPREY PETROMYZON MARINUS. + CANADA ENDOCRINOLOGY      4687F
ANADROMOUS HISTOLOGY CYTOL     LAMPREY PETROMYZON MARINUS. + CANADA NEW BRUNSWICK      5516F
AMMOCOETE ANADROMOUS CYTOL     LAMPREY PETROMYZON MARINUS. + CANADA NEW BRUNSWICK      5518F
ESPIRATION HISTOLOGY OSMORE    LAMPREY PETROMYZON MARINUS. + ONTARIO MORPHOLOGY R      5085F
NOLOGY METABOLISM REPRODUCT    LAMPREY PETROMYZON MARINUS. + U.S.A. ADULT ENDOCRI      4607F
PHOSIS NERVOUS SYSTEM TECHN    LAMPREY PETROMYZON MARINUS. + VISION ADULT METAMOR      5160F
BDOMINAL WINDOW IN THE SEA     LAMPREY PETROMYZON MARINUS. /SIMPLANTATION OF AN A      5613
DA ADULT AMMOCOETE BIOLOGY     LAMPREY PETROMYZON MARINUSIN NEW BRUNSWICK. + CANA      4710F
ATION MIGRATION /LT PIGMENT    LAMPREY PETROMYZONMARINUS L. + CANADA ADULT PIGMEN      4895F
FICATION AND PROPERTIES OF     LAMPREY PHOSPHORYLASE. /TIN LOWER VERTEBRATES PURI      4717
HBBS), OCTOBER 29-30, 1975.    LAMPREY RESEACH AT HAMMOND BAY BIOLOGICAL STATION(      4756F
THE SCHIFF REACTION OF THE     LAMPREY RETINA ENTOSPHNUS JAPONICUS. /       /       4658
NUS JAPONICUS BIOCHEMISTRY     LAMPREY SERUM IN ZONE ELECTROPHORESIS. + *ENTOSPHE      5074
E NERVOUS SYSTEM /AMMOCOET     LAMPREY SPINAL CORD. + *PETROMYZON MARINUS AMMOCOE      5069
AMMOCOETE ADULT MORPHOLOGY     LAMPREY SPINAL CORD. + *PETROMYZON MARINUS U.S.A.      4595F
Y /ROMYZON U.S.A. PHYSIOLOG    LAMPREY SPINAL CORD. + *PETROMYZON U.S.A. PHYSIOLO      5333F
        /GIANT INTERNEURONS OF LAMPREY SPINAL CORD. /      4560
20. + *LAMPETRA LAMOTTEI *P    LAMPREY SURVEY OF THE NIAGARA RIVER, 1973. PP. 19-      4890F
, 1975. PP. 7-12. + CANADA     LAMPREY SURVEYS OF STREAMS TRIBUTARY TO LAKE HURON      4605F
,1974.PP. 10-14. + *PETROMY    LAMPREY SURVEYS OF STREAMS TRIBUTARY TO LAKE HURON      4645F
,1973.PP.7-12. + *PETROMYZO    LAMPREY SURVEYS OF STREAMS TRIBUTARY TO LAKE HURON      4864F
. PP.9-11. + *PETROMYZON MA    LAMPREY SURVEYS OF STREAMS TRIBUTARY TO LAKE HURON      5249F
IO, NEW YORK STATE, 1975. P    LAMPREY SURVEYS OF STREAMS TRIBUTARY TO LAKE ONTAR      4553F
IOR,1973. PP.6-7. + *PETROM    LAMPREY SURVEYS OF STREAMS TRIBUTARY TO LAKE SUPER      4865F
IOR,1974. PP. 6-9. + *PETRO    LAMPREY SURVEYS OF STREAMS TRIBUTARY TO LAKE SUPER      5111F
IOR, 1975. PP. 6-7. + CANAD    LAMPREY SURVEYS OF STREAMS TRIBUTARY TO LAKE SUPER      5116F
IOR. PP. 7-9. + *PETROMYZON    LAMPREY SURVEYS OF STREAMS TRIBUTARY TO LAKE SUPER      5248F
AN SIDE OF LAKE ONTARIO, 19    LAMPREY SURVEYS OF STREAMS TRIBUTARY TO THE CANADI      4547F
AN SIDE OF LAKE ONTARIO 197    LAMPREY SURVEYS OF STREAMS TRIBUTARY TO THE CANADI      4861F
AN SIDE OF LAKE ONTARIO. P.    LAMPREY SURVEYS OF STREAMS TRIBUTARY TO THE CANADI      5110F
AN SIDE OF LAKE ONTARIO. PP    LAMPREY SURVEYS OF STREAMS TRIBUTARY TO THE CANADI      5250F
RK SIDE OF LAKE ONTARIO, 19    LAMPREY SURVEYS OF STREAMS TRIBUTARY TO THE NEW YO      4867F
RK SIDE OF LAKE ONTARIO, 19    LAMPREY SURVEYS OF STREAMS TRIBUTARY TO THE NEW YO      5109F
```

MOCOETE LIFE CYCLE BLOOD GR
S OF THE ANADROMOUS LARVAL
LIFE IN THE ANADROMOUS SEA
ADULT EXCRETION HISTOLOGY
ADULT BIOCHEMISTRY HISTOLO
M MIGRATION. + CANADA ATLAN
PMENT AND METAMORPHOSIS. /O
OCRINOLOGY BLOOD /IQUES END
AMMOCOETE ENDOCRINOLOGY EX
OETE OSMOREGULATION ENDOCRI
UMENT HISTOLOGY /LOGY INTEG
POPHYSIAL CELLS OF THE SEA
OST PREFERENCES OF THE SEA
TICS OF THE ANADROMOUS SEA
TRATIONS IN LANDLOCKED SEA
/CYTOKINESIS IN THE SEA
/US, EMBRYOS. + EMBRYOLOGY
CHEMISTRY /S. + EMBRYOLOGY
OWTH RATES FROM THREE STREA
COLOGY LIFE CYCLE BLOOD GRO
GILL VENTILATION. /ON, AND
OUS SYSTEM SENSE RECEPTORS
VISION HISTOLOGY /S SYSTEM
/ L. + EXCRETION HISTOLOGY
GY /+ ADULT USA ENDOCRINOLO
-RELEASE TOXICANTS FOR SEA
FACTORY NERVE. + NERVOUS SY
ETION FEEDING GROWTH PARASI
ND BASS MICROPTERUS SASMOID
TICULAR REFERENCE TO AZUROP
ENERATION, BLOOD GLUCOSE RE
TH MAMMALIAN CORTICOTROPIN.
PERIOD OF SEXUAL MATURATION
S OF THEIR LIFE CYCLE.] + R
TACT MALE AND FEMALE RIVER
LT LIFE MATURITY /57. + ADU
MPERATURE /HAVIOUR LIGHT TE
/CRANIAL NERVES IN
*LAMPETRA PLANERI *EUDONTOM
IGRATION.] PP.127-133. + *L
*LAMPETRA PLANERI BELGIUM A
BRITISH COLUMBIA,CANADA. +
RI *LAMPETRA MARIAE U.S.S.R
(BLOCH). + U.K. AMMOCOETE
/LIS AND LAMPETRA PLANERI.
. + BLOOD LIFE CYCLE PHYSIO
STRESS ORGLUCOSE BEFORE AN
OCOETE ADULT BLOOD CYTOLOGY
ERA, FROM THE DELMARVA PENI
TE PARASITISM OF ADULT /COE
T HISTOLOGY /AMMOCOETE ADUL
ADULT ENDOCRINOLOGY HISTOLO
YROIDS OF POST METAMORPHIC
ECTED. + U.S.A. AMMOCOETE A
ILIES.] + *PETROMYZONINAE *
ISTRIBUTION AND BIOLOGY OF
BUTARY. + CANADA BIOLOGY /I
RIVER AND LAKES HURON AND
ECOLOGY FECUNDITY FEEDING
N ICHTHYOMYZON *TETRAPLEURO
UVIATILIS FINLAND ADULT ION
PTERA *LAMPETRA LAMOTTEI AM
S ENGLAND AMMOCOETE ADULT B
*LAMPETRA (LETHENTERON) LAM
ANEUS U.S.A AMMOCOETEMUSCLE
ERON) LAMOTTENII AMMOCOETE
LIS *LAMPETRA PLANERI*ICHTH
Y LOCOMOTION NERVOUS SYSTEM
/ /VENTILATION OF LARVAL
EATMENT FOR CONTROL OF SEA
SIS ADULT /OLOGY METAMORPHO
THE NUCLEAR DNA CONTENT OF
/ELECTRO RETINOGRAM OF
GREAT LAKES MANAGEMENT /A.
IEKOENSIS ADULT ANATOMY CIR
ORE AND DURING PERIOD OF SE
UITARY GLAND IN LARVAL SEA
LANERI(BLOCH). + GREAT BRIT
TEI (PISCES: PETROMYZONIDA
NUS NAMAYCUSH. + CANADA MOR
UM TO CAERULEIN. + *LAMPETR

LAMPREY, PETROMYZON MARINUS L. + U.S.A. ECOLOGY AM    4975F
LAMPREY, PETROMYZON MARINUS L. /RRING METAMORPHOSI    4695
LAMPREY, PETROMYZON MARINUS L. /TEY DURING LARVAL    4539
LAMPREY, PETROMYZON MARINUS L.. + CANADA AMMOCOETE    4898F
LAMPREY, PETROMYZON MARINUS L.. + CANADA AMMOCOETE    4899F
LAMPREY, PETROMYZON MARINUS L., DURING ITS UPSTREA    5539F
LAMPREY, PETROMYZON MARINUS L., DURING LARVALDEVEL    4698
LAMPREY, PETROMYZON MARINUS. + ADULT TECHNIQUES EN    5208
LAMPREY, PETROMYZON MARINUS. + CANADA LAKE ONTARIO    4639F
LAMPREY, PETROMYZON MARINUS. + NEW BRUNSWICK AMMOC    5604F
LAMPREY, PETROMYZON MARINUS. + U.S.A. BIOLOGY INTE    5235F
LAMPREY, PETROMYZON MARINUS. / FICATION OF ADENOHY    5567
LAMPREY, PETROMYZON MARINUS. /MPTION, GROWTH AND H    5610F
LAMPREY, PETROMYZON MARINUS. /SAND SPAWNING ENERGE    5507
LAMPREY, PETROMYZON MARINUS. /TIC AND IONIC CONCEN    5344F
LAMPREY, PETROMYZON MARINUS, EMBRYO. + EMBRYOLOGY    5374F
LAMPREY, PETROMYZON MARINUS, EMBRYOS. + EMBRYOLOGY    5373
LAMPREY, PETROMYZON MARINUS, EMBRYOS. + EMBRYOLOGY    5375F
LAMPREY, PETROMYZON MARINUS, LARVAL (AMMOCOETE) GR    5133F
LAMPREY, PETROMYZON MARINUSL. + U.S.A. AMMOCOETE E    5616F
LAMPREY, PETROMYZON, DURING SUCTION, OLFACTION, AN    5461
LAMPREY,ENTOSPHENUS JAPONICUS. + ADULT ANIMAL NERV    4745F
LAMPREY,PETROMYZON MARINUS + U.S.A. NERVOUS SYSTEM    4785F
LAMPREY,PETROMYZON MARINUS L. + EXCRETION HISTOLOG    4880F
LAMPREY,PETROMYZON MARINUS. + ADULT USA ENDOCRINOL    4780F
LAMPREYCONTROL. + MANAGEMENT /R OF POSSIBLE BOTTOM    5432F
LAMPREYLAMPETRA FLUVIATILIS FROM STIMULATION OF OL    4739
LAMPREYPETROMYZON MARINUS. + CANADA DIGESTION EXCR    4673F
LAMPREYPETROMYZON MARINUS, TROUT SALMO GAIRDNERI A    4727F
LAMPREYS (LAMPETRA FLUVIATILIS (L.) GRAY) WITH PAR    5131F
LAMPREYS (LAMPETRA FLUVIATILIS L.): EFFECTS ON REG    5509
LAMPREYS (LAMPETRA FLUVIATILIS) AFTER TREATMENT WI    5527
LAMPREYS (LAMPETRA FLUVIATILIS) BEFORE AND DURING    5234
LAMPREYS (LAMPETRA FLUVIATILIS) ON DIFFERENT STAGE    5323F
LAMPREYS (LAMPETRA FLUVIATILIS). /UTOSTERONE ON IN    5607F
LAMPREYS (LAMPETRA FLUVIATILIS). PP. 156-157. + AD    4868F
LAMPREYS (PETROMYZON MARINUS). + BEHAVIOUR LIGHT T    5298F
LAMPREYS AND CEPHALASPIDS. /E    5510
LAMPREYS AND FISHES OF SLOVAKIA CZECHOSLOVAKIA. +    5175F
LAMPREYS ANDFISHES ON VARIOUS STAGES OF SPAWNING M    5361F
LAMPREYS IN AN AQUARIUM.] + *LAMPETRA FLUVIATILIS    5357F
LAMPREYS IN THE SALMONAND CERTAIN OTHER RIVERS IN    4974F
LAMPREYS IN WATERS OF THE BSSR.] + *LAMPETRA PLANE    5358F
LAMPREYS LAMPETRA FLUVIATILIS AND LAMPETRA PLANERI    4877F
LAMPREYS LAMPETRA FLUVIATILIS AND LAMPETRA PLANERI    4991
LAMPREYS LAMPETRA FLUVIATILIS AND LAMPETRA PLANERI    5301F
LAMPREYS LAMPETRA FLUVIATILIS TREATED WITH INSULIN    5062F
LAMPREYS LAMPETRA FLUVIATILIS. + GREAT BRITIAN AMM    4649F
LAMPREYS LAMPETRA LAMOTTENII ANDOKKELBERGIA AEPYPT    4720F
LAMPREYS LAMPETRA PLANERI (BLOCH). + SWEDEN AMMOCO    5467F
LAMPREYS LAMPETRA REISSNERI. + JAPAN AMMOCOETE ADU    5180
LAMPREYS LAMPETRAFLUVIATILIS (L.) GRAY. + DENMARK    4592F
LAMPREYS OF LAMPETR LAMPETRA FLUVIATILIS. /Y IN TH    5598
LAMPREYS OF THE FAMILY PETROMYZONIDAE MUST BE PROT    4600F
LAMPREYS OF THE FAMILY PETROMYZONIDAE INTO3 SUBFAM    4753F
LAMPREYS OF THE GENUS LAMPETRA IN OREGON. /MION, D    5554
LAMPREYS PETROMYZON MARINUS FROM A LAKE ONTARIO TR    4759F
LAMPREYS PETROMYZON MARINUS TAGGED IN THE ST.MARYS    4746F
LAMPREYS PETROMYZON MARINUS. + CANADA ADULT ANIMAL    5140F
LAMPREYS PETROMYZONIDAE. + *CASPIOMYZON *PETROMYZO    5391F
LAMPREYS UNDER NATURAL CONDITIONS.] + *LAMPETRA FL    4952F
LAMPREYS. + *ICHTHYOMYZON UNICUSPIS *LAMPETRA AEPY    5466F
LAMPREYS. + *LAMPETRA PLANERI *LAMPETRA FLUVIATILI    4584F
LAMPREYS. + *LAMPETRA PLANERI *PETROMYZON MARINUS    4565F
LAMPREYS. + *PETROMYZON MARINUS *ICHTHYOMYZON CAST    4674F
LAMPREYS. + *PETROMYZON MARINUS *LAMPETRA (LETHENT    4998F
LAMPREYS. + *PETROMYZON MARINUS *LAMPETRA FLUVIATI    4713
LAMPREYS. + *PETROMYZON MARINUS AMMOCOETE HISTOLOG    5026F
LAMPREYS. + *PETROMYZON MARINUS U.S.A. RESPIRATION    4787F
LAMPREYS. + ANIMAL CHEMISTRY ECOLOGY MANAGEMENT /R    5051F
LAMPREYS. + NERVOUS SYSTEM ENDOCRINOLOGY METAMORPH    4927
LAMPREYS. /    /    4696
LAMPREYS. /I    4956
LAMPREYS. PP. 211-216. + *PETROMYZON MARINUS U.S.A    5501F
LAMPREYS. 8-13PP. + *MYXINE GLUTINOSA *MAYOMYZON P    4843F
LAMPREYS.RIVER LAMPREYS (LAMPETRA FLUVIATILIS) BEF    5234
LAMPREYS(PETROMYZON MARINUS). /LRUCTURE OF THE PIT    5339
LAMPREYS, LAMPETRA FLUVIATILIS (L.) AND LAMPETRA P    5142F
LAMPREYS, OKKELBERGIA AEPYPTERA AND LAMPETRA LAMOT    5139F
LAMPREYS, PETROMYZON MARINUS ON LAKE TROUT SALVELI    4769F
LAMPREYS: IMMUNOFLUORESCENCE REACTION WITH ANTISER    5562F

NERVOUS SYSTEM TECHNIQUES
E NERVOUS SYSTEM HISTOLOGY
EMPERATURE. + CANADA ADULT
ULT CIRCADIAN DISTRIBUTION
HAVIOUR CIRCULATORY SYSTEM
ARINUS AMMOCOETE HISTOLOGY
  AND TEMPERATURE. + CANADA
AMPREY (LAMPETRA FLUVIATILI
ATILIS (L.) DURING THE SPAW
ILIS ANIMAL MUSCLE NERVOUS
NA KHOLINORETSEPTIVNYE MEM
TRY TECHNIQUES /            /
RETIDAE FROM SOUTH AUSTRALI
MYZON MARINUS SPAWNING MIGR
N. + *MYXINE GLUTINOSA UNIT
CIES USED 1950-1975.] + *PE
LUTINOSA BIOLOGY HISTOCHEMI
THE VERTEBRATE BRAIN. / OF
                            /
70. /STERN EUROPE.] PP. 63-
OMYZON GAGEI DISTRIBUTION P
A (MARTENS) CYTOLOGY METABO
THE HYPOPHYSIAL REGION IN
  + *MYXINE GLUTINOSA *LAMPE
MPREY PHOSPHORYLASE. /OF LA
LADDER CONTRACTION IN SOME
PHYLOGENY OF THE EARLIEST
ATING HORMONE SECRETION IN
E EC-2.4.1.1. IN MUSCLE OF
SEA ADULT ENDOCRINOLOGY TE
BIOCHEMISTRY BIOLOGY CYTOL
R. ANIMAL BIOCHEMISTRY HIST
CARBOHYDRATE METABOLISMIN
MPETRA JAPONICA (MARTENS) C
HE ECOLOGY OF THE PIKE ESOX
E.] PP. 63-70. /STERN EUROP
ROMETHYL-4-NITROPHENOL) BY
/EIR SYSTEMATIC POSITION.]
S AND THOSE OF THE COELACAN
CIRCULATORY SYSTEM /LUTION
CRINOLOGY PHYSIOLOGY / ENDO
TATRETUS STOUTII *PETROMYZO
ON AND CHARACTERIZATION OF
INOSA + *MYXINE GLUTINOSA S
            /ARE THERE
HISTOLOGY /                 /
/PHYLOGENETIC EMERGENCE OF
MIT AUSNAHME VON SAUGERN.
PREY LAMPETRA AEPYPTERA IN
*LAMPETRA TRIDENTATUS PISC
ELEOSTEI) UND LAMPETRA PLAN
LIDE (BAYER 73) ON BENTHIC
EMISTRY MANAGEMENT ANIMAL /
PLANERI CYCLOSTOMATA.] + H
XCRETION TECHNIQUES CYTOLOG
A FLUVIATILIS. + ADULT HIST
D MYRIOPHYLLUM SPICATUM L.)
TERS OF THE MAGALLANES CHIL
LT HEARING ANATOMY HISTOLOG
ID IN THE MUDPUPPYNECTURUS
HESE ORGANS IN BDELLOSTOMA
ER BLUE HOLE SANDUSKY COUNT
MYZONIDAE IN WATERS OF THE
EMISTRY /OLISM ANIMAL BIOCH
TABOLISM EXCRETION /LOGY ME
MISTRY /ANIMAL BLOOD BIOCHE
S). /S (LAMPETRA FLUVIATILI
(L.) GRAY. + DENMARK ADULT
00. + *PETROMYZON MARINUS C
OGLOBINS AND HEMOGLOBINS. /
MARINUS ENDOCRINOLOGY /ZON
            /RECRUITMENT IN
RESUMPTIVE INTERRENAL CELLS
ILIS) AFTER TREATMENT WITH
CRINOLOGY /ISTRY BLOOD ENDO
TIONS SPLIT BY LAMPREY AND
US STOUTII U.S.A. ENDOCRINO
ARK BIOCHEMISTRY TECHNIQUES
OF NANTES. /ATURAL HISTORY
RTEBRATES, VERTEBRATES AND
  /TRY MANAGEMENT TECHNIQUES

| | |
|---|---|
| LOCOMOTION /E TRANSECTION IN LARVAL SEA LAMPREY. + | 5579F |
| LOCOMOTION ANATOMY / ETRA FLUVIATILIS ANIMAL MUSCL | 5079F |
| LOCOMOTION MIGRATION /KIN RELATION TO WEIGHT AND T | 4593F |
| LOCOMOTION MIGRATION BEHAVIOUR /UIATILIS SWEDEN AD | 5141F |
| LOCOMOTION NERVOUS SYSTEM RESPIRATION /RTATA. + BE | 5464 |
| LOCOMOTION NERVOUS SYSTEM TECHNIQUES /(ETROMYZON M | 5026F |
| LOCOMOTION RESPIRATION /MN RELATION TO BODY WEIGHT | 4755F |
| LOCOMOTOR ACTIVITY AND LENGTH/WEIGHT OF THERIVER L | 5141F |
| LOCOMOTOR ACTIVITY OF RIVER LAMPREY LAMPETRA FLUVI | 5304F |
| LOCOMOTOR ORGAN IN SOME FISH.] + *LAMPETRA FLUVIAT | 5079F |
| LOKIRUYUSHCHEE DEISTVIE POLIPEPTIDOV ZMEINOGO YADA | 4636F |
| LONG RANGE ORDER IN GLOBULAR PROTEINS. + BIOCHEMIS | 4813 |
| LONGIPINNIS NEW SPECIES A NEW HAGFISH FAMILY EPTAT | 4992 |
| LOOK AT THE EVOLUTION OF THE ISLET ORGAN. + *PETRO | 5083F |
| LOW RESOLUTION CRYSTAL STRUCTURE OF HAGFISH INSULI | 4629 |
| LOWER AND MIDDLE ELBE RIVER, WEST GERMANY. THE SPE | 5506F |
| LOWER CHORDATES. + *LAMPETRA FLUVIATILIS *MYXINE G | 5265F |
| LOWER CHORDATESTELL US ABOUT THE EARLY EVOLUTION O | 5198 |
| LOWER DEVONIAN AGNATHANS OF YUNNAN AND SICHUAN. /I | 5417 |
| LOWER DEVONIAN DEPOSITS IN EASTERN EUROPE.] PP. 63 | 5087 |
| LOWER OUACHITA RIVER SYSTEM IN ARKANSAS. + *ICHTHY | 5202F |
| LOWER PART OF THE RIVER UMBRA. + *LAMPETRA JAPONIC | 4622F |
| LOWER VERTEBRATES (CYCLOSTOMES AND FISHES). / N OF | 4979 |
| LOWER VERTEBRATES (HAGFISH, LAMPREY AND CHIMAERA). | 5019F |
| LOWER VERTEBRATES PURIFICATION AND PROPERTIES OF L | 4717 |
| LOWER VERTEBRATES. + BIOCHEMISTRY ENDOCRINOLOGY /H | 5048 |
| LOWER VERTEBRATES. /RCARTILAGE AND BONE DURING THE | 4879 |
| LOWER VERTEBRATES. PP. 51-62. /H MELANOCYTE STIMUL | 5534 |
| LOWER VERTEBRATES. PP.547-566. /IOGEN PHOSPHORYLAS | 5179 |
| LOWER VERTEBRATES.] + *LAMPETRA FLUVIATILIS BALTIC | 4721F |
| LOWER VERTEBRATES.] + *LAMPETRA FLUVIATILIS RUSSIA | 5278 |
| LOWER VERTEBRATES.] + *LAMPETRA FLUVIATILIS U.S.S. | 4803F |
| LOWER VERTEBRATES.] /IYKH. [HORMONAL REGULATION OF | 4772F |
| LUCIUS IN THE LOWER PART OF THE RIVER UMBRA. + *LA | 4622F |
| LUCIUS L. NIZOV'EV R. UMBY. [MATERIAL CONCERNING T | 4622F |
| LUDLOW AND LOWER DEVONIAN DEPOSITS IN EASTERN EURO | 5087 |
| LUMINESCENCE SPECTROPHOMETRY. /I OF TFM (3-TRIFLUO | 5381 |
| LUNGFISHES (DIPNOI) AND THEIR SYSTEMATIC POSITION. | 4567F |
| LUNGFISHES LEPIDOSIREN NEOCERATODUS AND PROTOPTERU | 5156F |
| LYMPH VESSELS. + HISTOLOGY ENDOCRINOLOGY EVOLUTION | 4640F |
| LYMPHATIC SYSTEM OF MYXINE. PP. 57-64. + BLOOD END | 4857F |
| LYMPHO-HEMATOPOIETIC TISSUES OF CYCLOSTOMES. + *EP | 4550F |
| LYMPHOCYTE SUBPOPULATIONS. /              /IDENTIFICATI | 5565 |
| LYMPHOCYTE-LIKE CELLSIN THE BLOOD FROM MYXINE GLUT | 4945F |
| LYMPHOCYTES IN HAGFISH? / | 4957 |
| LYMPHOID CELLS IN THE HAGFISH. + BLOOD IMMUNOLOGY | 5114F |
| LYMPHOID TISSUES AND CELLS. PP. 149-202. / | 5568 |
| LYMPHOZYTEN DES PERIPHEREN BLUTES BEI WIRBELTIEREN | 5119F |
| LYNN CREEK WAYNE COUNTY WEST VIRGINIA U.S.A. / LAM | 5284 |
| M.V. CALIGUS JULY-AUGUST, 1974. + *LAMPETRA AYRESI | 5364F |
| MACRO-NEUROVEN VON SALMO IRIDEUS (GIBBONS 1855) (T | 4700F |
| MACROINVERTEBRATES IN A LAKE ENVIRONMENT. /ACYLANI | 5459 |
| MACROINVERTEBRATES. + CHEMISTRY MANAGEMENT ANIMAL | 5387F |
| MACRONEURONS OF SALMO IRIDEUS TELEOST AND LAMPETRA | 4700F |
| MACROPHAGES IN THE RENAL URINARY SPACE. + ANIMAL E | 5593 |
| MACROPHTHALMIA STAGE OF THE RIVER LAMPREY LAMPETR | 5101F |
| MACROPHYTES (ELODEA CANADENSIS (MICHX) PLANCHON AN | 5421F |
| MACROSTOMUS NEW RECORD MYXINE PETROMYZONIDAE IN WA | 5571 |
| MACULAE OFTHE LAMPREY ENTOSPHENUS JAPONICUS. + ADU | 4663F |
| MACULOSUS. + *MYXINE GLUTINOSA BIOCHEMISTRY /O FLU | 4627F |
| MADE BY BASHFORD DEAN. PP.67-102. /O DRAWINGS OF T | 5584 |
| MADE DURING A 1 DAY SCUBA INVESTIGATIONOF THE MILL | 4637F |
| MAGALLANES CHILE REGION. /INEW RECORD MYXINE PETRO | 5571 |
| MAINE USA MARINE SPECIES. + METABOLISM ANIMAL BIOC | 4964F |
| MAINE USA. + ANIMAL ADULT BIOCHEMISTRY HISTOLOGY M | 4604F |
| MAJOR CLASSES OF VERTEBRATES. + ANIMAL BLOOD BIOCH | 4549 |
| MALE AND FEMALE RIVER LAMPREYS (LAMPETRA FLUVIATIL | 5607F |
| MALE AND FEMALE RIVER LAMPREYS LAMPETRAFLUVIATILIS | 4592F |
| MALE SEA LAMPREY IN SABLE RIVER LAKE SUPERIOR. P.1 | 5263F |
| MAMMALIAN AND PISCINE MYOGLOBINS AND HEMOGLOBINS. | 4936 |
| MAMMALIAN AND SUBMAMMALIAN CHORDATES. + *PETROMYZO | 5056F |
| MAMMALIAN BILE DUCT FORMATION. + HISTOLOGY /A | 5392 |
| MAMMALIAN CORTICOTROPHIN ON THE ULTRASTRUCTURE OFP | 4899F |
| MAMMALIAN CORTICOTROPIN. /GPREYS (LAMPETRA FLUVIAT | 5527 |
| MAMMALIAN INSULIN. + CANADA BIOCHEMISTRY BLOOD END | 5326F |
| MAMMALIAN THROMBINS. + ADULT BIOCHEMISTRY BLOOD /I | 5146F |
| MAMMALIAN THYROID-STIMULATING HORMONE. + *EPTATRET | 5328F |
| MAMMALIANCORTICOTROPIN. + *PETROMYZON MARINUS DENM | 4574F |
| MAMMALS CONSERVED AT THE MUSEUM OF NATURAL HISTORY | 5008 |
| MAN.] + ANIMAL BLOOD BIOCHEMISTRY /OHANISMS ININVE | 5372F |
| MANAGEMENT + PISCES CHEMISTRY MANAGEMENT TECHNIQUE | 5397F |

| | | |
|---|---|---|
| MANAGEMENT GROWTH AMMOCOET | MARINUS *LAMPETRA LAMOTTEI *ICHTHYOMYZON SP CANADA | 5110F |
| MPETRA TRIDENTATA *MYXINE G | MARINUS *LAMPETRA PLANERI *LAMPETRA FLUVIATILIS*LA | 4705F |
| CEPTORS /OLFACTION SENSE RE | MARINUS *LAMPETRA PLANERIANATOMY OLFACTION SENSE R | 5047F |
| BOLISM MIGRATION /DULT META | MARINUS *LAMPETRA TRIDENTATA RESPIRATION ADULT MET | 5138F |
| I ENDOCRINOLOGY HISTOCHEMIS | MARINUS *MORDACIA MORDAX *GEOTRIA *LAMPETRA PLANER | 4706F |
| LOGY /DULT ANIMAL ENDOCRINO | MARINUS *PETROMYZON DORSATUS ADULT ANIMAL ENDOCRIN | 5066F |
| L INJECTION. + *PETROMYZON | MARINUS ADULT ENDOCRINOLOGY CANADA HISTOLOGY /AERO | 4881F |
| YOMYZON FOSSOR *PETROMYZON | MARINUS ADULT HISTOCHEMISTR /A 266:69-89. + *ICHTH | 5206F |
| LOGY /ENSE RECEPTORS PHYSIO | MARINUS ADULT NERVOUS SYSTEM SENSE RECEPTORS PHYSI | 4643F |
| SEA LAMPREY. + *PETROMYZON | MARINUS ADULT VISION NERVOUS SYSTEM /RN THE ADULT | 4798F |
| OOD HISTOLOGY /EVOLUTION BL | MARINUS AMMOCOETE ADULT HISTOCHEMISTRY EVOLUTION B | 4743F |
| RA FLUVIATILIS *PETROMYZON | MARINUS AMMOCOETE ADULTMUSCLE ANATOMY HISTOLOGY /E | 4840F |
| S). PP. 5-7. + *PETROMYZON | MARINUS AMMOCOETE ANATOMY SENSE RECEPTORS /PENTATU | 4731F |
| LAMPRICIDE. + *PETROMYZON | MARINUS AMMOCOETE CHEMISTRY MANAGEMENT TECHNIQUE / | 5053F |
| EM TECHNIQUES /NERVOUS SYST | MARINUS AMMOCOETE HISTOLOGY LOCOMOTION NERVOUS SYS | 5026F |
| SPINAL CORD. + *PETROMYZON | MARINUS AMMOCOETE NERVOUS SYSTEM /SSECTED LAMPREY | 5069 |
| LOGY / NERVOUS SYSTEM HISTO | MARINUS ANADROMOUS LARVA ADULT NERVOUS SYSTEM HIST | 5020F |
| CRYATALLINS. + *PETROMYZON | MARINUS ANIMAL EVOLUTION PHYSIOLOGY /LEBRAE ALPHA | 5024F |
| ANADA U.S.A. ADULT AMMOCOET | MARINUS AT DIFFERENT STAGES IN ITS LIFE CYCLE. + C | 4758F |
| ATRETUS STOUTI *PETROMYZON | MARINUS BIOCHEMISTRY DIGESTION /PNE GLUTINOSA *EPT | 4910F |
| RANSFERRINS. + *PETROMYZON | MARINUS BIOCHEMISTRY IMMUNOLOGY /PHER VERTEBRATE T | 5004F |
| AND FIBRIN. + *PETROMYZON | MARINUS BLOOD HISTOCHEMISTRY / LAMPREY FIBRINOGEN | 4960F |
| . PP. 66-67. + *PETROMYZON | MARINUS CANADA /EA LAMPREY IN ST. MARYS RIVER,1974 | 5103F |
| PP. 171-172. + *PETROMYZON | MARINUS CANADA ADULT / DAM ON THE SAUGEEN RIVER. | 5485F |
| 972. P. 166. + *PETROMYZON | MARINUS CANADA ADULT / AMPREY IN ST. MARYS RIVER,1 | 5488F |
| 1973. P. 88. + *PETROMYZON | MARINUS CANADA ADULT /SMPREY TAKEN AT DENN'S DAM, | 4848F |
| . PP. 80-81. + *PETROMYZON | MARINUS CANADA ADULT /TREY IN ST. MARY'S RIVER1973 | 4906F |
| HISTOLOGY CYTOLOGY /NOLOGY | MARINUS CANADA ADULT AMMOCOETE ANIMAL ENDOCRINOLOG | 4779F |
| RY MANAGEMENT /GROWTH HISTO | MARINUS CANADA ADULT AMMOCOETE BIOLOGY GROWTH HIST | 4908F |
| . PP. 81-84. + *PETROMYZON | MARINUS CANADA ADULT BIOLOGY /+S: LAKE HURON, 1973 | 4905F |
| 1973. P. 85. + *PETROMYZON | MARINUS CANADA ADULT BIOLOGY /MMMERCIALFISHERMEN, | 4902F |
| . PP. 85-87. + *PETROMYZON | MARINUS CANADA ADULT BIOLOGY /PE HUMBERRIVER, 1973 | 4900F |
| TE /RY MORTALITY OF AMMOCOE | MARINUS CANADA ADULT CHEMISTRY MORTALITY OF AMMOCO | 5237F |
| 972. P. 171. + *PETROMYZON | MARINUS CANADA ADULT MANAGEMENT /M HUMBER RIVER, 1 | 5486F |
| 1973. PP. 5. + *PETROMYZON | MARINUS CANADA ADULT MORTALITY OF /OR OPERATIONS: | 4866F |
| . PP. 95-96. + *PETROMYZON | MARINUS CANADA ADULT MORTALITY OF MANAGEMENT /C972 | 5481 |
| 973. PP. 77. + *PETROMYZON | MARINUS CANADA AMMOCOETE /+REAM TREATMENTS, 1971-1 | 4901F |
| 973. PP. 78. + *PETROMYZON | MARINUS CANADA AMMOCOETE /EREAM TREATMENTS, 1971-1 | 4904F |
| PETRA LAMOTTEI *PETROMYZON | MARINUS CANADA AMMOCOETE /M1973. PP. 19-20. + *LAM | 4890F |
| PP. 101-103. + *PETROMYZON | MARINUS CANADA AMMOCOETE /ORN RIVER SYSTEM, 1972. | 5477F |
| 1973. PP.15. + *PETROMYZON | MARINUS CANADA AMMOCOETE BIOLOGY / F LAKE ONTARIO | 4861F |
| R.PP. 88-91. + *PETROMYZON | MARINUS CANADA AMMOCOETE BIOLOGY /TST. MARY'S RIVE | 4826F |
| 3.PP. 21-31. + *PETROMYZON | MARINUS CANADA AMMOCOETE CHEMISTRY MORTALITY OF /M | 4889F |
| TER SYSTEMS. + *PETROMYZON | MARINUS CANADA AMMOCOETE HISTORY. /CIN SEDIMENT-WA | 4736F |
| . PP. 47-52. + *PETROMYZON | MARINUS CANADA AMMOCOETE MORTALITY OF CHEMISTRY /H | 4893F |
| 3. PP.32-46. + *PETROMYZON | MARINUS CANADA AMMOCOETE MORTALITY OF CHEMISTRY /S | 4888F |
| HURON. P.93. + *PETROMYZON | MARINUS CANADA GREAT LAKES ADULT DISTRIBUTION /AE | 5258F |
| MISTRY / EXCRETION HISTOCHE | MARINUS CANADA GREAT LAKES ADULT EXCRETION HISTOCH | 5045F |
| O. PP.98-99. + *PETROMYZON | MARINUS CANADA GREAT LAKES ADULT MORPHOMETRY / ARI | 5262F |
| TRIBUTION MANAGEMENT MORTAL | MARINUS CANADA GREAT LAKES AMMOCOETE CHEMISTRY DIS | 5253F |
| TRIBUTION MANAGEMENT MORTAL | MARINUS CANADA GREAT LAKES AMMOCOETE CHEMISTRY DIS | 5254F |
| TRIBUTION MANAGEMENT MORTAL | MARINUS CANADA GREAT LAKES AMMOCOETE CHEMISTRY DIS | 5256F |
| TRIBUTION MANAGEMENT MORTAL | MARINUS CANADA GREAT LAKES AMMOCOETE CHEMISTRY DIS | 5257F |
| ANAGEMENT /E DISTRIBUTION M | MARINUS CANADA GREAT LAKES AMMOCOETE DISTRIBUTION | 5248F |
| ANAGEMENT /E DISTRIBUTION M | MARINUS CANADA GREAT LAKES AMMOCOETE DISTRIBUTION | 5250F |
| ALITY OF /TE MANAGEMENTMORT | MARINUS CANADA GREAT LAKES AMMOCOETE MANAGEMENTMOR | 5252F |
| ION /NADA MANAGEMENT MIGRAT | MARINUS CANADA GREAT LAKES CANADA MANAGEMENT MIGRA | 5246F |
| ANAGEMENT MORTALITY OF /N M | MARINUS CANADA GREAT LAKES CHEMISTRY DISTRIBUTION | 5249F |
| SIS HISTOCHEMISTRY /ADOGENE | MARINUS CANADA GREAT LAKES ENDOCRINOLOGY GONADOGEN | 5470F |
| GEMENT MORTALITY OF /T MANA | MARINUS CANADA GREAT LAKES LAKE SUPERIOR ADULT MAN | 5247F |
| COMMISSION. + *PETROMYZON | MARINUS CANADA MANAGEMENT /LHE GREAT LAKES FISHERY | 5194F |
| 1974. P. 5. + *PETROMYZON | MARINUS CANADA MANAGEMENT /NOPERATIONS,LAKE HURON, | 4907F |
| /ADULT AMMOCOETE CHEMISTRY | MARINUS CANADA MANAGEMENT ADULT AMMOCOETE CHEMISTR | 5107F |
| /ADULT AMMOCOETE CHEMISTRY | MARINUS CANADA MANAGEMENT ADULT AMMOCOETE CHEMISTR | 5108F |
| /CHEMISTRY AMMOCOETE ADULT | MARINUS CANADA MANAGEMENT CHEMISTRY AMMOCOETE ADUL | 5106F |
| PP. 160-170. + *PETROMYZON | MARINUS CANADA MORPHOMETRY /MS, LAKE HURON, 1972. | 5487 |
| 3. PP 71-74. + *PETROMYZON | MARINUS CANADA MORTALITY OF CHEMISTRY AMMOCOETE /T | 4892F |
| CHEMISTRY ENDOCRINOLOGY MAN | MARINUS CANADA U.S.A. GREAT LAKES ADULT AMMOCOETE | 5263F |
| SEA LAMPREY. + *PETROMYZON | MARINUS CHEMISTRY AMMOCOETE MANAGEMENT /VS OF THE | 5050F |
| RTICOTROPIN. + *PETROMYZON | MARINUS DENMARK BIOCHEMISTRY TECHNIQUES BLOOD / CO | 4574F |
| COPY AND AUTO RADIOGRAPHY W | MARINUS DURING METAMORPHOSIS. PART 1: LIGHT MICROS | 5174F |
| HEMISTRY GONADOGENESIS ADUL | MARINUS DURING THEIR PARASTIC LIFE STAGE. + HISTOC | 4659F |
| N CHORDATES. + *PETROMYZON | MARINUS ENDOCRINOLOGY /P MAMMALIAN AND SUBMAMMALIA | 5056F |
| ATRETUS STOUTI *PETROMYZON | MARINUS FRANCE BIOLOGY EVOLUTION BIOCHEMISTRY / PT | 5337F |
| INS CHAINS.] + *PETROMYZON | MARINUS FRANCE BIOLOGY EVOLUTION BIOCHEMISTRY / UL | 4602F |
| LOGY /IBUTARY. + CANADA BIO | MARINUS FROM A LAKE ONTARIO TRIBUTARY. + CANADA BI | 4759F |
| OFF EUROPE. + *PETROMYZON | MARINUS GERMANY DISTRIBUTION ADULT /OY IN MIDWATER | 4733F |
| ERMEN. P.98. + *PETROMYZON | MARINUS GREAT LAKES ADULT MORPHOMETRY / RCIAL FISH | 5261F |
| A AMMOCOETE ADULT MANAGEMEN | MARINUS IN LAKE SUPERIOR 1953-1970. + NORTH AMERIC | 4630F |
| ANADA ADULT LOCOMOTION MIGR | MARINUS IN RELATION TO WEIGHT AND TEMPERATURE. + C | 4593F |
| ERS IN 1973. + *PETROMYZON | MARINUS IRELAND ADULT /EISHES TAKEN FROM IRISH WAT | 4686F |

| | | |
|---|---|---|
| S ENDOCRINOLOGY GONADOGENES | MARINUS L. + *PETROMYZON MARINUS CANADA GREAT LAKE | 5470F |
| G METAMORPHOSIS /ION FEEDIN | MARINUS L. + BLOOD ADULT AMMOCOETE MIGRATION FEEDI | 5145 |
| HE SEA LAMPREY, PETROMYZON | MARINUS L. + BLOOD BIOCHEMISTRY ENDOCRINOLOGY /  T | 5325F |
| BLOOD ENDOCRINOLOGY FEEDIN | MARINUS L. + CANADA ATLANTIC OCEAN ADULT AMMOCOETE | 5303F |
| LT SEA LAMPREY, PETROMYZON | MARINUS L. + CANADA ENDOCRINOLOGY HISTOLOGY /R ADU | 4781F |
| ON /OGY NERVOUS SYSTEM VISI | MARINUS L. + CYTOLOGY HISTOLOGY NERVOUS SYSTEM VIS | 5245F |
| THE SEA LAMPREY,PETROMYZON | MARINUS L. + EXCRETION HISTOLOGY /NHRIC KIDNEY OF | 4880F |
| ENDOCRINOLOGY EVOLUTION HI | MARINUS L. + HISTOLOGY EXCRETION ADULT GREAT LAKES | 4897F |
| HEMISTRY /OGY CYTOLOGY BIOC | MARINUS L. + IMMUNOLOGY ENDOCRINOLOGY CYTOLOGY BIO | 5538F |
| US SEA LAMPREY, PETROMYZON | MARINUS L. + METAMORPHOSIS GROWTH MORPHOMETRY / MO | 5537F |
| LOOD GROWTH METABOLISMMETAM | MARINUS L. + U.S.A. ECOLOGY AMMOCOETE LIFE CYCLE B | 4975F |
| LARVAL LAMPREY, PETROMYZON | MARINUS L. /RRING METAMORPHOSIS OF THE ANADROMOUS | 4695 |
| F THE ATLANTIC, PETROMYZON | MARINUS L. /SAKES AND THE ANADROMOUS SEA LAMPREY O | 4726 |
| US SEA LAMPREY, PETROMYZON | MARINUS L. /TEY DURING LARVAL LIFE IN THE ANADROMO | 4539 |
| ISTOLOGY HISTOCHEMISTRY / H | MARINUS L.. + CANADA AMMOCOETE ADULT BIOCHEMISTRY | 4899F |
| OLOGY HISTOCHEMISTRY / HIST | MARINUS L.. + CANADA AMMOCOETE ADULT EXCRETION HIS | 4898F |
| S MEERNEUNAGES (PETROMYZON | MARINUS L.) IM GOTTIN-SEE BEI POTSDAM. /TFANG EINE | 5580F |
| A ATLANTIC OCEAN ADULT HIST | MARINUS L., DURING ITS UPSTREAM MIGRATION. + CANAD | 5539F |
| SIS. /OPMENT AND METAMORPHO | MARINUS L., DURING LARVALDEVELOPMENT AND METAMORPH | 4698 |
| PP.100-101. + *PETROMYZON | MARINUS MANAGEMENT /. TO SEA LAMPREY BARRIER DAMS. | 5264F |
| N. PP.94-97. + *PETROMYZON | MARINUS MANAGEMENT MORPHOMETRY / RRIERS, LAKE HURO | 5260F |
| IQUES /LOGY HISTOLOGY TECHN | MARINUS METAMORPHOSIS ENDOCRINOLOGY HISTOLOGY TECH | 5174F |
| DA MORTALITY BY PARASITISM | MARINUS ON LAKE TROUT SALVELINUS NAMAYCUSH. + CANA | 4769F |
| STRY MANAGEMENT TECHNIQUES | MARINUS PISCES CHEMISTRY MANAGEMENT + PISCES CHEMI | 5397F |
| DULT PARASITISM BY BEHAVIOU | MARINUS PREDATION ON FRESHWATERTELEOST. + CANADA A | 4754F |
| BLOOD /RINOLOGY PHYSIOLOGY | MARINUS SPAWNING MIGRATION ENDOCRINOLOGY PHYSIOLOG | 5083F |
| ON AND MICHIGAN, 1963-67. + | MARINUS TAGGED IN THE ST.MARYS RIVER AND LAKES HUR | 4746F |
| OGY /E NERVOUS SYSTEM CYTOL | MARINUS U.S.A. ADULT AMMOCOETE NERVOUS SYSTEM CYTO | 4625F |
| LYCOPEPTIDE. + *PETROMYZON | MARINUS U.S.A. ADULT BLOOD HISTOCHEMISTRY /SIS A G | 4954F |
| LOGY /LOGY EVOLUTION IMMUNO | MARINUS U.S.A. ADULT BLOODCYTOLOGY EVOLUTION IMMUN | 4550F |
| HEMISTRY MORTALITY OF /BIOC | MARINUS U.S.A. AMMOCOETE ADULT ANIMAL CHEMISTRYBIO | 4751F |
| YSTEM /MORPHOLOGY NERVOUS S | MARINUS U.S.A. AMMOCOETE ADULT MORPHOLOGY NERVOUS | 4595F |
| 3. PP.15-18. + *PETROMYZON | MARINUS U.S.A. AMMOCOETE BIOLOGY /AKE ONTARIO, 197 | 4867F |
| BUTION ECOLOGY SYSTEMATICS | MARINUS U.S.A. AMMOCOETEADULT METAMORPHOSIS DISTRI | 4676F |
| / LANDLOCKED ENDOCRINOLOGY | MARINUS U.S.A. ANADROMOUS LANDLOCKED ENDOCRINOLOGY | 5289F |
| UR CHEMISTRY ENDOCRINOLOGY | MARINUS U.S.A. GREAT LAKES ADULT AMMOCOETE BEHAVIO | 5366F |
| RY DISTRIBUTION MANAGEMENT | MARINUS U.S.A. GREAT LAKES ADULT AMMOCOETE CHEMIST | 5368F |
| RY DISTRIBUTION MANAGEMENT | MARINUS U.S.A. GREAT LAKES ADULT AMMOCOETE CHEMIST | 5502F |
| SIOLOGY /NERVOUS SYSTEM PHY | MARINUS U.S.A. GREAT LAKES ADULT NERVOUS SYSTEM PH | 5608F |
| IOLOGY / NERVOUS SYSTEMPHYS | MARINUS U.S.A. GREAT LAKES ADULT NERVOUS SYSTEMPHY | 5283F |
| HYSIOLOGY /NERVOUS SYSTEM P | MARINUS U.S.A. GREAT LAKES ANATOMY NERVOUS SYSTEM | 5280F |
| ANAGEMENT MORTALITY OF /N M | MARINUS U.S.A. GREAT LAKES CHEMISTRY DISTRIBUTION | 5255F |
| PP. 211-216. + *PETROMYZON | MARINUS U.S.A. GREAT LAKES MANAGEMENT /TLAMPREYS. | 5501F |
| THE LAMPREY. + *PETROMYZON | MARINUS U.S.A. GREAT LAKES NERVOUS SYSTEM /HNS OF | 5578F |
| OTREMO STOUTII *PETROMYZON | MARINUS U.S.A. IMMUNOLOGYANIMAL PISCES /I+ *POLIST | 4566F |
| TION CULTURE /IMAL DISTRIBU | MARINUS U.S.A. MANAGEMENT CHEMISTRY ANIMAL DISTRIB | 4756F |
| . PP. 52-70. + *PETROMYZON | MARINUS U.S.A. MORTALITY OF CHEMISTRY AMMOCOETE /Y | 4894F |
| SEA LAMPREY. + *PETROMYZON | MARINUS U.S.A. NERVOUS SYSTEM CYTOLOGY PHYSIOLOG / | 4542F |
| AL LAMPREYS. + *PETROMYZON | MARINUS U.S.A. RESPIRATION /  /VENTILATION OF LARV | 4787F |
| CANADA OSMOREGULATION AMMO | MARINUS. + *LAMPETRA PLANERI *LAMPETRA FLUVIATILIS | 4599F |
| THE SEA LAMPREY PETROMYZON | MARINUS. + ADULT AMMOCOETE /Y ON THE ORAL DISC OF | 4702F |
| IN A CYCLOSTOME PETROMYZON | MARINUS. + ADULT HISTOLOGY /*TYPES OF ISLET CELLS | 5205F |
| HE SEA LAMPREY, PETROMYZON | MARINUS. + ADULT TECHNIQUES ENDOCRINOLOGY BLOOD /I | 5208 |
| OF THE LAMPREY,PETROMYZON | MARINUS. + ADULT USA ENDOCRINOLOGY /YIPHERAL BLOOD | 4780F |
| RENTS IN LARVAL PETROMYZON | MARINUS. + AMMOCOETE NERVOUS SYSTEM /AR NERVE AFFE | 4951F |
| EEDING IONIC REGULATION PAR | MARINUS. + CANADA ADULT ANIMAL ECOLOGY FECUNDITY F | 5140F |
| TH PARASITISM BY PHYSIOLOGY | MARINUS. + CANADA DIGESTION EXCRETION FEEDING GROW | 4673F |
| ATION / HISTOLOGY OSMOREGUL | MARINUS. + CANADA ENDOCRINOLOGY HISTOLOGY OSMOREGU | 4687F |
| OGY EXCRETION /E ENDOCRINOL | MARINUS. + CANADA LAKE ONTARIO AMMOCOETE ENDOCRINO | 4639F |
| US CYTOLOGY DIGESTION HISTO | MARINUS. + CANADA NEW BRUNSWICK AMMOCOETE ANADROMO | 5518F |
| Y CYTOLOGY DIGESTION /TOLOG | MARINUS. + CANADA NEW BRUNSWICK ANADROMOUS HISTOLO | 5516F |
| NDOCRINOLOGY /OREGULATION E | MARINUS. + NEW BRUNSWICK AMMOCOETE OSMOREGULATION | 5604F |
| OSMOREGULATION / HISTOLOGY | MARINUS. + ONTARIO MORPHOLOGY RESPIRATION HISTOLOG | 5085F |
| PRODUCTION /Y METABOLISM RE | MARINUS. + U.S.A. ADULT ENDOCRINOLOGY METABOLISM R | 4607F |
| HE SEA LAMPREY, PETROMYZON | MARINUS. + U.S.A. BIOLOGY INTEGUMENT HISTOLOGY /LT | 5235F |
| ORPHOSIS MOUTH /OLOGY METAM | MARINUS. + U.S.A. GREAT LAKES ADULT HISTOLOGY META | 5348F |
| M TECHNIQUES HISTOLOGY /STE | MARINUS. + VISION ADULT METAMORPHOSIS NERVOUS SYST | 5160F |
| HE SEA LAMPREY, PETROMYZON | MARINUS. / FICATION OF ADENOHYPOPHYSIAL CELLS OF T | 5567 |
| HE SEA LAMPREY, PETROMYZON | MARINUS. /MPTION, GROWTH AND HOST PREFERENCES OF T | 5610F |
| US SEA LAMPREY, PETROMYZON | MARINUS. /SAND SPAWNING ENERGETICS OF THE ANADROMO | 5507 |
| THE SEA LAMPREY PETROMYZON | MARINUS. /SIMPLANTATION OF AN ABDOMINAL WINDOW IN | 5613 |
| ED SEA LAMPREY, PETROMYZON | MARINUS. /TIC AND IONIC CONCENTRATIONS IN LANDLOCK | 5344F |
| + GREAT LAKES GROWTH MORTA | MARINUS) AND THE ASSOCIATEDRATE OF HOST MORTALITY. | 5340F |
| US SEA LAMPREY (PETROMYZON | MARINUS) IN RELATION TO AMBIENT SALINITY. /AADROMO | 5611F |
| HE SEA LAMPREY (PETROMYZON | MARINUS). + ANIMAL ADULT NERVOUS SYSTEM VISION /IT | 5130F |
| T SEA LAMPREYS (PETROMYZON | MARINUS). + BEHAVIOUR LIGHT TEMPERATURE /E IN ADUL | 5298F |
| AL SEA LAMPREYS(PETROMYZON | MARINUS). /LRUCTURE OF THE PITUITARY GLAND IN LARV | 5339 |
| HE SEA LAMPREY, PETROMYZON | MARINUS, EMBRYO. + EMBRYOLOGY / /CYTOKINESIS IN T | 5374F |
| IN SEA LAMPREY, PETROMYZON | MARINUS, EMBRYOS. + EMBRYOLOGY /HE ON DEVELOPMENT | 5373 |
| HE SEA LAMPREY, PETROMYZON | MARINUS, EMBRYOS. + EMBRYOLOGY CHEMISTRY /SNT IN T | 5375F |
| E STREAMS TRIBUTARY TO LAKE | MARINUS, LARVAL (AMMOCOETE) GROWTH RATES FROM THRE | 5133F |

INGDOM ADULT ENDOCRINOLOGY
Y ENDOCRINOLOGY MANAGEMENT
AT LAKES CANADA MANAGEMENT
 ECOLOGY GROWTH LIFE CYCLE
ON. + SWEDEN ADULT ECOLOGY
 1975. P.90 + CANADA ADULT
 + CANADA ADULT LOCOMOTION
ADULT BIOLOGY DISTRIBUTION
. PP.70-73. + CANADA ADULT
 ADULT BLOOD ENDOCRINOLOGY
LUTION BLOOD GONADOGENESIS
 CANADA ADULT PIGMENTATION
SPIRATION ADULT METABOLISM
SASMOIDES. + CANADA PISCES
S SEA LAMPREY, PETROMYZON M
OLOGY REPRODUCTIONCOMOTION
AN DISTRIBUTION LOCOMOTION
 + JAPAN BEHAVIOUR FEEDING
ETROMYZON MARINUS SPAWNING
L. + BLOOD ADULT AMMOCOETE
Y AT KETTLE FALLS. + ADULT
TION PARASITISM BY FEEDING
ION GROWTH /      /SEASONAL
MOUS SPAWNING MIGRATION. +
TH LIFECYCLE METAMORPHOSIS
LUVIATILIS. + RUSSIA ADULT
BIOCHEMISTRY ENDOCRINOLOGY
+ ENDOCRINOLOGY PHYSIOLOGY
ENDOCRINOLOGY IMMUNOLOGY /
STOLOGY /TION PHYSIOLOGY HI
SSIA ADULT BLOOD ENDOCRINOL
ATION OF THE KIDNEY AND PRO
PETRA FLUVIATILIS. [MICRO D
Y AND PROXIMAL REABSORPTION
I BIKARBONATA NATRIYA V PRO
 CHEMISTRY /O USA. + U.S.A.
 *LAMPETRA TRIDENTATA *LAMP
/COETE ANATOMY MORPHOMETRY
ENT /LOGY CHEMISTRY MANAGEM
 IN THE BLOOD OF DIADROMOUS
YNTHETASE ACTIVITY OF THE L
RONMENTAL NA22 IN STUDIES O
TION OF FRESHWATER LAMPREYS
D ELECTRONMICROSCOPIC STUDI
DY ON THE NEPHRON IN THE LA
N FROMGLYCEROL IN ISOLATED
HRINE ONAMYLASE ACTIVITY IN
 OF SNAKE VENOM POLYPEPTIDE
RCUS I TSYPLYAT. [THE EFFEC
V VERKHNEGO IRTYSHA. [MORPH
N SODIUM CHLORIDE AND SODIU
AZHENII. [SYNAPTIC ACTIONS
OGICAL AND FUNCTIONAL PROPE
PTION RATE IN THE PROXIMAL
THE LAMPREY, LAMPETRA FLUVI
HE HISTOLOGY AND HISTOCHEMI
RELATION TO THE PSEUDOBRANC
*LAMPETRA FLUVIATILISADULT
. + *LAMPETRA AEPYPTERA U.S
EEK WATERSHED IN SOUTHEAST
+ *ICHTHYOMYZON CASTANEUS *
OETE ANIMAL BLOOD /RI AMMOC
F ALKYLISOCYANIDES WITH LAM
ETHYL-4-NITROPHENOL (TFM).
 OXIDATIVEPHOSPHORYLATION A
LIS BIOCHEMISTRY EVOLUTION
LAMPETRA FLUVIATILIS RUSSIA
FATTY ACIDS IN BRAIN MYELIN
HE FISH FAUNA OF THE LOWER
SPECIES FROM COASTAL MAINE
GEMENT /ORY STUDIES. + MANA
73) TO SEVERAL BIRD SPECIES
O EGGS AND FRY OF COHO SALM
T /SCES CHEMISTRY MANAGEMEN
IMAL /EMISTRY MANAGEMENT AN
IMAL ECOLOGY /MANAGEMENT AN
ROMETHYL-4-NITROPHENOL) IN
.] /TIONARY NEUROPHYSIOLOGY
 NERVOUS SYSTEM. + *LAMPETR
 TYPE-A,PRICE CURRENT METER
101. + *PETROMYZON MARINUS

| | |
|---|---|
| MIGRATION / EY, LAMPETRA FLUVIATILIS L. + UNITED K | 5176F |
| MIGRATION /A. GREAT LAKES ADULT AMMOCOETE CHEMISTR | 5263F |
| MIGRATION /C P.5. + *PETROMYZON MARINUS CANADA GRE | 5246F |
| MIGRATION /CPETRA FLUVIATILIS UNITED KINGDOM ADULT | 5009 |
| MIGRATION /CVIATILIS (L.) DURING THE SPAWNING SEAS | 5304F |
| MIGRATION /EDATA ON SEA LAMPREY FROM HUMBER RIVER, | 5227F |
| MIGRATION /KIN RELATION TO WEIGHT AND TEMPERATURE. | 4593F |
| MIGRATION /LROMYZON MARINUS *LAMPETRA FLUVIATILIS | 5506F |
| MIGRATION /N-RECAPTURE STUDIES, LAKE ONTARIO, 1974 | 5135F |
| MIGRATION /027-133. + *LAMPETRA FLUVIATILIS RUSSIA | 5361F |
| MIGRATION /R GREAT BRITAIN ADULT ENDOCRINOLOGY EVO | 5065F |
| MIGRATION /STHE SEA LAMPREY PETROMYZONMARINUS L. + | 4895F |
| MIGRATION /YOMYZON MARINUS *LAMPETRA TRIDENTATA RE | 5138F |
| MIGRATION ADULT ANIMAL GROWTH /N BASS MICROPTERUS | 4727F |
| MIGRATION AND SPAWNING ENERGETICS OF THE ANADROMOU | 5507 |
| MIGRATION BEHAVIOUR /RFIC OCEAN ENDOCRINOLOGY HIST | 5152F |
| MIGRATION BEHAVIOUR /UIATILIS SWEDEN ADULT CIRCADI | 5141F |
| MIGRATION CIRCADIAN /VE HAGFISHEPTATRETUS BURGERI. | 4699F |
| MIGRATION ENDOCRINOLOGY PHYSIOLOGY BLOOD /IN. + *P | 5083F |
| MIGRATION FEEDING METAMORPHOSIS /RROMYZON MARINUS | 5145 |
| MIGRATION MANAGEMENT /                  /THE LAMPRE | 5129F |
| MIGRATION MANAGEMENT /RORTH AMERICA ADULT DISTRIBU | 4746F |
| MIGRATION OF EPTATRETUS BURGERI. + JAPAN DISTRIBUT | 4943F |
| MIGRATION OSMOREGULATION PHYSIOLOGY HISTOLOGY /GRO | 5115F |
| MIGRATION PARASITISM BY MORPHOMETRY /TFEEDING GROW | 4710F |
| MIGRATION PHYSIOLOGY //S OF THE LAMPREY LAMPETRA F | 5218F |
| MIGRATION PHYSIOLOGY /T + *LAMPETRA PLANERI ADULT | 4886F |
| MIGRATION TECHNIQUES ADULT /OAMPETRA FLUVIATILIS. | 5005 |
| MIGRATION. + CANADA ATLANTIC OCEAN ADULT HISTOLOGY | 5539F |
| MIGRATION. + MIGRATION OSMOREGULATION PHYSIOLOGY H | 5115F |
| MIGRATION.] PP.127-133. + *LAMPETRA FLUVIATILIS RU | 5361F |
| MIKRODISSEKCHII I MIKROPUNKCHII). [NEPHRON ORGANIZ | 5064F |
| MIKRODISSEKTSIONNE ISSLEDOVANIE NEFRONA MINOGI LAM | 4694F |
| MIKROPUNKCHII). [NEPHRON ORGANIZATION OF THE KIDNE | 5064F |
| MIKROPUNKTSIONNOE IZUCHENIE REABSORBTSII KHLORIDA | 5507 |
| MILLER BLUE HOLE SANDUSKY COUNTY OHIO USA. + U.S.A | 4637F |
| MINIMA, A DWARFED PARASITIC LAMPREY FROM OREGON. + | 5555 |
| MINIMUS U.S.A. ADULT AMMOCOETE ANATOMY MORPHOMETRY | 4750F |
| MINNOWS EXPOSED TO TFM. + BIOLOGY CHEMISTRY MANAGE | 5379F |
| MINOG I RYB V PERIOD NERESTOVOIMIGRATSII. [INSULIN | 5361F |
| MINOG I SKORPEM. [EFFECT OF HORMONES ON GLYCOGEN S | 4598F |
| MINOG V ESTESTVENNYKH USLOVIYAKH. [THE USE OF ENVI | 4952F |
| MINOG V VODEMAKH BSSR. [THE DISTRIBUTION AND EVOLU | 5358F |
| MINOGI LAMPETRA FLUVIATILIS. [LIGHT MICROSCOPIC AN | 5173 |
| MINOGI LAMPETRA FLUVIATILIS. [MICRO DISSECTION STU | 4694F |
| MINOGI LAMPETRA FLUVIATILIS. [SYNTHESIS OF GLYCOGE | 5360F |
| MINOGI LAMPETRA FLUVIATILIS. [THE EFFECT OF EPINEP | 4561F |
| MINOGI LAMPETRA FLUVIATILIS.[THE INHIBITORY EFFECT | 4636F |
| MINOGI LAMPETRA FLUVIATILIS, SKORPENY SCORPAENA PO | 5267F |
| MINOGI LAMPETRA JAPONICA KESSLERI (ANIKIN) VODOEMO | 4623F |
| MINOGI LAMPETRAFLUVIATILIS. [MICROPUNCTURE STUDY O | 5507 |
| MINOGI PRI SUPRASPINAL'NOM 1 INTRASPINAL'NOM RAZDR | 4601F |
| MINOGI V REZNYE PERIODY ZHIZNENNOGO TSIKLA. [CYTOL | 5323F |
| MINOGI. [THE TEMPERATURE DEPENDENCE OF THE REABSOR | 5354F |
| MINOGI. [ULTRASTRUCTURE OF THE NEUROHYPOPHYSIS IN | 5071F |
| MIOTUBO IN LAMPETRA PLANERI (BLOCH). [RESEARCHON T | 4619F |
| MIRABILE OF THE FISH EYE. PART 2:DISTRIBUTION AND | 4961F |
| MIRABILE. + *MYXINE GLUTINOSA *PETROMYZON MARINUS | 4961F |
| MISSISSIPPI WITH NOTES ONHABITATS AND DISTRIBUTION | 5089F |
| MISSOURI AND NORTHEAST ARKANSAS U.S.A. /IE CANE CR | 5287 |
| MISSOURI RIVER GAVINS-POINT DAM TO RULO NEBRASKA. | 5197F |
| MIT AUSNAHME VON SAUGERN. + *LAMPETRA PLANERI AMMO | 5119F |
| MITNEUNAUGEN HAMOGLOBIN.[STUDIES ON THE REACTION O | 4760F |
| MITOCHONDRIA BY THE LAMPREY LARVICIDE 3-TRIFLUOROM | 5077 |
| MITOCHONDRIA FROM THE LAMPREY LAMPETRA FLUVIATILIS | 4613F |
| MITOCHONDRIA IN VERTEBRATES.] + *LAMPETRA FLUVIATI | 4741F |
| MITOCHONDRIAL MEMBRANES IN LOWER VERTEBRATES.] + * | 5278 |
| MITOKHONDRII V RYADUPOZVONOCHNYKH. [SPHINGOMYELIN | 4741F |
| MITTLEREN ELBE: DIE GENUTZTEN ARTEN, 1950-1975. [T | 5506F |
| MIXED FUNCTION OXIDASE ACTIVITY OF SEVERAL MARINE | 4604F |
| MIXTURE ASLAMPRICIDES IN LABORATORY STUDIES. + MAN | 5463F |
| MIXTURE OF TFM AND CLONITRALID (BAYLUSCIDE, BAYER | 5430F |
| MIXTURE TO FINGERLINGS OF SEVEN FISH SPECIES AND T | 5095F |
| MIXTURES OF CHEMICALS. + PISCES CHEMISTRY MANAGEME | 5396F |
| MODEL STREAM COMMUNITIES. + CHEMISTRY MANAGEMENT A | 5386F |
| MODEL STREAM COMMUNITIES. + CHEMISTRY MANAGEMENT A | 5419F |
| MODEL STREAM COMMUNITIES. /ERICIDE (TFM: 3-TRIFLUO | 5388 |
| MODERN ACHIEVEMENTS OF EVOLUTIONARY NEUROPHYSIOLOG | 4642F |
| MODES OF TRANSMISSION AND EVOLUTION OF THE CENTRAL | 5505 |
| MODIFICATION OF THE LISTENING DEVICE OF THE SMALL, | 5494F |
| MODIFICATIONS TO SEA LAMPREY BARRIER DAMS. PP.100- | 5264F |

| | | |
|---|---|---|
| PISCINE MYOGLOBINS AND HEMO | **MOLECULAR** CHARACTERISTICS OF CERTIN MAMMALIAN AND | 4936 |
| MINO TRANSFERASES IN MUSCLE | **MOLECULAR** EVOLUTION OF GYCOGEN PHOSPHORYLASE AND A | 5305 |
| CORD A PHYLOGENETIC SYNTHES | **MOLECULAR** EVOLUTION OF MYOGLOBIN AND THE FOSSIL RE | 4737 |
| -12 BINDING CAPACITIES AND | **MOLECULAR** SIZES OF THE BINDING PROTEINS. /IA FOR B | 5188 |
| *MYXINE GLUTINOSA *LAMPETRA | **MOLECULAR** STRUCTURE ANDBIOSYNTHESIS OF INSULIN. + | 4725F |
| A AND PHYSIOLOGICAL ACTIVIT | **MOLECULARSPECULAR** SPECTROSCOPY. PART I  NMR SPECTR | 4930 |
| RY ANALYSIS OF THE INSULIN | **MOLECULE.** / ARCH AIMING AT A FUNCTIONAL EVOLUTIONA | 5192 |
| APPENDIX STATISTICAL ANALY | **MOLLUSCA** FROM BORNU PROVINCE NORTHERN NIGERIA WITH | 4544 |
| R 2353 IN AQUATIC INVERTEBR | **MOLLUSCICIDE** BAYER 73 AND RESIDUE DYNAMICS OF BAYE | 5444F |
| AQUATIC SYSTEMS. PP. 109-11 | **MOLLUSCICIDES** ON THE MICROFLORA AND MICROFAUNA OF | 5441 |
| MUTAGENIC ACTIVITY OF SOME | **MOLLUSCICIDES.** /                      /SCREENING FOR | 5451 |
| SM IN YOUNG FISH.] /ETABOLI | **MOLODI** RYB. [METHODS OF DETERMINING ACTIVE METABOL | 4924 |
| TRA REISSNERI MONGOLIA DIST | **MONGOLEI.** [ON THE FISH FAUNA OF MONGOLIA. + *LAMPE | 4912F |
| LIA. + *LAMPETRA REISSNERI | **MONGOLIA** DISTRIBUTION / ON THE FISH FAUNA OF MONGO | 4912F |
| ON /ERI MONGOLIA DISTRIBUTI | **MONGOLIA.** + *LAMPETRA REISSNERI MONGOLIA DISTRIBUT | 4912F |
| STOCHEMISTRY MANAGEMENT /HI | **MONITORING** AID FOR SOME WATERBOURNE CHEMICALS. + H | 5382F |
| ANALYSIS. POSSIBLE AID IN | **MONITORING** WATER QUALITY /              /FISH BILE | 5402 |
| THE LAMPREY USING A NEW FI | **MONO** AMINES IN THE HYPOTHALAMUS AND SPINAL CORD OF | 4940F |
| NEUROHYPOPHYSIS OF THE HAGF | **MONOAMINE** OXIDASE AND ACETYLCHOLINESTERASE IN THE | 4591F |
| HYPOTHALAMO-HYPOPHYSIAL SYS | **MONOAMINE** OXIDASE AND ACETYLCHOLINESTERASE IN THE | 4882F |
| S FIBRES IN ATLANTIC HAGFIS | **MONOAMINERGICINNERVATION** OF SLOW NON-TWITCH MUSCLE | 5335 |
| CELLS OF THE VERTEBRATE HYP | **MONOAMINES** IN CEREBROSPINAL FLUID - CONTACT NERVE | 5000 |
| ELLS OF THE LAMPREY LAMPETR | **MONOAMINES** IN PERIVENTRICULAR HYPOTHALAMIC NERVE C | 4973 |
| F THE LAMPREY, LAMPETRA JAP | **MONOAMINES** IN THE HYPOTHALAMO-HYPOPHYSIAL REGION O | 5029F |
| THE HYPOTHALAMUS IN VERTEBR | **MONOAMINES** IN THE LIQUOR CONTACTING NERVECELLS IN | 4606F |
| THE HYPOTHALAMUS OF THE LA | **MONOAMINES** IN THE LIQUOR-CONTACTING NERVE CELLS IN | 4941F |
| OTALAMUSAU POZVONOCHNYKH. [ | **MONOAMINY** VLIKVOR-KOTAKTNYKH NERVNYKH KLETKAKH GIP | 4606F |
| SOIDS OF HAGFISH. + CYTOLOG | **MONONUCLEAR** PHAGOCYTES KUPFFER CELLS IN LIVER SINU | 5591 |
| . UPTAKE OF RADIO LABELLED | **MONOSACCARIDES** AND AMINO ACIDS. /IEPATRETUS-STOUTI | 5166 |
| ON OSMOREGULATION /REGULATI | **MONSTROSA.** + NORWAY PHYSIOLOGY BLOOD IONIC REGULAT | 4811F |
| NOLOGY HISTOCHEMISTRY EVOLU | **MORDACIA** MORDAX *GEOTRIA *LAMPETRA PLANERI ENDOCRI | 4706F |
| TOCHEMISTRY EVOLUTION / HIS | **MORDAX** *GEOTRIA *LAMPETRA PLANERI ENDOCRINOLOGY HI | 4706F |
| SKOI MINOGI LAMPETRA JAPONI | **MORFOLOGICHESKIE** OSOBENNOSTI I IZMENCHIVOST' SIBIR | 4623F |
| HECHBYKH KLETOK ARTERIAL'NY | **MORFOLOGICHESKIE** OSOBENNOSTI PINOTSITOZA RLADKOMYS | 5322F |
| EM PHYSIOLOGY ANIMAL BIOCHE | **MORPHINE-LIKE** PEPTIDE "ENKEPHALIN". + NERVOUS SYST | 5158 |
| FORE AND AFTER METAMORPHOSI | **MORPHOGENESIS** AND BIOSYNTHESIS OF THYROGLOBULIN BE | 5559 |
| NEPHROS DURING METAMORPHOSI | **MORPHOGENESIS** AND GROWTH OF THE DEFINITIVE OPISTHO | 5350F |
| NEY IN THEANADROMOUS SEA LA | **MORPHOGENESIS** AND GROWTH OF THE OPISTHONEPHRIC KID | 4698 |
| ESIA INLAMPREY LIVER. + *LA | **MORPHOLOGIC** AND HISTOCHEMICAL STUDY OF BILIARY ATR | 4734F |
| S GASTROPODA PROSOBRANCHIA | **MORPHOLOGICAL** AND ANALYTICAL DATA. /HOSTOMA-ELEGAN | 5149 |
| Y /ICAL STUDIES. + HISTOLOG | **MORPHOLOGICAL** AND IMMUNOLOGICAL STUDIES. + HISTOLO | 5059 |
| INLAMPREY SPINAL CORD. + *P | **MORPHOLOGICAL** CORRELATES OF SYNAPTIC TRANSMISSION | 5020F |
| SEA LAMPREY, PETROMYZON MA | **MORPHOLOGICAL** DEVELOPMENT OF THE INTEGUMENT OF THE | 5235F |
| CRO TUBULES IN AXONAL TRANS | **MORPHOLOGICAL** EVIDENCE FOR THE PARTICIPATION OF MI | 4818 |
| LL AND TASTE. /        /NEW | **MORPHOLOGICAL** FOUNDATIONS OF THE PHYSIOLOGY OF SME | 5297 |
| IBERIAN LAMPREY LAMPETRA JA | **MORPHOLOGICAL** PECULIARITIES AND CHANGEABILITY OF S | 4623F |
| TH MUSCLE CELLS IN ARTERIAL | **MORPHOLOGICAL** PECULIARITIES OF PINOCYTOSIS OF SMOO | 5322F |
| ES. + IMMUNOLOGY HISTOCHEMI | **MORPHOLOGICAL,** HISTOCHEMICAL AND IMMUNOLOGIC STUDI | 5058 |
| DULT MUSCLE HISTOCHEMISTRY | **MORPHOLOGY** /RES. + *LAMPETRA FLUVIATILIS FINLAND A | 4876F |
| TORY SYSTEM HISTOCHEMISTRY | **MORPHOLOGY** CYTOLOGY /SLAMOTTENII AMMOCOETE CIRCULA | 4998F |
| OID SYNTHESIZING CELLULAR S | **MORPHOLOGY** HISTOCHEMISTRY AND BIOCHEMISTRY OF STER | 5190 |
| NUS U.S.A. AMMOCOETE ADULT | **MORPHOLOGY** NERVOUS SYSTEM /GRD. + *PETROMYZON MARI | 4595F |
| CT OF THELARVAL LAMPREY PET | **MORPHOLOGY** OF THE EPITHELIUM IN THE ALIMENTARY TRA | 5518F |
| ULT SEA LAMPREY PETROMYZON | **MORPHOLOGY** OF THE GILLS OF LARVAL AND PARASITIC AD | 5085F |
| .MYXINE GLUTINOSA. + ATLANTI | **MORPHOLOGY** OF THE KIDNEY OF THE ATLANTIC HAGFISH, | 4568 |
| LIS L. + POLAND ADULT HISTO | **MORPHOLOGY** OF THE RIVER LAMPREY, LAMPETRA FLUVIATI | 4778F |
| ROMYZON MARINUS. + ONTARIO | **MORPHOLOGY** RESPIRATION HISTOLOGY OSMOREGULATION /N | 5085F |
| HISTOLOGY /        /GENERAL | **MORPHOLOGY.** PP. 74-82. + AMMOCOETE SENSE RECEPTORS | 5057F |
| THE AMERICAN BROOK LAMPREY | **MORPHOMETRIC** AND MERISTIC STUDY OF A POPULATION OF | 5553 |
| ULES IN THE NEUROHYPOPHYSIS | **MORPHOMETRIC** CLASSIFICATION OF NEUROSECRETORY GRAN | 4538F |
| ULES IN THE NEUROHYPOPHYSIS | **MORPHOMETRIC** CLASSIFICATION OF NEUROSECRETORY GRAN | 5078F |
| AND LAMPETRA PLANERI (BLOC | **MORPHOMETRICS** OF THE LAMPREYS LAMPETRA FLUVIATILIS | 4877F |
| S CANADA GREAT LAKES ADULT | **MORPHOMETRY** / ARIO. PP.98-99. + *PETROMYZON MARINU | 5262F |
| L. + METAMORPHOSIS GROWTH | **MORPHOMETRY** / MOUS SEA LAMPREY, PETROMYZON MARINUS | 5537F |
| MARINUS GREAT LAKES ADULT | **MORPHOMETRY** / RCIAL FISHERMEN. P.98. + *PETROMYZON | 5261F |
| ROMYZON MARINUS MANAGEMENT | **MORPHOMETRY** / RRIERS, LAKE HURON. PP.94-97. + *PET | 5260F |
| TARIO. PP. 68-69. + CANADA | **MORPHOMETRY** /AD BY COMMERCIAL FISHERMEN IN LAKE ON | 4962F |
| . PP.86-89. + CANADA ADULT | **MORPHOMETRY** /ASSESSMENT BARRIERS, LAKE HURON, 1975 | 5226F |
| BUTION ECOLOGY SYSTEMATICS | **MORPHOMETRY** /G AMMOCOETEADULT METAMORPHOSIS DISTRI | 4676F |
| . + *LETHENTERON JAPONICUM | **MORPHOMETRY** /HRON LAMOTTEI (LE SEUR), FROM ONTARIO | 5553 |
| RON, 1974. P. 68. + CANADA | **MORPHOMETRY** /MCTRICAL ASSESSMENT BARRIERS, LAKE HU | 5088F |
| *PETROMYZON MARINUS CANADA | **MORPHOMETRY** /MS, LAKE HURON, 1972. PP. 160-170. + | 5487 |
| IS MIGRATION PARASITISM BY | **MORPHOMETRY** /TFEEDING GROWTH LIFECYCLE METAMORPHOS | 4710F |
| A. ADULT AMMOCOETE ANATOMY | **MORPHOMETRY** /VRIDENTATUS *ENTOSPHENUS MINIMUS U.S. | 4750F |
| Y FECUNDITY FEEDING GROWTH | **MORPHOMETRY** BEHAVIOUR PARASITISM LIFE CYCLE /LOLOG | 4974F |
| TOMY MUSCLE HISTOCHEMISTRY | **MORPHOMETRY** BIOCHEMISTRY /A PP. 21-22. + ADULT ANA | 4841F |
| HOPHAGA U.S.A. SYSTEMATICS | **MORPHOMETRY** MUSCLE /PETRA TRIDENTATA *LAMPETRA LET | 5555 |
| OSA. P. 23. + ADULT VISION | **MORPHOMETRY** NERVOUS SYSTEM HISTOLOGY / XINE GLUTIN | 4845F |
| TAIN AMMOCOETE SYSTEMATICS | **MORPHOMETRY** PIGMENTATION MUSCLE /O LETHENTERON BRI | 4713 |
| AMMOCOETE ADULT FECUNDITY | **MORPHOMETRY** SPAWNING /A, U.S.) + U.S.A. LIFE CYCLE | 5139F |
| ATOMY DISTRIBUTION ECOLOGY | **MORPHOMETRY** SYSTEMATICS /MASIN.] + RUSSIA ADULT AN | 4623F |

214

| | | |
|---|---|---|
| LT ENDOCRINOLOGY HISTOLOGY | **NERVOUS** SYSTEM /*MPETRA FLUVIATILIS.] + RUSSIA ADU | 5071F |
| MYZON MARINUS. + AMMOCOETE | **NERVOUS** SYSTEM /AR NERVE AFFERENTS IN LARVAL PETRO | 4951F |
| ANIMAL CIRCULATORY SYSTEM | **NERVOUS** SYSTEM /ATOMES. PP. 123-136. + *PETROMYZON | 5117F |
| EM. + JAPAN ADULT CYTOLOGY | **NERVOUS** SYSTEM /CE LAMPREY USING A NEW FILTER SYST | 4940F |
| ISTOCHEMISTRY BIOCHEMISTRY | **NERVOUS** SYSTEM /DTRA FLUVIATILIS. + SWEDEN ADULT H | 4735F |
| AMMOCOETE HISTOLOGY MUSCLE | **NERVOUS** SYSTEM /E. + *LAMPETRA FLUVIATILIS RUSSIA | 5317 |
| DOCRINOLOGY HISTOCHEMISTRY | **NERVOUS** SYSTEM /EETUS BURGERI (GIRARD). + JAPAN EN | 4782F |
| LAMOTTEI U.S.A. PHYSIOLOGY | **NERVOUS** SYSTEM /FIS *LAMPETRA AEPYPTERA *LAMPETRA | 5606F |
| ODIDE. + PHYSIOLOGY ANIMAL | **NERVOUS** SYSTEM /FTS OF BRAIN BARRIER SYSTEMS FOR I | 5513F |
| AN ENDOCRINOLOGY HISTOLOGY | **NERVOUS** SYSTEM /GHAGFISH EPTATRETUS BURGERI. + JAP | 5078F |
| AMMOCOETE ADULT MORPHOLOGY | **NERVOUS** SYSTEM /GRD. + *PETROMYZON MARINUS U.S.A. | 4595F |
| MARINUS U.S.A. GREAT LAKES | **NERVOUS** SYSTEM /HNS OF THE LAMPREY. + *PETROMYZON | 5578F |
| GY HISTOLOGY METAMORPHOSIS | **NERVOUS** SYSTEM /MANERII ADULT CYTOLOGY ENDOCRINOLO | 4747F |
| LIS BIOCHEMISTRY EVOLUTION | **NERVOUS** SYSTEM /MERTEBRATES.] + *LAMPETRA FLUVIATI | 4741F |
| + CANADA HISTOLOGY ANATOMY | **NERVOUS** SYSTEM /MPTATRETUS STOUTI (CYCLOSTOMATA) | 4922F |
| LT ENDOCRINOLOGY HISTOLOGY | **NERVOUS** SYSTEM /O LAMPETRA TRIDENTATA. + JAPAN ADU | 5042F |
| IOLOGY EVOLUTION OLFACTION | **NERVOUS** SYSTEM /ONOSA *PETROMYZON ANATOMY ANIMAL B | 4771F |
| LT ENDOCRINOLOGY HISTOLOGY | **NERVOUS** SYSTEM /PON DANFORDI REGAN.] + ROMANIA ADU | 5072F |
| FLUVIATILIS ANATOMY ANIMAL | **NERVOUS** SYSTEM /RENERAL INTRODUCTION. + *LAMPETRA | 4697F |
| MYZON MARINUS ADULT VISION | **NERVOUS** SYSTEM /RN THE ADULT SEA LAMPREY. + *PETRO | 4798F |
| TROMYZON MARINUS AMMOCOETE | **NERVOUS** SYSTEM /SSECTED LAMPREY SPINAL CORD. + *PE | 5069 |
| T ENDOCRINOLOGY TECHNIQUES | **NERVOUS** SYSTEM BIOCHEMISTRY /PLUTINOSA CANADA ADUL | 4558F |
| PTATRETUS BURGERI. + JAPAN | **NERVOUS** SYSTEM CIRCULATION HISTOLOGY /OE HAGFISH E | 4928F |
| GY ADULT /PHYSIOLOGY CYTOLO | **NERVOUS** SYSTEM CIRCULATORY SYSTEM PHYSIOLOGY CYTOL | 4874 |
| NUS U.S.A. ADULT AMMOCOETE | **NERVOUS** SYSTEM CYTOLOGY /OPREY. + *PETROMYZON MARI | 4625F |
| *PETROMYZON MARINUS U.S.A. | **NERVOUS** SYSTEM CYTOLOGY PHYSIOLO /DEA LAMPREY. + | 4542F |
| FLUVIATILIS SWEDEN VISION | **NERVOUS** SYSTEM CYTOLOGYSWEDEN ADULT /O + *LAMPETRA | 5039 |
| . + ADULT HISTOLOGY MUSCLE | **NERVOUS** SYSTEM DIGESTION /M GLUTINOSA L. PP. 24-25 | 4839F |
| . + U.S.A. ADULT HISTOLOGY | **NERVOUS** SYSTEM ENDOCRINOLOGY /ELAMPETRA TRIDENTATA | 4577F |
| RA JAPONICA. + JAPAN ADULT | **NERVOUS** SYSTEM ENDOCRINOLOGY /STHE LAMPREY, LAMPET | 4882F |
| HE MIDBRAIN OF LAMPREYS. + | **NERVOUS** SYSTEM ENDOCRINOLOGY METAMORPHOSIS ADULT / | 4927 |
| LAMPETRA FLUVIATILIS ADULT | **NERVOUS** SYSTEM EVOLUTION / EBRATES EVOLUTION.] + * | 4540F |
| HINA.] + ANATOMY EVOLUTION | **NERVOUS** SYSTEM EVOLUTION / ER DEVONIAN OF YUNNAN C | 4709F |
| + AMMOCOETE ADULT ANATOMY | **NERVOUS** SYSTEM EVOLUTION /A AND COMPARATIVE STUDY. | 4559F |
| NE GLUTINOSA NORWAY MUSCLE | **NERVOUS** SYSTEM HISTOLOGY /  MUSCLE FIBERS. + *MYXI | 5524F |
| + ADULT VISION MORPHOMETRY | **NERVOUS** SYSTEM HISTOLOGY / XINE GLUTINOSA. P. 23. | 4845F |
| L. CYCLOSTOMATA.] + NORWAY | **NERVOUS** SYSTEM HISTOLOGY /BIS IN MYXINE GLUTINOSA | 4809F |
| JAPONICA. + ADULT CYTOLOGY | **NERVOUS** SYSTEM HISTOLOGY /G THE LAMPREY, LAMPETRA | 5596 |
| NUS ANADROMOUS LARVA ADULT | **NERVOUS** SYSTEM HISTOLOGY /MORD. + *PETROMYZON MARI | 5020F |
| LUVIATILIS. + RUSSIA ADULT | **NERVOUS** SYSTEM HISTOLOGY /T THE LAMPREY LAMPETRA F | 4973 |
| VERTEBRATE HYPOTHALAMUS. + | **NERVOUS** SYSTEM HISTOLOGY BIOLOGY /CE CELLS OF THE | 5000 |
| FLUVIATILIS ANIMAL MUSCLE | **NERVOUS** SYSTEM HISTOLOGY LOCOMOTION ANATOMY / ETRA | 5079F |
| MYXINE GLUTINOSA. + U.S.A. | **NERVOUS** SYSTEM HISTOLOGY TECHNIQUES /ER SYSTEM OF | 4805F |
| + *MYXINE GLUTINOSA NORWAY | **NERVOUS** SYSTEM MUSCLE /IO FIBROBLAST CELL NUCLEI. | 5171 |
| ITS COMPARATIVE ANATOMY WIT | **NERVOUS** SYSTEM OF VERTEBRATES: A GENERALSURVEY OF | 4730 |
| NERVES. + SWEDEN HISTOLOGY | **NERVOUS** SYSTEM OLFACTION /HSTRUCTURE OF OLFACTORY | 4543F |
| + *CYCLOSTOMATA EVOLUTION | **NERVOUS** SYSTEM PHYSIOLOGY / AMMALIAN VERTEBRATES.] | 5181F |
| TEBRATE CENTRAL SYNAPSE. + | **NERVOUS** SYSTEM PHYSIOLOGY / ANSFER CURVES AT A VER | 4714 |
| VIATILIS RUSSIA METABOLISM | **NERVOUS** SYSTEM PHYSIOLOGY / RATES. + *LAMPETRA FLU | 5281F |
| TES. + BLOOD ENDOCRINOLOGY | **NERVOUS** SYSTEM PHYSIOLOGY /,YSTEMS FOR NONELECTROL | 5515 |
| S RUSSIA BIOLOGY HISTOLOGY | **NERVOUS** SYSTEM PHYSIOLOGY /D+ *LAMPETRA FLUVIATILI | 5505 |
| U.S.A. GREAT LAKES ANATOMY | **NERVOUS** SYSTEM PHYSIOLOGY /L+ *PETROMYZON MARINUS | 5280F |
| S U.S.A. GREAT LAKES ADULT | **NERVOUS** SYSTEM PHYSIOLOGY /O. + *PETROMYZON MARINU | 5608F |
| PETRA FLUVIATILIS U.S.S.R. | **NERVOUS** SYSTEM PHYSIOLOGY /OL STIMULATION.] + *LAM | 4601F |
| TION OF OLFACTORY NERVE. + | **NERVOUS** SYSTEM PHYSIOLOGY /OUVIATILIS FROM STIMULA | 4739 |
| ANIMAL CIRCULATORY SYSTEM | **NERVOUS** SYSTEM PHYSIOLOGY /S *LAMPETRA FLUVIATILIS | 5162 |
| BIOCHEMISTRY ENDOCRINOLOGY | **NERVOUS** SYSTEM PHYSIOLOGY /UJAPAN ADULT HISTOLOGY | 4591F |
| NEUS U.S.A AMMOCOETEMUSCLE | **NERVOUS** SYSTEM PHYSIOLOGY ADULT /GHTHYMYZON CASTA | 4674F |
| PETRA FLUVIATILIS.] + USSR | **NERVOUS** SYSTEM PHYSIOLOGY ANIMAL /RTHE LAMPREY LAM | 4636F |
| KE PEPTIDE "ENKEPHALIN". + | **NERVOUS** SYSTEM PHYSIOLOGY ANIMAL BIOCHEMISTRY /ELI | 5158 |
| STOMATA.] + HISTOCHEMISTRY | **NERVOUS** SYSTEM PHYSIOLOGY CYTOLOGY PISCES /E CYCLO | 4700F |
| TILATION IN THE LAMPREY. + | **NERVOUS** SYSTEM PHYSIOLOGY RESPIRATION /IROL OF VEN | 5269F |
| (LAMPETRA FLUVIATILIS). + | **NERVOUS** SYSTEM PHYSIOLOGY VISION /TE RIVER LAMPREY | 5299F |
| CULATORY SYSTEM LOCOMOTION | **NERVOUS** SYSTEM RESPIRATION /RTATA. + BEHAVIOUR CIR | 5464 |
| JAPONICUS. + ADULT ANIMAL | **NERVOUS** SYSTEM SENSE RECEPTORS HISTOLOGY /TSPHENUS | 4745F |
| *PETROMYZON MARINUS ADULT | **NERVOUS** SYSTEM SENSE RECEPTORS PHYSIOLOGY /OREY. + | 4643F |
| ER AT A VERTEBRATE CENTRAL | **NERVOUS** SYSTEM SYNAPSE. + *LAMPETRA BIOCHEMISTRY / | 4580F |
| LT ANATOMY EVOLUTION MOUTH | **NERVOUS** SYSTEM SYSTEMATICS /IR DU SPITSBERG. + ADU | 5346F |
| COETE HISTOLOGY LOCOMOTION | **NERVOUS** SYSTEM TECHNIQUES /(ETROMYZON MARINUS AMMO | 5026F |
| LAMPETRA JAPONICA. + ADULT | **NERVOUS** SYSTEM TECHNIQUES /EURONS OF THE LAMPREY, | 5572F |
| IN LAMPREY INTERNEURONS. + | **NERVOUS** SYSTEM TECHNIQUES /OAND GABA CONDUCTANCES | 5112F |
| VISION ADULT METAMORPHOSIS | **NERVOUS** SYSTEM TECHNIQUES HISTOLOGY /S MARINUS. + | 5160F |
| N IN LARVAL SEA LAMPREY. + | **NERVOUS** SYSTEM TECHNIQUES LOCOMOTION /E TRANSECTIO | 5579F |
| US L. + CYTOLOGY HISTOLOGY | **NERVOUS** SYSTEM VISION /AE LAMPREY PETROMYZON MARIN | 5245F |
| N MARINUS). + ANIMAL ADULT | **NERVOUS** SYSTEM VISION /ITHE SEA LAMPREY (PETROMYZO | 5130F |
| ETROMYZON MARINUS + U.S.A. | **NERVOUS** SYSTEM VISION HISTOLOGY /ELARVAL LAMPREY,P | 4785F |
| LFACTION /YTOLOGY BIOLOGY O | **NERVOUS** SYSTEM VISION TECHNIQUES CYTOLOGY BIOLOGY | 5586 |
| LOGY HISTOLOGY NERVOUS SYST | **NERVOUS** SYSTEM. + *LAMPETRA FLUVIATILIS RUSSIA BIO | 5505 |
| RY NEUROPHYSIOLOGY.] /TIONA | **NERVOUS** SYSTEMAND MODERN ACHIEVEMENTS OF EVOLUTION | 4642F |
| S U.S.A. GREAT LAKES ADULT | **NERVOUS** SYSTEMPHYSIOLOGY /CY. + *PETROMYZON MARINU | 5283F |
| LIS SWEDEN ADULT HISTOLOGY | **NERVOUS** SYSTEMVISION /ASTOMI. + *LAMPETRA FLUVIATI | 5214F |

INOSA L. /GFISH MYXINE GLUT
/DIE
N REPRESENTATIVES. /LEOSTEA
+ *PETROMYZON MARINUS CANA
. + JAPAN ADULT ENDOCRINOLO
MPETRA TRIDENTATA. + JAPAN
+ U.S.A. ADULT HISTOLOGY NE
ND MYOGLOBIN PHYLOGENIES. P
ERTAIN... + BIOCHEMISTRY /C
SE IN THE CYCLOSTOMEBRAIN.
+ JAPAN ADULT  HISTOLOGY CY
OLFACTION SENSE RECEPTORS H
ENS) CYTOLOGY METABOLISM /T
RA FLUVIATILIS SWEDEN ADULT
GLOBULIN. /NATION OF THYRO
HE UPPER PRIPYAT RIVER. + L
E SOUTH AUSTRALIAN MUSEUM.
N AT AN INVARIANT RESIDUE.
   IODINE-125. + *PETROMYZON
AND ELECTRON MICROSCOPY. /
EPTORS BY ELECTRON MICROGRA
E IN THE CYCLOSTOME BRAIN.
   CANADA GREAT LAKES ADULT E
ANIMAL BLOOD BIOCHEMISTRY
CH AND TO THE SWIMBLADDER R
ARD TISSUE) IN THE DERMAL S
*PETROMYZON MARINUS *LAMPET
NS FROM LITTLE SKATE RAJAER
G PROTEIN. + *ENTOSPHENUS J
BINS.] + *LAMPETRA FLUVIATI
/ULES IN AXONAL TRANSPORT.
A IN ADULTO.[RESEARCHES ON
OOD HISTOLOGY HISTOCHEMISTR
LARVA INTO THE ADULT.] + IT
UND MYOGLOBINEN. [DETERMINA
Y SYSTEM IN ONTO- AND PHYLO
WING STIMULATION OF DORSAL
N HISTOLOGY NERVOUS SYSTEM
YCH I RYB. + POLAND BIBLIOG
/YB. + POLAND BIBLIOGRAPHY
   PULP TO A CORTICO STEROID
A L.). PP. 67-69. + NORWAY
TINOSA. PP.70-72. + SWEDEN
P. 76-79. + SWEDEN DENMARK
   THE HAGFISH. PP. 73-75. +
RIVER LAMPREY LAMPETRA FLUV
OF A FACILITATED DIFFUSION
/FOLDING
ATUM. + *BDELLESTOMA CIRRHA
(TFM) TO BROOK TROUT (LALV
AND HUMAN FACTOR XIII. /II
HE SEA LAMPREY PETROMYZON M
ACTION OF THE INTRINSIC LAM
EFFECT OF EPINEPHRINE ONAM
MONES ON GLYCOGEN SYNTHETAS
EPITHELIAL CELLS OF LAMPREY
Y LAMPETRA JAPONICA KESSLER
S IN ARTERIAL VESSELS OF CY
APESPORINE OR AREMYXINE IN
MMOCOETE ADULT FECUNDITY MO
RIBUTION /ES). + ADULT DIST
DULT METAMORPHOSIS DISTRIBU
RESHWATER FISHES OF THE SUS
IONAL STRUCTUREOF THE POLY
NIMAL BIOCHEMISTRY /OLOGY A
IDERGIC NEUROSECRETION.] +
REYS. + NERVOUS SYSTEM ENDO
ADULT CYTOLOGY ENDOCRINOLOG
IONS SPLIT BY LAMPREY AND M
BRINOGEN FIBRINAND FIBRINO
E SUPERIOR STREAM TREATMENT
E HURON STREAM TREATMENTS,
EROTILAPIA MULTISPINOSA (GU
S IN RELATION TO WEIGHT AND
ZON MARINUS. + NEW BRUNSWIC
OF DIADROMOUS LAMPREYS ANDF
PHYSIOLOGY / ENDOCRINOLOGY
IATILIS) BEFORE AND DURING
LIS DURING THE PRESPAWNING
ERIVER LAMPREY (LAMPETRA FL
IONAL PROPERTIES OF ISLET T

S BURGERI JAPAN BLOOD HISTO          PERIPHERAL BLOOD CELLS OF THE HAGFISH + *EPTATRETU       4796F
. + ADULT USA ENDOCRINOLOGY         PERIPHERAL BLOOD OF THE LAMPREY,PETROMYZON MARINUS        4780F
N SAUGERN. + *LAMPETRA PLAN         PERIPHEREN BLUTES BEI WIRBELTIEREN MIT AUSNAHME VO        5119F
MPREY LAMPETRA FLUVIATILIS.         PERIVENTRICULAR HYPOTHALAMIC NERVE CELLS OF THE LA        4973
TRIBUTARIES. PP. 79-80. + C         PERMANENT STAFF GAUGES ON LAKES HURON AND ONTARIO        4903F
Y /NERVE. + VISION HISTOLOG         PEROXIDASE ALONG THE OPTIC NERVE. + VISION HISTOLO       5279
  EPTATRETUS BURGERI. + CYTO        PEROXIDASE IN HEPATIC PARENCHYMAL CELLS OF HAGFISH       5590
  LAMPREY. PART 2: THE TUBUL        PEROXIDASE IN THE OPISTHONEPHRIC KIDNEY OF THE SEA        5045F
EPTATRETUS BURGERI (GIRARD)         PEROXIDASE INTO THE THIRD VENTRICLEOF THE HAGFISH        4782F
  PP. 53-56 + *LAMPETRA PLAN        PEROXIDE IN BLOOD CELLS OF FISHES AND CYCLOSTOMES.       4859F
M) IN AQUATIC ENVIRONMENTS.         PERSISTENCE OF 3-TRIFLUOROMETHYL-4-NITROPHENOL (TF        5384
E LAMPREY IN A COMPARATIVE          PERSPECTIVE. PP.97-145. /          /THE BRAIN OF TH       5592
-95. + *PETROMYZON MARINUS          PERSPECTIVES IN COMPARATIVE ISLET RESEARCH. PP. 83       5289F
WITH AN INTRODUCTION TOTHE          PERTIN....NEW YORK; ACADEMIC PRESS, 3V. /TANATOMY        4730
TIVE DIETARY TOXICITIES OF          PESTICIDES TO BIRDS. + MANAGEMENT /        /COMPARA       5416F
ON NERVOUS SYSTEM / OLFACTI         PETROMYZON ANATOMY ANIMAL BIOLOGY EVOLUTION OLFACT       4771F
/ORY SYSTEM NERVOUS SYSTEM          PETROMYZON ANIMAL CIRCULATORY SYSTEM NERVOUS SYSTE       5117F
LGERI *PETROMYZON MARINUS *         PETROMYZON DORSATUS ADULT ANIMAL ENDOCRINOLOGY / U       5066F
ISTRY MANAGEMENT /A. + CHEM         PETROMYZON FLUVIATILIS AND MYXINE GLUTINOSA. + CHE       5447
  TECHNIQUES GENETICS /ISTRY        PETROMYZON FLUVIATILIS. + SWEDEN ADULT BIOCHEMISTR       4562F
  /DIE PARIETALORGANE VON           PETROMYZON FLUVIATILIS. /D                               5549
S *LAMPETRA *EUDONTOMYZON *          PETROMYZON ICHTHYOMYZON *TETRAPLEURODON *ENTOSPENU       5391F
ES BASED ON THE TYPE GENUS          PETROMYZON LINNAEUS,1758. Z.N.(S.) + SYSTEMATICS /       4766F
CRETION HISTOCHEMISTRY / EX         PETROMYZON MARINUS + CANADA ATLANTIC OCEAN ADULT E       5350F
IOLOGY / CANADA AMMOCOETE B         PETROMYZON MARINUS *ICHTHYOMYZON CANADA AMMOCOETE        4862F
IOLOGY / CANADA AMMOCOETE B         PETROMYZON MARINUS *ICHTHYOMYZON CANADA AMMOCOETE        4863F
IOLOGY / CANADA AMMOCOETE B         PETROMYZON MARINUS *ICHTHYOMYZON CANADA AMMOCOETE        4864F
IOLOGY / CANADA AMMOCOETE B         PETROMYZON MARINUS *ICHTHYOMYZON CANADA AMMOCOETE        4865F
MMOCOETEMUSCLE NERVOUS SYST         PETROMYZON MARINUS *ICHTHYOMYZON CASTANEUS U.S.A A       4674F
OTTEI CANADA AMMOCOETE DIST         PETROMYZON MARINUS *ICHTHYOMYZON SP. *LAMPETRA LAM       4645F
OTTII CANADA AMMOCOETE GROW         PETROMYZON MARINUS *ICHTHYOMYZON SP. *LAMPETRA LAM       5104F
OTTII CANADA AMMOCOETE GROW         PETROMYZON MARINUS *ICHTHYOMYZON SP. *LAMPETRA LAM       5105F
OTTEI CANADA AMMOCOETE MANA         PETROMYZON MARINUS *ICHTHYOMYZON SP. *LAMPETRA LAM       5109F
OTTICANADA AMMOCOETE CHEMIS         PETROMYZON MARINUS *ICHTHYOMYZON SP. *LAMPETRA LAM       5111F
S AMMOCOETE CHEMISTRY MANAG         PETROMYZON MARINUS *ICHTHYOMYZON SP. *LAMPETRA LAM       5491F
HENUS TRIDENTATUS *TETRAPLE         PETROMYZON MARINUS *ICHTHYOMYZON U.S.A. GREAT LAKE       5251F
NII *ICHTHYOMYZON FOSSOR AM         PETROMYZON MARINUS *ICHTHYOMYZON UNICUSPIS *ENTOSP       4752F
NII AMMOCOETE CIRCULATORY S         PETROMYZON MARINUS *LAMPETRA (LETHENTERON) LAMOTTE       4565F
RATIONNERVOUS SYSTEM /RESPI         PETROMYZON MARINUS *LAMPETRA (LETHENTERON) LAMOTTE       4998F
ERON JAPONICUM (LAMPETRA JA         PETROMYZON MARINUS *LAMPETRA AEPYPTERA U.S.A. RESP       4788F
  PLANERIGREAT BRITIAN CANAD        PETROMYZON MARINUS *LAMPETRA FLUVIATILIS * LETHENT       4770F
  PLANERI*ICHTHYOMYZON *LAMP        PETROMYZON MARINUS *LAMPETRA FLUVIATILIS *LAMPETRA       4651F
  PLANERI*LETHENTERON JAPONI        PETROMYZON MARINUS *LAMPETRA FLUVIATILIS *LAMPETRA       4713
  PLANERI*LETHENTERON JAPONI        PETROMYZON MARINUS *LAMPETRA FLUVIATILIS *LAMPETRA       5395F
LOGY DISTRIBUTION MIGRATION         PETROMYZON MARINUS *LAMPETRA FLUVIATILIS *LAMPETRA       5469F
BLOOD CIRCULATORY SYSTEM BI         PETROMYZON MARINUS *LAMPETRA FLUVIATILIS ADULT BIO       5506F
ULTNERVOUS SYSTEM SENSE REC         PETROMYZON MARINUS *LAMPETRA FLUVIATILIS L. ADULT        4760F
N EVOLUTION /LISADULT VISIO         PETROMYZON MARINUS *LAMPETRA FLUVIATILIS SWEDEN AD       4949F
Y DISTRIBUTION /EAST GERMAN         PETROMYZON MARINUS *LAMPETRA FLUVIATILISADULT VISI       4961F
AMPETRA JAPONICA (KESSLERI)         PETROMYZON MARINUS *LAMPETRA FLUVIATILISEAST GERMA       5186F
N SP CANADA MANAGEMENT GROW         PETROMYZON MARINUS *LAMPETRA JAPONICA (MARTENS) *L       5321F
VIATILIS*LAMPETRA TRIDENTAT         PETROMYZON MARINUS *LAMPETRA LAMOTTEI *ICHTHYOMYZO       5110F
ON SENSE RECEPTORS /OLFACTI         PETROMYZON MARINUS *LAMPETRA PLANERI *LAMPETRA FLU       4705F
N ADULT METABOLISM MIGRATIO         PETROMYZON MARINUS *LAMPETRA PLANERIANATOMY OLFACT       5047F
L ENDOCRINOLOGY /DULT ANIMA         PETROMYZON MARINUS *LAMPETRA TRIDENTATA RESPIRATIO       5138F
LOGY /RINOLOGY CANADA HISTO         PETROMYZON MARINUS *PETROMYZON DORSATUS ADULT ANIM       5066F
9. + *ICHTHYOMYZON FOSSOR *         PETROMYZON MARINUS ADULT ENDOCRINOLOGY CANADA HIST       4881F
TORS PHYSIOLOGY /ENSE RECEP         PETROMYZON MARINUS ADULT HISTOCHEMISTR /A 266:69-8       5206F
  THE ADULT SEA LAMPREY. + *        PETROMYZON MARINUS ADULT NERVOUS SYSTEM SENSE RECE       4643F
VOLUTION BLOOD HISTOLOGY /E         PETROMYZON MARINUS ADULT VISION NERVOUS SYSTEM /RN       4798F
STOLOGY /TMUSCLE ANATOMY HI         PETROMYZON MARINUS AMMOCOETE ADULT HISTOCHEMISTRY        4743F
S /E ANATOMY SENSE RECEPTOR         PETROMYZON MARINUS AMMOCOETE ADULTMUSCLE ANATOMY H       4840F
ECHNIQUE /STRY MANAGEMENT T         PETROMYZON MARINUS AMMOCOETE ANATOMY SENSE RECEPTO       4731F
ERVOUS SYSTEM TECHNIQUES /N         PETROMYZON MARINUS AMMOCOETE CHEMISTRY MANAGEMENT        5053F
ED LAMPREY SPINAL CORD. + *         PETROMYZON MARINUS AMMOCOETE HISTOLOGY LOCOMOTION        5026F
YSTEM HISTOLOGY / NERVOUS S         PETROMYZON MARINUS AMMOCOETE NERVOUS SYSTEM /SSECT       5069
BRAE ALPHA CRYATALLINS. + *         PETROMYZON MARINUS ANADROMOUS LARVA ADULT NERVOUS        5020F
  CYCLE. + CANADA U.S.A. ADU        PETROMYZON MARINUS ANIMAL EVOLUTION PHYSIOLOGY /LE       5024F
TINOSA *EPTATRETUS STOUTI           PETROMYZON MARINUS AT DIFFERENT STAGES IN ITS LIFE       4758F
ERTEBRATE TRANSFERRINS. + *         PETROMYZON MARINUS BIOCHEMISTRY DIGESTION /PNE GLU       4910F
  FIBRINOGEN AND FIBRIN. + *        PETROMYZON MARINUS BIOCHEMISTRY IMMUNOLOGY /PHER V       5004F
  RIVER,1974. PP. 66-67. + *        PETROMYZON MARINUS BLOOD HISTOCHEMISTRY / LAMPREY        4960F
EEN RIVER. PP. 171-172. + *         PETROMYZON MARINUS CANADA /EA LAMPREY IN ST. MARYS       5103F
RYS RIVER,1972. P. 166. + *         PETROMYZON MARINUS CANADA ADULT / DAM ON THE SAUG        5485F
ENN'S DAM, 1973. P. 88. + *         PETROMYZON MARINUS CANADA ADULT / AMPREY IN ST. MA       5488F
S RIVER1973. PP. 80-81. + *         PETROMYZON MARINUS CANADA ADULT /SMPREY TAKEN AT D       4848F
NDOCRINOLOGY HISTOLOGY CYTO         PETROMYZON MARINUS CANADA ADULT /TREY IN ST. MARY'       4906F
ROWTH HISTORY MANAGEMENT /G         PETROMYZON MARINUS CANADA ADULT AMMOCOETE ANIMAL E       4779F
HURON, 1973. PP. 81-84. + *         PETROMYZON MARINUS CANADA ADULT AMMOCOETE BIOLOGY        4908F
FISHERMEN, 1973. P. 85. + *         PETROMYZON MARINUS CANADA ADULT BIOLOGY /+s: LAKE        4905F
                                    PETROMYZON MARINUS CANADA ADULT BIOLOGY /MMERCIAL        4902F

| | | |
|---|---|---|
| RIVER, 1973. PP. 85-87. + * | PETROMYZON MARINUS CANADA ADULT BIOLOGY /PE HUMBER | 4900F |
| OF AMMOCOETE /RY MORTALITY | PETROMYZON MARINUS CANADA ADULT CHEMISTRY MORTALIT | 5237F |
| ER RIVER, 1972. P. 171. + * | PETROMYZON MARINUS CANADA ADULT MANAGEMENT /M HUMB | 5486F |
| PERATIONS: 1973. PP. 5. + * | PETROMYZON MARINUS CANADA ADULT MORTALITY OF /OR O | 4866F |
| MENT /T MORTALITY OF MANAGE | PETROMYZON MARINUS CANADA ADULT MORTALITY OF MANAG | 5481 |
| NTS, 1971-1973. PP. 77. + * | PETROMYZON MARINUS CANADA AMMOCOETE /+REAM TREATME | 4901F |
| NTS, 1971-1973. PP. 78. + * | PETROMYZON MARINUS CANADA AMMOCOETE /EREAM TREATME | 4904F |
| -20. + *LAMPETRA LAMOTTEI * | PETROMYZON MARINUS CANADA AMMOCOETE /M1973. PP. 19 | 4890F |
| TEM, 1972. PP. 101-103. + * | PETROMYZON MARINUS CANADA AMMOCOETE /ORN RIVER SYS | 5477F |
| KE ONTARIO 1973. PP.15. + * | PETROMYZON MARINUS CANADA AMMOCOETE BIOLOGY / F LA | 4861F |
| MARY'S RIVER.PP. 88-91. + * | PETROMYZON MARINUS CANADA AMMOCOETE BIOLOGY /TST. | 4826F |
| LITY OF /TE CHEMISTRY MORTA | PETROMYZON MARINUS CANADA AMMOCOETE CHEMISTRY MORT | 4889F |
| SEDIMENT-WATER SYSTEMS. + * | PETROMYZON MARINUS CANADA AMMOCOETE HISTORY. /CIN | 4736F |
| EMISTRY /TE MORTALITY OF CH | PETROMYZON MARINUS CANADA AMMOCOETE MORTALITY OF C | 4888F |
| EMISTRY /TE MORTALITY OF CH | PETROMYZON MARINUS CANADA AMMOCOETE MORTALITY OF C | 4893F |
| UTION / LAKES ADULT DISTRIB | PETROMYZON MARINUS CANADA GREAT LAKES ADULT DISTRI | 5258F |
| ON HISTOCHEMISTRY / EXCRETI | PETROMYZON MARINUS CANADA GREAT LAKES ADULT EXCRET | 5045F |
| ETRY /T LAKES ADULT MORPHOM | PETROMYZON MARINUS CANADA GREAT LAKES ADULT MORPHO | 5262F |
| EMISTRY DISTRIBUTION MANAGE | PETROMYZON MARINUS CANADA GREAT LAKES AMMOCOETE CH | 5253F |
| EMISTRY DISTRIBUTION MANAGE | PETROMYZON MARINUS CANADA GREAT LAKES AMMOCOETE CH | 5254F |
| EMISTRY DISTRIBUTION MANAGE | PETROMYZON MARINUS CANADA GREAT LAKES AMMOCOETE CH | 5256F |
| EMISTRY DISTRIBUTION MANAGE | PETROMYZON MARINUS CANADA GREAT LAKES AMMOCOETE CH | 5257F |
| TRIBUTION MANAGEMENT /E DIS | PETROMYZON MARINUS CANADA GREAT LAKES AMMOCOETE DI | 5248F |
| TRIBUTION MANAGEMENT /E DIS | PETROMYZON MARINUS CANADA GREAT LAKES AMMOCOETE DI | 5250F |
| AGEMENTMORTALITY OF /TE MAN | PETROMYZON MARINUS CANADA GREAT LAKES AMMOCOETE MA | 5252F |
| MENT MANAGE MIGRATION /NADA MANAGE | PETROMYZON MARINUS CANADA GREAT LAKES CANADA MANAG | 5246F |
| STRIBUTION MANAGEMENT MORTA | PETROMYZON MARINUS CANADA GREAT LAKES CHEMISTRY DI | 5249F |
| Y GONADOGENESIS HISTOCHEMIS | PETROMYZON MARINUS CANADA GREAT LAKES ENDOCRINOLOG | 5470F |
| R ADULT MANAGEMENT MORTALIT | PETROMYZON MARINUS CANADA GREAT LAKES LAKE SUPERIO | 5247F |
| KES FISHERY COMMISSION. + * | PETROMYZON MARINUS CANADA MANAGEMENT /LHE GREAT LA | 5194F |
| LAKE HURON, 1974. P. 5. + * | PETROMYZON MARINUS CANADA MANAGEMENT /NOPERATIONS, | 4907F |
| E CHEMISTRY /ADULT AMMOCOET | PETROMYZON MARINUS CANADA MANAGEMENT ADULT AMMOCOE | 5107F |
| E CHEMISTRY /ADULT AMMOCOET | PETROMYZON MARINUS CANADA MANAGEMENT ADULT AMMOCOE | 5108F |
| COETE ADULT /CHEMISTRY AMMO | PETROMYZON MARINUS CANADA MANAGEMENT CHEMISTRY AMM | 5106F |
| RON, 1972. PP. 160-170. + * | PETROMYZON MARINUS CANADA MORPHOMETRY /MS, LAKE HU | 5487 |
| MOCOETE /TY OF CHEMISTRY AM | PETROMYZON MARINUS CANADA MORTALITY OF CHEMISTRY A | 4892F |
| AMMOCOETE CHEMISTRY ENDOCR | PETROMYZON MARINUS CANADA U.S.A. GREAT LAKES ADULT | 5263F |
| STRY AMMOCOETE MANAGEMENT / | PETROMYZON MARINUS CHEMISTRY AMMOCOETE MANAGEMENT | 5050F |
| BLOOD /HEMISTRY TECHNIQUES | PETROMYZON MARINUS DENMARK BIOCHEMISTRY TECHNIQUES | 4574F |
| IGHT MICROSCOPY AND AUTO RA | PETROMYZON MARINUS DURING METAMORPHOSIS. PART 1: L | 5174F |
| E. + HISTOCHEMISTRY GONADOG | PETROMYZON MARINUS DURING THEIR PARASTIC LIFE STAG | 4659F |
| SUBMAMMALIAN CHORDATES. + * | PETROMYZON MARINUS ENDOCRINOLOGY /P MAMMALIAN AND | 5056F |
| ISTRY /GY EVOLUTION BIOCHEM | PETROMYZON MARINUS FRANCE BIOLOGY EVOLUTION BIOCHE | 4602F |
| ISTRY /GY EVOLUTION BIOCHEM | PETROMYZON MARINUS FRANCE BIOLOGY EVOLUTION BIOCHE | 5337F |
| CANADA BIOLOGY /IBUTARY. + | PETROMYZON MARINUS FROM A LAKE ONTARIO TRIBUTARY. | 4759F |
| IN MIDWATER OFF EUROPE. + * | PETROMYZON MARINUS GERMANY DISTRIBUTION ADULT /OY | 4733F |
| ERCIAL FISHERMEN. P.98. + * | PETROMYZON MARINUS GREAT LAKES ADULT MORPHOMETRY / | 5261F |
| ORTH AMERICA AMMOCOETE ADUL | PETROMYZON MARINUS IN LAKE SUPERIOR 1953-1970. + N | 4630F |
| RATURE. + CANADA ADULT LOCO | PETROMYZON MARINUS IN RELATION TO WEIGHT AND TEMPE | 4593F |
| M IRISH WATERS IN 1973. + * | PETROMYZON MARINUS IRELAND ADULT /EISHES TAKEN FRO | 4686F |
| GREAT LAKES ENDOCRINOLOGY | PETROMYZON MARINUS L. + *PETROMYZON MARINUS CANADA | 5470F |
| ATION FEEDING METAMORPHOSIS | PETROMYZON MARINUS L. + BLOOD ADULT AMMOCOETE MIGR | 5145 |
| OLOGY /IOCHEMISTRY ENDOCRIN | PETROMYZON MARINUS L. + BLOOD BIOCHEMISTRY ENDOCRI | 5325F |
| T AMMOCOETE BLOOD ENDOCRINO | PETROMYZON MARINUS L. + CANADA ATLANTIC OCEAN ADUL | 5303F |
| OGY /A ENDOCRINOLOGY HISTOL | PETROMYZON MARINUS L. + CANADA ENDOCRINOLOGY HISTO | 4781F |
| SYSTEM VISION /OGY NERVOUS | PETROMYZON MARINUS L. + CYTOLOGY HISTOLOGY NERVOUS | 5245F |
| GREAT LAKES ENDOCRINOLOGY E | PETROMYZON MARINUS L. + HISTOLOGY EXCRETION ADULT | 4897F |
| TOLOGY BIOCHEMISTRY /OGY CY | PETROMYZON MARINUS L. + IMMUNOLOGY ENDOCRINOLOGY C | 5538F |
| METRY /PHOSIS GROWTH MORPHO | PETROMYZON MARINUS L. + METAMORPHOSIS GROWTH MORPH | 5537F |
| IFE CYCLE BLOOD GROWTH META | PETROMYZON MARINUS L. + U.S.A. ECOLOGY AMMOCOETE L | 4975F |
| ANADROMOUS LARVAL LAMPREY, | PETROMYZON MARINUS L. /RRING METAMORPHOSIS OF THE | 4695 |
| A LAMPREY OF THE ATLANTIC, | PETROMYZON MARINUS L. /SAKES AND THE ANADROMOUS SE | 4726 |
| HE ANADROMOUS SEA LAMPREY, | PETROMYZON MARINUS L. /TEY DURING LARVAL LIFE IN T | 4539 |
| OCHEMISTRY HISTOLOGY HISTOC | PETROMYZON MARINUS L.. + CANADA AMMOCOETE ADULT BI | 4899F |
| CRETION HISTOLOGY HISTOCHEM | PETROMYZON MARINUS L.. + CANADA AMMOCOETE ADULT EX | 4898F |
| M GOTTIN-SEE BEI POTSDAM. / | PETROMYZON MARINUS L.) IM GOTTIN-SEE BEI POTSDAM. | 5580F |
| ON. + CANADA ATLANTIC OCEAN | PETROMYZON MARINUS L., DURING ITS UPSTREAM MIGRATI | 5539F |
| METAMORPHOSIS. /OPMENT AND | PETROMYZON MARINUS L., DURING LARVALDEVELOPMENT AN | 4698 |
| RRIER DAMS. PP.100-101. + * | PETROMYZON MARINUS MANAGEMENT /. TO SEA LAMPREY BA | 5264F |
| , LAKE HURON. PP.94-97. + * | PETROMYZON MARINUS MANAGEMENT MORPHOMETRY / RRIERS | 5260F |
| OLOGY TECHNIQUES /LOGY HIST | PETROMYZON MARINUS METAMORPHOSIS ENDOCRINOLOGY HIS | 5174F |
| USH. + CANADA MORTALITY BY | PETROMYZON MARINUS ON LAKE TROUT SALVELINUS NAMAYC | 4769F |
| ISCES CHEMISTRY MANAGEMENT | PETROMYZON MARINUS PISCES CHEMISTRY MANAGEMENT + P | 5397F |
| + CANADA ADULT PARASITISM | PETROMYZON MARINUS PREDATION ON FRESHWATERTELEOST. | 4754F |
| PHYSIOLOGY BLOOD /RINOLOGY | PETROMYZON MARINUS SPAWNING MIGRATION ENDOCRINOLOG | 5083F |
| D LAKES HURON AND MICHIGAN, | PETROMYZON MARINUS TAGGED IN THE ST.MARYS RIVER AN | 4746F |
| YSTEM CYTOLOGY /E NERVOUS S | PETROMYZON MARINUS U.S.A. ADULT AMMOCOETE NERVOUS | 4625F |
| Y /DULT BLOOD HISTOCHEMISTR | PETROMYZON MARINUS U.S.A. ADULT BLOOD HISTOCHEMIST | 4954F |
| TION IMMUNOLOGY /LOGY EVOLU | PETROMYZON MARINUS U.S.A. ADULT BLOODCYTOLOGY EVOL | 4550F |
| HEMISTRYBIOCHEMISTRY /MORTAL | PETROMYZON MARINUS U.S.A. AMMOCOETE ADULT ANIMAL C | 4751F |
| Y NERVOUS SYSTEM /MORPHOLOG | PETROMYZON MARINUS U.S.A. AMMOCOETE ADULT MORPHOLO | 4595F |

234

| | | |
|---|---|---|
| VER LAMPREY KIDNEY.] + *LAM | REABSORPTION RATE IN THE PROXIMAL TUBULE OF THE RI | 5354F |
| H SPECIES FROM THE HANFORD | REACH OF THE COLUMBIA RIVER. / VE ABUNDANCE OF FIS | 5569 |
| IN.] + *PETROMYZON MARINUS | REACTION OF ALKYLISOCYANIDES WITH LAMPREY HEMOGLOB | 4760F |
| S. / /THE SCHIFF | REACTION OF THE LAMPREY RETINA ENTOSPHNUS JAPONICU | 4658 |
| FLUVIATILIS *LAMPETRA PLANE | REACTION WITH ANTISERUM TO CAERULEIN. + *LAMPETRA | 5562F |
| PETRA TRIDENTATUS U.S.A. BL | REACTIONS OF LAMPREY FIBRINOGEN AND FIBRIN. + *LAM | 4594F |
| ONEPHRIC KIDNEY OF THESEA L | REACTIVITY IN THE TUBULAR EPITHELIUM OF THE OPISTH | 4898F |
| BIN.[STUDIES ON THE REACTIO | REAKTION VON ALKYLISOZYANIDEN MITNEUNAUGEN HAMOGLO | 4760F |
| FROM SEVERAL SPECIES OF FIS | REASSOCIATION AND HYBRIDIZATION PROPERTIES OF DNA | 5147 |
| ON MARINUS TAGGED IN THE ST | RECAPTURE OF PARASITIC PHASE SEA LAMPREYS PETROMYZ | 4746F |
| ND BIOCHEMISTRY OF STEROID | RECENT ADVANCES IN THE MORPHOLOGY HISTOCHEMISTRY A | 5190 |
| /EARLY AND | RECENT PRIMITIVE BRAIN FORMS. PP. 87-96. /R | 5594 |
| AORDINARY CONSERVATION OF B | RECEPTOR AND INSULIN OF THE ATLANTIC HAGFISH. EXTR | 5286 |
| ROM THE ATLANTIC HAGFISH MY | RECEPTOR BINDING AFFINITY AND POTENCY OF INSULIN F | 5154F |
| INOLOGY BIOCHEMISTRY /NDOCR | RECEPTOR BINDING REGION OF INSULIN. + ANIMAL ENDOC | 5601 |
| LIVER CELLS SPECIFICITY OF | RECEPTOR BINDING. / OF INSULIN WITH ISOLATED RAT | 5090 |
| RAJUNCTIONAL ACETYLCHOLINE | RECEPTOR FROM RAT MUSCLE. /I AND IODINATION OF EXT | 5014 |
| RNICA. /FROM TORPEDO CALIFO | RECEPTOR-RICH MEMBRANE VESICLES FROM TORPEDO CALIF | 5164 |
| OF BIO MEMBRANES AND PHOTO | RECEPTORDIFFERENTIATION. /SFORM CREATING FUNCTION | 4914 |
| ERIANATOMY OLFACTION SENSE | RECEPTORS /ALIS *PETROMYZON MARINUS *LAMPETRA PLAN | 5047F |
| L EVOLUTION CYTOLOGY SENSE | RECEPTORS /ETE: THE VERTEBRATE MEDIAN EYE. + ANIMA | 4555F |
| ADULTNERVOUS SYSTEM SENSE | RECEPTORS /HN MARINUS *LAMPETRA FLUVIATILIS SWEDEN | 4949F |
| OLFACTIONPHYSIOLOGY SENSE | RECEPTORS /ITINOSA CANADA ATLANTIC OCEAN BEHAVIOUR | 5219F |
| AGEMENT MORTALITY BY SENSE | RECEPTORS /OCHEMISTRY ENDOCRINOLOGY INNUNOLOGY MAN | 5366F |
| US AMMOCOETE ANATOMY SENSE | RECEPTORS /PENTATUS). PP. 5-7. + *PETROMYZON MARIN | 4731F |
| OLFACTION BEHAVIOUR SENSE | RECEPTORS /TTHE HAGFISH MYXINE GLUTINOSA. + SWEDEN | 5100F |
| AMPETRA FLUVIATILIS CYCLOST | RECEPTORS AND ASSOCIATED NEURONS INTHE RETINA OF L | 5001F |
| IES. /FIC AND ABLATION STUD | RECEPTORS BY ELECTRON MICROGRAFIC AND ABLATION STU | 4536 |
| . 74-82. + AMMOCOETE SENSE | RECEPTORS HISTOLOGY / /GENERAL MORPHOLOGY. PP | 5057F |
| DEN NORWAY OLFACTION SENSE | RECEPTORS HISTOLOGY /D OLFACTORY EPITHELIUM. + SWE | 4808F |
| UDY.] + NORWAY JAPAN SENSE | RECEPTORS HISTOLOGY /DRASTERELECTRONMICROSCOPIC ST | 5543F |
| NIMAL NERVOUS SYSTEM SENSE | RECEPTORS HISTOLOGY /TSPHENUS JAPONICUS. + ADULT A | 4745F |
| ADULT NERVOUS SYSTEM SENSE | RECEPTORS PHYSIOLOGY /OREY. + *PETROMYZON MARINUS | 4643F |
| ALS FROM LAMPREY OLFACTORY | RECEPTORS. / /INTRA CELLULAR POTENTI | 4728 |
| GLYCOGEN FROMGLYCEROL IN I | RECHNOI MINOGI LAMPETRA FLUVIATILIS. [SYNTHESIS OF | 5360F |
| Y EFFECT OF SNAKE VENOM POL | RECHNOI MINOGI LAMPETRA FLUVIATILIS.[THE INHIBITOR | 4636F |
| REABSORPTION RATE IN THE P | RECHNOI MINOGI. [THE TEMPERATURE DEPENDENCE OF THE | 5354F |
| F MYOGLOBIN AND THE FOSSIL | RECORD A PHYLOGENETIC SYNTHESIS. /SLAR EVOLUTION O | 4737 |
| LANES CHILE REGION. / MAGAL | RECORD MYXINE PETROMYZONIDAE IN WATERS OF THE MAGA | 5571 |
| YPTERA, (PISCES: PETROMYZON | RECORD OF THE LEAST BROOK LAMPREY, OKKELBERGIA AEP | 5523F |
| MATODA CUCULLANICAE) FROM S | RECORDS OF CUCULLANUS TRUTTAE (FABRICUS, 1794) (NE | 5467F |
| STEM IN ARKANSAS. + *ICHTHY | RECORDS OF FISHES FROM THE LOWER OUACHITA RIVER SY | 5202F |
| /1ST ARKANSAS U.S.A. | RECORDS OF LAMPETRA SPP. (PETROMYZONIDAE). /S | 5055 |
| SPECIES TO THE ICHTHYO FAU | RECORDS OF SOME ARKANSAS FISHES WITH ADDITION OF 3 | 4944F |
| TENII ANDOKKELBERGIA AEPYPT | RECORDS OF THE FRESH WATER LAMPREYS LAMPETRA LAMOT | 4720F |
| YPTERA (ABBOTT), FROM THE D | RECORDS OF THE LEAST BROOK LAMPREY OKKELBERGIA AEP | 4676F |
| CTION IN LARVAL SEA LAMPREY | RECOVERY AND REGENERATION AFTER SPINAL CORD TRANSE | 5579F |
| STOLOGY / / | RECRUITMENT IN MAMMALIAN BILE DUCT FORMATION. + HI | 5392 |
| OBRANCH AND MARINE TELEOST | RED CELLS. /TND IRON NUCLEOTIDE COMPLEXES IN ELASM | 5168 |
| ECHELLI, 1910) (NEMATODA: C | REDESCRIPTION OF TRUTTAEDACNITIS STELMIOIDES (VESS | 5327F |
| Y HISTOCHEMISTRY / HISTOLOG | REFERENCE TO AZUROPHIL LEUCOCYTES. + BLOOD HISTOLO | 5131F |
| 42-45. + *LAMPETRA FLUVIATI | REFERENCE TO THE STRUCTURE OF HEMOGLOBIN III. PP. | 4834F |
| THE ADULT.] + ITALY METAMOR | REFERENCE TO THE TRANSFORMATION OF THE LARVA INTO | 4807F |
| ARINUS ADULT VISION NERVOUS | REFLEXES IN THE ADULT SEA LAMPREY. + *PETROMYZON M | 4798F |
| HYPOPHYSIAL NEUROSECRETORY | REGAN. [EFFECT OFCHLORPROMAZINE OF HYPOTHALAMO AND | 5072F |
| X OF EUDONTOMYZON DANFORDI | REGAN. /UNS IN THE HYPOTHALAMIC HYPOPHYSIAL COMPLE | 4878F |
| VOUS SYSTEM / HISTOLOGY NER | REGAN.] + ROMANIA ADULT ENDOCRINOLOGY HISTOLOGY NE | 5072F |
| UND EUDONTOMYZON DANFORDI ( | REGAN). /TCHEN AMMOCOTEN LAMPETRA PLANERI (BLOCH) | 5331 |
| ER BASINS. / AND HORNAD RIV | REGARD TO POPULATION FROM THE POPRAD AND HORNAD RI | 5582 |
| AL SEA LAMPREY. + NERVOUS S | REGENERATION AFTER SPINAL CORD TRANSECTION IN LARV | 5579F |
| PINAL TRANSECTION IN LARVAL | REGENERATION OF MUELLER AND MAUTHNER AXONS AFTER S | 5026F |
| *PETROMYZON MARINUS AMMOCOE | REGENERATION OF TRANSECTED LAMPREY SPINAL CORD. + | 5069 |
| OGENESIS. + DENMARK ADULT E | REGENERATION, BLOOD GLUCOSE REGULATION, AND VITELL | 5509 |
| ). /(CYCLOSTOMES AND FISHES | REGION IN LOWER VERTEBRATES (CYCLOSTOMES AND FISHE | 4979 |
| INGERIKE GROUP OF THE OSLO | REGION NORWAY. /RAPHY AND FACIES ANALYSIS OF THE R | 5031 |
| TRY /NDOCRINOLOGY BIOCHEMIS | REGION OF INSULIN. + ANIMAL ENDOCRINOLOGY BIOCHEMI | 5601 |
| TRA JAPONICA JAPAN ADULT HI | REGION OF THE LAMPREY, LAMPETRA JAPONICA. + *LAMPE | 5029F |
| US. + CANADA LAKE ONTARIO A | REGION OF THE LARVAL SEA LAMPREY, PETROMYZON MARIN | 4639F |
| LT ECOLOGY GROWTH LIFE CYCL | REGION. + *LAMPETRA FLUVIATILIS UNITED KINGDOM ADU | 5009 |
| RS OF THE MAGALLANES CHILE | REGION. /INEW RECORD MYXINE PETROMYZONIDAE IN WATE | 5571 |
| ST OF FISHES IN THE GORKII | REGION.] + *PETROMYZONIDAE U.S.S.R. DISTRIBUTION / | 4650F |
| OF DORSAL PARTS OF THE SPIN | REGIONS OF THE LAMPREY BRAINFOLLOWING STIMULATION | 4612 |
| 976. PP. 189-199. + CHEMIST | REGISTRATION-ORIENTED RESEARCH ON LAMPRICIDES IN 1 | 5367F |
| 975. PP. 189-195. + CHEMIST | REGISTRATION-ORIENTED RESEARCH ON LAMPRICIDES IN 1 | 5500F |
| 310. + CHEMISTRY MANAGEMENT | REGISTRATION-ORIENTED RESEARCH ON TFM IN 1972. P. | 5472F |
| S OF THE HYPOTHALAMO-HYPOPH | REGULARITIES OF THE DEVELOPMENT OF NEUROHEMAL PART | 4556F |
| OLOGY OSMOREGULATION IONIC | REGULATION / F THE BLOOD AND MUSCLE CELLS. + PHYSI | 4767F |
| A FLUVIATILIS ANIMAL IONIC | REGULATION /AIN THE VERTEBRATE KIDNEY.] + *LAMPETR | 4670F |
| RA FLUVIATILIS BLOOD IONIC | REGULATION /EBLOOD SERUM OF VERTEBRATES. + *LAMPET | 5204 |
| ATILIS FINLAND ADULT IONIC | REGULATION /SATURAL CONDITIONS.] + *LAMPETRA FLUVI | 4952F |
| IOLOGY METAMORPHOSIS IONIC | REGULATION /SOCOETE ADULT SPAWNING LIFE CYCLE PHYS | 4599F |

Y LAMPETRA FLUVIATILIS.] +
SYSTEM / HISTOLOGY NERVOUS
/NERVOUS SYSTEM PHYSIOLOGY
2. + *LAMPETRA FLUVIATILIS
.] + *LAMPETRA FLUVIATILIS
THE KIDNIN THE LAMPREY.] +
INFLUENCE OF DIURETICS.] +
S. + *LAMPETRA FLUVIATILIS
FIBERS OF THE LAMPREY.] +
REZA. [COMBINED CHROMATOGRA
F ENVIRONMENTAL NA22 IN STU
ES OF CYCLOSTOMATA AND BONY
BRAIN MYELIN AND MITOCHONDR
LUNGFISHES (DIPNOI) AND TH
HA CYCLOSTOMATE (CYCLOSTOMA
RKII REGION.] + *PETROMYZON
LOSTOMATA) AND IN FISH(PISC
BLOOD OF DIADROMOUS LAMPREY
OF CERTAIN FISH.] + BIOCHEM
LISM / OF FISHES.] + METABO
.] + U.S.S.R. ANIMAL GENETI
OUNG FISH.] /ETABOLISM IN Y
MERCIAL FISHES OF THE USSR.
I. PASOZYTY KRAGLOUSTYCH I
(TEXT TO THE ATLAS OF COLO
RISUNKOV RYB). [COMMERCIAL
R, 1975. PP.96-98. + SEA LA
RINUS CANADA U.S.A. GREAT L
ANATOMY OLFACTION /ON SPP.
NIQUE. /ENOLS STAINING TECH
IN IMMATURE ADULTS OF THE L
IN LANDLOCKED SEA LAMPREY,
US) IN RELATION TO AMBIENT
CANADA PISCES MIGRATION ADU
E STAGES OF RAINBOW TROUT (
AYER 73) IN RAINBOW TROUT (
AYER 73) IN RAINBOW TROUT (
TRA PLANERI (BLOCH 1784) (C
OMATA.] + HISTOCHEMISTRY NE
TO 3-TRIFLUOROMETHYL-4-NITR
HARENGUS PALLASI) IN THE ST
DING GROWTH /STRIBUTION FEE
Y MANAGEMENT /IMAL CHEMISTR
UDY OF FISH SEGREGATION IN
ND TO EGGS AND FRY OF COHO
                    /ATLANTIC
,CANADA. + *LAMPETRA PLANER
TOXICANTS TO GREEN EGGS OF
73) TO 4 BIRD SPECIES. /ER
*EPTATRETUS STOUTI *PETROMY
ND PROTOPTERUS AND THOSE OF
OPHENOL (TFM). + CHEMISTRY
ISM BY /ORTALITY BY PARASIT
ION /MATICS ADULT DISTRIBU
TION MIDDLE DEVONIAN GASPE
IONOF THE MILLER BLUE HOLE
CES DU SYSTEMECIRCULATOIRE
.A. /        /FISHES OF THE
. + *MYXINE GLUTINOSA NORWA
IATILIS U.S.S.R. ANIMAL BIO
IN CYLLOSTOMES AND FISHES.
ANDVENTRICLE OF THE CYCLOST
F SOME MARINE AND FRESH WAT
MYXINE GLUTINOSA L. + NORWA
AKTIVNOSTI U KRUGLOROTYKH 1
8. [FRACTIONATION OF PROTEI
ROWTH /ATION ADULT ANIMAL G
ROMYZONIDAE. + *CASPIOMYZON
ANADA ADULT /YZON MARINUS C
/RI AMMOCOETE ANIMAL BLOOD
Y COMMISSION. /LAKES FISHER
RY COMMISSION. /LAKES FISHE
ERY COMMISSION. + SEA LAMPR
ERY COMMISSION. + *PETROMYZ
UTION /SA SWEDEN BLOOD EVOL
LED ENDOTHELIUM OF THE HAGF
LBIOLOGICAL TISSUES IN THE
F THELAMPREY, LAMPETRA FLUV
IN INFISH TEETH. + HISTOLOG
R WALL INTHE LAMPREY LAMPET
RINOLOGY HISTOLOGY BIOCHEMI

242

RUSSIA BIOCHEMISTRY METABOLISM MUSCLE /ATHE LAMPRE    5360F
RUSSIA BIOLOGY CIRCULATORY SYSTEM HISTOLOGY NERVOU    5504
RUSSIA BIOLOGY HISTOLOGY NERVOUS SYSTEM PHYSIOLOGY    5505
RUSSIA ENDOCRINOLOGY METABOLISM MUSCLE /+PP.345-36    5123F
RUSSIA EXCRETION /PULE OF THE RIVER LAMPREY KIDNEY    5354F
RUSSIA EXCRETION PHYSIOLOGY /BPROXIMAL TUBULES OF     5507
RUSSIA IONIC REGULATION /UA FLUVIATILIS UNDER THE     5210F
RUSSIA METABOLISM NERVOUS SYSTEM PHYSIOLOGY / RATE    5281F
RUSSIA MUSCLE /ECHOLINE CONTRACTURES INFAST MUSCLE    5094F
RY' SOCHETANIEM METODOV KHROMATOGRAFII 1 ELEKTROFO    4896F
RY' 1 MINOG V ESTESTVENNYKH USLOVIYAKH. [THE USE O    4952F
RY'. [PHOSPHORYLASE ISOENZYMES FROM SKELETAL MUSCL    4887F
RYADUPOZVONOCHNYKH. [SPHINGOMYELIN FATTY ACIDS IN     4741F
RYB (DIPNOI) I IKH POLOZHENIEV SISTEME. [STATUS OF    4567F
RYB (PISCES). [GLYCOGEN CONTENT IN ORGANS OF AGNAT    5122F
RYB GOR'KOVSKOI OBLASTI. [LIST OF FISHES IN THE GO    4650F
RYB PISCES. [LEVEL OF GLYCEMIA IN CYCLOSTOMES (CYC    4681F
RYB V PERIOD NERESTOVOIMIGRATSII. [INSULIN IN THE     5361F
RYB. [CHARACTER OF THE NUCLEOTIDE SEQUENCE OF DNA     4548F
RYB. [CHLORIDE SECRETING CELLS OF FISHES.] + METAB    4616F
RYB. [GENOSYTEMATICS AND EVOLUTION OF FISH GENOMES    4802F
RYB. [METHODS OF DETERMINING ACTIVE METABOLISM IN     4924
RYB. [TEKST K ATLASU TSVETNYKH RISUNKOV RYB). [COM    5321F
RYB. + POLAND BIBLIOGRAPHY /EICZEJ POLSKI. CZESC I    4688F
RYB). [COMMERCIAL FISHES OF THE USSR. DESCRIPTION.    5321F
RYBY SSSR OPISENIIA RYB. (TEKST K ATLASU TSVETNYKH    5321F
SABLE RIVER EXPERIMENTAL MECHANICAL ASSESSMENT WEI    5230F
SABLE RIVER LAKE SUPERIOR. P.100. + *PETROMYZON MA    5263F
SAC NASAL CHEZ LES OSTEOSTRACI. + *PETROMYZON SPP.    4692F
SACCHARIDESIN THE PRESENCE OF PHENOLS STAINING TEC    5040
SALINITY ON PITUITARY THYROID AND INTERRENALCELLS     4687F
SALINITY ON SERUM OSMOTIC AND IONIC CONCENTRATIONS    5344F
SALINITY. /AADROMOUS SEA LAMPREY (PETROMYZON MARIN    5611F
SALMO GAIRDNERI AND BASS MICROPTERUS SASMOIDES. +     4727F
SALMO GAIRDNERI). + CHEMISTRY MANAGEMENT /LRLY LIF    5080F
SALMO GAIRDNERI). / LORO-4'-NITROSALICYLANILIDE (B    5438
SALMO GAIRDNERI). /HLORO-4'-NITROSALICYLANILIDE (B    5440
SALMO IRIDEUS (GIBBONS 1855) (TELEOSTEI) UND LAMPE    4700F
SALMO IRIDEUS TELEOST AND LAMPETRA PLANERI CYCLOST    4700F
SALMON (ONCORHYNCHUS KISUTCH) AFTERACUTE EXPOSURE     5422F
SALMON (ONCORHYNCHUS) AND PACIFIC HERRING (CLUPEA     5343F
SALMON AND HERRING STOCKS. + ADULT DISTRIBUTION FE    4633F
SALMON FROM THE UPPER GREAT LAKES + IMAL CHEMIST      5189F
SALMON SPAWNINGSTREAMS. /        /A PRELIMINARY ST    4638
SALMON. + MANAGEMENT /FNGS OF SEVEN FISH SPECIES A    5095F
SALMON: THEIR RESPONSES TO FISHERY CHEMICALS. /A      5429
SALMONAND CERTAIN OTHER RIVERS IN BRITISH COLUMBIA    4974F
SALMONIDS. + CHEMISTRY MANAGEMENT / ICITY OF FOUR     5449
SALT OF 2',5-DICHLORO-4'-NITROSALICYLANILIDE (BAYE    5456
SALTS IN FISHES. PP. 595-612. + *MYXINE GLUTINOSA     4910F
SALTS OF THE LUNGFISHES LEPIDOSIREN NEOCERATODUS A    5156F
SALVELINUS FONTINALIS) TO 3-TRIFLUOROMETHYL-4-NITR    5577F
SALVELINUS NAMAYCUSH. + CANADA MORTALITY BY PARASI    4769F
SAME GENUS. + AMMOCOETE SYSTEMATICS ADULT DISTRIBU    5542F
SANDSTONE CANADA.] + *ENTOSPHENUS JAPONICUS /SORMA    5022F
SANDUSKY COUNTY OHIO USA. + U.S.A. CHEMISTRY /HGAT    4637F
SANGUIN DES OSTEOSTRACES. /S SPECIALISATIONS PRECO    4959
SANTA-CLARA RIVER SYSTEM, SOUTHERN CALIFORNIA, U.S    5075
SARCOLEMMA IN THE SUBNEURALREGION OF MUSCLE FIBERS    5524F
SARCOPLASM IN LOWER VERTEBRATES.] + *LAMPETRA FLUV    4803F
SARCOPLASMIC PROTEINS AND THEIR ENZYMIC PROPERTIES    4896F
SARCOPLASMIC RETICULUM  IN THE PORTAL VEIN, HEART     4597F
SARCOPLASMIC RETICULUM AND THE SYSTEM IN MUSCLES O    4583
SARCOPLASMIC RETICULUM INTHE CARDIAC VENTRICLE OF     4792F
SARKOPLAZMATICHESKIKH 'ELKOV 1 IKH FERMENTATIVNOI     4896F
SARKOPLAZME MYSHTS NIZSHIKH POZVONOCHNYKH. PP.50-5    4803F
SASMOIDES. + CANADA PISCES MIGRATION ADULT ANIMAL     4727F
SATELLITE SPECIES AMONG THE HOLARCTIC LAMPREYS PET    5391F
SAUGEEN RIVER. PP. 171-172. + *PETROMYZON MARINUS     5485F
SAUGERN. + *LAMPETRA PLANERI AMMOCOETE ANIMAL BLOO    5119F
SAULT STE. MARIE, ONTARIO TO THE GREAT LAKES FISHE    4783F
SAULT STE. MARIE, ONTARIO, TO THE GREAT LAKES FISH    4784F
SAULT STE. MARIE, ONTARIO, TO THE GREAT LAKES FISH    4844F
SAULT STE. MARIE, ONTARIO, TO THE GREAT LAKES FISH    5194F
SCAGERAC SEA. + *MYXINE GLUTINOSA SWEDEN BLOOD EVO    5063F
SCANNING ELECTRON MICROSCOPE STUDY ON THE HIGH WAL    5293F
SCANNING ELECTRON MICROSCOPE.PP. 287-294. / NTERNA    4977
SCANNING ELECTRON MICROSCOPIC STUDY OF THE GILLS O    5587
SCANNING ELECTRON MICROSCOPY OF ENAMELOID AND DENT    4875F
SCANNING ELECTRON MICROSCOPY OF THE 3RD VENTRICULA    5187F
SCANNING ELECTRON MICROSCOPY. + SWEDEN ADULT ENDOC    4762F

INGERIKE GROUP STAGE 10 OF
LOGII. [THE IDEAS OF A.A. U
TRA LAMOTTEI *ICHTHYOMYZON
N GROWTH /COETE DISTRIBUTIO
GROWTH /MOCOETE MANAGEMENT
GROWTH /NAGEMENT AMMOCOETE
WTH MANAGEMENT /EMISTRY GRO
ISTRY /MMOCOETE GROWTH CHEM
ISTRY /MMOCOETE GROWTH CHEM
ESIS CYTOLOGY /RI GONADOGEN
TRIFLUOROMETHYL-4-NITROPHEN
PHS OF MAYFLIES (HEXAGENIA
O CONFORMERS, HYPO OSMO REG
HAGES IN THE RENAL URINARY
S /TERON ANATOMY SYSTEMATIC
COETE ADULT BIOLOGY GROWTH
DULT FECUNDITY MORPHOMETRY
  ANIMAL BLOOD BIOCHEMISTRY
Y METAMORPHOSIS LIFE CYCLE
MPETRA AEPYPTERA. + U.S.A.
PETROMYZON MARINUS. /PREY,
REGULATION /ORPHOSIS IONIC
RINOLOGY PHYSIOLOGY BLOOD /
IOLOGY HISTOLOGY /TION PHYS
ATILIS RUSSIA ADULT BLOOD E
M.] + *LAMPETRA FLUVIATILIS
GHT OF THERIVER LAMPREY (LA
T ENDOCRINOLOGY METABOLISM
GREAT BRITAIN ADULT HISTOLO
ENTATA RESPIRATION ADULT ME
N ADULT ECOLOGY MIGRATION /
E EASTERNBALTIC 1973. /    /
  EASTERN BALTIC. /         /
FISH SEGREGATION IN SALMON
II. PP. 42-45. + *LAMPETRA
LIFE HISTORY, 1960-72. PP.
GUIN DES OSTEOSTRACES. /SAN
  SYSTEME CIRCULATOIRE DES O
            /SOME ENDOCRINE
RCULATORY SYSTEM MUSCLE ANI
TATRETUS BURGERI SWEDEN CAN
TH AUSTRALIA WITH A KEY TO
E. + *CASPIOMYZON *PETROMYZ
UBLE FRACTIONS FROM LITTLE
  ALDEHYDEREDUCTASE. + ANIMA
EMENT /COHO SALMON. + MANAG
BENTHIC FISH FAUNA OF THE S
  PRIPYAT RIVER. + LAMPETRA
ULAR SIZES OF THE BINDING P
CHEMISTRY HISTOLOGY METABOL
R. /CH OF THE COLUMBIA RIVE
ONSIDERATION OF ENDANGERED
-4-NITROPHENOL (TFM) TO 10
(MICHX) PLANCHON AND MYRIOP
  JAPAN. /   /NOTES ON SOME
OM KLAMATH RIVER SYSTEM, CA
TRO BIOTRANSFORMATION. + MA
L BIOCHEMISTRY PHYSIOLOGY /
VER DRAINAGE ABOVE CONOWING
E FROM EASTERN TRIBUTARIES
ZONIDAE FROM SOUTH CENTRAL
ON MARINUS *LAMPETRA (LETHE
CREASER AND HUBBS, 1922, (
ADULT DISTRIBUTION /MATICS
  STATISTICAL ANALYSES OF 2
  U.S.A. SYSTEMATICS / GAGEI
MPETRA FLUVIATILIS ADULT BI
RY IMMUNOLOGY /L BIOCHEMIST
  BAYER 73) TO SEVERAL BIRD
N COASTAL MAINE USA MARINE
BIPHENLS IN VARIOUS MARINE
ILIDE (BAYER 73) TO 4 BIRD
  *EPTATRETUS DEANI *EPTATRE
LPHA-1 3 COLLAGENS. + ANIMA
HENUS JAPONICUS JAPAN CIRCU
                           /
  *LAMPETRA FLUVIATILIS ANIM
IMITIVE VERTEBRATE. /ERY PR
AN APPRAISAL OF THE PROBLE
H ISOLATED RAT LIVER CELLS
PARTICULAR REFERENCE TO THE

SOUTHERN NORWAY. / /MARINE CALCARENITES FROM THE R   4679
SOVREMENNYE DOSTIZHENIYA EVOLYUTSIONNOI NEIROFIZIO   4642F
SP CANADA MANAGEMENT GROWTH AMMOCOETE /NNUS *LAMPE   5110F
SP. *LAMPETRA LAMOTTEI CANADA AMMOCOETE DISTRIBUTI   4645F
SP. *LAMPETRA LAMOTTEI CANADA AMMOCOETE MANAGEMENT   5109F
SP. *LAMPETRA LAMOTTEI CANADA MANAGEMENT AMMOCOETE   5111F
SP. *LAMPETRA LAMOTTICANADA AMMOCOETE CHEMISTRY GR   5491F
SP. *LAMPETRA LAMOTTII CANADA AMMOCOETE GROWTH CHE   5104F
SP. *LAMPETRA LAMOTTII CANADA AMMOCOETE GROWTH CHE   5105F
SP. *MYXINE GLUTINOSA *EPTATRETUS BURGERI GONADOGE   4704F
SP.) AS AN INDICATOR OF ENVIRONMENTAL LEVELS OF 3-   5242F
SP.). + MANAGEMENT /9ROMETHYL-4-NITROPHENOL TO NYM   5137F
SPACE AND INTRA CELLULAR ION CONCENTRATIONS IN OSM   4931
SPACE. + ANIMAL EXCRETION TECHNIQUES CYTOLOGY /EOP   5593
SPADICEUS *LAMPETRA *LETHENTERON ANATOMY SYSTEMATI   4752F
SPAWNING /A, LAMPETRA PLANERI (BLOCH,1784). + AMMO   4764F
SPAWNING /A, U.S.) + U.S.A. LIFE CYCLE AMMOCOETE A   5139F
SPAWNING /TTTY-ACIDS IN THE BLOOD OF MARINE FISH +   5015
SPAWNING ADULT /±THE LARVA INTO THE ADULT.] + ITAL   4807F
SPAWNING ADULT /IT CONSTRUCTION OF THE LAMPREY, LA   4653F
SPAWNING ENERGETICS OF THE ANADROMOUS SEA LAMPREY,   5507
SPAWNING LIFE CYCLE PHYSIOLOGY METAMORPHOSIS IONIC   4599F
SPAWNING MIGRATION ENDOCRINOLOGY PHYSIOLOGY BLOOD   5083F
SPAWNING MIGRATION. + MIGRATION OSMOREGULATION PHY   5115F
SPAWNING MIGRATION.] PP.127-133. + *LAMPETRA FLUVI   5361F
SPAWNING OF RIVER AND BROOK LAMPREYS IN AN AQUARIU   5357F
SPAWNING PERIOD, LOCOMOTOR ACTIVITY AND LENGTH/WEI   5141F
SPAWNING REPRODUCTION /ELLOGENESIS. + DENMARK ADUL   5509
SPAWNING RIVER LAMPREY, LAMPETRA FLUVIATILIS L. +   5169F
SPAWNING RUN. + *PETROMYZON MARINUS *LAMPETRA TRID   5138F
SPAWNING SEASON. + SWEDEN ADULT ECOLOGY MIGRATION   5304F
SPAWNING STOCK AND FISHERY FOR RIVER LAMPREY IN TH   5012
SPAWNING STOCK AND FISHERY OF RIVER LAMPREY IN THE   4814
SPAWNINGSTREAMS. /         /A PRELIMINARY STUDY OF   4638
SPECIAL REFERENCE TO THE STRUCTURE OF HEMOGLOBIN I   4834F
SPECIAL REPORT ON THE STUDIES OF SEA LAMPREY EARLY   5473F
SPECIALISATIONS PRECOCES DU SYSTEMECIRCULATOIRE SA   4959
SPECIALISATIONS PRECOCES ET CARATERES PRIMITIFS DU   5547
SPECIALIZATION OF THE BIRD. /F                      5027
SPECIALIZATIONSIN VERTEBRATE CARDIAC MUSCLES. + CI   5602
SPECIES + *MYXINE GLUTINOSA *EPTATRETUS STOUTI *EP   4763F
SPECIES A NEW HAGFISH FAMILY EPTATRETIDAE FROM SOU   4992
SPECIES AMONG THE HOLARCTIC LAMPREYS PETROMYZONIDA   5391F
SPECIES AND PARTIAL CHARACTERIZATION IN HEPATICSOL   4972
SPECIES AND TISSUE DISTRIBUTION OF NADPH-DEPENDANT   5570
SPECIES AND TO EGGS AND FRY OF COHO SALMON. + MANA   5095F
SPECIES COMPOSITION AND ZOOGEOGRAPHIC ANALYSIS OF   5597
SPECIES COMPOSITION OF FISH POPULATION INTHE UPPER   4873
SPECIES DATA FOR B-12 BINDING CAPACITIES AND MOLEC   5188
SPECIES FROM COASTAL MAINE USA. + ANIMAL ADULT BIO   4604F
SPECIES FROM THE HANFORD REACH OF THE COLUMBIA RIV   5569
SPECIES IN LAND USE DECISIONS. /         /THE C     5319
SPECIES OF ALGAE. + MANAGEMENT /E3-TRIFLUOROMETHYL   5118F
SPECIES OF AQUATIC MACROPHYTES (ELODEA CANADENSIS   5421F
SPECIES OF COLOURLESS EUGLENOPHYCEAE FROM HOKKAIDO   4938
SPECIES OF ENTOSPHENUS GILL,1862 PETROMYZONIDAE FR   4750F
SPECIES OF FISH WITH PRELIMINARY NOTESON ITS IN VI   5404F
SPECIES OF FISH. + ANIMAL BIOCHEMISTRY PHYSIOLOGY   5147
SPECIES OF FRESHWATER FISHES OF THE SUSQUEHANNA RI   5068F
SPECIES OF LAMPREY GENUS LETHENTERON PETROMYZONIDA   5150
SPECIES OF LAMPREY OF THE GENUS ENTOSPENUS PETROMY   4821
SPECIES OF LAMPREYS. + *LAMPETRA PLANERI *PETROMYZ   4565F
SPECIES OF THE HOLARCTIC LAMPREY GENUS LETHENTERON   5542F
SPECIES OF THE SAME GENUS. + AMMOCOETE SYSTEMATICS   5542F
SPECIES PILA WERNEI AND ASPATHRIA COMPLANATA. /NIX   4544
SPECIES TO THE ICHTHYO FAUNA. + *ICHTHYOMYZON GAGE   4944F
SPECIES USED 1950-1975.] + *PETROMYZON MARINUS *LA   5506F
SPECIES. + *ENTOSPHENUS JAPONICUS ANIMAL BIOCHEMIS   5574F
SPECIES. + CHEMISTRY MANAGEMENT /PLID (BAYLUSCIDE,   5430F
SPECIES. + METABOLISM ANIMAL BIOCHEMISTRY / YMES I   4964F
SPECIES. / T-RELATED COMPOUNDS AND POLYCHLORINATED   4935
SPECIES. /OSALT OF 2',5-DICHLORO-4'-NITROSALICYLAN   5456
SPECIESOF HAGFISH MYXINIDAE. + *EPTATRETUS STOUTII   4704F
SPECIFIC CLEAVAGE OF TYPES ALPHA-1 2,ALPHA-2 AND A   4571
SPECIFIC GRANULES IN THE LAMPREY ATRIUM. + *ENTOSP   5266F
SPECIFIC GRANULES IN THE LAMPREY ATRIUM. /C         5531
SPECIFIC VOLUMES OF HEMOGLOBINS AND MYOGLOBINS.] +   5128F
SPECIFICITY AND NEGATIVE COOPERATIVITY IN A VERY P   5286
SPECIFICITY AND THE EMBRYONIC DERIVATION OF ORGANS   5093
SPECIFICITY OF RECEPTOR BINDING. / OF INSULIN WIT   5090
SPECIMENS AT DIFFERENT STAGES OF DEVELOPMENT WITH   4807F

FISHES. + ANTARCTICA ANIMAL    SPECIMENS IN THE SOUTH AUSTRALIAN MUSEUM. PART 1:    4718F
BIOCHEMISTRY /CERTAIN... +    SPECTRA AND PHYSIOLOGICAL ACTIVITY OF CERTAIN... +    4930
TROPHENOL) BY LUMINESCENCE    SPECTROPHOMETRY. /I OF TFM (3-TRIFLUOROMETHYL-4-NI    5381
S STOUTI /TINOSA *EPTATRETU    SPECTROSCOPY. P. 46. + *MYXINE GLUTINOSA *EPTATRET    4832F
L ACTIVITY OF CERTAIN... +    SPECTROSCOPY. PART I  NMR SPECTRA AND PHYSIOLOGICA    4930
ACROSOMAL SYSTEM IN EARLY    SPERMATIDS OF MYXINE GLUTINOSA. + HISTOLOGY /T THE    5296F
(CYCLOSTOMATA). + *MYXINE    SPERMATOGENETISCHEN STADIEN VON MYXINEGLUTINOSA L.    5418
H MYXINIDAE. + *EPTATRETUS    SPERMIOGENESIS IN EASTERN PACIFIC SPECIESOF HAGFIS    4704F
INEN. [DETERMINATION OF PAR    SPEZIFISCHER VOLUMINA VON HAMOGLOBINEN UND MYOGLOB    5128F
CHONDRIA IN VERTEBRATES.] +    SPHINGOMYELIN FATTY ACIDS IN BRAIN MYELIN AND MITO    4741F
PLANCHON AND MYRIOPHYLLUM    SPICATUM L.). + CHEMISTRY MANAGEMENT ECOLOGY /NHX)    5421F
EM. + JAPAN ADULT CYTOLOGY    SPINAL CORD OF THE LAMPREY USING A NEW FILTER SYST    4940F
ADULT NERVOUS SYSTEM SENSE    SPINAL CORD OF THE LAMPREY. + *PETROMYZON MARINUS    4643F
NUS U.S.A. NERVOUS SYSTEM C    SPINAL CORD OF THE SEA LAMPREY. + *PETROMYZON MARI    4542F
NUS U.S.A. ADULT AMMOCOETE    SPINAL CORD OF THE SEA LAMPREY. + *PETROMYZON MARI    4625F
ERVOUS SYSTEM TECHNIQUES LO    SPINAL CORD TRANSECTION IN LARVAL SEA LAMPREY. + N    5579F
S SYSTEM / AMMOCOETE NERVOU    SPINAL CORD. + *PETROMYZON MARINUS AMMOCOETE NERVO    5069
A ADULT NERVOUS SYSTEM HIST    SPINAL CORD. + *PETROMYZON MARINUS ANADROMOUS LARV    5020F
E ADULT MORPHOLOGY NERVOUS    SPINAL CORD. + *PETROMYZON MARINUS U.S.A. AMMOCOET    4595F
T LARGE SYNAPSESIN LAMPREY    SPINAL CORD. + *PETROMYZON U.S.A. PHYSIOLOGY /CN A    5333F
NT INTERNEURONS OF LAMPREY    SPINAL CORD. /    /GIA    4560
ION OF DORSAL PARTS OF THE    SPINAL CORD. /+THE LAMPREY BRAINFOLLOWING STIMULAT    4612
CE. PP. 109-122. /AN EMINEN    SPINAL FLUID BY EPENDYMAL CELLS OFTHE MEDIAN EMINE    4984
.S.A. GREAT LAKES ANATOMY N    SPINAL NEURONS IN LAMPREY. + *PETROMYZON MARINUS U    5280F
.R. NERVOUS SYSTEM PHYSIOLO    SPINAL STIMULATION.] + *LAMPETRA FLUVIATILIS U.S.S    4601F
ON MARINUS AMMOCOETE HISTOL    SPINAL TRANSECTION IN LARVAL LAMPREYS. + *PETROMYZ    5026F
OD FROM MYXINE GLUTINOSA +    SPINDLE CELLS INTO LYMPHOCYTE-LIKE CELLSIN THE BLO    4945F
RATS, CHINESE HAMSTERS AND    SPINY MICE. + MANAGEMENT / D NEWBORN GUINEA PIGS,    5365F
THE GORKII REGION.] + *PET    SPISOK RYB GOR'KOVSKOI OBLASTI. [LIST OF FISHES IN    4650F
SYSTEM SYSTEMATICS /ERVOUS    SPITSBERG. + ADULT ANATOMY EVOLUTION MOUTH NERVOUS    5346F
E DU DEVONIEN INFERIEUR DU    SPITSBERG. /ICYCLOSTOMI, OSTEOSTRACI), CEPHALASPID    4578
TOPOIESIS OF THE PRIMITIVE    SPLEEN OF HAGFISH.] /    /[HEMA    4980F
BLOOD HISTOLOGY CYTOLOGY /    SPLEEN OF THE HAGFISH. + *EPTATRETUS BURGERI JAPAN    4795F
IOCHEMISTRY BLOOD / ADULT B    SPLIT BY LAMPREY AND MAMMALIAN THROMBINS. + ADULT    5146F
U.S.A. RECORDS OF LAMPETRA    SPP. (PETROMYZONIDAE). /    /1ST ARKANSAS    5055
975, P.86. + *ICHTHYOMYZON    SPP. ADULT BIOLOGY / ED BY COMMERCIAL FISHERMEN, 1    5225F
OSTEOSTRACI. + *PETROMYZON    SPP. ANATOMY OLFACTION /MME DU SAC NASAL CHEZ LES    4692F
ENT /OETE CHEMISTRY MANAGEM    SPP. SEA LAMPREY CANADA AMMOCOETE CHEMISTRY MANAGE    5223F
RSE TUBULAR SYSTEM IN HAGFI    SPREAD OF THE ACTION POTENTIAL THROUGH THE TRANSVE    5006F
TUBULAR SYSTEM IN HAGFISH    SPREAD OF THE JUNCTION POTENTIAL IN THE TRANSVERSE    5007
NKOV RYB). [COMMERCIAL FISH    SSSR OPISENIIA RYB. (TEKST K ATLASU TSVETNYKH RISU    5321F
FAUNA OF THE USSR.] + SYST    SSSR. [METACERCARIA OFTHE GENUS DIPLOSTOMUM IN THE    5153
RINUS CANADA GREAT LAKES AD    ST MARYS RIVER, LAKE HURON. P.93. + *PETROMYZON MA    5258F
ANADA AMMOCOETE BIOLOGY / C    ST. MARY'S RIVER.PP. 88-91. + *PETROMYZON MARINUS    4826F
NUS CANADA ADULT /YZON MARI    ST. MARY'S RIVER1973. PP. 80-81. + *PETROMYZON MAR    4906F
GE AREA FROM ROOT RIVER TO    ST. MARYS ISLAND. P. 174. /    /TRANSFER OF STORA    5497F
EMENT AMMOCOETE /NADA MANAG    ST. MARYS RIVER, 1971-74. PP. 91-93. + CANADA MANA    5132F
RINUS *ICHTHYOMYZON CANADA    ST. MARYS RIVER, 1973. PP. 12-15. + *PETROMYZON MA    4862F
CHEMISTRY /ANADA AMMOCOETE    ST. MARYS RIVER, 1973. PP.75-76. + CANADA AMMOCOET    4891F
TRIBUTION /CANADA ADULT DIS    ST. MARYS RIVER, 1975. PP.83-85. + CANADA ADULT DI    5224F
CANADA ADULT /YZON MARINUS    ST. MARYS RIVER,1972. P. 166. + *PETROMYZON MARINU    5488F
NUS CANADA /PETROMYZON MARI    ST. MARYS RIVER,1974. PP. 66-67. + *PETROMYZON MAR    5103F
67. + NORTH AMERICA ADULT D    ST.MARYS RIVER AND LAKES HURON AND MICHIGAN, 1963-    4746F
ASE FROM THE MUSCLES OF ECT    STABILITY OF GLYCERALDEHYDE 3 PHOSPHATE DEHYDROGEN    5522
SE FROM MUSCLES OF ECTOTHER    STABILITY OF GYCERALDEHYDE-3-PHOSPHATE DEHYDROGENA    5612
EIRSIGNIFICANCE IN ANALYZIN    STABLE CYTOLOGICAL AND HISTOLOGICAL INDICES AND TH    4701F
OSTOMATA). LICHT-UND ELEKTR    STADIEN DER OOGENESE BEI MYXINE GLUTINOSA L. (CYCL    4589F
MYXINE GLUTINOSA REPRODUCTI    STADIEN VON MYXINEGLUTINOSA L. (CYCLOSTOMATA). + *    5418
ASFORMAZIONE DELLA LARVA IN    STADIO DI SVILUPPO CON PARTICOLARE RIGUARDO ALLATR    4807F
S. PP. 79-80. + CANADA TECH    STAFF GAUGES ON LAKES HURON AND ONTARIO TRIBUTARIE    4903F
+ ADULT HISTOLOGY OSMOREGUL    STAGE OF THE  RIVER LAMPREY LAMPETRA FLUVIATILIS.    5101F
S FROM THE RINGERIKE GROUP    STAGE 10 OF SOUTHERN NORWAY. / /MARINE CALCARENITE    4679
DURING THEIR PARASTIC LIFE    STAGE. + HISTOCHEMISTRY GONADOGENESIS ADULT /VNUS    4659F
MOCOETE ENDOCRINOLOGY HISTO    STAGES IN ITS LIFE CYCLE. + CANADA U.S.A. ADULT AM    4758F
MARINUS *LAMPETRA FLUVIATI    STAGES IN THE LIFE CYCLE OFLAMPREYS. + *PETROMYZON    4651F
THE TRANSFORMATION OF THE    STAGES OF DEVELOPMENT WITH PARTICULAR REFERENCE TO    4807F
TRY MANAGEMENT /). + CHEMIS    STAGES OF RAINBOW TROUT (SALMO GAIRDNERI). + CHEMI    5080F
ETRA FLUVIATILIS RUSSIA ADU    STAGES OF SPAWNING MIGRATION.] PP.127-133. + *LAMP    5361F
ARINUS L. + U.S.A. ECOLOGY    STAGES OF THE LANDLOCKED SEA LAMPREY, PETROMYZON M    4975F
ARINUSL. + U.S.A. AMMOCOETE    STAGES OF THE LANDLOCKED SEA LAMPREY, PETROMYZON M    5616F
STOMATA). LIGHT- AND ELECTR    STAGES OF THE OOGENESIS IN MYXINE GLUTINOSA (CYCLO    4589F
HEMISTRY AMMOCOETE MANAGEME    STAGES OF THE SEA LAMPREY. + *PETROMYZON MARINUS C    5050F
TA (TREWAVAS, 1933) AND HER    STAGES OF THE VARIEGATED PERCHES TILAPIA LEUCOSTIC    5448
NOLOGY /USSIA ADULT ENDOCRI    STAGES OF THEIR LIFE CYCLE.] + RUSSIA ADULT ENDOCR    5323F
IN THE PRESENCE OF PHENOLS    STAINING TECHNIQUE. /LH ACID MUCO POLY SACCHARIDES    5040
ETABOLISM /ION PHYSIOLOGY M    STANDARD METABOLIC RATE. + RESPIRATION PHYSIOLOGY    5182F
ENY AND PHYLOGENY FROM THE    STANDP... PP. 97-112. /AIN VERTEBRATA DURING ONTOG    5290
OFIZAROI NEIROSEKRETORNOI S    STANOVLENIYA NEIROGEMAL'NYKHOTDELOV GIPOTALAMO-GIP    4556F
AMPREY U.S.A. CHEMISTRY ADU    STATE SIDE OFLAKE ONTARIO, 1975. PP.55-68. + SEA L    5221F
ION MANAGEMENT /E DISTRIBUT    STATE, 1975. PP.14-16. + U.S.A. AMMOCOETE DISTRIBU    4553F
+ *PETROMYZON MARINUS *ICHT    STATES (NEW YORK) SIDE OF LAKE ONTARIO. PP.14-15.    5251F

258

| | | |
|---|---|---|
| E AQUATIC MIDGE CHIRONOMUS | **TENTANS.** + CHEMISTRY MANAGEMENT /K BY LARVAE OF TH | 4647F |
| LICYLANILIDE BY CHIRONOMUS | **TENTANS.** + MANAGEMENT /AE 2',5-DICHLORD-4'-NITROSA | 5453F |
| VAEOF THE MIDGE CHIRONOMUS | **TENTANS.** + MANAGEMENT /OLANILIDE (BAYER 73) TO LAR | 5120F |
| E AQUATIC MIDGE CHIRONOMUS | **TENTANS.** /NFM) IN SUBLETHALLY ESPOSED LARVAE OF TH | 5457 |
| ONOMID LARVAE (CHIRONOMOUS | **TENTANS).** /CTRIFLUOROMETHYL-4-NITROPHENOL) IN CHIR | 5380 |
| AMPREY, LAMPETRA JAPONICA. | **TERMINAL** COUPLING' IN THE SKELETAL MUSCLE OF THE L | 4644F |
| SCLES. + ADULT AMMOCOETE AN | **TERMINAL** COUPLINGS" IN SOME VERTEBRATE SKELETAL MU | 4757F |
| E ATLANTIC HAGFISH MYXINE G | **TERMINALS** ON DIFFERENT TYPESOF MUSCLE FIBERS IN TH | 4610F |
| ITCH MUSCLES FIBRES IN ATLA | **TERMINALS.** MONOAMINERGICINNERVATION OF SLOW NON-TW | 5335 |
| / | **TERRESTRIAL** GASTROPODS FROM THE PORTUGUESE FAUNA. | 5172 |
| SOMERSET ISLAND NORTHWEST | **TERRITORIES** CANADA. + EVOLUTION SYSTEMATICS /LN OF | 4995 |
| ZING CELLULAR SITES IN THE | **TESTES** O... PP.99-136. /TISTRY OF STEROID SYNTHESI | 5190 |
| RETUS STOUTI. + CANADA PACI | **TESTICULAR** INTERSTITIAL TISSUE OF THEHAGFISH EPTAT | 5152F |
| OMATA). + *LAMPETRA FLUVIAT | **TESTIS** OF THE HAGFISH, EPTATRETUS BURGERI (CYCLOST | 5347F |
| CYCLOSTOMATA). + JAPAN ENDO | **TESTIS** OF THE HAGFISH, EPTATRETUS BURGERI GIRARD ( | 5525F |
| NTACT MALE AND FEMALE RIVER | **TESTOSTERONE** AND OESTRADIOL ON GONADECTOMIZEDAND I | 4592F |
| AMPREY, LAMPETRA FLUVIATILI | **TESTOSTERONE** IMPLANTATION IN THE MIGRATING RIVER L | 5176F |
| ETUS BURGERI. + SWEDEN ENDO | **TESTOSTERONE** IN HEPATIC TISSUE OF A HAGFISH EPTATR | 5199F |
| EPTATRETUS BURGERI ENDOCRIN | **TESTOSTERONE** IN THE HAGFISH EPATRETUS BURGERI. + * | 4990F |
| EYS (LAMPETRA FLUVIATILIS). | **TESTOSTERONE** ON INTACT MALE AND FEMALE RIVER LAMPR | 5607F |
| TRA FLUVIATILIS. + GREAT BR | **TESTOSTERONE** ON THE MIGRATING RIVER LAMPREY, LAMPE | 5065F |
| O NONTARGET FISH IN STATIC | **TESTS.** + MANAGEMENT /OROMETHL-4-NITROPHENOL(TFM) T | 4712F |
| ARGET FISH IN FLOW-THROUGH | **TESTS.** + MANAGEMENT /TL-4-NITROPHENOL(TFM) TO NONT | 4691F |
| *LETHENTERON SYSTEMATICS / | **TETRAPLEURODON** *ENTOSPENUS *LAMPETRA *EUDONTOMYZON | 5391F |
| TOMY SYSTEMATICS /TERON ANA | **TETRAPLEURODON** SPADICEUS *LAMPETRA *LETHENTERON AN | 4752F |
| ISHES.)] PP. 14-23. + *CASP | **TEXT** TO THE ATLAS OF COLOUR ILLUSTRATIONS OF THE F | 5321F |
| OW TROUT (SALMO GAIRDNERI). | **TFM** (LAMPRICIDE) TO SIX EARLY LIFE STAGES OF RAINB | 5080F |
| N. + CHEMISTRY MANAGEMENT A | **TFM** (3-TRIFLUOROMETHYL-4-NITROPHENOL) AND ANTIMYCI | 5403F |
| CE SPECTROPHOMETRY. /NESCEN | **TFM** (3-TRIFLUOROMETHYL-4-NITROPHENOL) BY LUMINESCE | 5381 |
| D LARVAE (CHIRONOMOUS TENTA | **TFM** (3-TRIFLUOROMETHYL-4-NITROPHENOL) IN CHIRONOMI | 5380 |
| RAL BIRD SPECIES. + CHEMIST | **TFM** AND CLONITRALID (BAYLUSCIDE, BAYER 73) TO SEVE | 5430F |
| TFM GLUCURONIDE IN BILE OF | **TFM** EXPOSED RAINBOW TROUT. /UND IDENTIFICATION OF | 5415 |
| / | **TFM** FIELD FORMULATION. + CHEMISTRY MANAGEMENT /E | 5377F |
| T. /FM EXPOSED RAINBOW TROU | **TFM** GLUCURONIDE IN BILE OF TFM EXPOSED RAINBOW TRO | 5415 |
| DYNAMICS OF QUINALDINE AND | **TFM** IN RAINBOW TROUT. + CHEMISTRY MANAGEMENT /IUE | 4660F |
| ATION-ORIENTED RESEARCH ON | **TFM** IN 1972. P. 310. + CHEMISTRY MANAGEMENT /DISTR | 5472F |
| ANTIMYCIN A, BAYER 73, AND | **TFM** TO THE OSTRACOD CYPRETTA KAWATAI. /TCITIES OF | 5454F |
| OW TROUT. + *PETROMYZON MAR | **TFM** UPTAKE AND CONJUGATION IN SEALAMPREY AND RAINB | 4751F |
| FATHEAD MINNOWS EXPOSED TO | **TFM.** + BIOLOGY CHEMISTRY MANAGEMENT /+ODUCTION OF | 5379F |
| / | **TFM.** + CHEMISTRY MANAGEMENT /I | 5378F |
| /MICROBIAL DEGRADATION OF | **TFM.** /E | 5376 |
| THE AQUATIC MIDGE CHIRONOM | **TFM)** AND BAYER 73 ON IN VIVO OXYGEN CONSUMPTION OF | 4937F |
| 4'-NITROSALICYLANILIDE (BAY | **TFM)** AND THE 2-AMINOETHANOL SALT OF 2',5-DICHLORO- | 5456 |
| ER 73) TO LARVAEOF THE MIDG | **TFM)** AND 2' 5-DICHLORO-4'-NITROSALICYLANILIDE (BAY | 5120F |
| ETE CHEMISTRY MANAGEMENT TE | **TFM)** AS A LAMPRICIDE. + *PETROMYZON MARINUS AMMOCO | 5053F |
| TANS. + CHEMISTRY MANAGEMEN | **TFM)** BY LARVAE OF THE AQUATIC MIDGE CHIRONOMUS TEN | 4647F |
| M UPTAKE AND CONJUGATION IN | **TFM)** FOR THE SEA LAMPREY;COMPARATIVE ASPECTS OF TF | 4751F |
| TROL OF SEA LAMPREYS. + ANI | **TFM)** IN A STREAM ECOSYSTEM AFTER TREATMENT FOR CON | 5051F |
| LUOROMETHYL-4-NITROPHENOL ( | **TFM)** IN AQUATIC ENVIRONMENTS. /HSISTENCE OF 3-TRIF | 5384 |
| LUOROMETHYL-4-NITROPHENOL ( | **TFM)** IN AQUATIC INVERTEBRATES. + MANAGEMENT /HTRIF | 5144F |
| S CHEMISTRY MANAGEMENT + PI | **TFM)** IN RAINBOW TROUT. + *PETROMYZON MARINUS PISCE | 5397F |
| IDGE CHIRONOMUS TENTANS. /M | **TFM)** IN SUBLETHALLY ESPOSED LARVAE OF THE AQUATIC | 5457 |
| *PETROMYZON MARINUS CHEMIS | **TFM)** ON DEVELOPMENTAL STAGES OF THE SEA LAMPREY. + | 5050F |
| LUOROMETHYL-4-NITROPHENOL ( | **TFM)** TO BROOK TROUT (LALVELINU FONTINALIS). / TRIF | 5406 |
| LUOROMETHYL-4-NITROPHENOL ( | **TFM)** TO 10 SPECIES OF ALGAE. + MANAGEMENT /E3-TRIF | 5118F |
| ROM THE UPPER GREAT LAKES + | **TFM)** UNDETECTED IN LAKE TROUT AND CHINOOK SALMON F | 5189F |
| LUOROMETHYL-4-NITROPHENOL ( | **TFM).** + ANIMAL CHEMISTRY TOXICITY MANAGEMENT /ORIF | 5077 |
| LUOROMETHYL-4-NITROPHENOL ( | **TFM).** + CHEMISTRY MANAGEMENT /YNTINALIS) TO 3-TRIF | 5577F |
| LUOROMETHYL-4-NITROPHENOL ( | **TFM).** + U.S.A. MANAGEMENT ANIMAL CHEMISTRY /Y-TRIF | 5242F |
| 3) AND A 98:2 MIXTURE ASLAM | **TFM),** 2', 5-DICHLORO-4'-NITROSALICYLANIDE (BAYER 7 | 5463F |
| 73), AND A 98:2 MIXTURE TO | **TFM),** 2',5-DICHLORO-4'-NITROSALICYLANILIDE (BAYER | 5095F |
| MODEL STREAM COMMUNITIES. | **TFM,** 3-TRIFLUOROMETHYL-4-NITROPHENOL). RESIDUES IN | 5419F |
| AM COMMUNITIES. /MODEL STRE | **TFM:** 3-TRIFLUOROMETHYL-4-NITROPHENOL) IN MODEL STR | 5388 |
| D PRODUCTION OF TWO SPECIES | **TFM:** 3-TRIFLUOROMETHYL-4-NITROPHENOL) ON GROWTH AN | 5421F |
| ACROINVERTEBRATES. + CHEMIS | **TFM:** 3-TRIFLUOROMETHYL-4-NITROPHENOL) TO SELETED M | 5387F |
| F MERCURY CADMIUM LEAD AND | **THALLIUM** IN A EUTROPHIC LAKE. / /DISTRIBUTION O | 4614F |
| EVOLUTION CYTOLOGY /NATOMY | **THAT** OF OTHER CHORDATES. PP. 14-16. + ADULT ANATOM | 4842F |
| DURING LARVALDEVELOPMENT AN | **THEANADROMOUS** SEA LAMPREY, PETROMYZON MARINUS L., | 4698 |
| STRIATED MUSCLE FIBERS IN | **THEATLANTIC** HAGFISH MYXINE GLUTINOSA. /T TYPES OF | 4588 |
| TISSUES OF THE LAMPREY, LAM | **THEENDOCRINE** PANCREAS, INTERRENAL, AND CHROMAFFIN | 5002F |
| AND LAMPETRA PLANERI. /LIS | **THEGILLS** OF AMMOCOETE LAMPREYS LAMPETRA FLUVIATILI | 4991 |
| C STUDIES OFTHE THYROID GLA | **THEHAGFISH** EPTATRETUS BUGERI A PART OF PHYLOGENETI | 4690F |
| ON THE ADENOHYPOPHYSIS OF | **THEHAGFISH** EPTATRETUS BURGERI. /ACROSCOPIC STUDIES | 4608 |
| AN ENDOCRINOLOGY HISTOLOGY | **THEHAGFISH** EPTATRETUS STOUTI. + CANADA PACIFIC OCE | 5152F |
| REGULATION OSMOREGULATION | **THEHAGFISH** MYXINE GLUTINOSA. + GERMANY BLOOD IONIC | 5351F |
| AMPREY AMMOCOETE. + ANIMAL | **THEINTRA** MEDULLARY PRIMARY AFFERENT CELLS OF THE L | 5295F |
| ATEPHYLOGENESIS AND SYSTEMA | **THEIRSIGNIFICANCE** IN ANALYZING PROBLEMS IN VERTEBR | 4701F |
| UNA OF THE TISZA BASIN.] + | **THEISSBECKENS.[TOWARD** THE KNOWLEDGE OF THE FISH FA | 5127F |
| PAN ADULT RESPIRATION HISTO | **THELAMPREY** GILL FILAMENTS. + *LAMPETRA JAPONICA JA | 5540F |
| OPIC STUDY OF THE GILLS OF | **THELAMPREY,** LAMPETRA FLUVIATILIS (L.) /SON MICROSC | 5587 |
| BRUNSWICK AMMOCOETE ANADRO | **THELARVAL** LAMPREY PETROMYZON MARINUS. + CANADA NEW | 5518F |

```
15. + *PETROMYZON MARINUS
.11-14. + *PETROMYZON MARIN
75. PP. 12-14. + CANADA AMM
73. PP.15-18. + *PETROMYZON
74. PP. 15-17. + *PETROMYZO
LAKE ONTARIO. PP.14-15. + *
F LAKE ONTARIO. PP.66-78. +
ONTARIO. PP. 66-78. + *PETR
ARINUS FROM A LAKE ONTARIO
OF PHYTOPHAGOUS LARVAE OF
MORPHOMETRY MUSCLE /MATICS
AMMOCOETE BIOCHEMISTRY META
RAMYXINE ATAMI ADULT AMMOCO
N VERTEBRATES. + *LAMPETRA
DULT METABOLISM MIGRATION /
ION NERVOUS SYSTEM RESPIRAT
ERVOUS SYSTEM / HISTOLOGY N
ENDOCRINOLOGY /VOUS SYSTEM
YSIOLOGY /US BIOCHEMISTRYPH
ETHENTERON ANATOMY SYSTEMAT
OPARD FROG. + *ENTOSPHENUS
TO FRICTION. + *ENTOSPENUS
A TRIDENTATUS *ENTOSPHENUS
TY FEEDING GROWTH MORPHOMET
*LAMPETRA AYRESI *LAMPETRA
ASMOBRANCH. + *ENTOSPHENUS
EN AND FIBRIN. + *LAMPETRA
OETE ANATOMY SENSE RECEPTOR
OSIS IN LAMPREYS. + *ICHTHY
OMYZON MARINUS U.S.A. GREAT
ID AND SERUM OF THE SEA LAM
 MUSCLE CELLS. + PHYSIOLOGY
S GRAY. + AUSTRALIA AMMOCOE
INUS ADULT ENDOCRINOLOGY CA
L. + MANAGEMENT /          /
ORD DE LA FAMILLE PETROMYZO
TRY MUSCLE PHYSIOLOGY /EMIS
/IMMUNOCHEMICAL STUDIES OF
) OF THE TROPONIN COMPLEX A
RACTION. + BIOCHEMISTRY MUS
NITROPHENOL (TFM) TO BROOK
RLY LIFE STAGES OF RAINBOW
LIDE (BAYER 73) IN RAINBOW
LIDE (BAYER 73) IN RAINBOW
-4-NITROPHENOL (TFM). + CHE
Y CONJUGATION AND EXCRETION
S + ANIMAL CHEMISTRY MANAGE
ERBOURNE CHEMICALS. + HISTO
OSALICYLANILIDE IN RAINBOW
UOROMETHYL-4-NITROPHENOL. /
TION TO THE FISHERY THE SEA
ES. + CANADA PISCES MIGRATI
ARASITISM BY /ORTALITY BY P
GEMENT + PISCES CHEMISTRY M
T ANIMAL CHEMISTRYBIOCHEMIS
NALDINE AND TFM IN RAINBOW
L-4-NITROPHENOL IN RAINBOW
ILE OF TFM EXPOSED RAINBOW
G THE COURSE OF EVOLUTION?
ATILIS *MYXINE GLUTINOSA BI
LUVIATILIS RUSSIA AMMOCOETE
POPULATION FROM THE POPRAD
ROM SWEDISH BROOK LAMPREYS
YSTED LARVAE OF CUCULLANUS
EMATODA: CUCULLANIDAE) FROM
EMATODA: CUCULLANIDAE) IN A
ISLET TISSUE IN LAMPREYS (L
DROGENAZ V KLETKAKH NEFRONA
OVLOV POLZHELYDOCHNOI ZHELE
 USSR. DESCRIPTION. (TEXT T
SE UPTAKE BY MUSCLES OF THE
 THESEA LAMPREY, PETROMYZON
T LAKES ADULT EXCRETION HIS
OTENTIAL IN THE TRANSVERSE
ATLANTIC ADULT MUSCLE PHYSI
LUVIATILIS RUSSIA EXCRETION
THE PARTICIPATION OF MICRO
BNEURALREGION OF MUSCLE FIB
E OF AGRANULAR CYTOPLASMIC
ONICA JAPAN HISTOLOGY CYTOL
ICARBONATE IN THE PROXIMAL
```

```
TRIBUTARY TO THE CANADIAN SIDE OF LAKE ONTARIO. P.      5110F
TRIBUTARY TO THE CANADIAN SIDE OF LAKE ONTARIO. PP      5250F
TRIBUTARY TO THE CANADIAN SIDE OF LAKE ONTARIO, 19      4547F
TRIBUTARY TO THE NEW YORK SIDE OF LAKE ONTARIO, 19      4867F
TRIBUTARY TO THE NEW YORK SIDE OF LAKE ONTARIO, 19      5109F
TRIBUTARY TO THE UNITED STATES (NEW YORK) SIDE OF       5251F
TRIBUTARY TO THE UNITED STATES (NEW YORK) WATERS O      5255F
TRIBUTARY TOTHE CANADIAN (ONTARIO) WATERS OF LAKE       5254F
TRIBUTARY. + CANADA BIOLOGY /IAMPREYS PETROMYZON M      4759F
TRICHOPTERA. /IOF LAMPETRA PLANERI IN THE PRESENCE      5016F
TRIDENTATA *LAMPETRA LETHOPHAGA U.S.A. SYSTEMATICS      5555
TRIDENTATA *LAMPETRA PLANERI UNITED KINGDOM ADULT       4678F
TRIDENTATA *MYXINE GLUTINOSA *EPTATRETUS STOUTI*PA      4705F
TRIDENTATA BIOCHEMISTRY /HRILLARY ACIDIC PROTEIN I      4570F
TRIDENTATA RESPIRATION ADULT METABOLISM MIGRATION       5138F
TRIDENTATA. + BEHAVIOUR CIRCULATORY SYSTEM LOCOMOT      5464
TRIDENTATA. + JAPAN ADULT ENDOCRINOLOGY HISTOLOGY       5042F
TRIDENTATA. + U.S.A. ADULT HISTOLOGY NERVOUS SYSTE      4577F
TRIDENTATUS *ENTOSPHENUS TRIDENTATUS BIOCHEMISTRYP      4955F
TRIDENTATUS *TETRAPLEURODON SPADICEUS *LAMPETRA *L      4752F
TRIDENTATUS ADULT ANIMALS ENDOCRINOLOGY / N THE LE      5003F
TRIDENTATUS ANIMAL INTEGUMENT /CTEBRATE EPIDERMIS       4693F
TRIDENTATUS BIOCHEMISTRYPHYSIOLOGY / S. + *LAMPETR      4955F
TRIDENTATUS CANADA ADULT AMMOCOETE ECOLOGY FECUNDI      4974F
TRIDENTATUS PISCES PARASITISM BY /3UGUST, 1974. +       5364F
TRIDENTATUS U.S.A. ANIMAL BIOCHEMISTRY ADULT /C EL      4729F
TRIDENTATUS U.S.A. BLOOD ANIMAL PHYSIOLOGY / RINOG      4594F
TRIDENTATUS). PP. 5-7. + *PETROMYZON MARINUS AMMOC      4731F
TRIGEMINAL MOTORNUCLEUS BEFORE AND AFTER METAMORPH      5466F
TRIGEMINAL SENSORY NEURONS OF THE LAMPREY. + *PETR      5578F
TRIIODOTHYRONINE AND THYROXINE LEVELS IN THE THYRO      5208
TRIMETHYLAMINE OXIDE AND POTASSIUMOF THE BLOOD AND      4767F
TRITATED MELATONIN IN THE LAMPREY GEOTRIA-AUSTRALI      5209F
TRITIATED CHOLESTEROL INJECTION. + *PETROMYZON MAR      4881F
TRITIUM LABELING OF 3-TRIFLUOROMETHYL-4-NITROPHENO      5413F
TROIS SOUS FAMILLES DES LAMPROIES DE L'HEMISPHEREN      4753F
TROPONIN COMPLEX AND THEIR INTERACTION. + BIOCHEMI      4628F
TROPONIN-C. PP. 65-67. + ANIMAL BIOCHEMISTRY /          4671F
TROPONIN-C) AND THE INHIBITORY PROTEIN (TROPONIN-I      4628F
TROPONIN-I) OF THE TROPONIN COMPLEX AND THEIR INTE      4628F
TROUT (LALVELINU FONTINALIS). / TRIFLUOROMETHYL-4-     5406
TROUT (SALMO GAIRDNERI). + CHEMISTRY MANAGEMENT /L     5080F
TROUT (SALMO GAIRDNERI). / LORO-4'-NITROSALICYLANI     5438
TROUT (SALMO GAIRDNERI). /HLORO-4'-NITROSALICYLANI     5440
TROUT (SALVELINUS FONTINALIS) TO 3-TRIFLUOROMETHYL     5577F
TROUT - EFFECT OFSALICYLAMIDE ON THE ACUTE TOXICIT     4652F
TROUT AND CHINOOK SALMON FROM THE UPPER GREAT LAKE     5189F
TROUT BILE: A PROPOSED MONITORING AID FOR SOME WAT     5382F
TROUT BY CARBARYL. //TER AND 2',5-DICHLORO-4'-NITR     5439
TROUT EXPOSED TO 3-TRIFLUOROMETHYL-4-NITROPHENOL.      5544
TROUT POPULATION OF SOUTHERN LAKE SUPERIOR IN RELA     4711
TROUT SALMO GAIRDNERI AND BASS MICROPTERUS SASMOID     4727F
TROUT SALVELINUS NAMAYCUSH. + CANADA MORTALITY BY      4769F
TROUT. + *PETROMYZON MARINUS PISCES CHEMISTRY MANA     5397F
TROUT. + *PETROMYZON MARINUS U.S.A. AMMOCOETE ADUL     4751F
TROUT. + CHEMISTRY MANAGEMENT /IUE DYNAMICS OF QUI     4660F
TROUT. /T ANDBILIARY EXCRETION OF 3-TRIFLUOROMETHY     5412F
TROUT. /UND IDENTIFICATION OF TFM GLUCURONIDE IN B     5415
TRUE HOMOLOGY OR ON ADAPTATION BY THE LATTER DURIN     4983
TRUNK MUSCLE OF LOWER CHORDATES. + *LAMPETRA FLUVI     5265F
TRUNK MUSCULATURE OF LARVAL LAMPREY. + *LAMPETRA F     5317
TRUNK MYOMERES IN LAMPETRA PLANERI WITH REGARD TO      5582
TRUTTAE (FABRICUS, 1794) (NEMATODA CUCULLANICAE) F     5467F
TRUTTAE IN THE BROOK LAMPREY, LAMPETRA PLANERI. /K     5270
TRUTTAEDACNITIS STELMIOIDES (VESSECHELLI, 1910) (N     5327F
TRUTTAEDACNITIS STELMIOIDES (VESSICHELLI, 1910) (N     5541F
TSIKLA. [CYTOLOGICAL AND FUNCTIONAL PROPERTIES OF      5323F
TSITOFOTOMETRICHESKOE ISSLEDOVANIE AKTIVNOSTI DEGI     4742F
TSITOLOGICHESKIE I FUNKTSIONAL'NYE OOBENNOSTI OSTR     5323F
TSVETNYKH RISUNKOV RYB). [COMMERCIAL FISHES OF THE     5321F
TSYPLYAT. [THE EFFECT OF VARIOUS INSULINS ON GLUCO     5267F
TUBULAR EPITHELIUM OF THE OPISTHONEPHRIC KIDNEY OF     4898F
TUBULAR NEPHRON. + *PETROMYZON MARINUS CANADA GREA     5045F
TUBULAR SYSTEM IN HAGFISH SLOW MUSCLE FIBERS. /F P     5007
TUBULAR SYSTEM IN HAGFISH TWITCH MUSCLE FIBERS. +      5006F
TUBULE OF THE RIVER LAMPREY KIDNEY.] + *LAMPETRA F     5354F
TUBULES IN AXONAL TRANSPORT. /NGICAL EVIDENCE FOR      4818
TUBULES INVAGINATING FROM THE SARCOLEMMA IN THE SU     5524F
TUBULES OF LAMPREY CELLS.HLORIDE CELLS. /OSTRUCTUR     4722
TUBULES OF LAMPREY CHLORIDE CELLS. + *LAMPETRA JAP     4603F
TUBULES OF THE KIDNEY IN THE LAMPREY. /ORIDE AND B     4569
```

TOUTII *PETROMYZON MARINUS
ETRY SPAWNING /DITY MORPHOM
HYL-4-NITROPHENOL (TFM). +
TURE /IMAL DISTRIBUTION CUL
-70. + *PETROMYZON MARINUS
REY. + *PETROMYZON MARINUS
TEM OF MYXINE GLUTINOSA. +
MPREY,PETROMYZON MARINUS +
STOLOGY NERVOUS SYSTEM / HI
SPINAL CORD. + *PETROMYZON
PYPTERA *LAMPETRA LAMOTTEI
             /1ST ARKANSAS
EYS. + *PETROMYZON MARINUS
ARINUS *LAMPETRA AEPYPTERA
REY, LAMPETRA AEPYPTERA. +
XON? + *LAMPETRA AEPYPTERA
UNA. + *ICHTHYOMYZON GAGEI
MOTTEI *LAMPETRA AEPYPTERA
NTATA *LAMPETRA LETHOPHAGA
YZON GAGEI IN AN EASTTEXAS
VERS OF EASTERN KAZAKH-SSR
, LAMPETRA FLUVIATILIS.] +
Y LAMPETRA FLUVIATILIS.] +
/TRY HISTOCHEMISTRY PISCES
LUTION OF FISH GENOMES.] +
E /STRY ENDOCRINOLOGY MUSCL
REGION.] + *PETROMYZONIDAE
(LINNE) *LAMPETRA PLANERI
.] + *LAMPETRA FLUVIATILIS
ETE DISTRIBUTION EVOLUTION
ONAL REGULATION OF CARBOHYD
CENTRAL NERVOUS SYSTEMAND M
RL'NOI NERVNOI SISTEMY I SO
IBUTION EVOLUTION SYSTEMATI
FISH POPULATION INTHE UPPE
TURE OF THE NEUROHYPOPHYSIS
LS IN THEGILLS OF AMMOCOETE
ING TISSUES IN THE SEXUALLY
RACTERIZATION OF THE INTEST
ETWEENTHE MYOCARDIAL GRANUL
IN MYXINE GLUTINOSA. P. 23
ONE LOCALIZATION IN THE ISL
ONE CONTAINING CELLS IN THE
AND ITS HOMOLOGOUS ORGAN.
VERTEBRATE CARDIAC MUSCLES.
NATOMY HISTOLOGY ADULT /+ A
NT TYPESOF MUSCLE FIBERS IN
AND RAT.SPECIALIZATIONS BE
OF THE LAMPREY LAMPETRA TR
THE PARSDISTALIS OF THE LA
NG NEURONS IN THE HYPOTHALA
LAMPETRA FLUVIATILIS (L.),
HAGFISH,EPTATRETUS STOUTI
EY, LAMPETRA FLUVIATILIS.]
TUS STOUTI + GONADOGENESIS
AMPREY PETROMYZON MARINUS A
IVER LAMPREY, LAMPETRA FLUV
UE OF THEHAGFISH EPTATRETUS
AND OF JAPANESE HAGFISHES P
RS IN THEATLANTIC HAGFISH M
HE OPISTHONEPHROS OF THE LA
OF METABOLISM FUNCTION AND
BOLISM /TENS) CYTOLOGY META
ESOX LUCIUS IN THE LOWER P
D HORMONES IN A CYCLOSTOME
R MITOCHONDRIA BY THE LAMPR
FINLAND ADULT IONIC REGULA
EGULATION /+ RUSSIA IONIC R
-234. /          /PROBLEMS OF
HE UPPER GREAT LAKES + ANIM
MDERIVATE DER CYCLOSTOMEN
THE BIOSYNTHESIS OF THYROG
EC LES CROISSANCES NATURELL
SPADICEUS *LAMPETRA *LETHE
US) TRIDENTATA *LAMPETRA PL
MMOCOETE ADULT FEEDING PHYS
.S.A. PHYSIOLOGY NERVOUS SY
ON CASTANEUS *ICHTHYOMYZON
M FORMATION A NEW SI-URIAN
LISM METAMORPHOSIS PHYSIOLO
SIS RESPIRATION /METAMORPHO

| Entry | Ref |
|---|---|
| U.S.A. IMMUNOLOGYANIMAL PISCES /I+ *POLISTOTREMO S | 4566F |
| U.S.A. LIFE CYCLE AMMOCOETE ADULT FECUNDITY MORPHO | 5139F |
| U.S.A. MANAGEMENT ANIMAL CHEMISTRY /Y-TRIFLUOROMET | 5242F |
| U.S.A. MANAGEMENT CHEMISTRY ANIMAL DISTRIBUTION CU | 4756F |
| U.S.A. MORTALITY OF CHEMISTRY AMMOCOETE /Y. PP. 52 | 4894F |
| U.S.A. NERVOUS SYSTEM CYTOLOGY PHYSIOLOG /DEA LAMP | 4542F |
| U.S.A. NERVOUS SYSTEM HISTOLOGY TECHNIQUES /ER SYS | 4805F |
| U.S.A. NERVOUS SYSTEM VISION HISTOLOGY /ELARVAL LA | 4785F |
| U.S.A. PACIFIC OCEAN ENDOCRINOLOGY GONADOGENESIS H | 5184F |
| U.S.A. PHYSIOLOGY /CN AT LARGE SYNAPSESIN LAMPREY | 5333F |
| U.S.A. PHYSIOLOGY NERVOUS SYSTEM /FIS *LAMPETRA AE | 5606F |
| U.S.A. RECORDS OF LAMPETRA SPP. (PETROMYZONIDAE). | 5055 |
| U.S.A. RESPIRATION /   /VENTILATION OF LARVAL LAMPR | 4787F |
| U.S.A. RESPIRATIONNERVOUS SYSTEM /,+ *PETROMYZON M | 4788F |
| U.S.A. SPAWNING ADULT /IT CONSTRUCTION OF THE LAMP | 4653F |
| U.S.A. SYSTEMATICS /OPETROMYZONIDAE) A DISTINCT TA | 4768F |
| U.S.A. SYSTEMATICS /YF 3 SPECIES TO THE ICHTHYO FA | 4944F |
| U.S.A. SYSTEMATICS ANIMAL /LAS USA. + *LAMPETRA LA | 5035F |
| U.S.A. SYSTEMATICS MORPHOMETRY MUSCLE /PETRA TRIDE | 5555 |
| U.S.A. WATERSHED. /IOUTHERN BROOK LAMPREY ICHYHYOM | 5207 |
| U.S.S.R. /UIBERIAN LAMPREY LAMPETRA KESSLERI IN RI | 5318 |
| U.S.S.R. ADULT BIOCHEMISTRY ENDOCRINOLOGY /RAMPREY | 4561F |
| U.S.S.R. ANATOMY EXCRETION /ONEPHRON IN THE LAMPRE | 4694F |
| U.S.S.R. ANIMAL BIOCHEMISTRY HISTOCHEMISTRY PISCES | 4803F |
| U.S.S.R. ANIMAL GENETICS /Y[GENOSYTEMATICS AND EVO | 4802F |
| U.S.S.R. BEHAVIOUR BIOCHEMISTRY ENDOCRINOLOGY MUSC | 4598F |
| U.S.S.R. DISTRIBUTION /LT OF FISHES IN THE GORKII | 4650F |
| U.S.S.R. FISHERIES SYSTEMATICS /NPETRA FLUVIATILIS | 5321F |
| U.S.S.R. NERVOUS SYSTEM PHYSIOLOGY /OL STIMULATION | 4601F |
| U.S.S.R. UKRAINE BALTIC SEA BLACK SEA ADULT AMMOCO | 5358F |
| UGLEVODNOGO OBMENA U NIZSHIKH POZVONOCHNYKH. [HORM | 4772F |
| UKHTOMSKII ONTHE HIERARCHICAL ORGANIZATION OF THE | 4642F |
| UKHTOMSKOGO NA IERARKHICHESKUYU ORGANIZATSIYUTSENT | 4642F |
| UKRAINE BALTIC SEA BLACK SEA ADULT AMMOCOETE DISTR | 5358F |
| UKRAINIAN-SSR USSR. PART 1. SPECIES COMPOSITION OF | 4873 |
| UL'TRASTUKTURA NEIROGIPOFIZA U MINOGI. [ULTRASTRUC | 5071F |
| ULTASTRUCTURE OF THE PRESUMED ION TRANSPORTING CEL | 4991 |
| ULTRASTRUCTURAL ANALYSES OFPRESUMED STEROID-PRODUC | 5470F |
| ULTRASTRUCTURAL AND FLUORESCENCE MICROSCOPICAL CHA | 4786F |
| ULTRASTRUCTURAL EVIDENCE FOR A DIRECT CONNECTION B | 4792F |
| ULTRASTRUCTURAL INVESTIGATION OF THE OPTICAL TRACT | 4845F |
| ULTRASTRUCTURAL INVESTIGATIONS ON POLYPEPTIDE HORM | 4617F |
| ULTRASTRUCTURAL INVESTIGATIONS ON POLYPEPTIDE HORM | 4870F |
| ULTRASTRUCTURE AND IODINEMETABOLISM OF THE THYROID | 4916 |
| ULTRASTRUCTURE OF CELL REMBRANE SPECIALIZATIONSIN | 5602 |
| ULTRASTRUCTURE OF LIVING CYCLOSTOME ECTOPROCTS. + | 5067F |
| ULTRASTRUCTURE OF MOTOR NERVE TERMINALS ON DIFFERE | 4610F |
| ULTRASTRUCTURE OF MYOTENDINOUS JUNCTIONS IN MYXINE | 4563F |
| ULTRASTRUCTURE OF PARS NERVOSA AND PARS INTERMEDIA | 5042F |
| ULTRASTRUCTURE OF THE ANTERIOR NEUROHYPOPHYSIS AND | 4577F |
| ULTRASTRUCTURE OF THE CEREBROSPINAL FLUID CONTACTI | 5596 |
| ULTRASTRUCTURE OF THE GILLS OF THE RIVER LAMPREY, | 5115F |
| ULTRASTRUCTURE OF THE NEUROHYPOPHYSIAL LOBE OF THE | 4922F |
| ULTRASTRUCTURE OF THE NEUROHYPOPHYSIS IN THE LAMPR | 5071F |
| ULTRASTRUCTURE OF THE OVARY OF THE HAGFISH EPTATRE | 5151F |
| ULTRASTRUCTURE OF THE PITUITARY GLAND IN THE SEA L | 4758F |
| ULTRASTRUCTURE OF THE PRO-ADENOHYPOPHYSIS OF THE R | 4646F |
| ULTRASTRUCTURE OF THE TESTICULAR INTERSTITIAL TISS | 5152F |
| ULTRASTRUCTURE OF THE THREAD CELLS IN THE SLIME GL | 5044F |
| ULTRASTRUCTURE OF 4 TYPES OF STRIATED MUSCLE  FIBE | 4588 |
| ULTRASTRUCTURE OFPRESUMPTIVE INTERRENAL CELLS IN T | 4899F |
| ULTRASTRUCTUREOF THE VERTEBRATE KIDNEY /DVE STUDY | 4684 |
| UMBRA. + *LAMPETRA JAPONICA (MARTENS) CYTOLOGY MET | 4622F |
| UMBY. [MATERIAL CONCERNING THE ECOLOGY OF THE PIKE | 4622F |
| UN CYCLOSTOME ET DES POISSONS. [CIRCULATING THYROI | 5193 |
| UNCOUPLING OF OXIDATIVE PHOSPORYLATION IN RAT LIVE | 5077 |
| UNDER NATURAL CONDITIONS.] + *LAMPETRA FLUVIATILIS | 4952F |
| UNDER THE INFLUENCE OF DIURETICS.] + RUSSIA IONIC | 5210F |
| UNDERSTANDING THE SUBSTRUCTURE OF SYNAPSES. PP.207 | 5213 |
| UNDETECTED IN LAKE TROUT AND CHINOOK SALMON FROM T | 5189F |
| UNDFISCHE. P. 223. /LKIEMEN DER ANAMNIER-KIEMENDAR | 5588 |
| UNE LAMPROIE ADULTE, LAMPETRA PLANERI (BLOCH). [ON | 5025F |
| UNE NOUVELLE METHODE D'ALIMENTATION. COMPARISON AV | 5352F |
| UNICUSPIS *ENTOSPHENUS TRIDENTATUS *TETRAPLEURODON | 4752F |
| UNICUSPIS *ICHTHYOMYZON HUBBSI *LAMPETRA (ENTOSPEN | 4678F |
| UNICUSPIS *LAMPETRA AEPYPTERA *LAMPETRA LAMOTTEI A | 5466F |
| UNICUSPIS *LAMPETRA AEPYPTERA *LAMPETRA LAMOTTEI U | 5606F |
| UNICUSPIS U.S.A. /ATO RULO NEBRASKA. + *ICHTHYOMYZ | 5197F |
| UNIT IN THE CANADIANARCTIC. /          /CAPE STOR | 4871 |
| UNITED KINGDOM ADULT AMMOCOETE BIOCHEMISTRY METABO | 4678F |
| UNITED KINGDOM ADULT AMMOCOETE EVOLUTION METAMORPH | 5300F |

VERTEBRATES (CYCLOSTOMES A
  (CYCLOSTOMATA). + HISTOLOG
D PHYLOGENESIS. + *LAMPETRA
POLISTOTREMA STOUTII. + *EP
IGRATING ANADROMOUS SEA LAM
EGIUM ADULT CYTOLOGY ENDOCR
DEN ADULT CIRCADIAN DISTRIB
ORY SYSTEM ENDOCRINOLOGY /T
S CIRCULATORY SYSTEM PHYSIO
GLUTINOSA. + NORWAY CIRCULA
  HISTOLOGY /CULATORY SYSTEM
OCHEMISTRY PHYSIOLOGY /M BI
MARINUS *LAMPETRA AEPYPTER
ETROMYZON MARINUS *ICHTHYOM
DU NORD. [THE VELAR TENTACL
AMERIQE DU NORD. [THE VELAR
F THE LAMPREY LAMPETRA FLUV
LOGY RESPIRATION /EM PHYSIO
NUS U.S.A. RESPIRATION  /
CTION, OLFACTION, AND GILL
AMPETRA FLUVIATILIS) DURING
IVER LAMPREY LAMPETRA FLUVI
AL ACTIVITY OF THE CARDIAC
ORY SYSTEM HISTOLOGY HISTOC
NOSA L. + NORWAY CIRCULATOR
). + JAPAN ENDOCRINOLOGY HI
BOUT THE EARLY EVOLUTION OF
ERVOUS SYSTEM HISTOLOGY TEC
              /THE CEREBRAL
ADULT HISTOLOGY /PONICA. +
MYXINE GLUTINOSA L. (CYCLOS
GEFABSYSTEM. /          /
D CHANGEABILITY OF SIBERIAN
ANIMAL EVOLUTION PHYSIOLOG
LOWER DEVONIAN OF YUNNAN CH
TANDP... PP. 97-112. /THE S
EVONIEN INTERIEUR DU YUNNAN
S JAPONICUS ANIMAL BIOCHEMI
OCHEMICAL EVOLUTION OF THE
THE EARLY EVOLUTION OF THE
                      /THE
USCLE ANIMAL TECHNIQUES PHY
TRA BIOCHEMISTRY / + *LAMPE
LOGY /NERVOUS SYSTEM PHYSIO
RE ABSORPTION IN DIFFERENT
IDENTATUS ANIMAL INTEGUMENT
N PATHWAY FOR AMIDESIN THE
SERIES NO. 4. PROBLEMS IN
  /CHROMOSOMAL CHANGES IN
AND LOWER DEVONIAN DEPOSIT
TERN CZECHOSLOVAKIA): THE R
ION PHYSIOLOGY /STRY EVOLUT
BIOLOGY / SYSTEM HISTOLOGY
MARINUS FRANCE BIOLOGY EVOL
N AND ULTRASTRUCTUREOF THE
IONIC REGULATION /S ANIMAL
OOD / /GLUCONEOGENESIS IN
              /KETOGENESIS IN
SENSE RECEPTORS / CYTOLOGY
DOGENESIS / PHYSIOLOGY GONA
AL HISTOLOGY /MMOCOETE ANIM
FUNCTIONAL REVIEW. /    /
+ SWEDEN METABOLISM BIOCHEM
HEMISTRY IMMUNOLOGY /S BIOC
MUSEUM. PART 1: FISHES. + A
TIVITY IN A VERY PRIMITIVE
  /AGFISH, MYXINE GLUTINOSA.
OF INSULIN IN A PRIMITIVE
/A L. + BIOCHEMISTRY BLOOD
E IN ANALYZING PROBLEMS IN
YPOPHYSIAL REGION IN LOWER
XINE GLUTINOSA *LAMPETRA FL
DIES.)] + *LAMPETRA FLUVIAT
ANIMAL BLOOD BIOCHEMISTRY /
YZON MARINUS *LAMPETRA FLUV
ULT NERVOUS SYSTEM EVOLUTIO
              /ORDOVICIAN
PHOSPHORYLASE. /OF LAMPREY
/ATA BIOLOGY ENDOCRINOLOGY
VOLUTION MUSCLE /HEMISTRY E
GULATION /IS BLOOD IONIC RE

| | |
|---|---|
| VASCULARIZATION OF THE HYPOPHYSIAL REGION IN LOWER | 4979 |
| VASKULARISATION DER KIEMEN VON MYXINE GLUTINOSA L. | 5575F |
| VASOMOTOR ADRENERGIC INNERVATION IN ONTOGENESIS AN | 5504 |
| VASOTOCIN IN THE PITUITARY OF THE PACIFIC HAGFISH | 4558F |
| VASOTOCIN ON PLASMA FREE FATTY ACID LEVEL IN THE M | 5096F |
| VASOTOCINERGIC SYSTEM OF LAMPETRA FLUVIATILIS. + B | 5312 |
| VASTERBOTTEN, SWEDEN.] + *LAMPETRA FLUVIATILIS SWE | 5141F |
| VEIN HEART OF MYXINE GLUTINOSA L. + NORWAY CIRCULA | 4789F |
| VEIN HEART OF MYXINE GLUTINOSA. + SWEDEN TECHNIQUE | 4573F |
| VEIN, HEART ANDVENTRICLE OF THE CYCLOSTOME MYXINE | 4597F |
| VEINS OF AMIA CALVA L.. + ANIMAL CIRCULATORY SYSTE | 4749F |
| VEINS OF MYXINE. PP. 39-41. + CIRCULATORY SYSTEM B | 4835F |
| VELAR MOTONEURONS OF LAMPREY LARVAE. + *PETROMYZON | 4788F |
| VELAR TENTACLES IN LAMPREY OF NORTH AMERICA.] + *P | 4752F |
| VELAR TENTACLES") CHEZ LES LAMPROIES DE L'AMERIQE | 4752F |
| VELUM ("VELAR TENTACLES") CHEZ LES LAMPROIES DE L' | 4752F |
| VENOM POLYPEPTIDES ON CHOLINORECEPTIVE MEMBRANES O | 4636F |
| VENTILATION IN THE LAMPREY. + NERVOUS SYSTEM PHYSI | 5269F |
| VENTILATION OF LARVAL LAMPREYS. + *PETROMYZON MARI | 4787F |
| VENTILATION. /N THE LAMPREY, PETROMYZON, DURING SU | 5461 |
| VENTILATORY FREQUENCY AND HEART RATE OFLAMPREYS (L | 5138F |
| VENTILATORY FREQUENCY ANDHEART RATE IN THE ADULT R | 4719F |
| VENTRICLE IN MYXINE GLUTINOSA. /EATION OF MECHANIC | 5046 |
| VENTRICLE OF MYXINE GLUTINOSA L. + NORWAY CIRCULAT | 4792F |
| VENTRICLEAND THE PORTAL VEIN HEART OF MYXINE GLUTI | 4789F |
| VENTRICLEOF THE HAGFISH EPTATRETUS BURGERI (GIRARD | 4782F |
| VENTRICULAR ORGANS OF THE LOWER CHORDATESTELL US A | 5198 |
| VENTRICULAR SYSTEM OF MYXINE GLUTINOSA. + U.S.A. N | 4805F |
| VENTRICULAR SYSTEM OF MYXINE-GLUTINOSA. /E | 4946 |
| VENTRICULAR WALL INTHE LAMPREY LAMPETRA JAPONICA. | 5187F |
| VERBINDUNG ZWISCHEN NEURO- UND ADENOHYPOPHYSE BEI | 4809F |
| VERGLEICHENOE ANATOMIE DER MYXOIDEN. III. UBER DAS | 5291 |
| VERKHNEGO IRTYSHA. [MORPHOLOGICAL PECULIARITIES AN | 4623F |
| VERTEBRAE ALPHA CRYATALLINS. + *PETROMYZON MARINUS | 5024F |
| VERTEBRATA CYCLOSTOMATA CEPHALASPIDOMORPHES OF THE | 4709F |
| VERTEBRATA DURING ONTOGENY AND PHYLOGENY FROM THE | 5290 |
| VERTEBRATA, CYCLOSTOMATA) CEPHALASPIDOMORPHES DU D | 4709F |
| VERTEBRATE AND INVERTEBRATE SPECIES. + *ENTOSPHENU | 5574F |
| VERTEBRATE BRAIN. + BIOCHEMISTRY /          /BI | 5512F |
| VERTEBRATE BRAIN. /A LOWER CHORDATESTELL US ABOUT | 5198 |
| VERTEBRATE BRAIN. /E | 4923 |
| VERTEBRATE CARDIAC MUSCLES. + CIRCULATORY SYSTEM M | 5602 |
| VERTEBRATE CENTRAL NERVOUS SYSTEM SYNAPSE. + *LAMP | 4580F |
| VERTEBRATE CENTRAL SYNAPSE. + NERVOUS SYSTEM PHYSI | 4714 |
| VERTEBRATE CLASSES. / HRON STRUCTURE AND PROXIMAL | 5165 |
| VERTEBRATE EPIDERMIS TO FRICTION. + *ENTOSPENUS TR | 4693F |
| VERTEBRATE ERYTHROCYTE. / F A FACILITATED DIFFUSIO | 4968 |
| VERTEBRATE EVOLUTION. /          /LINNEAN SOCIETY | 5369 |
| VERTEBRATE EVOLUTION. /Y | 5566 |
| VERTEBRATE FAUNA AND THE CORRELATION OF THE LUDLOW | 5087 |
| VERTEBRATE FAUNA OF NORTHWESTERN BOHEMIA (NORTHWES | 5186F |
| VERTEBRATE FIBRINOGEN. + ANIMAL BIOCHEMISTRY EVOLU | 5097F |
| VERTEBRATE HYPOTHALAMUS. + NERVOUS SYSTEM HISTOLOG | 5000 |
| VERTEBRATE IMMUNOGLOBULINS CHAINS.] + *PETROMYZON | 4602F |
| VERTEBRATE KIDNEY /DVE STUDY OF METABOLISM FUNCTIO | 4684 |
| VERTEBRATE KIDNEY.] + *LAMPETRA FLUVIATILIS ANIMAL | 4670F |
| VERTEBRATE LIVERS. + ANIMAL BIOCHEMISTRY MUSCLE BL | 5211 |
| VERTEBRATE LIVERS. /S | 5212 |
| VERTEBRATE MEDIAN EYE. + ANIMAL EVOLUTION CYTOLOGY | 4555F |
| VERTEBRATE OVIDUCT-UTERUS. + ANIMAL PHYSIOLOGY GON | 5390F |
| VERTEBRATE SKELETAL MUSCLES. + ADULT AMMOCOETE ANI | 4757F |
| VERTEBRATE SUBCOMMISSURAL ORGAN. A STRUCTURAL AND | 5272F |
| VERTEBRATE THE ATLANTIC HAGFISH MYXINE GLUTINOSA. | 4654F |
| VERTEBRATE TRANSFERRINS. + *PETROMYZON MARINUS BIO | 5004F |
| VERTEBRATE TYPE SPECIMENS IN THE SOUTH AUSTRALIAN | 4718F |
| VERTEBRATE. /NING SPECIFICITY AND NEGATIVE COOPERA | 5286 |
| VERTEBRATE, THE ATLANTIC HAGFISH, MYXINE GLUTINOSA | 5546 |
| VERTEBRATE, THE CYCLOSTOME MYXINE GLUTINOSA. /ISIS | 5330 |
| VERTEBRATEMYXINE GLUTINOSA L. + BIOCHEMISTRY BLOOD | 5244F |
| VERTEBRATEPHYLOGENESIS AND SYSTEMATICS. /INIFICANC | 4701F |
| VERTEBRATES (CYCLOSTOMES AND FISHES). / N OF THE H | 4979 |
| VERTEBRATES (HAGFISH, LAMPREY AND CHIMAERA). + *MY | 5019F |
| VERTEBRATES (MICRODISSECTION AND MICROPUNCTURE STU | 5064F |
| VERTEBRATES AND MAN.] + ANIMAL BLOOD BIOCHEMISTRY | 5372F |
| VERTEBRATES COMPLETE TO OCTOBER 1, 1976. + *PETROM | 5395F |
| VERTEBRATES EVOLUTION.] + *LAMPETRA FLUVIATILIS AD | 4540F |
| VERTEBRATES FROM ONTARIO CANADA. /L | 4615 |
| VERTEBRATES PURIFICATION AND PROPERTIES OF LAMPREY | 4717 |
| VERTEBRATES. + *CYCLOSTOMATA BIOLOGY ENDOCRINOLOGY | 5389F |
| VERTEBRATES. + *LAMPETRA FLUVIATILIS BIOCHEMISTRY | 5305 |
| VERTEBRATES. + *LAMPETRA FLUVIATILIS BLOOD IONIC R | 5204 |

LISM NERVOUS SYSTEM PHYSIOL
BRILLARY ACIDIC PROTEIN IN
N NERVOUS SYSTEM / EVOLUTIO
FROM THE MAJOR CLASSES OF
CONTRACTION IN SOME LOWER
NUS ANDTHE ANCESTRY OF THE
GENY OF THE EARLIEST LOWER
N OF PISCINE RETINOL BINDIN
/PALEOZOIC
HORMONE SECRETION IN LOWER
.4.1.1. IN MUSCLE OF LOWER
EMISTRY /MAL CYTOLOGY BIOCH
TEM PHYSIOLOGY /NERVOUS SYS
MISTRY VISION /NIMAL BIOCHE
DULT ENDOCRINOLOGY TECHNIQU
EVOLUTION NERVOUS SYSTEM /
EXCRETION ANIMAL /HEMISTRY
EMISTRY BIOLOGY CYTOLOGY MU
MAL BIOCHEMISTRY HISTOCHEMI
NOLOGY NERVOUS SYSTEM /OCRI
LLS OF THE HYPOTHALAMUS IN
HYDRATE METABOLISMIN LOWER
ATOMY WITH AN INTRODUCTION
/PHIBIA, REPTILIA. 244 PP.
NDOCRINES INTESTINALES DES
URING ECHINODERM HAEMOGGLUT
N PHYSIOLOGY METABOLISM /IO
EGATIVE COOPERATIVITY IN A
A FREEZE CLEAVE STUDYON THE
INE RECEPTOR-RICH MEMBRANE
NERVATION OF SLOW NON-TWITC
AMPETRA LAMOTTENII (LESUEUR
N MARINUS. + CANADA LAKE ON
H MUSCLE CELLS IN ARTERIAL
ATORY SYSTEM /LUTION CIRCUL
RICAN BROOK LAMPREY (LAMPET
D HISTOCHEMICAL STUDIES ON
MYZON MARINUS. + AMMOCOETE
. + *PETROMYZON MARINUS ADU
NIMAL OF SCIENTIFIC INTERES
PRIL, 1972.GOTEBORG:KUNGL.
HE POLYMERIZATION PRODUCTS
/CHEMICAL ZOOLOGY. VOL.
ISHES OF GAULEY RIVER WEST
NN CREEK WAYNE COUNTY WEST
CTIOUS PANCREATIC NECROSIS
ING LATIN NAMES. A.VERTEBRA
Y HISTOLOGY NERVOUS SYSTEM
T HISTOCHEMISTRY HISTOLOGY
NIMAL ADULT NERVOUS SYSTEM
AL BIOCHEMISTRY TECHNIQUES
.] + CYCLOSTOMES HISTOLOGY
ATILIS ANIMAL BIOCHEMISTRY
NERVOUS SYSTEM PHYSIOLOGY
S HISTOLOGY /STEM TECHNIQUE
BURGERI *MYXINE GLUTINOSA
*LAMPETRA FLUVIATILISADULT
UVIATILIS.] + RUSSIA ADULT
E ALONG THE OPTIC NERVE. +
US + U.S.A. NERVOUS SYSTEM
TILIS CYCLOSTOMI. + SWEDEN
DEGENERATIVE EVOLUTION. +
GLUTINOSA. P. 23. + ADULT
*PETROMYZON MARINUS ADULT
AMPETRA FLUVIATILIS SWEDEN
PLANERI. + NERVOUS SYSTEM
SMOBRANCHS.] + CYCLOSTOMES
IN CERTAIN VERTEBRATES.] +
MPETRA FLUVIATILIS ADULT NE
VERTEBRATES. ISOLATION AND
/COMPARATIVE ASPECTS OF
LUVIATILIS ADULT GONADOGENE
BOLISM SPAWNING REPRODUCTIO
4-NITROPHENOL (TFM) IN RAIN
RY OF THE HAGFISH, EPTATRET
PRELIMINARY NOTESON ITS IN
OF A HAGFISH EPTATRETUS BU
ITH PRELIMINARY NOTESON ITS
TUITARY THYROID TISSUES IN
BORG:KUNGL. VETENSKAPS-OCH
ESTERONE BY THE SEA LAMPREY

| | | |
|---|---|---|
| LAMPETRA FLUVIATILIS). PP. | 156-157. + ADULT LIFE MATURITY /O RIVER LAMPREYS ( | 4868F |
| ARINUS CANADA MORPHOMETRY / | 160-170. + *PETROMYZON MARINUS CANADA MORPHOMETRY | 5487 |
| CAN FRESHWATER FISHES. PP. | 1622-1627. / T OF CERTAIN PARASITES ON NORTH AMERI | 4662 |
| N ST. MARYS RIVER,1972. P. | 166. + *PETROMYZON MARINUS CANADA ADULT / AMPREY I | 5488F |
| S AMMOCOETE ADULTMUSCLE ANA | 17-20.. + *LAMPETRA FLUVIATILIS *PETROMYZON MARINU | 4840F |
| T AMMOCOETE CHEMISTRY /ADUL | 17-26. + *PETROMYZON MARINUS CANADA MANAGEMENT ADU | 5108F |
| /S CANADA ADULT MANAGEMENT | 171. + *PETROMYZON MARINUS CANADA ADULT MANAGEMENT | 5486F |
| ON THE SAUGEEN RIVER. PP. | 171-172. + *PETROMYZON MARINUS CANADA ADULT / DAM | 5485F |
| ER TO ST. MARYS ISLAND. P. | 174. / /TRANSFER OF STORAGE AREA FROM ROOT RIV | 5497F |
| A,PRICE CURRENT METER. PP. | 174-175. /ZHE LISTENING DEVICE OF THE SMALL, TYPE- | 5494F |
| ETRA FLUVIATILIS LINNAEUS, | 1758. + GREAT BRITAIN ADULT PARASITISM OF / M LAMP | 5061F |
| ANFORD *EUDONTOMYZON MARIAE | 1784) (CYCLOSTOMATA) IN ROMANIA. + *EUDONTOMYZON D | 4541F |
| OF THE NUCLEAR DNA IN MACRO | 1784) (CYCLOSTOMATA). [QUALITATIVE INVESTIGATIONS | 4700F |
| AMPREYS LAMPETRA PLANERI (B | 1794) (NEMATODA CUCULLANICAE) FROM SWEDISH BROOK L | 5467F |
| PETRA LAMOTTENII (LESUEUR, | 1827). /O, 1910) (NEMATODA: CUCULLANIDAE) FROM LAM | 5327F |
| ) (CYCLOSTOMATA). [QUALITAT | 1855) (TELEOSTEI) UND LAMPETRA PLANERI (BLOCH 1784 | 4700F |
| N LAMPRICIDES IN 1975. PP. | 189-195. + CHEMISTRY MANAGEMENT //ENTED RESEARCH O | 5500F |
| N LAMPRICIDES IN 1976. PP. | 189-199. + CHEMISTRY MANAGEMENT /TENTED RESEARCH O | 5367F |
| PIA MULTISPINOSA (GUNTNER, | 1898). /SLEUCOSTICTA (TREWAVAS, 1933) AND HEROTILA | 5448 |
| ADA AMMOCOETE / MARINUS CAN | 19-20. + *LAMPETRA LAMOTTEI *PETROMYZON MARINUS CA | 4890F |
| HE SUCTORIAN TENTACLE. PP. | 191-208. / /TRANSPORT OF MATERIALS IN T | 4965 |
| ENII (LESUEUR, 1827). /MOTT | 1910) (NEMATODA: CUCULLANIDAE) FROM LAMPETRA LAMOT | 5327F |
| AMPREY (LAMPETRA LAMOTTENII | 1910) (NEMATODA: CUCULLANIDAE) IN AMERICAN BROOK L | 5541F |
| A AEPYPTERA U.S.A. SYSTEMAT | 1922 (PETROMYZONIDAE) A DISTINCT TAXON? + *LAMPETR | 4768F |
| RICA WITH NOTES ON OTHER SP | 1922, (PETROMYZONIDAE) FROM NORTHWESTERN NORTH AME | 5542F |
| /ISPINOSA (GUNTNER, 1898). | 1933) AND HEROTILAPIA MULTISPINOSA (GUNTNER, 1898) | 5448 |
| *EPTATRETUS BURGERI*PETROM | 195-257. + *MYXINE GLUTINOSA *LAMPETRA FLUVIATILIS | 4706F |
| ELBE RIVER, WEST GERMANY. | 1950-1975. [THE FISH FAUNA OF THE LOWER AND MIDDLE | 5506F |
| TILIS ADULT BIOLOGY DISTRIB | 1950-1975.] + *PETROMYZON MARINUS *LAMPETRA FLUVIA | 5506F |
| ENT HISTORY MORTALITY OF MO | 1953-1970. + NORTH AMERICA ADULT MANAGEM | 4630F |
| AMPREY EARLY LIFE HISTORY, | 1960-72. PP. 223-271. /SRT ON THE STUDIES OF SEA L | 5473F |
| TISM BY FEEDING MIGRATION M | 1963-67. + NORTH AMERICA ADULT DISTRIBUTION PARASI | 4746F |
| OCOETE / MARINUS CANADA AMM | 1971-1973. PP. 77. + *PETROMYZON MARINUS CANADA AM | 4901F |
| OCOETE / MARINUS CANADA AMM | 1971-1973. PP. 78. + *PETROMYZON MARINUS CANADA AM | 4904F |
| NADA MANAGEMENT AMMOCOETE / | 1971-74. PP. 91-93. + CANADA MANAGEMENT AMMOCOETE | 5132F |
| SERVATIONS ON RARE FISH IN | 1972. / /ENGLISH OB | 4816 |
| ON RARE FISH IN FRANCE IN | 1972. / /OBSERVATIONS | 4815 |
| NAGEMENT /S CANADA ADULT MA | 1972. P. 171. + *PETROMYZON MARINUS CANADA ADULT M | 5486F |
| RIENTED RESEARCH ON TFM IN | 1972. P. 310. + CHEMISTRY MANAGEMENT /DISTRATION-O | 5472F |
| OCOETE / MARINUS CANADA AMM | 1972. PP. 101-103. + *PETROMYZON MARINUS CANADA AM | 5477F |
| YZON SP. *LAMPETRA LAMOTTIC | 1972. PP. 154-161. + *PETROMYZON MARINUS *ICHTHYOM | 5491F |
| PHOMETRY /ARINUS CANADA MOR | 1972. PP. 160-170. + *PETROMYZON MARINUS CANADA MO | 5487 |
| MORTALITY OF MANAGEMENT /T | 1972. PP. 95-96. + *PETROMYZON MARINUS CANADA ADUL | 5481 |
| ALLET. 104 PP. /ERHETS-SAMH | 1972.GOTEBORG:KUNGL. VETENSKAPS-OCH VITTERHETS-SAM | 4793F |
| ARIE, ONTARIO TO THE GREAT | 1973 OF THE SEA LAMPREY CONROL CENTRE SAULT STE. M | 4783F |
| TAKEN FROM IRISH WATERS IN | 1973. + *PETROMYZON MARINUS IRELAND ADULT /EISHES | 4686F |
| MPREY IN THE EASTERNBALTIC | 1973. / /SPAWNING STOCK AND FISHERY FOR RIVER LA | 5012 |
| LOGY /INUS CANADA ADULT BIO | 1973. P. 85. + *PETROMYZON MARINUS CANADA ADULT BI | 4902F |
| MPREY TAKEN AT DENN'S DAM, | 1973. P. 88. + *PETROMYZON MARINUS CANADA ADULT /S | 4848F |
| LITY OF CHEMISTRY AMMOCOETE | 1973. PP 71-74. + *PETROMYZON MARINUS CANADA MORTA | 4892F |
| ON CANADA AMMOCOETE BIOLOGY | 1973. PP. 12-15. + *PETROMYZON MARINUS *ICHTHYOMYZ | 4862F |
| ARINUS CANADA AMMOCOETE / M | 1973. PP. 19-20. + *LAMPETRA LAMOTTEI *PETROMYZON | 4890F |
| COETE MORTALITY OF CHEMISTR | 1973. PP. 47-52. + *PETROMYZON MARINUS CANADA AMMO | 4893F |
| TALITY OF /CANADA ADULT MOR | 1973. PP. 5. + *PETROMYZON MARINUS CANADA ADULT MO | 4866F |
| ALITY OF CHEMISTRY AMMOCOET | 1973. PP. 52-70. + *PETROMYZON MARINUS U.S.A. MORT | 4894F |
| BIOLOGY /INUS CANADA ADULT | 1973. PP. 81-84. + *PETROMYZON MARINUS CANADA ADUL | 4905F |
| BIOLOGY /INUS CANADA ADULT | 1973. PP. 85-87. + *PETROMYZON MARINUS CANADA ADUL | 4900F |
| BIOLOGY / CANADA AMMOCOETE | 1973. PP.15. + *PETROMYZON MARINUS CANADA AMMOCOET | 4861F |
| ETE BIOLOGY / U.S.A. AMMOCO | 1973. PP.15-18. + *PETROMYZON MARINUS U.S.A. AMMOC | 4867F |
| OETE MORTALITY OF CHEMISTRY | 1973. PP.32-46. + *PETROMYZON MARINUS CANADA AMMOC | 4888F |
| ATMENT OF ST. MARYS RIVER, | 1973. PP.75-76. + CANADA AMMOCOETE CHEMISTRY /ITRE | 4891F |
| OETE CHEMISTRY MORTALITY OF | 1973.PP. 21-31. + *PETROMYZON MARINUS CANADA AMMOC | 4889F |
| DULT CHEMISTRY MORTALITY OF | 1973-1974. PP. 1-V. + *PETROMYZON MARINUS CANADA A | 5237F |
| MARIE, ONTARIO, TO THE GREA | 1974 OF THE SEA LAMPREY CONTROL CENTRE SAULT STE. | 4784F |
| ES PARASITISM BY /ATUS PISC | 1974. + *LAMPETRA AYRESI *LAMPETRA TRIDENTATUS PIS | 5364F |
| /MARINUS CANADA MANAGEMENT | 1974. P. 5. + *PETROMYZON MARINUS CANADA MANAGEMEN | 4907F |
| MENT BARRIERS, LAKE HURON, | 1974. P. 68. + CANADA MORPHOMETRY /MCTRICAL ASSESS | 5088F |
| ON SP. *LAMPETRA LAMOTTEI C | 1974. PP. 15-17. + *PETROMYZON MARINUS *ICHTHYOMYZ | 5109F |
| GEMENT ADULT AMMOCOETE CHEM | 1974. PP. 17-26. + *PETROMYZON MARINUS CANADA MANA | 5108F |
| GEMENT ADULT AMMOCOETE CHEM | 1974. PP. 26-39. + *PETROMYZON MARINUS CANADA MANA | 5107F |
| N SP. *LAMPETRA LAMOTTII CA | 1974. PP. 51-58 + *PETROMYZON MARINUS *ICHTHYOMYZO | 5105F |
| ON SP. *LAMPETRA LAMOTTII C | 1974. PP. 59-65. + *PETROMYZON MARINUS *ICHTHYOMYZ | 5104F |
| URE STUDIES, LAKE ONTARIO, | 1974. PP.70-73. + CANADA ADULT MIGRATION /N-RECAPT | 5135F |
| ADULT AMMOCOETE BIOLOGY GRO | 1974-1975. PP. I-VI. + *PETROMYZON MARINUS CANADA | 4908F |
| MARIE, ONTARIO, TO THE GREA | 1975 OF THE SEA LAMPREY CONTROL CENTRE SAULT STE. | 4844F |
| ISTRY ANIMAL DISTRIBUTION C | 1975. + *PETROMYZON MARINUS U.S.A. MANAGEMENT CHEM | 4756F |
| /T MANAGEMENT MORTALITY OF | 1975. P. 5. + CANADA ADULT MANAGEMENT MORTALITY OF | 4969F |
| LAMPREY FROM HUMBER RIVER, | 1975. P.90 + CANADA ADULT MIGRATION /EDATA ON SEA | 5227F |
| NAGEMENT /E DISTRIBUTION MA | 1975. PP. 12-14. + CANADA AMMOCOETE DISTRIBUTION M | 4547F |
| RESEARCH ON LAMPRICIDES IN | 1975. PP. 189-195. + CHEMISTRY MANAGEMENT //ENTED | 5500F |
| GEMENT /E DISTRIBUTION MANA | 1975. PP. 6-7. + CANADA AMMOCOETE DISTRIBUTION MAN | 5116F |

273

# Source Index

```
(RUSS.)                                                              5079  F
      HYDROBIOL. J., 12:57-59.
   (ENG.)
   (ENG.)
ACAD. SCI. PARIS, C.R. HEBD. SEANCES, SER. D, 276:57-60.             5437
   (FR.)
ACAD. SCI. PARIS, C.R. HEBD. SEANCES, 272:2434-2436.                 4692  F
   (FR.)
ACAD. SCI. PARIS, C.R. HEBD. SEANCES, 273:2223-2224.                 4810  F
   (FR.)
ACTA ANAT., 86:353-375                                               4559  F
ACTA BIOCHIM. POL., 22:229-238.                                      4613  F
ACTA BIOL. ACAD. SCI. HUNG., 27:45-56.                               5131  F
ACTA BIOL. CRACOV., SER. ZOOL., 17:263-274.                          4778  F
ACTA BIOL. MED. GER., 28:K27-K30.                                    4775  F
   (GER.)
ACTA BIOL. MED. GER., 30:607-615.                                    4760  F
   (GER.)
   ENG. SUMM.
ACTA BIOL. MED. GER., 31:383-388.                                    5128  F
   (GER.)
   ENGL. SUMM.
ACTA CRYSTALLOGR., SECT. B, STRUCT. CRYSTALLOGR. CRYST. CHEM         5349
   ., 33:3198-3204.
ACTA ENDOCRINOL., SUPPL., 204.                                       5163
ACTA ENDOCRINOL., 86:561-569.                                        5371  F
ACTA ENDOCRINOL., 86:570-577.                                        5365  F
ACTA HAEMATOL. JAP., 36:231-232.                                     4980  F
   (JAP.)
ACTA HAEMATOL. JAP., 36:256-257.                                     4953
ACTA HAEMATOL. JAP., 36:329-330.                                     4957
ACTA HISTOCHEM. CYTOCHEM., 7:56.                                     4658
ACTA MED. & BIOL., 21:33-43.                                         5572  F
ACTA PARASITOL. LITH., 12:59-62.                                     4664
ACTA PHYSIOL. SCAND., SUPPL., 440:83.                                4874
   (ABSTR.)
ACTA PHYSIOL. SCAND., 68:142.                                        5336  F
ACTA PHYSIOL. SCAND., 90:501-504.                                    4740
   (FR.)
   ENG. SUMM.
ACTA PHYSIOL. SCAND., 91:13A-14A.                                    4945  F
ACTA PHYSIOL. SCAND., 91:430-432.                                    5100  F
ACTA PHYSIOL. SCAND., 96:11A-12A.                                    5090
ACTA PHYSIOL. SCAND., 96:29-49                                       5006  F
ACTA PHYSIOL. SCAND., 96:50-57                                       5007
ACTA ZOOL. PATHOL. ANTVERPIENSIA., 58:51-56.                         4702  F
ACTA ZOOL., (STOCKH.) 54:285-295.                                    4543  F
ACTA ZOOL., (STOCKH.) 55:173-177.                                    4649  F
ACTA ZOOL., (STOCKH.), 54:201-207.                                   4596  F
ACTA ZOOL., (STOCKH.), 56:189-198.                                   4704  F
ACTA ZOOL., (STOCKH.), 56:219-223.                                   4881  F
ACTA ZOOL., (STOCKH.), 56:61-66.                                     4669  F
ACTA ZOOL., (STOCKH.), 56:85-91.                                     4565  F
ACTA ZOOL., (STOCKH.), 57:103-112.                                   4877  F
ACTA ZOOL., (STOCKH.), 57:137-146.                                   4990  F
ACTA ZOOL., (STOCKH.), 57:167-173.                                   4942
ACTA ZOOL., (STOCKH.), 57:41-51.                                     4786  F
ACTA ZOOL., (STOCKH.), 57:89-102.                                    4800  F
ACTA ZOOL., (STOCKH.), 58:117-123.                                   5310  F
ACTA ZOOL., (STOCKH.), 58:17-25.                                     5152  F
ACTA ZOOL., (STOCKH.), 58:205-221.                                   5341  F
ACTA ZOOL., (STOCKH.), 58:27-40.                                     5151  F
ACTA ZOOL., (STOCKH.), 59:57-62.                                     5520  F
ACTA ZOOL., SUPPL., 1974:215 PP.                                     5234
ACTA ZOOL., 53:243-266.                                              4922  F
ACTA ZOOL., 54:271-284                                               4808  F
ACTA ZOOL., 55:245-254                                               4655  F
ACTA ZOOL., 55:61-69.                                                4651  F
ACTA ZOOL., 56:1-9.                                                  4648  F
   (FR.)
   ENGL. SUMM.
ACTA ZOOL., 56:199-204.                                              4703  F
ACTA ZOOL., 56:213-216.                                              4762  F
ACTA ZOOL., 56:265-269.                                              4782  F
ACTA ZOOL., 56:95-118                                                4646  F
ACTA ZOOL., 57:7-11.                                                 4763  F
ACTA ZOOL., 58:223.                                                  5347  F
AKAD. NAUK KAZ. SSR, ALMA-ALTA, IZV., SER. BIOL., 15:40-47.          5318
AKAD. NAUK SSSR, LENINGR., DOKL., SER. BIOL., 210:1230-1232.         4598  F
   .
   (RUSS.)
AKAD. NAUK SSSR, LENINGR., DOKL., SER. BIOL., 210:232-235.           4548  F
   (RUSS.)
      ACAD. SCI. USSR, PROC., 210:184-187
   (ENG.)
   (RUSS.)
   ENG. TRANS.
AKAD. NAUK SSSR, LENINGR., DOKL., SER. BIOL., 218:474-476.           4567  F
   (RUSS.)
AKAD. NAUK SSSR, LENINGR., DOKL., SER. BIOL., 231:605-607.           5317
   (ENG.)
```

```
AKAD. WISS. OSTERR., MAT.-NAT. KL., SITZ., 180:49-63.          4830
    180: 49-63.
AKAD. WISS. OSTERR., MAT.-NAT. KL., SITZ., 21:3 PP.            5543  F
    (GER.)
AKAD. WISS., BERLIN, ABH., 1841:1-131.                        5291
AM. ASSOC. ANAT., LOS ANGELES.                                4785  F
    (ABSTR.)
AM. FISH. SOC., TRANS., 102:829-831.                          5454  F
AM. FISH. SOC., TRANS., 103:355-358.                          4593  F
AM. FISH. SOC., TRANS., 103:551-556.                          4937  F
AM. FISH. SOC., TRANS., 105:119-123.                          5407  F
AM. FISH. SOC., TRANS., 105:322-326.                          5076  F
AM. J. ANAT., 138:235-252.                                    4899  F
AM. J. ANAT., 139:309-333.                                    4568
AM. J. ANAT., 145:207-223.                                    4993
AM. J. ANAT., 151:239-264.                                    5519
AM. J. ANAT., 151:319-336.                                    5540  F
AM. J. ANAT., 152:263-268.                                    5538  F
AM. J. PHYSIOL., 234:R51-R60.                                 5515
AM. J. PHYSIOL., 234:R61-R65.                                 5513  F
AM. OIL CHEM. SOC., CHICAGO, J., 51:521A.                     4967
AM. ZOOL. SUPPL., 15:255-270.                                 4744  F
AM. ZOOL., 13:567-590.                                        4920  F
AM. ZOOL., 13:591-604.                                        4918  F
AM. ZOOL., 13:625-638.                                        4765  F
AM. ZOOL., 13:823-838.                                        4919
AM. ZOOL., 13:933-936.                                        4921
AM. ZOOL., 14:1296.                                           4659  F
    (ABSTR.)
AM. ZOOL., 15:29-38.                                          4550  F
AM. ZOOL., 15:39-49.                                          4869  F
AM. ZOOL., 15:786.                                            5048
    (ABSTR.)
AM. ZOOL., 17:365-377.                                        5238  F
AM. ZOOL., 17:727-737.                                        5389  F
AM. ZOOL., 17:763-773.                                        5390  F
AM. ZOOL., 17:833-849.                                        5393
AM. ZOOL., 17:897.                                            5392
    (ABSTR.)
AM. ZOOL., 17:93-105.                                         5200
AM. ZOOL., 17:973.                                            5391  F
ANAT. REC., 175:400.                                          5245  F
ANAT. REC., 181:449-450.                                      4780  F
    (ABSTR.)
ANAT. REC., 182:321-338.                                      4644  F
ANAT. REC., 184:397-398.                                      5083  F
    (ABSTR.)
ANAT. REC., 187:383-403.                                      5185  F
ANN. BIOL., 12:139-184.                                       5337  F
ANN. BIOL., 29:177                                            4814
ANN. BIOL., 29:180-181.                                       4816
ANN. BIOL., 29:182-183.                                       4815
ANN. BIOL., 30:206-207.                                       5012
ANN. HYDROBIOL., 17:255-262.                                  5352  F
ANN. IMMUNOL., 125C:731-745.                                  4602  F
    (FR.)
    ENG. SUMM.
ANN. INST. MICHEL PACHA, 5:69-235.                            4624
ANN. INST. PATAGONIA, 7:211-214.                              5571
ANN. PALEONTOL., VERTEBR., 61:3-16.                           5047  F
    (FR.)
    ENG. SUMM.
ANN. PALEONTOL., VERTEBR., 63:1-32.                           5346  F
    (FR.)
ANN. SCI. NAT., ZOOL. BIOL. ANIM., 17:535-558.                5149
ANNU. REV. PHARMACOL., 14:47-55.                              5307  F
ANTHROPOLOGIE (PARIS), 79:451-481.                            5037
ARCH. BIOCHEM. BIOPHYS., 181:447-453.                         4717
ARCH. BIOL., 87:429-476.                                      5272  F
ARCH. HISTOL. JAP., 37:277-290.                               4690  F
ARCH. HISTOL. JAP., 40:391-406.                               5602
ARCH. HISTOL. JAP., 40:41-49.                                 5187  F
ARCH. INT. PHYSIOL. BIOCHIM., 85:865-870.                     5096  F
ARCH. ORAL BIOL., 20:635-640.                                 4875  F
ARCH. OTO-RHINO-LARYNGOL., 210:1-42.                          5297
ARKANSAS ACAD. SCI., FAYETTEVILLE, PROC., 27:27-29.           4926
ARKANSAS ACAD. SCI., FAYETTEVILLE, PROC., 28:22-26.           5033
ARKANSAS ACAD. SCI., FAYETTEVILLE, PROC., 28:59-64.           5032  F
ARKANSAS ACAD. SCI., FAYETTEVILLE, PROC., 28:65-70.           5035  F
ARKANSAS ACAD. SCI., FAYETTEVILLE, PROC., 29:37-39.           5203
ARKANSAS ACAD. SCI., FAYETTEVILLE, PROC., 29:54-56.           5202  F
ARKANSAS ACAD. SCI., FAYETTEVILLE, PROC., 30:100-104.         5287
ARKH. ANAT. GISTOL. EMBRIOL., 67:22-40.                       4634  F
    (RUSS.)
    ENG. SUMM.
ARKH. ANAT. GISTOL. EMBRIOL., 67:92-99.                       4616  F
    (RUSS.)
ARKH. ANAT. GISTOL. EMBRIOL., 70:68-74.                       5073
ARKH. PATOL., 35:71-72.                                       5040
ASSOC. CAN.-FR. AV. SCI., MONT., 39:148.                      4752  F
    (ABSTR.)
```

```
          ABSTRACT
ASSOC. CAN.-FR. AV. SCI., MONT., 39:148.                    4753  F
     (FR.)
     (ABSTR.)
ASSOC. OFF. ANAL. CHEM., J., 59:862-865.                    5408  F
ASSOC. SOUTHEAST. BIOL., BULL., 24:58.                      4994  F
     (ABSTR.)
ASSOC. SOUTHEAST. BIOL., BULL., 25:55.                      5564
ASTM, PROC., IN PROGRESS.                                   5455
AUST. ZOOL., 18:137-148.                                    4992
BASEL: S. KARGER.                                           5282
BEAVER, MAG. NORTH, AUT:18-19.                              5129  F
BERLIN. UNIV., ZOOL. MUS., MITT., 49:49-67.                 4912  F
     (GER.)
     RUSS. SUMM.
BIOCHEM. BIOPHYS. RES. COMMUN., 60:1090-1096.               4954  F
BIOCHEM. GENET., 10:57-67.                                  4562  F
BIOCHEM. J., 161:201-204.                                   5156  F
BIOCHEM. PHARMACOL., 23:2403-2410.                          4652  F
BIOCHIM. BIOPHYS. ACTA, 271:277-282.                        4960  F
BIOCHIM. BIOPHYS. ACTA, 338:254-264.                        4572
BIOCHIM. BIOPHYS. ACTA, 351:273-289.                        4628  F
BIOCHIM. BIOPHYS. ACTA, 359:415-420.                        4594  F
BIOCHIM. BIOPHYS. ACTA, 444:344-348.                        5028
BIOCHIM. BIOPHYS. ACTA, 453:426-438.                        5146  F
BIOCHIM. BIOPHYS. ACTA, 453:439-452.                        4715  F
BIOFIZIKA, 18:216-222.                                      4930
     (RUSS.)
        BIOPHYSICS, 18:222-229.
        (ENG.)
BIOKHIMIIA, 40:550-555.                                     4989
BIOKHIMIIA, 40:652-658.                                     4887  F
     (RUSS.)
     ENGL. SUMM.
BIOKHIMIIA, 42:1960-1964.                                   5522
BIOKHIMIIA, 42:1960-1964.                                   5612
     (RUSS.)
        BIOCHEMISTRY, 42:1545-1548.
        (ENG.)
BIOL. BULL., 146:137-156.                                   4961  F
BIOL. LISTY, 41:172-185.                                    5372  F
BIOL. ZH. ARM., 26:56-63.                                   4612
BIOLOGIA (BRATISL.), 31:641-647.                            5175  F
BIOLOGIA, (BRATISL.), 25:123-128.                           5582
BIOLOGIST, 59:73-84                                         5161  F
BRAIN BEHAV. EVOL., 10:121-129.                             4951
BRAIN RES., 111:204-211.                                    5158
BRAIN RES., 140:33-42.                                      5466  F
BRAIN RES., 61:279-293.                                     4570  F
BRAIN RES., 72:1-23                                         4595  F
BRAIN RES., 99:17-33                                        4805  F
BULL. ENVIRON. CONTAM. TOXICOL., 17:57-65.                  5421  F
CALIF. FISH GAME, 61:56-59.                                 5157  F
CAN. FED. BIOL. SOC., PROC., 19:150.                        5570
CAN. FED. BIOL. SOC., PROC., 19:24.                         5567
CAN. FIELD NAT., 87:235-239.                                4600  F
CAN. J. ZOOL. 50:1215-1223.                                 5461
CAN. J. ZOOL., 51:101-104.                                  4895  F
CAN. J. ZOOL., 51:769-799.                                  4639  F
CAN. J. ZOOL., 52:1047-1055.                                5553
CAN. J. ZOOL., 52:1447-1455.                                4726
CAN. J. ZOOL., 52:1585-1589.                                4743  F
CAN. J. ZOOL., 53:516-520.                                  5130  F
     ANIMAL ADULT NERVOUS SYSTEM VISION
CAN. J. ZOOL., 54:1449-1458.                                4539
CAN. J. ZOOL., 54:180-184.                                  4723  F
CAN. J. ZOOL., 54:421-425                                   4768  F
CAN. J. ZOOL., 54:843-851.                                  4749  F
CAN. J. ZOOL., 54:974-989.                                  4750  F
CAN. J. ZOOL., 55:469-473.                                  4695
CAN. J. ZOOL., 56:1420-1429.                                5541  F
CAN. J. ZOOL., 56:561-570.                                  5537  F
CAN. PET. GEOL., BULL., 23:67-83.                           4871
CAS. MORAV. MUS., 56-57:375-384.                            5581  F
CELL TISSUE RES., 150:505-520.                              4586
CELL TISSUE RES., 152:259-270.                              4641
CELL TISSUE RES., 154:109-119.                              4876  F
CELL TISSUE RES., 154:17-27.                                4882  F
CELL TISSUE RES., 157:1-6.                                  5043
CELL TISSUE RES., 157:141-164.                              4758  F
CELL TISSUE RES., 157:165-184.                              5042  F
CELL TISSUE RES., 157:503-516.                              5045  F
CELL TISSUE RES., 158:75-87.                                4790  F
CELL TISSUE RES., 159:109-120.                              5038
CELL TISSUE RES., 159:311-323.                              5044  F
CELL TISSUE RES., 161:25-32.                                5029
CELL TISSUE RES., 163:327-341.                              4991
CELL TISSUE RES., 163:353-363.                              4792  F
CELL TISSUE RES., 164:201-213.                              5086
CELL TISSUE RES., 166:145-157.                              5078  F
CELL TISSUE RES., 166:185-200.                              4779  F
```

```
CELL TISSUE RES., 168:433-444.                                          5101  F
CELL TISSUE RES., 171:317-330.                                          5205  F
CELL TISSUE RES., 172:487-502.                                          5266  F
CELL TISSUE RES., 173:271-277.                                          5115  F
CELL TISSUE RES., 174:427-430.                                          5296  F
CELL TISSUE RES., 177:281-286.                                          5517
CELL TISSUE RES., 177:317-323.                                          5312
CELL TISSUE RES., 178:385-396.                                          5355
CELL TISSUE RES., 178:477-482.                                          4872
CELL TISSUE RES., 180:1-10.                                             5169  F
CELL TISSUE RES., 181:73-79.                                            5524  F
CELL TISSUE RES., 187:473-478.                                          5539  F
CHESAPEAKE SCI., 15:154-155.                                            4676  F
CHESAPEAKE SCI., 16:70-72.                                              4720  F
CHICAGO NAT. HIST. MUS., FIELIANA: GEOL., 33:83-93.                     4537
CHILE. MUS. NAC. HIST. NAT., BOL., 33:53-63.                            5036
CLUJ, ROM., UNIV. BABES-BOLYAI, STUD., SER. BIOL., 1:105-113            5399  F
     (ITAL.)
       FR. AND RUSS. SUMM.
CLUJ, ROM., UNIV. BABES-BOLYAI, STUD., SER. BIOL., 20:43-47.            4878  F
     (ROM.)
COLL. INT. CENTR. NAT. RECH. SCI., PARIS.                               4959
COLL. INT. CENTR. NAT. RECH. SCI., PARIS, 218:15-31.                    5547
COMP. BIOCHEM. PHYSIOL., 45A:1009-1021.                                 4767  F
COMP. BIOCHEM. PHYSIOL., 48A:145-151.                                   4627  F
COMP. BIOCHEM. PHYSIOL., 48A:555-560.                                   4545
COMP. BIOCHEM. PHYSIOL., 48B:329-342.                                   4579
COMP. BIOCHEM. PHYSIOL., 49:273-280.                                    4621
COMP. BIOCHEM. PHYSIOL., 49A:677-688.                                   4599  F
COMP. BIOCHEM. PHYSIOL., 50A:379-382.                                   4687  F
COMP. BIOCHEM. PHYSIOL., 50A:753-757.                                   4673  F
COMP. BIOCHEM. PHYSIOL., 51A:521-522.                                   4573  F
COMP. BIOCHEM. PHYSIOL., 51A:723-726.                                   4656  F
COMP. BIOCHEM. PHYSIOL., 51B:139-142.                                   4620
COMP. BIOCHEM. PHYSIOL., 51B:403-408.                                   4609  F
COMP. BIOCHEM. PHYSIOL., 51B:517-520.                                   4584  F
COMP. BIOCHEM. PHYSIOL., 52:17-22.                                      5046
COMP. BIOCHEM. PHYSIOL., 52A:639-643.                                   4880  F
COMP. BIOCHEM. PHYSIOL., 52B:547-549.                                   5041
COMP. BIOCHEM. PHYSIOL., 53A:73-77.                                     4885
COMP. BIOCHEM. PHYSIOL., 53B:295-298.                                   5019  F
COMP. BIOCHEM. PHYSIOL., 53B:555-559.                                   5004  F
COMP. BIOCHEM. PHYSIOL., 54B:369-374.                                   5074
COMP. BIOCHEM. PHYSIOL., 54B:409-411.                                   5465  F
COMP. BIOCHEM. PHYSIOL., 55B:381-385.                                   5177  F
COMP. BIOCHEM. PHYSIOL., 55B:49-56.                                     5024  F
COMP. BIOCHEM. PHYSIOL., 55B:69-75.                                     5023  F
COMP. BIOCHEM. PHYSIOL., 56:305-309.                                    5188
COMP. BIOCHEM. PHYSIOL., 56A:457-460.                                   5182  F
COMP. BIOCHEM. PHYSIOL., 56B:81-85.                                     5147
COMP. BIOCHEM. PHYSIOL., 57:133-138.                                    5212
COMP. BIOCHEM. PHYSIOL., 57B:127-132.                                   5211
COMP. BIOCHEM. PHYSIOL., 57B:185-190.                                   5218  F
COMP. BIOCHEM. PHYSIOL., 57B:191-196.                                   5015
COMP. BIOCHEM. PHYSIOL., 60A:431-434.                                   5145
COMP. BIOCHEM. PHYSIOL., 60A:435-443.                                   5604  F
COMP. BIOCHEM. PHYSIOL., 60B:87-92.                                     5603
COMP. GEN. PHARMACOL., 3:160-166.                                       5397  F
COMP. GEN. PHARMACOL., 5:67-76.                                         5404  F
CONNECT. TISSUE. RES., 2:57-64.                                         4571
COPEIA, 1973:135-136.                                                   4653  F
COPEIA, 1973:568-574.                                                   5555
COPEIA, 1974:794-795.                                                   4727  F
COPEIA, 1975:136-137.                                                   4733  F
COPEIA, 1977:762-766.                                                   5348  F
COPEIA, 1977:767-768.                                                   5125  F
COPEIA, 1978:349-352.                                                   5613
DEL. CONSERV., 20:16-20.                                                5010
DIABETES, 26:322-340.                                                   5275
DIABETES, 26:353.                                                       5286
DIABETOLOGIA, 10:364.                                                   4725  F
     (ABSTR.)
DRUG METAB. DISPOS., 2:545-555.                                         4604  F
ELECTROENCEPHALOGR. CLIN. NEUROPHYSIOL., 36:438.                        4661
ENDOCRINOLOGY, 95:854-862.                                              5056  F
ENV. BIOL. FISH., 3:241-243.                                            5545  F
ENV. BIOL. FISH., 4.                                                    5507
     (RUSS.)
       ENG. SUMM.
ENVIRON. BIOL. FISHES., 2:103-120.                                      4975  F
ENVIRON. CAN., FISH. MAR. SERV., SEA LAMPREY CONTROL CENT.,             5231  F
     SAULT STE MARIE, REP., 14 PP.
ENVIRON. CAN., FISH. MAR. SERV., TECH. REP., 611:25PP.                  4633  F
EXPERIENTIA, 31:912-913.                                                4696
EXPERIENTIA, 32:1537-1538.                                              5447
EXPERIENTIA, 32:443-445.                                                5171
FAUNA (OSLO), 29:1-64.                                                  5395  F
     (NORW.)
FAUNA FLORA (STOCKH.), 72:15-17.                                        5277  F
     (SWED.)                              1977
FAUNA, 29:1-64.                                                         5469  F
```

                        (NORW.)
          FEBS LETT., 62:139-141.                                4813
          FEBS LETT., 65:92-95.                                  5070
          FEBS LETT., 78:279-283.                                5314   F
          FED. PROC., 31:606.                                    5415
          FED. PROC., 33:1474.                                   4738
            (ABSTR.)
          FED. PROC., 34: 281.                                   4683
          FED. PROC., 34:305.                                    4685
            (ABSTR.)
          FED. PROC., 34:345.                                    4682
            (ABSTR.)
          FED. PROC., 35:1126.                                   5014
          FED. PROC., 35:1131.                                   5013
          FED. PROC., 35:2145-2149                               5097   F
          FED. PROC., 35:321.                                    5018   F
            (ABSTR.)
          FED. PROC., 35:656.                                    5011
          FED. PROC., 36:1239.                                   5166
          FED. PROC., 36:2386-2389.                              5269   F
          FED. PROC., 36:528.                                    5168
          FED. PROC., 36:540.                                    5162
          FED. PROC., 36:642.                                    5164
          FED. PROC., 36:676.                                    5167
          FED. PROC., 37:929                                     5464
            (ABSTR.)
          FISH. RES. BOARD CAN., J. 34:159-163                   5235   F
          FISH. RES. BOARD CAN., J., 30:1047-1052.               5080   F
          FISH. RES. BOARD CAN., J., 30:1367-1370                4755   F
          FISH. RES. BOARD CAN., J., 30:1841-1846.               5411   F
          FISH. RES. BOARD CAN., J., 30:461-463.                 5414   F
          FISH. RES. BOARD CAN., J., 30:565-568.                 5343   F
          FISH. RES. BOARD CAN., J., 30:601-605.                 4754   F
          FISH. RES. BOARD CAN., J., 31:122-123.                 4759   F
          FISH. RES. BOARD CAN., J., 32:1455-1459.               5387   F
          FISH. RES. BOARD CAN., J., 32:1873-1876.               5422   F
          FISH. RES. BOARD CAN., J., 32:515-522.                 5440
          FISH. RES. BOARD CAN., J., 32:729-738.                 5219   F
          FISH. RES. BOARD CAN., J., 33:1198-1201.               5405   F
          FISH. RES. BOARD CAN., J., 33:2740-2746.               5386   F
          FISH. RES. BOARD CAN., J., 34:1373-1378.               5340   F
          FISH. RES. BOARD CAN., J., 34:276-281.                 5419   F
          FISH. RES. BOARD CAN., J., 35:1262-1265.               5462   F
          FISH. RES. BOARD CAN., J., 5:623-631.                  4769   F
          FISH. RES. BOARD CAN., MANUSCR. REP. SER., 1377: 37 PP.  5364   F
          FISH. RES. BOARD CAN., TECH. REP., 304:28P.            4761   F
          FIZIOL. BIOKHIM. NIZSHIKH POZVONO.,   :49-53.          5267   F
            (RUSS.)
            ENG. SUMM.
          FIZIOL. ZH. SSSR IM. I.M. SECHENOVA, 63:1195-1198.     5354   F
            (RUSS.)
            ENG. SUMM.
          FLA. MAR. RES. PUBL., 26:22-33.                        5316
          FLORA & FAUNA (MOSCOW), 1960:171-178.                  5573   F
          FLORENCE, 11TH INT. CANCER CONGR., ABSTR., 2:176.      5546
          FOLIA BIOCHEM. BIOL. GRAECA., 13:25-46.                4983
          FOLIA BIOL., 25:409-414.                               5598
          FOLIA HISTOCHEM. CYTOCHEM., 14:283-308.                5273
          FOLIA MORPHOL., 25:64-67.                              5265   F
          FOLIA PARASITOL., (PRAGUE), 24:323-329                 5467   F
          GEGENBAURS MORPHOL. JAHRB., 119:796-808.               5563
          GEGENBAURS MORPHOL. JAHRB., 119:823-856.               5215
          GEN. COMP. ENDOCRINOL., SUPPL., 2:510-521.             4979
          GEN. COMP. ENDOCRINOL., 19:56-68.                      4897   F
          GEN. COMP. ENDOCRINOL., 21:188-195.                    5315
          GEN. COMP. ENDOCRINOL., 21:214-215.                    4928   F
            (ABSTR.)
          GEN. COMP. ENDOCRINOL., 21:231-240.                    5328   F
          GEN. COMP. ENDOCRINOL., 21:451-460.                    5559
          GEN. COMP. ENDOCRINOL., 22;480-488.                    4558   F
          GEN. COMP. ENDOCRINOL., 22:312-314.                    4564   F
          GEN. COMP. ENDOCRINOL., 22:335-336.                    4927
            (ABSTR.)
          GEN. COMP. ENDOCRINOL., 22:372.                        4916
            (ABSTR.)
          GEN. COMP. ENDOCRINOL., 22:384.                        5607   F
            (ABSTR.)
          GEN. COMP. ENDOCRINOL., 22:391                         4915   F
            (ABSTR.)
          GEN. COMP. ENDOCRINOL., 23:311-324.                    4748   F
          GEN. COMP. ENDOCRINOL., 24:249-256.                    4591   F
          GEN. COMP. ENDOCRINOL., 24:305-313.                    4592   F
          GEN. COMP. ENDOCRINOL., 25:274-291.                    4617   F
          GEN. COMP. ENDOCRINOL., 25:487-508                     4577   F
          GEN. COMP. ENDOCRINOL., 26:2-15.                       5450
          GEN. COMP. ENDOCRINOL., 26:368-373.                    4607   F
          GEN. COMP. ENDOCRINOL., 26:420-422.                    4585   F
          GEN. COMP. ENDOCRINOL., 26:59-69.                      4729   F
          GEN. COMP. ENDOCRINOL., 26:96-99.                      4574   F
          GEN. COMP. ENDOCRINOL., 26:96-99.                      5527
          GEN. COMP. ENDOCRINOL., 27:179-192.                    4886   F

```
GEN. COMP. ENDOCRINOL., 27:320-349.                               5066  F
GEN. COMP. ENDOCRINOL., 27:495-508.                               4998  F
GEN. COMP. ENDOCRINOL., 27:517-526.                               4781  F
GEN. COMP. ENDOCRINOL., 28:184-204.                               5002  F
GEN. COMP. ENDOCRINOL., 28:213-227.                               4870  F
GEN. COMP. ENDOCRINOL., 28:228-246.                               4806  F
GEN. COMP. ENDOCRINOL., 28:358-364.                               5005
GEN. COMP. ENDOCRINOL., 28:365-367.                               5003  F
GEN. COMP. ENDOCRINOL., 28:473-480.                               5065  F
GEN. COMP. ENDOCRINOL., 29:1-13.                                  5062  F
GEN. COMP. ENDOCRINOL., 29:258.                                   5193
    (FR.)
    (ABSTR.)
GEN. COMP. ENDOCRINOL., 29:287.                                   5196
    (ABSTR.)
GEN. COMP. ENDOCRINOL., 29:288.                                   5192
    (ABSTR.)
GEN. COMP. ENDOCRINOL., 29:301-312.                               5021
GEN. COMP. ENDOCRINOL., 30:243-257.                               5174  F
GEN. COMP. ENDOCRINOL., 30:258-266.                               5199  F
GEN. COMP. ENDOCRINOL., 30:340-346.                               5176  F
GEN. COMP. ENDOCRINOL., 30:500-516.                               5184  F
GEN. COMP. ENDOCRINOL., 31:270-275.                               5209  F
GEN. COMP. ENDOCRINOL., 31:381-383.                               5208
GEN. COMP. ENDOCRINOL., 31:75-79.                                 5159  F
GEN. COMP. ENDOCRINOL., 31:91-100.                                5155
GEN. COMP. ENDOCRINOL., 33:423-427.                               5326  F
GEN. COMP. ENDOCRINOL., 33:53-60.                                 5306  F
GEN. COMP. ENDOCRINOL., 34.                                       5605  F
    (ABSTR.)
GEN. COMP. ENDOCRINOL., 34:26-37.                                 5470  F
GEN. COMP. ENDOCRINOL., 35:197-204.                               5509
GEN. COMP. ENDOCRINOL., 35:96-98.                                 5600
GEN. ELECTR., ENVIRON. SCI. LAB.:55 PP.                           5381
GEN. PHARMACOL., 6:15-18.                                         4660  F
GEOL. SURV. CAN., BULL., 222:35-46.                               5329
GREAT LAKES FISH. COMM., ANNU. REP., PP.1-93.                     4784  F
    RESTRICTED
GREAT LAKES FISH. COMM., ANNU. REP., 101 PP.                      5194  F
    RESTRICTED
GREAT LAKES FISH. COMM., ANNU. REP., 361 PP.                      5499  F
GREAT LAKES FISH. COMM., ANNU. REP., 91 PP.                       4783  F
    RESTRICTED
GREAT LAKES FISH. COMM., ANNU. REP., 98PP.                        4844  F
    RESTRICTED
GREAT LAKES FISH. COMM., TECH. REP.                               5425
GREAT LAKES FISH. COMM., TECH. REP.                               5459
GREAT LAKES FISH. COMM., TECH. REP., 18:1-16.                     4736  F
    RESTRICTED
GREAT LAKES FISH. COMM., TECH. REP., 26:1-60.                     4630  F
    RESTRICTED
GREAT LAKES FISH. COMM., TECH. REP., 27:1-19.                     4746  F
    RESTRICTED
GREAT LAKES FISH. COMM., TECH. REP., 28:34 PP.                    4711
GREAT LAKES FISH. COMM., TECH. REP., 8:21 PP.                     5552
GREAT LAKES FISH. COMM., TECH. REV., 18PP.                        4756  F
    RESTRICTED
HELMINTHOL. SOC. WASH., PROC., 45:238-245.                        5327  F
HISTOCHEMISTRY, 40:263-266.                                       4940  F
HISTOCHEMISTRY, 43:283-290.                                       4587  F
HOKKAIDO UNIV., SAPPORO, JAPAN, FAC. SCI., J., SER. VI            5170  F
    ZOOL., 20:277-287.
HYDROBIOLOGIA, 46:207-222.                                        4614  F
HYDROBIOLOGIA, 47:415-430.                                        5030
HYDROBIOLOGIA, 55:265-270.                                        5304  F
ILL. STATE ACAD. SCI., TRANS., 69:313-314.                        5523  F
IN  ANDREWS, S., R. MAHALA, S. MILES AND A.D. WALKER, EDS.,       5510
      LINNEAN SOCIETY SYMPOSIUM SERIES, NO. 4. PROBLEMS IN VERTEBR
      ATE EVOLUTION. LONDON: ACADEMIC PRESS.
IN  BOSMA, J.F., ED., SYMPOSIUM ON DEVELOPMENT OF THE BASICR      5615
      ANIUM. BETHESDA, MD., JUNE 23-25, 1975. BETHESDA: DEP. HEALT
      H EDUC. WELFARE.
IN  BOURNE, G.H. AND J.F. DANIELLI EDS. INTERNATIONAL REVIEW      5190
      OF CYTOLOGY, VOL. 47. NEW YORK: ACADEMIC PRESS.
IN  CHIARELLI, S.B. AND E. CAPANNA, EDS., CYTOTAXONOMY AND V      5560
      ERTEBRATE EVOLUTION. LONDON: ACADEMIC PRESS.
IN  COLLOQUE INT. CENT. NAT. RECH. SCI., PARIS, 104:13-29. P      5528
      ROBLEMES ACTUELS DE PALEONTOLOGIE - EVOLUTION DES VERTEBRES.
      PARIS: INT. CENT. NAT. RECH. SCI.
IN  COLLOQUE INTERNATIONAL C.N.R.S., PARIS 4-9 JUIN, PROBLEM      4771  F
      ES ACTUELS DE PALEONTOLOGIE - EVOLUTION DES VERTEBRES. FR.,
      CENT. NAT. RECH. SCI.
      (FR.)
      ENGL. SUMM.
IN  CORNER, M.A., AND D.F. SWAAB EDS. PROGRESS IN BRAIN RESE      5213
      ARCH. VOL. 45. PERSPECTIVES IN BRAIN RESEARCH. 9TH INTERNATI
      ONAL SUMMER SCHOOL. AMSTERDAM. NETHERLANDS, JULY 28-AUG.1,19
      75. AMSTERDAM: ELSEVIER SCIENTIFIC PUBLISHING COMP.
IN  COULSTON, F. AND F. KORTE EDS. ENVIRONMENTAL QUALITY AND      5183  F
      SAFETY, VOL.5.  GLOBAL ASPECTS OF CHEMISTRY TOXICOLOGY AND
      TECHNOLOGY AS APPLIED TO THE ENVIRONMENT., NEW YORK: ACADEMIC
```

PRESS.

IN  CSERR,H.F. ET.AL. EDS. FLUID ENVIRONMENT OF THE BRAIN. N        5117  F
       EW YORK: ACADEMIC PRESS.
IN  DAWE, C.J., D.G. SCARPELLI, AND S.R. WELLINGS, EDS., PRO         5285  F
       GRESS IN EXPERIMENTAL TUMOUR RESEARCH, VOL. 20. BASEL: S. KA
       RGER.
IN  DENTON,D.A. AND J.P.COGHLAN EDS. OLFACTION AND TASTE V.          5081
       PROCEEDING OF THE FIFTH INTERNATIONAL SYMPOSIUM. MELBOURNE,
       AUSTRALIA, OCT. 1974. NEW YORK: ACADEMIC PRESS.
IN  DIMOND, S.J. AND D.A. BLIZZARD, EDS., ANNALS OF THE NEW          5592
       YORK ACADEMY OF SCIENCES, V.299. EVOLUTION AND
       LATERALIZATION OF THE BRAIN. NEW YORK: N.Y. ACAD. SCI.
IN  DIMOND, S.J. AND D.A. BLIZZARD, EDS., ANNALS OF THE NEW          5594
       YORK ACADEMY OF SCIENCES, V. 299. EVOLUTION AND
       LATERALIZATION OF THE BRAIN. NEW YORK: N.Y. ACAD. SCI.
IN  DRABIKOWSKI,H., H. STRZELECKA-GOLASZEWSKA AND E.                 4671  F
       CARAFOLI, EDS., CALCIUM BINDING PROTEINS. WARSAW: PWN-POLISH
       SCIENTIFIC PUBLISHERS.
IN  EAKIN, R.M. THE THIRD EYE. BERKELEY: UNIV. OF CALIFORNIA         4731  F
       PRESS.
IN  EAKIN,R.M. THE THIRD EYE. BERKELEY: UNIV. OF CALIFORNIA          5057  F
       PRESS.
IN  ELEFTHERIOU, B.E. AND R.L. SPROTT, EDS., HORMONAL CORREL         5535
       ATES OF BEHAVIOUR, VOL. 1. A LIFESPAN VIEW. NEW YORK: PLENUM
IN  FANGE, R. ED. MYXINE GLUTINOSA. BIOCHEMISTRY, PHYSIOLOGY         4847  F
       AND STRUCTURE. REPORT FROM A SYMPOSIUM IN GOTEBORG 28-29
       APRIL, 1972. GOTEBORG:KUNGL. VETENSKAPS- OCH VITTERHETS-
       SAMHLLET.
IN  FANGE,R. ED. MYXINE GLUTINOSA. BIOCHEMISTRY. PHYSIOLOGY          4858  F
       AND STRUCTURE. REPORT FROM A SYMPOSIUM IN GOTEBORG 28-29 APR
       IL, 1972. GOTEBORG: KUNGL. VETENSKAPS- OCH VITTERHTS-SAMHLLT
       .
IN  FANGE,R. ED. MYXINE GLUTINOSA. BIOCHEMISTRY, PHYSIOLOGY          4582  F
       AND STRUCTURE. REPORT FROM A SYMPOSIUM IN GOTEBORG 28-29
       APRIL, 1972. GOTEBORG:KUNGL. VETENSKAPS-OCH VITTERHETS-
       SAMHLLT.
IN  FANGE,R. ED. MYXINE GLUTINOSA. BIOCHEMISTRY, PHYSIOLOGY          4804  F
       AND STRUCTURE. REPORT FROM A SYMPOSIUM IN GOTEBORG 28-29 APR
       IL,1972. GOTEBORG: KUNGL. VETENSKAPS- OCH VITTERHETS-SAMHLLT
       .
IN  FANGE,R. ED. MYXINE GLUTINOSA. BIOCHEMISTRY, PHYSIOLOGY          4832  F
       AND STRUCTURE. REPORT FROM A SYMPOSIUM IN GOTEBORG 28-29
       APRIL, 1972. GOTEBORG:KUNGL. VETENSKAPS- OCH
       VITTERHETS-SAMHLLT.
IN  FANGE,R. ED. MYXINE GLUTINOSA. BIOCHEMISTRY, PHYSIOLOGY          4833  F
       AND STRUCTURE. REPORT FROM A SYMPOSIUM IN GOTEBORG 28-29
       APRIL, 1972. GOTEBORG: KUNGL. VETENSKAPS- OCH VITTERHETS-
       SAMHLLT.
IN  FANGE,R. ED. MYXINE GLUTINOSA. BIOCHEMISTRY, PHYSIOLOGY          4834  F
       AND STRUCTURE. REPORT FROM A SYMPOSIUM IN GOTEBORG 28-29,APR
       IL, 1972. GOTEBORG: KUNGL. VETENSKAPS- OCH VITTERHETS-SAMHLL
       ET.
IN  FANGE,R. ED. MYXINE GLUTINOSA. BIOCHEMISTRY, PHYSIOLOGY          4835  F
       AND STRUCTURE. REPORT FROM A SYMPOSIUM IN GOTEBORG 28-29
       APRIL, 1972. GOTEBORG: KUNGL. VETENSKAPS- OCH VITTERHETS-
       SAMHLLET.
IN  FANGE,R. ED. MYXINE GLUTINOSA. BIOCHEMISTRY, PHYSIOLOGY          4836  F
       AND STRUCTURE. REPORT FROM A SYMPOSIUM IN GOTEBORG 28-29 APR
       IL, 1972. GOTEBORG:KUNGL. VETENSKAPS- OCH VITTERHETS-SAMHLLE
       T.
IN  FANGE,R. ED. MYXINE GLUTINOSA. BIOCHEMISTRY, PHYSIOLOGY          4837  F
       AND STRUCTURE. REPORT FROM A SYMPOSIUM IN GOTEBORG 28-29
       APRIL, 1972. GOTEBORG: KUNGL. VETENSKAPS- OCH VITTERHETS-
       SAMHLLET.
IN  FANGE,R. ED. MYXINE GLUTINOSA. BIOCHEMISTRY, PHYSIOLOGY          4838  F
       AND STRUCTURE. REPORT FROM A SYMPOSIUM IN GOTEBORG 28-29, AP
       RIL, 1972. GOTEBORG: KUNGL. VETENSKAPS- OCH VITTERHETS-SAMHL
       LET.
IN  FANGE,R. ED. MYXINE GLUTINOSA. BIOCHEMISTRY, PHYSIOLOGY          4839  F
       AND STRUCTURE. REPORT FROM A SYMPOSIUM IN GOTEBORG 28-29
       APRIL, 1972. GOTEBORG:KUNGL. VETENSKAPS- OCH VITTERHETS-
       SAMHLLET.
IN  FANGE,R. ED. MYXINE GLUTINOSA. BIOCHEMISTRY, PHYSIOLOGY          4840  F
       AND STRUCTURE. REPORT FROM A SYMPOSIUM IN GOTEBORG 28-29
       APRIL, 1972. GOTEBORG:KUNGL. VETENSKAPS- OCH VITTERHETS-
       SAMHLLET.
IN  FANGE,R. ED. MYXINE GLUTINOSA. BIOCHEMISTRY, PHYSIOLOGY          4841  F
       AND STRUCTURE. REPORT FROM A SYMPOSIUM IN GOTEBORG 28-29 APR
       IL, 1972. GOTEBORG:KUNGL. VETENSKAPS- OCH VITTERHETS-SAMHLLE
       T.
IN  FANGE,R. ED. MYXINE GLUTINOSA. BIOCHEMISTRY, PHYSIOLOGY          4842  F
       AND STRUCTURE. REPORT FROM A SYMPOSIUM IN GOTEBORG 28-29
       APRIL, 1972. GOTEBORG:KUNGLE. VETENSKAPS- OCH VITTERHETS-
       SAMHALLET.
IN  FANGE,R. ED. MYXINE GLUTINOSA. BIOCHEMISTRY, PHYSIOLOGY          4845  F
       AND STRUCTURE. REPORT FROM A SYMPOSIUM IN GOTEBORG 28-29
       APRIL, 1972. GOTEBORG:KUNGL. VETENSKAPS- OCH VITTERHETS-
       SAMHLLET.
IN  FANGE,R. ED. MYXINE GLUTINOSA. BIOCHEMISTRY, PHYSIOLOGY          4846  F
       AND STRUCTURE. REPORT FROM A SYMPOSIUM IN GOTEBORG 28-29 APR
       IL, 1972. GOTEBORG:KUNGL. VETENSKAPS- OCH VITTERHETS-SAMHLLE

                T.
IN   FANGE,R. ED. MYXINE GLUTINOSA. BIOCHEMISTRY, PHYSIOLOGY          4849  F
     AND STRUCTURE. REPORT FROM A SYMPOSIUM IN GOTEBORG 28-29
     APRIL, 1972. GOTEBORG:KUNGL. VETENSKAPS-OCH VITTERHETS-
     SAMHLLT.
IN   FANGE,R. ED. MYXINE GLUTINOSA. BIOCHEMISTRY, PHYSIOLOGY          4850  F
     AND STRUCTURE. REPORT FROM A SYMPOSIUM IN GOTEBORG 28-29 APR
     IL, 1972. GOTEBORG: KUNGL. VETENSKAPS- OCH VITTERHETS-SAMHLL
     T.
IN   FANGE,R. ED. MYXINE GLUTINOSA. BIOCHEMISTRY, PHYSIOLOGY          4851  F
     AND STRUCTURE. REPORT FROM A SYMPOSIUM IN GOTEBORG 28-29
     APRIL, 1972. GOTEBORG:KUNGL. VETENSKAPS- OCH VITTERHETS-
     SAMHLLT.
IN   FANGE,R. ED. MYXINE GLUTINOSA. BIOCHEMISTRY, PHYSIOLOGY          4852  F
     AND STRUCTURE. REPORT FROM A SYMPOSIUM IN GOTEBORG 28-29
     APRIL, 1972. GOTEBORG:KUNGL. VETENSKAPS- OCH VITTERHETS-
     SAMHLLT.
IN   FANGE,R. ED. MYXINE GLUTINOSA. BIOCHEMISTRY, PHYSIOLOGY          4853  F
     AND STRUCTURE. REPORT FROM A SYMPOSIUM IN GOTEBORG 28-29 APR
     IL, 1972. GOTEBORG:KUNGL. VETENSKAPS- OCH VITTERHETS-SAMHLLT

IN   FANGE,R. ED. MYXINE GLUTINOSA. BIOCHEMISTRY, PHYSIOLOGY          4854  F
     AND STRUCTURE. REPORT FROM A SYMPOSIUM IN GOTEBORG 28-29 APR
     IL, 1972. GOTEBORG: KUNGL. VETENSKAPS- OCH VITTERHETS-SAMHLL
     T.
IN   FANGE,R. ED. MYXINE GLUTINOSA. BIOCHEMISTRY, PHYSIOLOGY          4855  F
     AND STRUCTURE. REPORT FROM A SYMPOSIUM IN GOTEBORG 28-29
     APRIL, 1972. GOTEBORG: KUNGL. VETENSKAPS- OCH VITTERHETS-
     SAMHLLT.
IN   FANGE,R. ED. MYXINE GLUTINOSA. BIOCHEMISTRY, PHYSIOLOGY          4856  F
     AND STRUCTURE. REPORT FROM A SYMPOSIUM IN GOTEBORG 28-29 APR
     IL, 1972. GOTEBORG:KUNGL. VETENSKAPS- OCH VITTERHETS-SAMHLLT
     .
IN   FANGE,R. ED. MYXINE GLUTINOSA. BIOCHEMISTRY, PHYSIOLOGY          4857  F
     AND STRUCTURE. REPORT FROM SYMPOSIUM IN GOTEBORG 28-29
     APRIL, 1972. GOTEBORG: KUNGL. VETENSKAPS- OCH VITTERHETS-
     SAMHLLT.
IN   FANGE,R. ED. MYXINE GLUTINOSA. BIOCHEMISTRY, PHYSIOLOGY          4859  F
     AND STRUCTURE. REPORT FROM A SYMPOSIUM IN GOTEBORG 28-29
     APRIL, 1972. GOTEBORG: KUNGL. VETENSKAPS- OCH VITTERHETS-
     SAMHLLT.
IN   FANGE,R. ED. MYXINE GLUTINOSA. BIOCHEMISTRY, PHYSIOLOGY          4860  F
     AND STRUCTURE. REPORT FROM A SYMPOSIUM GOTEBORG 28-29 APRIL,
     1972. GOTEBORG:KUNGL. VETENSKAPS-OCH VITTERHETS-SAMHLLT.
IN   FANGE,R. ED. MYXINE GLUTINOSA. BIOCHEMISTRY,PHYSIOLOGY A         4843  F
     ND STRUCTURE. REPORT FROM A SYMPOSIUM IN GOTEBORG 28-29 APRI
     L, 1972. GOTEBORG:KUNGL. VETENSKAPS- OCH VITTERHETS-SAMHALLE
     T.
IN   FARVAR, M.T., AND J.P. MILTON, EDS., TECHNOLOGY, ECOLOGY         5441
     , AND INTERNATIONAL DEVELOPMENT. NEW YORK: DOUBLEDAY.
IN   FENTRESS, J.C., ED., SIMPLER NETWORKS AND BEHAVIOUR.             5532
     SUNDERLAND: SINAUER.
IN   FLORKIN,M. AND B.T.SCHEER EDS. CHEMICAL ZOOLOGY. VOL.VII         4705  F
     I: PRIMITIVE DEUTEROSTOMIANS, CYCLOSTOMATA, FISHES. NEW YORK
     :ACADEMIC PRESS.
IN   FLORKIN,M. AND BRADLEY,T.S. EDS. CHEMICAL ZOOLOGY,VOL.8.         4910  F
     DEUTEROSTOMIANS, CYCLOSTOMES AND FISHES. NEW YORK: ACADEMIC
     PRESS.
IN   FLORKIN,M. AND BRADLEY,T.S.(ED.). CHEMICAL ZOOLOGY, VOL.         4909  F
     8. DEUTEROSTOMIANS, CYCLOSTOMES AND FISHES. NEW YORK;
     ACADEMIC PRESS.
IN   FLORKIN,M. AND SCHEER,B.T. EDS. CHEMICAL ZOOLOGY. VOL.           4706  F
     VIII: PRIMATIVE DEUTEROSTOMIANS, CYCLOSTOMATA, FISHES. NEW
     YORK, ACADEMIC PRESS.
IN   GRAHAM, D.H.   A TREASURY OF NEW ZEALAND FISHES. WELLINGT        5232  F
     ON: A.H. AND A.W. REED.
IN   GREAT LAKES FISH. COMM., ANNU. REP.                             5366  F
     RESTRICTED
IN   GREAT LAKES FISH. COMM., ANNU. REP.                             5472  F
     RESTRICTED
IN   GREAT LAKES FISH. COMM., ANNU. REP., 1973.                      5473  F
     RESTRICTED
IN   GREAT LAKES FISH. COMM., ANNU. REP., 1976.                      5500  F
     RESTRICTED
IN   GREAT LAKES FISH. COMM., ANNU. REP., 1976.                      5501  F
     RESTRICTED
IN   GREAT LAKES FISH. COMM., ANNU. REP., 1976.                      5502  F
     RESTRICTED
IN   GREAT LAKES FISH. COMM., ANNU. REP., 1977.                      5368  F
     RESTRICTED
IN   GRILLO, T.A.I. AND LIEBSON, L. AND EPPLE, A., EDS. THE E         5126  F
     VOLUTION OF PANCREATIC ISLETS. OXFORD: PERGAMON PRESS.
IN   GRILLO, T.A.I., AND A. EPPLE, EDS., EVOLUTION OF PANCREA         5436
     TIC ISLETS. OXFORD: PERGAMMON.
IN   GRILLO, T.A.I., L. LEIBSON AND A. EPPLE, EDS., THE EVOLU         5289  F
     TION OF PANCREATIC ISLETS. OXFORD: PERGAMON PRESS.
IN   GRILLO, T.A.I., L. LEIBSON AND A. EPPLE, EDS., THE EVOLU         5292
     TION OF PANCREATIC ISLETS. OXFORD: PERGAMON.
IN   GRILLO, T.A.I., L. LEIBSON AND A.EPPLE, EDS., THE EVOLUT         5288  F
     ION OF PANCREATIC ISLETS. OXFORD: PERGAMON PRESS.
IN   GRILLO, T.A.I., L. LEIBSON, AND A. EPPLE, EDS., THE EVOL         5290

UTION OF PANCREATIC ISLETS. OXFORD: PERGAMON PRESS.

| | | | |
|---|---|---|---|
| IN | GRILLO, T.A.I., LEIBSON, L., EPPLE, A., EDS. THE EVOLUTI ON OF PANCREATIC ISLETS. OXFORD: PERGAMON PRESS. | 5123 | F |
| IN | GRILLO, T.A.I, L. LEIBSON, A. EPPLE EDS., THE EVOLUTION OF PANCREATIC ISLETS. OXFORD: PERGAMON PRESS. | 5124 | F |
| IN | GRZIMEK, B., ED., GRZIMEK'S ENCYCLOPEDIA OF EVOLUTION. N EW YORK: VAN NOSTRAND REINHOLD CO. | 5240 | |
| IN | GUDGER, E.W., ED., THE BASHFORD DEAN MEMORIAL VOLUME ARC HAIC FISHES. ART III. NEW YORK:AMERICAN MUSEUM NATURAL HISTO RY. | 5584 | |
| IN | HAMILTON, W.J. ED. TEXTBOOK OF HUMAN ANATOMY. SAINT LOUI S, MO.:MOSBY COMPANY. | 5201 | |
| IN | HANDBUCH DER VERGLEICHENDEN ANATOMIE DER WIRBELTIERE. BD. III. | 5588 | |
| IN | HANKE,W. AND M. LINDAUER (EDS.) FORTSCHRITTE DER ZOOLOGIE, BAND 22, HEFT 2/3. VERGLEICHENDE ENDOKRINOLOGIE 2. INTERNATIONALES SYMPOSIUM DER AKADEMIE DER WISSENSCHAFTEN UND DER LITERATURE ZU MAINZ VOM 3 BIS 5, APRIL, 1973. STUTTGART:GUSTAV FISHER VERLAG. | 4958 | |
| IN | HIRSCH,G.C., RUSKA, H. AND SITTE,P. EDS. GRUNDLAGEN DER CYTOLOGIE. JENA: G. FISCHER. | 4978 | |
| IN | HOUCK, J.C. ED. CHALONES. AMSTERDAM: NORTH-HOLLAND PUBLI SHING CO. | 5093 | |
| IN | HUREAU,J.C. AND MONOD,TH. EDS. CHECK LIST OF THE FISHES OF THE NORTH EASTERN ATLANTIC AND OF THE MEDITERRANEAN. PARI S:UNESCO. AGNATHA 1-6. | 4770 | F |
| IN | JOHARI,O.M. AND I. CORVIN EDS. SCANNING ELECTRON MICROSCOPY 1975. CHICAGO:IIT RESEARCH INSTITUTE. | 4977 | |
| IN | KEKENTHAL, W., ED., HANDBUCH DER ZOOLOGIE. VOL. 6. BERLI N: WALTER DE GRUYTER & CO. | 5556 | |
| IN | KNIGGE,K.M. ET AL. ED. BRAIN-ENDOCRINE INTERACTION II. THE VENTRICULAR SYSTEM IN NEUROENDOCRINE MECHANISMS. SYMPOSIUM. SHIZUOKA, JAPAN. OCT. 16-18. 1974. BASEL: S. KARGER. | 4984 | |
| IN | KNIGGE,K.M. ET.AL. EDS. BRAIN-ENDOCRINE INTERACTION 11. THE VENTRICULAR SYSTEM IN NEUROENDOCRINE MECHANISMS. SYMPOSI UM. SHIZUOKA, JAPAN. OCT. 16-18,1974. BASEL:S.KARGER. | 4986 | |
| IN | KNOBIL,E. AND W.H. SAWYER EDS. HANDBOOK OF PHYSIOLOGY, SECTION 7. ENDOCRINOLOGY, VOL. IV. THE PITUITARY GLAND AND NEUROENDOCRINE CONTROL, PART 1. WASHINGTON, AMERICAN PHYSIOL OGICAL SOCIETY. | 4988 | |
| IN | KNOBIL,E. AND W.H.SAWYER EDS. HANDBOOK OF PHYSIOLOGY, SE CTION,7. ENDOCRINOLOGY, VOL.IV. THE PITUITARY GLAND AND ITS NEUROENDOCRINE CONTROL, PART I. WASHINGTON, AMERICAN PHYSIOL OGICAL SOCIETY. | 5017 | |
| IN | KNOBIL,E. AND W.H.SAWYER EDS. HANDBOOK OF PHYSIOLOGY, SE CTION 7. ENDOCRINOLOGY, VOL.IV. THE PITUITARY GLAND AND ITS NEUROENDOCRINE CONTROL, PART 2. WASHINGTON:AMERICAN PHYSIOLO GICAL SOCIETY. | 5034 | F |
| IN | KRAYBILL, H.F. ET AL, EDS., ANNALS OF THE NEW YORK ACADEMY OF SCIENCES, V. 298. AQUATIC POLLUTANTS AND BIOLOGIC EFFECTS WITH EMPHASIS ON NEOPLASIS. NEW YORK: N.Y. ACAD. SCI. | 5595 | |
| IN | KREPSA, E.M. ED. FERMENTY V EVOLYUTSII ZIVOTNYKH, LENING RAD: NAUKA. (RUSS.) ENGL. SUMM. | 4803 | F |
| IN | KREPSA,E.M. ED. FERMENTY V EVOLYUTSII ZHIVOTNYKH. LENING RAD: NAUKA. (RUSS.) ENGL. SUMM. | 4896 | F |
| IN | L'ETUDE PHYLOGENIQUE ET ONTOGENIQUE DE LA REPONSE IMMUNI TAIRE ET SON APPORT A LA THEORY., PARIS,12-14 OCTOBRE. (PHYL OGENIC AND ONTOGENIC STUDY OF THE IMMUNE RESPONSE AND ITS CO NTRIBUTION TO THE IMMUNOLOGICAL THEORY.) PARIS: SOC. FRAN. I MMUNOL. | 4801 | F |
| IN | LAMPRECHT, I. AND B. SCHAARSCHMEDT, EDS., APPLICATION OF CALORIMETRY IN LIFE SCIENCES. BERLIN: WALTER DE GRUYTER. | 5511 | |
| IN | LOWENSTEIN,O. ED. ADVANCES IN COMPARATIVE PHYSIOLOGY AND BIOCHEMISTRY, VOL.4. NEW YORK: ACADEMIC PRESS. | 4657 | |
| IN | LUEKEN,B. AND J.-H.SCHARF EDS. NOVA ACTA LEOPOLDINA. VOL . 37/2. NO.208. HALLE:DEUTSCHE AKADEMIE DER NATURFORSCHER LE OPOLDINA. | 4950 | |
| IN | MARCHALONIS, J.J., ED., IMMUNOLOGY SERIES, V.5. THE LYMP HOCYTE: STRUCTURE AND FUNCTION. PART 1. NEW YORK: BASEL. | 5568 | |
| IN | MARKERT, C.L. ED., ISOZYMES II. PHYSIOLOGICAL FUNCTION. THIRD INTERNATIONAL CONFERENCE. NEW HAVEN, CONN., U.S.A. NEW YORK: ACADEMIC PRESS. | 5179 | |
| IN | MICH. DEP. NAT. RESOUR., MICH. FISH. CENTEN. REP., 1873 TO 1973. FISH MANAGE. REP., #6. | 5445 | |
| IN | MOTHES,K. AND JOACHIM-HERMANN,S. EDS. NOVA ACTA LEOPOLDI NA. ABHANDLUNGEN DER DEUTSCHEN AKADEMIE DER NATURFORSCHER LE OPOLDINA, BAND 38 NO.211. EAST GERMANY: DEUTSCHE AKADEMIE DE R NATURFORSCHER LEOPOLDINA. | 4949 | F |
| IN | OBRUCHEV, D.V. ED. [ESSAYS ON THE PHYLOGENY AND SYSTEMAT ICS OF FOSSIL FISH AND AGNATHA.] MOSCOW:NAUKA | 5087 | |
| IN | OBRUCHEV, D.V. ED. [ESSAYS ON THE PHYLOGENY AND SYSTEMAT ICS OF FOSSIL FISH AND AGNATHA.] MOSCOW:NAUKA. | 5091 | |
| IN | OBRUCHEV, D.V. ED. [ESSAYS ON THE PHYLOGENY AND SYSTEMAT ICS OF FOSSIL FISH AND AGNATH.] MOSCOW: NAUKA (RUSS.) | 5092 | F |

IN  OBRUCHEV, D.V., ED., [ESSAYS ON THE PHLOGENY AND SYSTEMA          5054
    TICS OF FOSSIL FISH AND AGNATHA.] MOSCOW: NAUKA.
    (RUSS.)
IN  PETERS,H. ED. THE DEVELOPMENT AND MATURATION OF THE               4868  F
    OVARY AND ITS FUNCTIONS. INTERNATIONAL CONGRESS SERIES NO.
    267. AMSTERDAM:EXCERPTA MEDICA AMSTERDAM.
IN  PLISETSKAYA, E. AND L. LEIBSON, EDS., EVOLUTIONARY ENDOC          5361  F
    RINOLOGY OF PANCREAS. LENINGRAD: AKAD. NAUK SSSR.
    (RUSS.)
    ENG. SUMM.
IN  SCHARF ANLAESSLICH DER JAHRESVERSAMMLUNG. HALLE,EAST GER          5049
    MANY, OCT. 11-14,1973.
    HALLE: DEUTSCHE AKADEMIE DER NATURFORSCHER LEOPOLDINA.
IN  SCHWARTZKOPFT,J. ED. ABHANDLUNGEN DER RHEINISCH-WESTFAEL          5052
    ISCHEN AKADEMIE DER WISSEN-SCHAFTEN,BAND 53. SYMPOSIUM. BOCH
    UM, WEST GERMANY, OCT. 14-18, 1973. OPLADEN:WESTDEUTSCHER VE
    RLAG.
IN  SLADECEK,V. ED., PROCEEDINGS OF THE INTERNATIONAL                 4662
    ASSOCIATION OF THEORETICAL AND APPLIED LIMNOLOGY. VOL. 18.
    LENINGRAD: NAUKA PUBLISHING HOUSE.
IN  SLEIGH,M.A. AND D.H.JENNINGS EDS. SYMPOSIA OF THE SOCIET          4965
    Y FOR EXPERIMENTAL BIOLOGY, NO.28. TRANSPORT AT THE CELLULAR
    LEVEL. LONDON, ENGLAND, AUG. 27-31, 1973. LONDON: CAMBRIDGE
    UNIV. PRESS.
IN  STEINER, D.F. ET AL, EARLY DIABETES IN EARLY LIFE. NEW Y          5400  F
    ORK: ACADEMIC PRESS.
IN  TALMAGE,R.V. OWEN,M. AND PARSONS,J.A. EDS. EXCERPTA              5099
    MEDICA INTERNATIONAL CONGRESS SERIES, NO.346. CALCIUM
    REGULATING HORMONES. PROCEEDINGS OF THE FIFTH PARATHYROID
    CONFERENCE, OXFORD, ENGLAND. JULY 21-26,1974. AMSTERDAM:
    EXCERPTA MEDICA.
IN  TIBBLES, J.J. ED. ANNUAL REPORT 1973 OF THE SEA LAMPREY          4894  F
    CONTROL CENTRE SAULT STE. MARIE, ONTARIO, TO THE GREAT LAKES
    FISHERY COMMISSION. GREAT LAKES FISH. COMM., ANNU. REP.
    RESTRICTED
IN  TIBBLES, J.J. ED. ANNUAL REPORT 1975 OF THE SE LAMPREY C         5236  F
    ONTROL CENTRE SAULT STE. MARIE, ONTARIO, TO GREAT LAKES FISH
    ERY COMMISSION. GREA LAKES FISH. COMM., ANNU. REP.
    RESTRICTED
IN  TIBBLES, J.J. ED. ANNUAL REPORT 1975 OF THE SEA LAMPREY          4969  F
    CONTROL CENTRE SAULT STE. MARIE, ONTARIO, TO THE GREAT LAKES
    FISHERY COMMISSION. GREAT LAKES FISH. COMM., ANNU. REP.
    RESTRICTED
IN  TIBBLES, J.J. ED. ANNUAL REPORT 1975 OF THE SEA LAMPREY          5222  F
    CONTROL CENTRE SAULT STE. MARIE, ONTARIO, TO THE GREAT LAKES
    FISHERY COMMISSION. GREAT LAKE FISH. COMM., ANNU. REP.
    RESTRICTED
IN  TIBBLES, J.J. ED. ANNUAL REPORT 1975 OF THE SEA LAMPREY          5223  F
    CONTROL CENTRE SAULT STE. MARIE, ONTARIO, TO THE GREAT LAKES
    FISH COMMISSION. GREAT LAKES FISH. COMM., ANNU. REP.
    RESTRICTED
IN  TIBBLES, J.J. ED., ANNUAL REPORT 1974 OF THE SEA LAMPREY         5132  F
    CONTROL CENTRE SAULT STE. MARIE, ONTARIO, TO THE GREAT LAKE
    S FISHERY COMMISSION. GREAT LAKES FISH. COMM., ANNU. REP.
    RESTRICTED
IN  TIBBLES, J.J. ED., ANNUAL REPORT 1974 OF THE SEA LAMPREY         5133  F
    CONTROL CENTRE SAULT STE. MARIE, ONTARIO, TO THE GREAT LAKE
    S FISHERY COMMISSION. GREAT LAKES FISH. COMM., ANNU. REP.
    RESTRICTED
IN  TIBBLES, J.J. ED., ANNUAL REPORT 1975 OF THE SEA LAMPREY         4547  F
    CONTROL CENTRE SAULT STE. MARIE, ONTARIO, TO THE GREAT LAKE
    S FISHERY COMMISSION. GREAT LAKES FISH. COMM., ANNU. REP.
    RESTRICTED
IN  TIBBLES, J.J. ED., ANNUAL REPORT 1975 OF THE SEA LAMPREY         4553  F
    CONTROL CENTRE SAULT STE. MARIE, ONTARIO, TO THE GREAT LAKE
    S FISHERY COMMISSION. GREAT LAKES FISH. COMM., ANNU. REP.
    RESTRICTED
IN  TIBBLES, J.J. ED., ANNUAL REPORT 1975 OF THE SEA LAMPREY         4554  F
    CONTROL CENTRE SAULT STE. MARIE, ONTARIO TO THE GREAT LAKES
    FISHERY COMMISSION. GREAT LAKES FISH. COMM., ANNU. REP.
    RESTRICTED
IN  TIBBLES, J.J. ED., ANNUAL REPORT 1975 OF THE SEA LAMPREY         4605  F
    CONTROL CENTRE, SAULT STE. MARIE, ONTARIO, TO THE GREAT LAK
    ES FISHERY COMMISSION. GREAT LAKES FISH. COMM., ANNU. REP.
    RESTRICTED
IN  TIBBLES, J.J. ED., ANNUAL REPORT 1975 OF THE SEA LAMPREY         5216  F
    CONTROL CENTRE SAULT STE. MARIE, ONTARIO, TO THE GREAT LAKE
    S FISHERY COMMISSION. GREAT LAKES FISH. COMM., ANNU. REP.
    RESTRICTED
IN  TIBBLES, J.J. ED., ANNUAL REPORT 1975 OF THE SEA LAMPREY         5220  F
    CONTROL CENTRE SAULT STE. MARIE, ONTARIO, TO THE GREAT LAKE
    S FISHERY COMMISSION. GREAT LAKES FISH. COMM., ANNU. REP.
    RESTRICTED
IN  TIBBLES, J.J. ED., ANNUAL REPORT 1975 OF THE SEA LAMPREY         5221  F
    CONTROL CENTRE SAULT STE. MARIE, ONTARIO, TO THE GREAT LAKE
    S FISHERY COMMISSION. GREAT LAKES FISH. COMM., ANNU. REP.
    RESTRICTED
IN  TIBBLES, J.J. ED., ANNUAL REPORT 1975 OF THE SEA LAMPREY         5224  F
    CONTROL CENTRE SAULT STE. MARIE, ONTARIO, TO THE GREAT LAKE
    S FISHERY COMMISSION. GREAT LAKES FISH. COMM., ANNU. REP.
    RESTRICTED

```
IN  TIBBLES, J.J. ED., ANNUAL REPORT 1975 OF THE SEA LAMPREY            5225  F
        CONTROL CENTRE SAULT STE. MARIE, ONTARIO, TO THE GREAT LAKE
        S FISHERY COMMISSION, GREAT LAKES FISH. COMM., ANNU. REP.
        RESTRICTED
IN  TIBBLES, J.J. ED., ANNUAL REPORT 1975 OF THE SEA LAMPREY            5226  F
        CONTROL CENTRE SAULT STE. MARIE, ONTARIO, TO THE GREAT LAKE
        S FISHERY COMMISSION. GREAT LAKES FISH. COMM., ANNU. REP.
        RESTRICTED
IN  TIBBLES, J.J. ED., ANNUAL REPORT 1975 OF THE SEA LAMPREY            5227  F
        CONTROL CENTRE SAULT STE. MARIE, ONTARIO, TO THE GREAT LAKE
        S FISHERY COMMISSION. GREAT LAKES FISH. COMM. ANNU. REP.
        RESTRICTED
IN  TIBBLES, J.J. ED., ANNUAL REPORT 1975 OF THE SEA LAMPREY            5228  F
        CONTROL CENTRE SAULT STE. MARIE, ONTARIO, TO THE GREAT LAKE
        S FISHERY COMMISSION. GREAT LAKES FISH. COMM., ANNU. REP.
        RESTRICTED
IN  TIBBLES, J.J. ED., ANNUAL REPORT 1975 OF THE SEA LAMPREY            5229  F
        CONTROL CENTRE SAULT STE. MARIE, ONTARIO, TO THE GREAT LAKE
        S FISHERY COMMISSION. GREAT LAKES FISH. COMM. ANNU. REP.
        RESTRICTED
IN  TIBBLES, J.J. ED., ANNUAL REPORT 1975 OF THE SEA LAMPREY            5230  F
        CONTROL CENTRE SAULT STE  MARIE, ONTARIO, TO THE GREAT LAKE
        S FISHERY COMMISSION. GREAT LAKES FISH. COMM., ANNU. REP.
        RESTRICTED
IN  TIBBLES, J.J. ED., ANNUAL REPORT 1976 OF THE SEA LAMPREY            5246  F
        CONTROL CENTRE SAULT STE. MARIE, ONTARIO, TO THE GREAT LAKE
        S FISHERY COMMISSION. GREAT LAKES FISH. COMM., ANNU. REP.
        RESTRICTED
IN  TIBBLES, J.J. ED., ANNUAL REPORT 1976 OF THE SEA LAMPREY            5247  F
        CONTROL CENTRE SAULT STE. MARIE, ONTARIO, TO THE GREAT LAKE
        S FISHERY COMMISSION. GREAT LAKES FISH. COMM., ANNU. REP.
        RESTRICTED
IN  TIBBLES, J.J. ED., ANNUAL REPORT 1976 OF THE SEA LAMPREY            5248  F
        CONTROL CENTRE SAULT STE. MARIE, ONTARIO, TO THE GREAT LAKE
        S FISHERY COMMISSION. GREAT LAKES FISH. COMM., ANNU. REP.
        RESTRICTED
IN  TIBBLES, J.J. ED., ANNUAL REPORT 1976 OF THE SEA LAMPREY            5249  F
        CONTROL CENTRE SAULT STE. MARIE, ONTARIO, TO THE GREAT LAKE
        S FISHERY COMMISSION. GREAT LAKES FISH. COMM., ANNU. REP.
        RESTRICTED
IN  TIBBLES, J.J. ED., ANNUAL REPORT 1976 OF THE SEA LAMPREY            5250  F
        CONTROL CENTRE SAULT STE. MARIE, ONTARIO, TO THE GREAT LAKE
        S FISHERY COMMISSION. GREAT LAKES FISH. COMM., ANNU. REP.
        RESTRICTED
IN  TIBBLES, J.J. ED., ANNUAL REPORT 1976 OF THE SEA LAMPREY            5251  F
        CONTROL CENTRE SAULT STE. MARIE, ONTARIO, TO THE GREAT LAKE
        S FISHERY COMMISSION. GREAT LAKES FISH. COMM., ANNU. REP.
        RESTRICTED
IN  TIBBLES, J.J. ED., ANNUAL REPORT 1976 OF THE SEA LAMPREY            5252  F
        CONTROL CENTRE SAULT STE. MARIE, ONTARIO, TO THE GREAT LAKE
        S FISHERY COMMISSION. GREAT LAKES FISH. COMM., ANNU. REP.
        RESTRICTED
IN  TIBBLES, J.J. ED., ANNUAL REPORT 1976 OF THE SEA LAMPREY            5253  F
        CONTROL CENTRE SAULT STE. MARIE, ONTARIO, TO THE GREAT LAKE
        S FISHERY COMMISSION. GREAT LAKES FISH. COMM., ANNU. REP.
        RESTRICTED
IN  TIBBLES, J.J. ED., ANNUAL REPORT 1976 OF THE SEA LAMPREY            5254  F
        CONTROL CENTRE SAULT STE. MARIE, ONTARIO, TO THE GREAT LAKE
        S FISHERY COMMISSION. GREAT LAKES FISH. COMM., ANNU. REP.
        RESTRICTED
IN  TIBBLES, J.J. ED., ANNUAL REPORT 1976 OF THE SEA LAMPREY            5255  F
        CONTROL CENTRE SAULT STE. MARIE, ONTARIO, TO THE GREAT LAKE
        S FISHERY COMMISSION. GREAT LAKES FISH. COMM., ANNU. REP.
        RESTRICTED
IN  TIBBLES, J.J. ED., ANNUAL REPORT 1976 OF THE SEA LAMPREY            5256  F
        CONTROL CENTRE SAULT STE. MARIE, ONTARIO, TO THE GREAT LAKE
        S FISHERY COMMISSION. GREAT LAKES FISH. COMM., ANNU. REP.
        RESTRICTED
IN  TIBBLES, J.J. ED., ANNUAL REPORT 1976 OF THE SEA LAMPREY            5257  F
        CONTROL CENTRE SAULT STE. MARIE, ONTARIO, TO THE GREAT LAKE
        S FISHERY COMMISSION. GREAT LAKES FISH. COMM., ANNU. REP.
        RESTRICTED
IN  TIBBLES, J.J. ED., ANNUAL REPORT 1976 OF THE SEA LAMPREY            5258  F
        CONTROL CENTRE SAULT STE. MARIE, ONTARIO, TO THE GREAT LAKE
        S FISHERY COMMISSION. GREAT LAKES FISH. COMM., ANNU. REP.
        RESTRICTED
IN  TIBBLES, J.J. ED., ANNUAL REPORT 1976 OF THE SEA LAMPREY            5259  F
        CONTROL CENTRE SAULT STE. MARIE, ONTARIO, TO THE GREAT LAKE
        S FISHERY COMMISSION. GREAT LAKES FISH. COMM., ANNU. REP.
        RESTRICTED
IN  TIBBLES, J.J. ED., ANNUAL REPORT 1976 OF THE SEA LAMPREY            5260  F
        CONTROL CENTRE SAULT STE. MARIE, ONTARIO, TO THE GREAT LAKE
        S FISHERY COMMISSION. GREAT LAKES FISH. COMM., ANNU. REP.
        RESTRICTED
IN  TIBBLES, J.J. ED., ANNUAL REPORT 1976 OF THE SEA LAMPREY            5261  F
        CONTROL CENTRE SAULT STE. MARIE, ONTARIO, TO THE GREAT LAKE
        S FISHERY COMMISSION. GREAT LAKES FISH. COMM., ANNU. REP.
        RESTRICTED
IN  TIBBLES, J.J. ED., ANNUAL REPORT 1976 OF THE SEA LAMPREY            5262  F
        CONTROL CENTRE SAULT STE. MARIE, ONTARIO, TO THE GREAT LAKE
        S FISHERY COMMISSION. GREAT LAKES FISH. COMM., ANNU. REP.
```

```
          RESTRICTED
IN   TIBBLES, J.J. ED., ANNUAL REPORT 1976 OF THE SEA LAMPREY          5263  F
       CONTROL CENTRE SAULT STE. MARIE, ONTARIO, TO THE GREAT LAKE
       S FISHERY COMMISSION. GREAT LAKES FISH. COMM., ANNU. REP.
          RESTRICTED
IN   TIBBLES, J.J. ED., ANNUAL REPORT 1976 OF THE SEA LAMPREY          5264  F
       CONTROL CENTRE SAULT STE. MARIE, ONTARIO, TO THE GREAT LAKE
       S FISHERY COMMISSION. GREAT LAKES FISH. COMM., ANNU. REP.
          RESTRICTED
IN   TIBBLES, J.J., ED., ANNUAL REPORT 1972 OF THE SEA LAMPRE          5477  F
       Y CONTROL CENTRE SAULT STE. MARIE, ONTARIO, TO THE GREAT LAK
       ES FISHERY COMMISSION. GREAT LAKES FISH. COMM., ANNU. REP.
          RESTRICTED
IN   TIBBLES, J.J., ED., ANNUAL REPORT 1972 OF THE SEA LAMPRE          5481
       Y CONTROL CENTRE SAULT STE. MARIE, ONTARIO, TO THE GREAT LAK
       ES FISHERY COMMISSION. GREAT LAKES FISH. COMM., ANNU. REP.
          RESTRICTED
IN   TIBBLES, J.J., ED., ANNUAL REPORT 1972 OF THE SEA LAMPRE          5485  F
       Y CONTROL CENTRE SAULT STE. MARIE, ONTARIO, TO THE GREAT LAK
       ES FISHERY COMMISSION. GREAT LAKES FISH. COMM., ANNU. REP.
          RESTRICTED
IN   TIBBLES, J.J., ED., ANNUAL REPORT 1972 OF THE SEA LAMPRE          5486  F
       Y CONTROL CENTRE SAULT STE. MARIE, ONTARIO, TO THE GREAT LAK
       ES FISHERY COMMISSION. GREAT LAKES FISH. COMM., ANNU. REP.
          RESTRICTED
IN   TIBBLES, J.J., ED., ANNUAL REPORT 1972 OF THE SEA LAMPRE          5487
       Y CONTROL CENTRE SAULT STE. MARIE, ONTARIO, TO THE GREAT LAK
       ES FISHERY COMMISSION. GREAT LAKES FISH. COMM., ANNU. REP.
          RESTRICTED
IN   TIBBLES, J.J., ED., ANNUAL REPORT 1972 OF THE SEA LAMPRE          5488  F
       Y CONTROL CENTRE SAULT STE. MARIE, ONTARIO, TO THE GREAT LAK
       ES FISHERY COMMISSION. GREAT LAKES FISH. COMM., ANNU. REP.
          RESTRICTED
IN   TIBBLES, J.J., ED., ANNUAL REPORT 1972 OF THE SEA LAMPRE          5491  F
       Y CONTROL CENTRE SAULT STE. MARIE, ONTARIO, TO THE GREAT LAK
       ES FISHERY COMMISSION. GREAT LAKES FISH. COMM., ANNU. REP.
          RESTRICTED
IN   TIBBLES, J.J., ED., ANNUAL REPORT 1972 OF THE SEA LAMPRE          5494  F
       Y CONTROL CENTRE SAULT STE. MARIE, ONTARIO, TO THE GREAT LAK
       ES FISHERY COMMISSION. GREAT LAKES FISH. COMM., ANNU. REP.
          RESTRICTED
IN   TIBBLES, J.J., ED., ANNUAL REPORT 1972 OF THE SEA LAMPRE          5497  F
       Y CONTROL CENTRE SAULT STE. MARIE, ONTARIO, TO THE GREAT LAK
       ES FISHERY COMMISSION. GREAT LAKES FISH. COMM., ANNU. REP.
          RESTRICTED
IN   TIBBLES, J.J., ED., ANNUAL REPORT 1973 OF THE SEA LAMPRE          4863  F
       Y CONTROL CENTRE SAULT STE. MARIE, ONTARIO, TO THE GREAT LAK
       ES FISHERY COMMISSION. GREAT LAKES FISH. COMM., ANNU. REP.
          RESTRICTED
IN   TIBBLES, J.J., ED., ANNUAL REPORT 1973 OF THE SEA LAMPRE          4889  F
       Y CONTROL CENTRE, SAULT STE. MARIE, ONTARIO, TO THE GREAT LA
       KES FISHERY COMMISSION. GREAT LAKES FISH. COMM., ANNU. REP.
          RESTRICTED
IN   TIBBLES, J.J., ED., ANNUAL REPORT 1973 OF THE SEA LAMPRE          5237  F
       Y CONTROL CENTRE STE. MARIE, ONTARIO, TO THE GREAT LAKES FIS
       HERY COMMISSION. GREAT LAKES FISH. COMM., ANNU. REP.
          RESTRICTED
IN   TIBBLES,J.J. ANNUAL REPORT 1973 OF THE SEA LAMPREY               4826  F
       CONTROL CENTRE SAULT STE. MARIE, ONTARIO, TO THE GREAT LAKES
        FISHERY COMMISSION. GREAT LAKES FISH. COMM., ANNU. REP.
          RESTRICTED
IN   TIBBLES,J.J. ANNUAL REPORT 1973 OF THE SEA LAMPREY               4848  F
       CONTROL CENTRE SAULT STE. MARIE, ONTARIO, TO THE GREAT LAKES
        FISHERY COMMISSION. GREAT LAKES FISH. COMM., ANNU. REP.
          RESTRICTED
IN   TIBBLES,J.J. ANNUAL REPORT 1973 OF THE SEA LAMPREY               4866  F
       CONTROL CENTRE SAULT STE. MARIE, ONTARIO, TO THE GREAT LAKES
        FISHERY COMMISSION.GREAT LAKES FISH. COMM., ANNU. REP.
          RESTRICTED
IN   TIBBLES,J.J. ANNUAL REPORT 1973 OF THE SEA LAMPREY               4900  F
       CONTROL CENTRE SAULT STE. MARIE, ONTARIO. TO THE GREAT LAKES
        FISHERY COMMISSION. GREAT LAKES FISH. COMM., ANNU. REP.
          RESTRICTED
IN   TIBBLES,J.J. ANNUAL REPORT 1973 OF THE SEA LAMPREY               4902  F
       CONTROL CENTRE SAULT STE. MARIE, ONTARIO, TO THE GREAT LAKES
        FISHERY COMMISSION. GREAT LAKES FISH. COMM., ANNU. REP.
          RESTRICTED
IN   TIBBLES,J.J. ANNUAL REPORT 1973 OF THE SEA LAMPREY               4904  F
       CONTROL CENTRE SAULT STE. MARIE, ONTARIO. TO THE GREAT LAKES
        FISHERY COMMISSION. GREAT LAKES FISH. COMM., ANNU. REP.,
          RESTRICTED
IN   TIBBLES,J.J. ANNUAL REPORT 1973 OF THE SEA LAMPREY               4905  F
       CONTROL CENTRE SAULT STE. MARIE, ONTARIO, TO THE GREAT
       LAKES FISHERY COMMISSION. GREAT LAKES FISH. COMM., ANNU. REP
          RESTRICTED
IN   TIBBLES,J.J. ANNUAL REPORT 1973 OF THE SEA LAMPREY               4906  F
       CONTROL CENTRE SAULT  STE. MARIE, ONTARIO, TO THE GREAT
       LAKES FISHERY COMMISSION. GREAT LAKES FISH. COMM., ANNU.REP.
          RESTRICTED
IN   TIBBLES,J.J. ANNUAL REPORT 1973 OF THE SEA LAMPREY CONTR         4903  F
       OL CENTRE SALUT STE. MARIE, ONTARIO, TO THE GREAT LAKES FISH
```

```
        ERY COMMISSION.
        RESTRICTED
IN   TIBBLES,J.J. ED. ANNUAL REPORT 1973 OF THE SEA LAMPREY              4892  F
        CONTROL CENTRE SAULT STE. MARIE, ONTARIO, TO THE GREAT LAKES
        FISHERY COMMISSION. GREAT LAKES FISH. COMM., ANNU. REP.
        RESTRICTED
IN   TIBBLES,J.J. ED. ANNUAL REPORT 1973 OF THE SEA LAMPREY              4901  F
        CONTROL CENTRE SAULT STE. MARIE, ONTARIO, TO THE GREAT LAKES
        FISHERY COMMISSION. GREAT LAKES FISH. COMM., ANNU. REP.
        RESTRICTED
IN   TIBBLES,J.J. ED. ANNUAL REPORT 1973 OF THE SEA LAMPREY C            4861  F
        ONTROL CENTRE SAULT STE. MARIE, ONTARIO, TO THE GREAT LAKES
        FISHERY COMMISSION. GREAT LAKES FISH. COMM., ANNU. REP.
        RESTRICTED
IN   TIBBLES,J.J. ED. ANNUAL REPORT 1973 OF THE SEA LAMPREY C            4862  F
        ONTROL CENTRE SAULT STE. MARIE, ONTARIO, TO THE LAKES FISHER
        Y COMMISSION. GREAT LAKES FISH. COMM., ANNU. REP.
        RESTRICTED
IN   TIBBLES,J.J. ED. ANNUAL REPORT 1973 OF THE SEA LAMPREY C            4864  F
        ONTROL CENTRE SAULT STE. MARIE, ONTARIO, TO THE GREAT LAKES
        FISHERY COMMISSION. GREAT LAKES FISH. COMM., ANNU. REP.
        RESTRICTED
IN   TIBBLES,J.J. ED. ANNUAL REPORT 1973 OF THE SEA LAMPREY C            4865  F
        ONTROL CENTRE SAULT STE. MARIE, ONTARIO, TO THE GREAT LAKES
        FISHERY COMMISSION GREAT LAKES FISH. COMM., ANNU. REP.
        RESTRICTED
IN   TIBBLES,J.J. ED. ANNUAL REPORT 1973 OF THE SEA LAMPREY C            4890  F
        ONTROL CENTRE SAULT STE. MARIE, ONTARIO, TO THE GREAT LAKES
        FISHERY COMMISSION. GREAT LAKES FISH. COMM., ANNU. REP.
        RESTRICTED
IN   TIBBLES,J.J. ED. ANNUAL REPORT 1974 OF SEA LAMPREY CONTR            5134  F
        OL CENTRE SAULT STE. MARIE, ONTARIO, TO THE GREAT LAKES FISH
        ERY COMMISSION. GREAT LAKES FISH. COMM., ANNU. REP.
        RESTRICTED
IN   TIBBLES,J.J. ED. ANNUAL REPORT 1974 OF SEA LAMPREY CONTR            5135  F
        OL CENTRE SAULT STE. MARIE, ONTARIO, TO THE GREAT LAKES FISH
        ERY COMMISSION GREAT LAKES FISH. COMM., ANNU. REP.
        RESTRICTED
IN   TIBBLES,J.J. ED. ANNUAL REPORT 1974 OF THE SEA LAMPREY              4907  F
        CONTROL CENTRE SAULT STE. MARIE, ONTARIO, TO THE GREAT LAKES
        FISHERY COMMISSION. GREAT LAKES FISH. COMM., ANNU. REP.
        RESTRICTED
IN   TIBBLES,J.J. ED. ANNUAL REPORT 1974 OF THE SEA LAMPREY C            4645  F
        ONTROL CENTRE SAULT STE. MARIE, ONTARIO, TO THE GREAT LAKES
        FISHERY COMMISSION. GREAT LAKES FISH. COMM., ANNU. REP.
        RESTRICTED
IN   TIBBLES,J.J. ED. ANNUAL REPORT 1974 OF THE SEA LAMPREY C            4908  F
        ONTROL CENTRE SAULT STE. MARIE, ONTARIO, TO THE GREAT LAKES
        FISHERY COMMISSION. GREAT LAKES FISH. COMM., ANNU. REP.
        RESTRICTED
IN   TIBBLES,J.J. ED. ANNUAL REPORT 1974 OF THE SEA LAMPREY C            4962  F
        ONTROL CENTRE SAULT STE. MARIE, ONTARIO, TO THE GREAT LAKES
        FISHERY COMMISSION. GREAT LAKES FISH. COMM., ANNU. REP.
        RESTRICTED
IN   TIBBLES,J.J. ED. ANNUAL REPORT 1974 OF THE SEA LAMPREY C            5088  F
        ONTROL CENTRE SAULT STE. MARIE, ONTARIO, TO THE GREAT LAKES
        FISHERY COMMISSION. GREAT LAKES FISH. COMM., ANNU. REP.
        RESTRICTED
IN   TIBBLES,J.J. ED. ANNUAL REPORT 1974 OF THE SEA LAMPREY C            5103  F
        ONTROL CENTRE SAULT STE. MARIE, ONTARIO, TO THE GREAT LAKES
        FISHERY COMMISSION. GREAT LAKES FISH. COMM., ANNU. REP.
        RESTRICTED
IN   TIBBLES,J.J. ED. ANNUAL REPORT 1974 OF THE SEA LAMPREY C            5104  F
        ONTROL CENTRE SAULT STE. MARIE, ONTARIO, TO THE GREAT LAKES
        FISHERY COMMISSION. GREAT LAKES FISH. COMM., ANNU. REP.
        RESTRICTED
IN   TIBBLES,J.J. ED. ANNUAL REPORT 1974 OF THE SEA LAMPREY C            5105  F
        ONTROL CENTRE SAULT STE. MARIE, ONTARIO, TO THE GREAT LAKES
        FISHERY COMMISSION. GREAT LAKES FISH. COMM., ANNU. REP.
        RESTRICTED
IN   TIBBLES,J.J. ED. ANNUAL REPORT 1974 OF THE SEA LAMPREY C            5106  F
        ONTROL CENTRE SAULT STE. MARIE, ONTARIO, TO THE GREAT LAKES
        FISHERY COMMISSION. GREAT LAKES FISH. COMM., ANNU. REP.
        RESTRICTED
IN   TIBBLES,J.J. ED. ANNUAL REPORT 1974 OF THE SEA LAMPREY C            5107  F
        ONTROL CENTRE SAULT STE. MARIE, ONTARIO, TO THE GREAT LAKES
        FISHERY COMMISSION. GREAT LAKES FISH. COMM., ANNU. REP.
        RESTRICTED
IN   TIBBLES,J.J. ED. ANNUAL REPORT 1974 OF THE SEA LAMPREY C            5108  F
        ONTROL CENTRE SAULT STE. MARIE, ONTARIO, TO THE GREAT LAKES
        FISHERY COMMISSION. GREAT LAKES FISH. COMM., ANNU. REP.
        RESTRICTED
IN   TIBBLES,J.J. ED. ANNUAL REPORT 1974 OF THE SEA LAMPREY C            5109  F
        ONTROL CENTRE SAULT STE. MARIE, ONTARIO, TO THE GREAT LAKES
        FISHERY COMMISSION. GREAT LAKES FISH. COMM. ANNU. REP.
        RESTRICTED
IN   TIBBLES,J.J. ED. ANNUAL REPORT 1974 OF THE SEA LAMPREY C            5110  F
        ONTROL CENTRE SAULT STE. MARIE, ONTARIO, TO THE GREAT LAKES
        FISHERY COMMISSION. GREAT LAKES FISH. COMM., ANNU. REP.
        RESTRICTED
IN   TIBBLES,J.J. ED. ANNUAL REPORT 1974 OF THE SEA LAMPREY C            5111  F
```

ONTROL CENTRE SAULT STE. MARIE, ONTARIO, TO THE GREAT LAKES  
FISHERY COMMISSION. GREAT LAKES FISH. COMM., ANNU. REP.  
RESTRICTED

IN  TIBBLES,J.J. ED. ANNUAL REPORT 1974 OF THE SEA LAMPREY C       5136  F  
ONTROL CENTRE SAULT STE. MARIE, ONTARIO, TO THE GREAT LAKES  
FISHERY COMMISSION. GREAT LAKES FISH. COMM., ANNU. REP.  
RESTRICTED

IN  TIBBLES,J.J.ED. ANNUAL REPORT 1973 OF THE LAMPREY       4891  F  
CONTROL CENTRE SAULT STE. MARIE, ONTARIO, TO THE GREAT LAKES  
FISHERY COMMISSION. GREAT LAKES FISH. COMM., ANNU. REP.  
RESTRICTED

IN  TIBBLES,J.J.ED. ANNUAL REPORT 1973 OF THE SEA LAMPREY       4888  F  
CONTROL CENTRE SAULT STE. MARIE, ONTARIO, TO THE GREAT  
LAKES FISHERY COMMISSION. GREAT LAKES FISH. COMM., ANNU.REP.  
RESTRICTED

IN  TIBBLES,J.J.ED. ANNUAL REPORT 1973 OF THE SEA LAMPREY       4893  F  
CONTROL CENTRE SAULT STE. MARIE, ONTARIO, TO THE GREAT  
LAKES FISHERY COMMISSION. GREAT LAKES FISH. COMM., ANNU.REP.  
RESTRICTED

IN  TIBBLES,J.J.ED. ANNUAL REPORT 1973 OF THE SEA LAMPREY CO      4867  F  
NTROL CENTRE SAULT STE. MARIE, ONTARIO, TO THE GREAT LAKES F  
ISHERY COMMISSION. GREAT LAKES FISH. COMM., ANNU. REP.  
RESTRICTED

IN  TILDERS, F.J.H., D.F. SWAAB AND T.B. VAN WILMERSMA       5534  
GREIDANUS, EDS., FRONTIERS OF HORMONE RESEARCH, VOL. 4.  
MELANOCYTE STIMULATING HORMONE: CONTROL, CHEMISTRY AND  
EFFECTS. BASEL: S. KARGER.

IN  TROSHIN, A.S., ED., BIOKHIMICHESKAYA GENETIKA RYB. MATER      4802  F  
IALY 1-GO VSESOYUZ. SOVESHCH. LENINGRAD, 6-9 FEVR. 1973 G. L  
ENINGRAD: VSESOYUZ. SOVESHCH. PO BIOKHIM. GENET. RYB.  
(RUSS.)  
ENGL. SUMM.

IN  USSING,H.H. AND N.A. THORN (EDS.) PROCEEDINGS OF THE       4934  
ALFRED BENZON SYMPOSIUM, NO. 5. TRANSPORT MECHANISMS IN  
EPITHELIA. COPENHAGEN, DENMARK, SEPTEMBER 10-14, 1972. NEW  
YORK: ACADEMIC PRESS.

IN  UYEDA, R., ELECTRON MICROSCOPY. VOL. II. TOKYO: MARUZEN.    5529  F  
IN  WESSELLS,N.K. ED. READINGS FROM SCIENTIFIC AMERICAN. VER   4932  
TEBRATE STRUCTURES AND FUNCTIONS. SAN FRANCISCO:FREEMAN AND  
COMPANY.

IN  WILLIAMS,W.D. ED. MONOGRAPHIAE BIOLOGICAE, VOL. 25.       4680  
BIOGEOGRAPHY AND ECOLOGY IN TASMANIA. NETHERLANDS,THE HAGUE.

IN GREAT LAKES FISH. COMM., ANNU. REP., 1977.       5367  F  
INDIANA ACAD. SCI., PROC., 85:147-155.       5268  
INST. NATL. SANTE RECH. MED., PARIS, COLLOQ., 30:182.     4684  
INT. COMM. ZOOL NOMENCL., BULL., 30:198-199.       4766  F  
INT. CONG. ANAT., 10TH., TOKYO, 327P.       4757  F  
INT. MSCHR. ANAT. HISTOL., 3:8-21.       5530  
INT. REV. HYDROBIOL., 44:395-429.       5331  
INT. SOC. THROMB. HAEMOST., J., 38:429-437.       5302  F  
FR. AND GER. SUMM.  
IR. NAT. J., 18:57-65.       4686  F  
IRCS MED. SCI., 13-1-1.       5562  F  
IRCS MED. SCI., 13-1-5.       5561  F  
J. ANIM. ECOL., 45:699-712.       5142  F  
J. ANIM. ECOL., 45:81-90       4776  F  
J. BIOL. CHEM., 252:602-608.       5154  F  
J. BIOL. CHEM.,250:5183-5191.       4654  F  
J. COLL. SCI., IMP. UNIV. TOKYO, 19:1-23.       5583  
J. COMP. NEUROL., 154:189-206       4542  F  
J. COMP. NEUROL., 154:207-224.       4625  F  
J. COMP. NEUROL., 156:255-276.       4697  F  
J. COMP. NEUROL., 165:1-15.       4884  
J. COMP. NEUROL., 168:545-554.       5026  F  
J. COMP. NEUROL., 171:465-480.       5160  F  
J. COMP. PHYSIOL., 104:175-183.       4788  F  
J. COMP. PHYSIOL., 104:185-203.       4787  F  
J. COMP. PHYSIOL., 112:159-164.       4798  F  
J. COMP. PHYSIOL., 123A:329-334.       5578  F  
J. COMP. PHYSIOL., 94A:57-68.       4674  F  
J. CONCHOL., 28:81-94.       4544  
J. DENT. RES., SUPPL., 1966:75.       5373  
J. DENT. RES., 1965:116.       5374  F  
J. DENT. RES., 55:538       4827  
J. ELECTRONMICROSC., 22:113.       5243  F  
J. ELECTRONMICROSC., 23:299.       4663  F  
J. ELECTRONMICROSC., 23:51-55.       4603  F  
J. ELECTRONMICROSC., 24:195-196.       5098  
(ABSTR.)  
J. ELECTRONMICROSC., 25:119       5180  
(ABSTR.)  
J. ELECTRONMICROSC., 25:217.       5531  
J. ELECTRONMICROSC., 26:228.       5591  
(ABSTR.)  
J. ELECTRONMICROSC., 26:228-229.       5590  
(ABSTR.)  
J. ELECTRONMICROSC., 26:264.       5596  
(ABSTR.)  
J. ELECTRONMICROSC., 26:275.       5593  
(ABSTR.)  
J. ELECTRONMICROSC., 26:72       5294  F

```
J. ELECTRONMICROSC., 26:72.                                       5293  F
   (ABSTR.)
J. EMBRYOL. EXP. MORPHOL., 42:219-235.                            5350  F
J. ENDOCRINOLOGY, 59.                                             5339
J. EVOL. BIOCHEM. PHYSIOL., 12:104-109.                           5165
   (ENG.)
J. EXP. BIOL., 63:193-206.                                        5138  F
J. EXP. BIOL., 65:449-458.                                        4678  F
J. EXP. BIOL., 69:187-198.                                        5300  F
J. EXP. BIOL., 73:261.                                            5536  F
J. EXP. ZOOL., 202:27-32.                                         5303  F
J. EXP. ZOOL., 202:431-437.                                       5298  F
J. FISH BIOL., 10:473-480.                                        5424
J. FISH BIOL., 7:539-564.                                         4708  F
J. FISH BIOL., 7:95-104.                                          4638
J. FISH BIOL., 8:441-448.                                         5063  F
J. FISH BIOL., 9:425-440.                                         5113  F
J. GEN. PHYSIOL., 61:254.                                         5333  F
   (ABSTR.)
J. ICHTHYOL., 13:929-933.                                         4622
   (RUSS.)
J. INFECT. DIS., 129:21-27.                                       4566  F
J. INVEST. DERMATOL., 65:39-44.                                   4693  F
J. LABELLED COMPD., 8:499-504.                                    5413  F
J. LIPID RES., 18:679-691.                                        5320
J. MICROSC. BIOL. CELL., 26:43-46.                                5148  F
J. MOL. BIOL., 105:39-74.                                         4982
J. MOL. BIOL., 75:13-31.                                          4626  F
J. MOL. BIOL., 82:231-265.                                        4552
J. MOL. BIOL., 87:23-30.                                          4629
J. MOL. BIOL., 89:245-248.                                        4971  F
J. MOL. CELL CARDIOL., 5:433-439.                                 4597  F
J. MORPHOL., 140:119-134.                                         4898  F
J. MORPHOL., 149:73-104.                                          5085
J. MORPHOL., 155:193-218.                                         5518  F
J. MORPHOL., 155:219-236.                                         5516  F
J. NEURO-VISC. REL. 31:308-333.                                   5586
   (FR.)
   ENG. SUMM.
J. NEUROBIOL., 5:443-462.                                         4536
J. NEUROPATHOL. EXP. NEUROL., 35:349.                             5069
   (ABSTR.)
J. NEUROPHYSIOL., 39:197-212.                                     5020  F
J. ORNITHOL., 117:257-278.                                        5027
J. PALEONTOL., 50:350-353.                                        5067  F
J. PHYSIOL. (LOND.), 270:115-132.                                 5283  F
J. PHYSIOL. (LOND.), 270:89-114.                                  5280  F
J. PHYSIOL., 242:84P-86P.                                         4714
J. PHYSIOL., 251:395-407.                                         4643  F
J. PHYSIOL., 251:409-426.                                         4580  F
J. PHYSIOL., 266:69-89.                                           5206  F
J. PHYSIOL., 277:395-408.                                         5579  F
J. PHYSIOL., 279:231-252.                                         5606  F
J. PHYSIOL., 279:551-567.                                         5608  F
   J. PHYSIOL., 279:551-567.
J. STEROID BIOCHEM., 8:1249-1252.                                 5325  F
J. SUBMICROSC. CYTOL., 8:243                                      5058
   (ABSTR.)
J. SUBMICROSC. CYTOL., 8:258                                      5059
   (ABSTR.)
J. ULTRASTRUCT. RES., 43:1-17.                                    4745  F
J. ZOOL., 171:239-250.                                            4719  F
J. ZOOL., 176:311-329.                                            4713
J. ZOOL., 177:57-72.                                              4710  F
J. ZOOL., 178:261-277.                                            4811  F
J. ZOOL., 178:305-317.                                            5233  F
J. ZOOL., 178:319-340.                                            4716  F
J. ZOOL., 181:113-130.                                            5140  F
J. ZOOL., 183:111-123.                                            5301  F
JAP. J. GENET., 53:91-102                                         5574  F
JAP. J. ICHTHYOL., 19:191-194.                                    4943  F
   (JAP.)
   ENG. SUMM.
JAP. SOC. COMP. ENDOCRINOL., GIFU, PROC., 1976:1 P.               5526  F
   (ABSTR.)
JAP. SOC. PHYCOL., BULL., 24:62-67.                               4938
JENA: GUSTAV FISHER VERLAG.                                       4822
   (GER.)
KAIBOGAKA ZASSHI. ACTA ANAT. NIPPON, 48:406.                      4722
KAT. FAUNY. POL., 24:1-253.                                       5084
KOREAN J. BIOCHEM., 9:58.                                         5614
LENINGRAD: NAUKA. 215PP.                                          4772  F
   (RUSS.)
LINN. SOC. LOND., J. (BIOL.),                                     4611
LINN. SOC. LOND., J. (BIOL.), 8:346.                              5239
LINN. SOC. LOND., SYMP. SER., 4:411.                              5369
LONDON: ACADEMIC PRESS, 738 PP.                                   4829
LOS ANGELES. NAT. HIST. MUS., CONTRIB. SCI., (295):1-20.          5075
LYON MED., 231:79-81.                                             4976
MAR. BIOL. ASSOC. U.K., J., 58:81-86                              5514  F
MAR. BIOL., 27:351-356.                                           4699  F
```

```
MAR. POLLUT. BULL., 5:12-18.                                     4933
MED. COLL. WIS., IN PRESS.                                       5452
METAB. CLIN. EXP., 25:1491-1494.                                 5195
MEXICO: PROC. 4TH. INT. CONGR. HORMONAL STEROIDS.                5370
   (ABSTR.)
MICHIGAN STATE UNIV., M.SC. THESIS:162 PP.                       5388
MICRON. 7:205-211.                                               5587
MIDWEST REG. CONF. DEV. BIOL., MAY 1968.                         5375   F
MISS. ACAD. SCI., J., 19:128:134.                                5089   F
MOL. BIOL., 8:536-542.                                           4632   F
   (RUSS.)
      MOL. BIOL., 8:427-433.
      (ENG.)
MOL. BIOL., 9:252-274.                                           5401   F
   (RUSS.)
      ENG. TRANSL.
MOL. PHARMACOL., 13:181-184.                                     5308
MOSCOW: NAUKA PRESS.                                             5363
MOSCOW: NAUKA PRESS, 214 PP.                                     4823
MOSCOW: VSES. NAUCHNO-ISSLED. INST. MORSK. RYBN. (VNIRO)         5321   F
   (RUSS.)
MT. DESERT ISL. BIOL. LAB., BULL., 11:11-15.                     4773   F
MT. DESERT ISL. BIOL. LAB., BULL., 12:105-108.                   4936
MT. DESERT ISL. BIOL. LAB., BULL., 12:43-45.                     4794   F
MT. DESERT ISL. BIOL. LAB., BULL., 12:6-9.                       4935
MT. DESERT ISL. BIOL. LAB., BULL., 12:99-104.                    4931
MT. DESERT ISL. BIOL. LAB., BULL., 13:120-121                    4799   F
MT. DESERT ISL. BIOL. LAB., BULL., 13:4-8.                       4972
MT. DESERT ISL. BIOL. LAB., BULL., 13:62-63.                     4968
MT. DESERT ISL. BIOL. LAB., BULL., 13:84-85.                     4946
MT. DESERT ISL. BIOL. LAB., BULL., 13:87-88                      4963
MT. DESERT ISL. BIOL. LAB., BULL., 13:94-98.                     4964   F
MT. DESERT ISL. BIOL. LAB., BULL., 16:1-2.                       5351   F
N.Y. ACAD. SCI., ANN., 127:443-458.                              5503
N.Y. ACAD. SCI., ANN., 253:472-506.                              4818
NAT. CAN., 103:111-118.                                          5022   F
   (FR.)
      ENG. SUMM.
NATL. MUS. NAT. SCI., OTTAWA, PUBL. ZOOL., 12:1-36.              5150
NATURALIST (LEEDS), 102:105-108.                                 5533
NATURE (LOND.), 243:229-231.                                     5009
NATURE (LOND.), 246:389-395.                                     4737
NATURE (LOND.), 251:239-240.                                     4966   F
NATURE (LOND.), 256:66-68.                                       5244   F
NATURE (LOND.), 273:504-509.                                     5601
NATURWISS. VER., HAMBURG, ABH. VERH., 20:185-222.                5506   F
   (GERM.)
NAUCHN. SOV. ELEKTRON. MIKROSK. S.S.S.R., KONFERENTSIYA, 197     5322   F
   I.
   (RUSS.)
NEIROFIZIOLOGIYA, 6:629-635.                                     4601   F
   (RUSS.)
      ENGL. SUMM.
NEIROFIZIOLOGIYA, 7:12-20.                                       4540   F
   (RUSS.)
      ENGL. SUMM.
NEIROFIZIOLOGIYA, 9:390-396.                                     5353
NEIROFIZIOLOGIYA, 9:512-517.                                     5521
NEUROSCI. ABSTR., 2:1124.                                        5112   F
   (ABSTR.)
NEUROSCI. ABSTR., 2:786.                                         5116   F
      CONTROL CENTRE SAULT STE. MARIE, ONTARIO, TO THE GREAT LAKE
      S FISHERY COMMISSION. GREAT LAKES FISH. COMM., ANNU. REP.
      RESTRICTED
NEUROSCI. LETT., 5:39-44.                                        5295   F
NEW YORK: ACADEMIC PRESS.                                        4689
NEW YORK: ACADEMIC PRESS.                                        4707   F
NEW YORK: ACADEMIC PRESS.                                        4730
NEW YORK: ACADEMIC PRESS, 746 PP.                                4923
NEW YORK: UNIPUB, 786 PP.                                        4828
NEW YORK: WILEY, 416 PP.                                         4831
NOR. GEOL. TIDSSKR., 54:1-12.                                    4679
NOR. GEOL. UNDERS. BULL., (36):1-87.                             4911
NOR. GEOL. UNDERS. BULL., 27:101-131.                            5031
NORTHWEST SCI., 50:76-86.                                        5082
NORTHWEST SCI., 51:208-215.                                      5569
NORW. J. ZOOL., 22:81-93.                                        4589   F
NORW. J. ZOOL., 23:111-120.                                      4667   F
   (GER.)
      ENGL. SUMM.
NORW. J. ZOOL., 23:297-306.                                      4809   F
   (GER.)
OHIO J. SCI., 74:330-331.                                        4637   F
OKAJIMAS FOL. ANAT. JAP., 54:25-60.                              5356   F
OREG. ACAD. SCI., PROC., 11:47.                                  5319
OREGON STATE UNIV., PH.D. THESIS.                                5544
P. & M. CURIE UNIV., THESIS: 157 PP.                             4578
PA. ACAD. SCI., HARRISBURG, PROC., 49:82-88.                     5068   F
PALAEONTOLOGY (LOND.), 19:1-5.                                   4995
PARAZITOLOGIYA (LENINGR.), 10:346-351.                           5153
   (RUSS.)
```

ENG. SUMM.
PARIS. MUS. NATL. HIST. NAT., BULL., SCI. TERRE, 41:1-16.        4709  F
    (FR.)
PARIS. MUS. NATL. HIST. NAT., BULL., ZOOL., 177:1469-1490.       4546
PEST. BIOCHEM. PHYSIOL., 6:363-366.                              5077
PFLANZENSCHUTS-NACHR., 15:50-70.                                 5550
PHARMACOLOGIST, 16:327.                                          5438
PHYSIOL. SOC. JAP., J., 35:505.                                  4560
PHYSIOL. SOC. JAP., J., 35:528.                                  4956
PRIMER CENTENARIO DE LA R.SOC. ESPANOLA DE HIST.NAT.,313-        4764  F
    323.
PROBL. ENDOKRINOL., 20:104-111.                                  4635
PROG. FISH CULT., 33:32-36.                                      5426  F
PROG. FISH CULT., 36:122-128.                                    4970  F
PROG. FISH CULT., 37:143-147.                                    5449
PROG. FISH CULT., 38:197.                                        5189  F
PROG. FISH CULT., 39:127-128.                                    5241  F
R. SOC. EDINB., PROC., SECT. B (BIOL. SCI.), 199:377-397.        5566
R. SOC. EDINB., PROC., SECT. B (NAT. ENVIRON.), 75:223-232.      5178
R. SOC. LOND., PHILOS. TRANS., SER. B, BIOL. SCI., 282-205-2     5121
    20.
R. SOC. TROP. MED. HYG., TRANS., 69:167-168.                     5451
RECH. BIOL. CONTEMP., 4:155-162.                                 5558
    (FR.)
    ENG. SUMM.
REV. SUISSE ZOOL., 82:35-40.                                     4640  F
RIV. BIOL., 65:331-345.                                          4619  F
    (ITAL.)
    ENGL. TRANSL.
RIV. BIOL., 66:270-291.                                          4807  F
    (ITAL.)
    ENGL. TRANSL.
S.D. ACAD. SCI., PROC., 54:194-222.                              5197  F
SARSIA, 51:97-106.                                               4789  F
SCAND. J. IMMUNOL., 7:245-250.                                   5338
SCIENCE, 184:72-73.                                              4575  F
SCIENCE, 193:680-681.                                            5382  F
SCIENCE, 200:529-531.                                            5599
SCR. FAC. SCI. NAT. UNIV. PURKYNIANAE BRUN BIOL., 6:17-20.       5270
SEMIN. ARTHRITIS RHEUM., 7:245-277.                              5565
SHIZUOKA. UNIV. FAC. SCI., REP., 9:67-78.                        4608
SMITHSONIAN CONTRIB. ZOOL., 206:1-45                             4819  F
SOC. BIOL., PARIS, C.R. HEBD. SEANCES, 170:59-64.                5025  F
    (FR.)
    ENG. SUMM.
SOC. BIOL., PARIS, C.R. HEBD. SEANCES, 171:1302-1305.            5468  F
    (FR.)
    ENG. SUMM.
SOC. BIOL., PARIS, C.R. HEBD. SEANCES, 171:308-313.              5342  F
    (FR.)
    ENG. SUMM.
SOC. PORT. CIENC. NAT., BOL., 16:21-46                           5172
SOC. R. ZOOL. BELG., ANN., 51:151-162.                           5357  F
SOC. SCI. NAT. OUEST FR., BULL., 73:1-6.                         5008
    (FR.)
SOC. ZOOL. FR., PARIS, BULL., 101:135-136.                       5016  F
SOC. ZOOL. FR., PARIS, BULL., 98:532-539.                        4665
SOUTH AUST. MUS., ADELAIDE, REC., 17:169-175.                    4718  F
SOUTH. CALIF. ACAD. SCI., LOS ANGELES, BULL., 75:60-67.          4821
SOUTH. CALIF. ACAD. SCI., LOS ANGELES, BULL., 75:99-111.         5139  F
SOUTHWEST. NAT., 19:220-223.                                     4944  F
SOUTHWEST. NAT., 20:414-416.                                     5055
SOUTHWEST. NAT., 22:107-114.                                     5207
STOCKHOLM, 9TH INT. CONGR. BIOCHEM., ABSTR.                      5330
STOCKHOLM: UNIV. OF STOCKHOLM. 11P.                              4791  F
STUD. CERCET. BIOL., SER. BIOL. ANIM., 27:101-104.               5072  F
TENN. ACAD. SCI., J. 49:81-87.                                   4590  F
TENN. ACAD. SCI., J., 51:66-67.                                  4812  F
TEX. J. SCI., 25:122-123.                                        4724
    (ABSTR.)
THROMB. DIATH. HAEMORRH., 29:313-338.                            4549
THROMB. RES., 3:419-424.                                         4576
TISCIA, 7:69-77.                                                 5127  F
    (GER.)
TOXICOL. & APPL. PHARMACOL., 25:430-434.                         5402
TOXICOL. & APPL. PHARMACOL., 25:542-552.                         5412  F
TOXICOL. & APPL. PHARMACOL., 31:150-158.                         4751  F
TOXICOL. & APPL. PHARMACOL., 36:281-296.                         5439
TOXICOLOGY.                                                      5457
TOXICOLOGY, 4:183-194.                                           4647  F
TRANSPLANT PROC., 6:47-50.                                       4929
TSITOLOGIYA, 15:1338-1344.                                       4583
TSITOLOGIYA, 17:219-237.                                         4883  F
TSITOLOGIYA, 19:1238-1244.                                       5323  F
U.S. DEP. INT., FISH WILDL. SERV., DENVER WILDL. RES. CENT.,     5377  F
    INT. REP. SER. PHARMACOL.: 4 PP.
    RESTRICTED
U.S. DEP. INT., FISH WILDL. SERV., DENVER WILDL. RES. CENT.,     5378  F
    INT. REP. SER. PHARMACOL.: 4 PP.
    RESTRICTED
U.S. DEP. INT., FISH WILDL. SERV., DENVER WILDL. RES. CENT.:     5430  F

```
                5 PP.
                RESTRICTED
U.S. DEP. INT., FISH WILDL. SERV., FISH CONTROL LAB.: 5 PP.        5429
U.S. DEP. INT., FISH WILDL. SERV., FISH CONTROL LAB.: 5 PP.        5433
U.S. DEP. INT., FISH WILDL. SERV., FISH CONTROL LAB.: 6 PP.        5432  F
                RESTRICTED
U.S. DEP. INT., FISH WILDL. SERV., FISH CONTROL LAB.: 7 PP.        5434
U.S. DEP. INT., FISH WILDL. SERV., HAMMOND BAY BIOL. STN.: 4       5428
                PP.
U.S. DEP. INT., FISH WILDL. SERV., INVEST. FISH CONTROL, IN        5406
                PROGRESS.
U.S. DEP. INT., FISH WILDL. SERV., INVEST. FISH CONTROL, IN        5456
                PROGRESS.
U.S. DEP. INT., FISH WILDL. SERV., INVEST. FISH CONTROL, 56:       5118  F
                1-17.
U.S. DEP. INT., FISH WILDL. SERV., INVEST. FISH CONTROL, 57:       5120  F
                7 PP.
U.S. DEP. INT., FISH WILDL. SERV., INVEST. FISH CONTROL, 58:       5137  F
                7 PP.
U.S. DEP. INT., FISH WILDL. SERV., INVEST. FISH CONTROL, 59:       5144  F
                1-9.
U.S. DEP. INT., FISH WILDL. SERV., INVEST. FISH CONTROL, 60:       4712  F
                1-27.
U.S. DEP. INT., FISH WILDL. SERV., INVEST. FISH CONTROL, 61:       4691  F
                3-9.
U.S. DEP. INT., FISH WILDL. SERV., INVEST. FISH CONTROL, 62:       4557  F
                3-7.
U.S. DEP. INT., FISH WILDL. SERV., INVEST. FISH CONTROL, 63:       5053  F
                3-11.
U.S. DEP. INT., FISH WILDL. SERV., INVEST. FISH CONTROL, 64:       5050  F
                3-8.
U.S. DEP. INT., FISH WILDL. SERV., INVEST. FISH CONTROL, 65:       4985  F
                3-10.
U.S. DEP. INT., FISH WILDL. SERV., INVEST. FISH CONTROL, 66:       5051  F
                3-8.
U.S. DEP. INT., FISH WILDL. SERV., INVEST. FISH CONTROL, 67:       5396  F
                1-8.
U.S. DEP. INT., FISH WILDL. SERV., INVEST. FISH CONTROL, 69:       5095  F
                1-9.
U.S. DEP. INT., FISH WILDL. SERV., INVEST. FISH CONTROL, 70:       5242  F
                1-5.
U.S. DEP. INT., FISH WILDL. SERV., INVEST. FISH CONTROL, 77:       5463  F
                11 PP.
U.S. DEP. INT., FISH WILDL. SERV., INVEST. FISH CONTROL, 78:       5444  F
                7 PP.
U.S. DEP. INT., FISH WILDL. SERV., INVEST. FISH CONTROL, 79:       5453  F
                8 PP.
U.S. DEP. INT., FISH WILDL. SERV., INVEST. FISH CONTROL, 84:       5577  F
                6 PP.
U.S. DEP. INT., FISH WILDL. SERV., INVEST. FISH CONTROL, 85:       5442  F
                5 PP.
U.S. DEP. INT., FISH WILDL. SERV., LIT. REV., 74-02: 6 PP.        5458
U.S. DEP. INT., FISH WILDL. SERV., SOUTHEAST. FISH CONTROL L       5379  F
                AB.:11 PP.
                RESTRICTED
U.S. DEP. INT., FISH WILDL. SERV., SOUTHEAST. FISH CONTROL L       5427
                AB.: 10 PP.
U.S. DEP. INT., FISH WILDL. SERV., SOUTHEAST. FISH CONTROL L       5431
                AB: 10 PP.
U.S. DEP. INT., FISH WILDL. SERV., SPEC. SCI. REP., 152:1-57       5416  F
                RESTRICTED
U.S. DEP. INT., FISH WILDL. SERV., SPEC. SCI. REP., 191:1-61       5498  F
U.S. DEP. INT., FISH WILDL. SERV., 2 PP.                           5143  F
                RESTRICTED
UMEA UNIV. MED. DISS., NEW SER., 15:41 PP.                         4774  F
UNIV. ALASKA  BIOL. PAP., 19:74 PP.                                5542  F
UNIV. B.C., DEP. ZOOL.                                             4974  F
UNIV. GUELPH, M. SC. THESIS: 67 PP.                                5611  F
                RESTRICTED
UNIV. GUELPH, M.SC. THESIS: 44 PP.                                 5344  F
                RESTRICTED
UNIV. GUELPH, M.SC. THESIS: 85 PP.                                 5616  F
                RESTRICTED
UNIV. GUELPH, PH.D. THESIS: 76 PP.                                 5610  F
                RESTRICTED
UNIV. LAVAL, QUE., PH.D. THESIS.                                   5435
UNIV. MICH., MUS. PALEONTOL., CONTRIB., 24:23-30.                  4615
UNIV. OREGON STATE, PH.D. THESIS.                                  5554
UNIV. TORONTO, PH.D. THESIS: 184 PP.                               4698
UNIV. TUBINGEN, PH.D. THESIS: 70 PP.                               5448
UNIV. WIS., M.SC. THESIS.                                          5403  F
UNIV. WIS., M.SC. THESIS: 32 PP.                                   5460
UNIV. WIS., M.SC. THESIS: 39 PP.                                   5409  F
UNIV. WIS., PH.D. THESIS:163 PP.                                   5384
UNIV. WIS.: 1P.                                                    5376
UOPR. IKHTIOL., 17:843-861.                                        5597
USP. SOVREM. BIOL., 77:348-359.                                    4914
USP. SOVREM. BIOL., 79:302-310.                                    4701  F
                (RUSS.)
VERT. PALASIAT., 13:202-216.                                       5417
VESTN. CESK. SPOL. ZOOL., 3:179-182.                               5119  F
```

```
        (GER.)
VESTN. CESK. SPOL. ZOOL., 38:95-97.                              4541  F
VESTN. ZOOL., 9(4):9-15.                                         4873
        (RUSS.)
        ENG. SUMM.
VISION RES., 14:137-140.                                         4555  F
VISION RES., 15:253-2599                                         4631  F
VISION RES., 16:237-239.                                         5039
VISION RES., 16:659-662.                                         5001  F
VISION RES., 17:715-717.                                         5299  F
VISION RES., 17:719-722.                                         5214  F
VOPR. ICHTHYOL., 15:369-370.                                     4924
        (RUSS.)
        J. ICHTHYOL., 15:334-337.
        (ENG.)
VOPR. IKHTIOL., 11:1077-1087.                                    4681  F
        (RUSS.)
        J. ICHTHYOL., 11:1061-1070.
        (ENG.)
VOPR. IKHTIOL., 12:297-306.                                      5122  F
        (RUSS.)
VOPR. IKHTIOL., 14:218-230.                                      4623  F
        (RUSS.)
        J. ICHTHIOL., 14:192-202.
        (ENG.)
VOPR. IKHTIOL., 14:34-40.                                        4650  F
        (RUSS.)
VOPR. IKHTIOL., 5: 7 PP.                                         5358  F
VOPR. MED. KHIM., 24:227-232.                                    4996
W. VA. ACAD. SCI., PROC., 47:150-153.                            5284
W.H.O., BULL., 27:95-98.                                         5551
WARSAW: PANSTW. WYDAWN. NAUK.                                    4688  F
        (POL.)
WATER RESOUR. BULL., 12:1233-1243.                               5410  F
WET. MEDED. K.N.N.V.(K.NED. NATUURHIST. VER.), 108:1-44.         4825
WIS., VITERBO COLL.:24 PP.                                       5380
WISS. Z., MATH-NATURWISS. ZEIHE, (POTSDAM), 3:141-144.           5580  F
Z. ANAT. ENTWICKLUNGSGESCH., 142:91-101.                         4563  F
Z. PARASITENK., 49:233-242.                                      5061  F
Z. WISS. ZOOL., 11:1-112.                                        5549
Z. ZELLFORSCH MIKROSK. ANAT., 140:425-432.                       5335
Z. ZELLFORSCH. MIKROSK ANAT., 143:273-290.                       4588
Z. ZELLFORSCH. MIKROSK. ANAT., 147:87-106.                       4610  F
Z. ZELLFORSCH., 136:85-96.                                       4734  F
Z. ZELLFORSCH., 141:33-54.                                       4735  F
Z. ZELLFORSCH., 142:329-345.                                     4747  F
        (GERM.)
        ENGL. SUMM.
Z. ZELLFORSCH., 356:1-5                                          4941  F
Z. ZELLFORSCH., 58:638-640.                                      5334
Z. ZELLFORSCH., 88:67-79.                                        4700  F
        (GER.)
        ENGL. SUMM.
ZH. EVOL. BIOCHEM. PHYSIOL., 13:106-113.                         5305
        J. EVOL. BIOCHEM. PHYSIOL., 13:106-113.
        (ENG.)
ZH. EVOL. BIOKHIM. FIZIOL., 10:223-231.                          4947  F
        (RUSS.)
        J. EVOL. BIOCHEM. PHYSIOL., 10:201-208.
        (ENG.)
ZH. EVOL. BIOKHIM. FIZIOL., 10:300-302.                          4955  F
        (RUSS.)
        J. EVOL. BIOCHEM. PHYSIOL., 10:266-268
        (ENG.)
ZH. EVOL. BIOKHIM. FIZIOL., 10:325-329.                          4741  F
        (RUSS.)
        ENGL. SUMM.
ZH. EVOL. BIOKHIM. FIZIOL., 10:524-526.                          4636  F
        (RUSS.)
        ENGL. SUMM.
ZH. EVOL. BIOKHIM. FIZIOL., 11:187-190.                          4606  F
        (RUSS.)
ZH. EVOL. BIOKHIM. FIZIOL., 11:187-190.                          5000
        (RUSS.)
        J. EVOL. BIOCHEM. PHYSIOL., 11:151-155.
        (ENG.)
ZH. EVOL. BIOKHIM. FIZIOL., 11:20-27.                            4668
        (RUSS.)
        J. EVOL. BIOCHEM. PHYSIOL., 11:14-19.
        (ENG.)
ZH. EVOL. BIOKHIM. FIZIOL., 11:218-224.                          4642  F
        (RUSS.)
ZH. EVOL. BIOKHIM. FIZIOL., 11:346-352.                          4817  F
        (RUSS.)
        ENG. SUMM.
        J. EVOL. BIOCHEM. PHYSIOL., 11:304-309.
        (ENG.)
ZH. EVOL. BIOKHIM. FIZIOL., 11:45-52.                            4670  F
        (RUSS.)
        ENGL. SUMM.
ZH. EVOL. BIOKHIM. FIZIOL., 11:567-572.                          5094  F
```

```
        (RUSS.)
        ENG. SUMM.
ZH. EVOL. BIOKHIM. FIZIOL., 11:605-611.                    5071  F
        (RUSS.)
        ENG. SUMM.
ZH. EVOL. BIOKHIM. FIZIOL., 11:88-90.                      4694  F
        (RUSS.)
        ENGL. SUMM.
ZH. EVOL. BIOKHIM. FIZIOL., 12:113-119.                    5064  F
        (RUSS.)
        ENG. SUMM.
ZH. EVOL. BIOKHIM. FIZIOL., 12:189-192.                    5060
        (RUSS.)
        J. EVOL. BIOCHEM. PHYSIOL., 12:174-176.
        (ENG.)
ZH. EVOL. BIOKHIM. FIZIOL., 12:282-284.                    5360  F
        (RUSS.)
        ENG. SUMM.
ZH. EVOL. BIOKHIM. FIZIOL., 12:358-361.                    5173
        (RUSS.)
ZH. EVOL. BIOKHIM. FIZIOL., 12:369-372.                    5204
        (RUSS.)
        J. EVOL. BIOCHEM. PHYSIOL., 12:336-338.
        (ENG.)
ZH. EVOL. BIOKHIM. FIZIOL., 12:75-77                       5191
        (RUSS.)
        J. EVOL. BIOCHEM. PHYSIOL., 12:65-67
        (ENG.)
ZH. EVOL. BIOKHIM. FIZIOL., 13:118-124.                    5278
        (RUSS.)
        J. EVOL. BIOCHEM. PHYSIOL., 13:99-105.
        (ENG.)
ZH. EVOL. BIOKHIM. FIZIOL., 13:146-151.                    5276
        (RUSS.)
        J. EVOL. BIOCHEM. PHYSIOL., 13:124-129.
        (ENG.)
ZH. EVOL. BIOKHIM. FIZIOL., 13:162-166.                    5274
        (RUSS.)
ZH. EVOL. BIOKHIM. FIZIOL., 13:340-343.                    5281  F
ZH. EVOL. BIOKHIM. FIZIOL., 13:405-407.                    5279
        (RUSS.)
        J. EVOL. BIOCHEM. PHYSIOL., 13:256-258.
        (ENG.)
ZH. EVOL. BIOKHIM. FIZIOL., 13:503-505.                    5313
        (RUSS.)
        J. EVOL. BIOCHEM. PHYSIOL., 13:343-345.
        (ENG.)
ZH. EVOL. BIOKHIM. FIZIOL., 13:556-569                     5512  F
ZH. EVOL. BIOKHIM. FIZIOL., 13:607-613.                    5362
ZH. EVOL. BIOKHIM. FIZIOL., 13:614-620.                    5504
        (RUSS.)
        J. EVOL. BIOCHEM. PHYSIOL., 13:429-433.
        (ENG.)
ZH. EVOL. BIOKHIM. FIZIOL., 13:621-632.                    5505
        (RUSS.)
        J. EVOL. BIOCHEM. PHYSIOL., 13:434-443.
        (ENG.)
ZH. EVOL. BIOKHIM. FIZIOL., 13:642-644.                    4569
        (RUSS.)
        J. EVOL. BIOCHEM. PHYSIOL., 13:455-456
        (ENG.)
ZH. EVOL. BIOKHIM. FIZIOL., 13:82-83.                      5210  F
        (RUSS.)
        J. EVOL. BIOCHEM. PHYSIOL., 13:63-64.
        (ENG.)
ZH. EVOL. BIOKHIM. FIZIOL., 8:324-332                      5181  F
        (RUSS.)
        ENG. SUMM.
ZH. EVOL. BIOKHIM. FIZIOL., 8:499-505.                     4797  F
        (RUSS.)
        J. EVOL. BIOCHEM. PHYSIOL., 8:440-444.
        (ENG.)
ZH. EVOL. BIOKHIM. FIZIOL., 8:558-560.                     4739
        (RUSS.)
        J. EVOL. BIOCHEM. PHYSIOL., 8:494-560.
        (ENG.)
ZH. EVOL. BIOKHIM. FIZIOL., 9:123-129.                     4952  F
        (RUSS.)
        J. EVOL. BIOCHEM. PHYSIOL., 9:107-112.
        (ENG.)
ZH. EVOL. BIOKHIM. FIZIOL., 9:209-211.                     4742  F
        (RUSS.)
        ENGL. SUMM.
ZH. EVOL. BIOKHIM. FIZIOL., 9:307-308.                     4973
        (RUSS.)
        J. EVOL. BIOCHEM. PHYSIOL., 9:269-271.
        (ENG.)
ZH. EVOL. BIOKHIM. FIZIOL., 9:355-363.                     4556  F
        (RUSS.)
        ENG. SUMM.
ZH. EVOL. BIOKHIM. FIZIOL., 9:611-613.                     4561  F
```

```
        (RUSS.)
        ENGL. SUMM.
ZH. EVOL. BIOKHIM. PIZIOL., 10:623-625.                          4721  F
        (RUSS.)
        J. EVOL. BIOCHEM. PHYSIOL., 10:567-569.
        (ENG.)
ZH. OBSHCH. BIOL., 36:173-188.                                   4997
ZH. OBSHCH. BIOL., 36:361-372.                                   4879
ZH. OBSHCH. BIOL., 36:483-491.                                   4987
ZOOL. ABH. MUS. TIERK. DRESDEN, SUPPL., 33:1-150.                5186  F
        (GERM.)
ZOOL. ANZ., 199:371-380.                                         5525  F
ZOOL. MAG. (TOKYO), 82:133-136.                                  5114  F
ZOOL. MAG. (TOKYO), 82:211-214.                                  4796  F
ZOOL. MAG. (TOKYO), 82:215-217.                                  4795  F
ZOOL. MAG. (TOKYO), 82:269.                                      4925
ZOOL. MAG. (TOKYO), 82:381.                                      4948
ZOOL. MAG. (TOKYO), 84:151-155.                                  4618  F
ZOOL. MAG. (TOKYO), 84:340.                                      4728
ZOOL. MAG. (TOKYO), 84:390                                       4538  F
        (JAP.)
ZOOL. MAG. (TOKYO), 84:405.                                      4581
ZOOL. MAG. (TOKYO), 84:510.                                      5198
ZOOL. POL., 23:263-267.                                          4777  F
        (POL.)
ZOOL. REVY., 35:131-134.                                         4672  F
        (SWED.)
        ENGL. SUMM.
ZOOL. REVY., 36:41-48                                            5141  F
        (SWED.)
        ENGL. SUMM.
ZOOL. REVY., 38:113-118.                                         5271  F
ZOOL. SCR., 3:193-200.                                           4675  F
ZOOL. SCR., 3:91-99.                                             4939  F
ZOOL. SCR., 5:35-47.                                             5102
ZOOL. SCR., 6:113-126.                                           5418
        (GER.)
        ENG. SUMM.
ZOOL. SCR., 6:331-341.                                           5575  F
        (GER.)
        ENG. SUMM.
ZOOL. ZH., 54:526-531.                                           5324  F
```